Neglected Viruses

They are the silent threats lurking in the shadows of global health. Until they are not.

While the world focuses on the pandemics of the day, a group of dangerous viruses remains dangerously underexplored. Chikungunya. Lassa. Nipah. Hanta. Dengue. These pathogens, capable of causing severe neurological damage, hemorrhagic fevers, and devastating outbreaks, represent a critical gap in our scientific and public health defenses.

This essential volume, drawing on the expertise of top researchers, emerges as a vital response to this neglect. It consolidates the latest research into a single, coherent resource, offering a comprehensive analysis of these viruses from their molecular structures and epidemiological patterns to their clinical management and the societal factors that allow them to persist.

> *"A very important step in the right direction... a powerful tool against ignorance and a strong basis for life in tomorrow."*
>
> **Prof. Paul Pumpens, from the Foreword**

More than just a compendium of knowledge, this book is a compelling call to action. It is an indispensable guide for virologists, epidemiologists, and public health professionals, and an accessible entry point for students and non-specialists, urging the scientific community to prioritize these latent threats before the next crisis strikes.

Key Features

- **Fills a Critical Void:** Provides a thorough examination of the world's most overlooked and impactful viral pathogens.
- **In-Depth Analysis:** Offers a unique comparative study of Chikungunya, Lassa, Nipah, Hantaviruses, and Dengue.
- **From Lab to Clinic:** Bridges the gap between cutting-edge research and practical clinical implications.
- **Timely and Urgent:** A crucial resource in a post-COVID world, advocating for proactive viral surveillance.
- **For a Broad Audience:** Written for seasoned scientists while remaining accessible to students and professionals in related fields.

Neglected Viruses

Edited by
Alaa A. A. Aljabali

CRC Press
Taylor & Francis Group
Boca Raton London New York

CRC Press is an imprint of the
Taylor & Francis Group, an **informa** business

Designed cover image: Shutterstock

First edition published 2026
by CRC Press
2385 NW Executive Center Drive, Suite 320, Boca Raton FL 33431

and by CRC Press
4 Park Square, Milton Park, Abingdon, Oxon, OX14 4RN

CRC Press is an imprint of Taylor & Francis Group, LLC

ISBN: 978-1-041-00016-7 (hbk)
ISBN: 978-1-041-00017-4 (pbk)
ISBN: 978-1-003-60781-6 (ebk)

DOI:10.1201/9781003607816

Typeset in Times
by KnowledgeWorks Global Ltd.

Contents

Foreword

Diseases desperate grown,
By desperate appliance are relieved,
Or not at all.
William Shakespeare, *Hamlet*, Act IV, Scene 3

Dear Reader,

Not all that bad in fact, though Shakespeare's pessimism should generally be considered. We certainly fight rather successfully and win over not only global diseases, but also neglected ones, the *Silent Threats* presented by the Author. But it's not quite simple to *Dive Deep*. That's why we need this Book that you, dear Reader, have just opened. Believe me, this Book came out just in time. Because often you can hear such an opinion about the present time – as in his famous monologue, *Hamlet*, he says:

Tis now the very witching time of night,
When churchyards yawn, and hell itself breathes out
Contagion to this world.
William Shakespeare, *Hamlet*, Act III, Scene 2

But again, this pessimism is excessive. It is more than simply helpful that the novel classification of viruses was established in 2019–2020, thanks to the tireless efforts of many highly enthusiastic and prospective virologists, first, Eugene Coonin. For the first time, the novel taxonomy included viruses into the global biological Linnaean classification, allowed its traditional 15-rank classification hierarchy, and recognized therefore the global virosphere as an integral part of wildlife.

For the first time in years, we have been able to correctly calculate the number of viruses we know, more precisely, how many viruses are already officially accounted for by our science. It turned out that there are 14,690 species to date. And few of them deny themselves the pleasure of causing some unpleasant malady in anyone, whether humans, animals, plants, or even microorganisms themselves. And this is only a small portion of what is actually present in nature.

Honestly to say, we are not paying adequate attention to this obstacle until the time. What we know is actually far from everything. And that our attention is insufficient. And when we are forced to turn such attention, it can be a little late. These are the circumstances that give rise to the *Silent Threats*, namely the neglected viruses and neglected diseases – the subject of the Book you opened.

Recently I was struck by an article presenting a "complete" list of bacteria, considered as the cause of human infections: Bartlett A, Padfield D, Lear L, Bendall R, Vos M. A comprehensive list of bacterial pathogens infecting humans. *Microbiology* (Reading). 2022;168:10.1099/mic.0.001269. The authors counted exactly 1513 harmful microbes, which were at least three times recorded as the cause of disease in humans. How many viruses cause diseases in humans and animals, yet no one has counted.

I am absolutely convinced that this Book represents a very important step in the right direction. The neglected viruses are just the kind of danger that is not yet fully understood by the general public. It's still out there, not me. It is not yet fully understood by the people that this danger is already here and now, but even more so it is a danger of tomorrow.

The sad and protracted history of COVID-19 has shown what happens when people ignore potential danger, and do not look in time for tomorrow. So, you can provoke the situation that made Kent say in *King Lear*:

Kill thy physician, and thy fee bestow upon the foul disease.
William Shakespeare, *King Lear*, Act I, Scene 1

It's too bad that this sad story has taught so little to many of us. I suspect that this is one of the arguments why the Author has conceived this Book, which lies before you.

The idea and main investment of the Author is that he offers not just a specific medicine, not just vaccines, but a way to think and know, to understand the diversity of viruses, to understand how new neglected diseases arise and old ones emerge. What is common and what is different about infections that cause both very similar and completely different viruses. This is also a powerful tool against the conspiracy and mystical theories that are so popular in our time and spread with mad speed on mass media and social networks.

And here we will again need the wise advice of the Bard of Avon, which is clearly followed by the Author of the Book:

Invest me in my motley. Give me leave
To speak my mind, and I will through and through
Cleanse the foul body of th' infected world If they will patiently receive my medicine.
William Shakespeare, *As You Like It*, Act II, Scene 7

This is Jacques de Boys, *the second son*. By the way, the same Shakespeare's character who said one of the most famous and quoted phrases today: *All the world's a stage.*

I am sure that this Book will be important not only for professionals, but also for a wide range of readers, especially young people, students of any specialization. This Book will be a real tool of knowledge against ignorance, truthful information against rumors and manipulations. And, most importantly, the strong basis for life in tomorrow.

Here's where you should not be afraid of the paradoxical advice of Oscar Wilde:

Thinking is the most unhealthy thing in the world, and people die of it just as they die of any other disease.
Oscar Wilde, *Beautiful and Impossible Things*

No, here the Reader will not be deceived, not disappointed. Here the Book is written in this way, and all the content is to avoid a situation where, according to Gloucester:

Tis the times' plague, when madmen lead the blind.
William Shakespeare, *King Lear*, Act IV, Scene 1

Read this Book, written by a wise and experienced Author, and think, think about the future. This is the real way to think and make informed decisions that will keep us alive from infectious diseases that are caused by the *Silent Threats.*

And if you, dear Reader, will be surprised by the abundance of quotes from Shakespeare, then there is a direct parallel. Because the Author's Book that lies before you is the same drama, with the main five characters. They are called Chikungunya, Lassa, Nipah, Hanta, and Dengue, and these five – by his own deep intention or by his involuntary whim – the Author made the actors of his play. To present their true essence to the *Urbi et Orbi* against the backdrop of a viral crowd of thousands, which remains in the shadows until the cut of the Book but will certainly fight for the place of the main characters in the Author's next Book.

Go, Reader, go! You will not regret it!

Paul Pumpens
Riga

Preface

In a world where new infectious threats frequently arise, our understanding of infectious diseases is rapidly advancing. However, as attention shifts toward the latest threats, certain viral pathogens remain underexplored. This book addresses a critical gap in our collective knowledge by identifying neglected pathogens whose potential risk to public health is significant but often overlooked. The viruses discussed herein range from the neurological impact of the Nipah virus to the often-overlooked transmission of Hantaviruses, each of which highlights a pathogen deserving of more focused scientific inquiry.

Gathering insights from top researchers, this volume explores both the biological mechanisms underlying these viruses and the societal factors that contribute to their relative neglect. By covering topics ranging from molecular structure to therapeutic strategies, our goal is to create an essential link between laboratory findings and clinical implications.

This work serves dual purposes: as a comprehensive resource for researchers and clinicians, and as a call to action for heightened awareness in public health and scientific research. With the shifting dynamics of climate, urbanization, and global interconnectivity affecting viral transmission, the timing of this book could not be more pressing. Through in-depth analyses of viral evolution, host-pathogen relationships, and novel diagnostic and treatment methods, the book underscores why these neglected viruses require urgent and focused attention.

We hope that this compilation inspires the scientific community to prioritize these latent threats, moving toward a proactive stance on viral surveillance and control.

Prof. Alaa A. A. Aljabali

About the Editor

Prof. Alaa A. A. Aljabali is Distinguished Professor of Nanoscience and Biotechnology at Yarmouk University, Jordan. A seasoned scientist, he holds a Ph.D. in Bionanoscience from the University of East Anglia and the John Innes Centre, following an M.Res. in Biotechnology from the University of Essex.

His extensive research career includes prestigious postdoctoral work as a Research Scientist at the University of Oxford and a role as a Bionanotechnologist at the John Innes Centre. Beyond his research, Prof. Aljabali has held significant leadership positions, including Vice Dean of the Faculty of Pharmacy at Yarmouk University.

He is an active contributor to the global scientific community, serving as an editorial board member and ad hoc reviewer for several high-impact journals, including *Viruses*, *Clinical Cancer Research*, and *PLOS One* among others.

Contributors

Muhamed Adilović
Department of Genetics and Bioengineering
International University of Sarajevo
Sarajevo, Bosnia and Herzegovina

Lina Al-Ebbini
Department of Biomedical Systems and Informatics Engineering
Yarmouk University
Irbid, Jordan

Alaa A. A. Aljabali
Department of Pharmaceutics and Pharmaceutical Technology
Faculty of Pharmacy, Yarmouk University
Irbid, Jordan

Almuthanna K. Alkaraki
Department of Biological Sciences
Faculty of Science, Yarmouk University
Irbid, Jordan

Abdelrahim Alqudah
Department of Clinical Pharmacy and Pharmacy Practice
Faculty of Pharmaceutical Sciences, The Hashemite University
Zarqa, Jordan

Alaa Alqudah
Department of Pharmaceutics and Pharmaceutical Technology
Faculty of Pharmacy, Yarmouk University
Irbid, Jordan

Rawan H. Alsharedeh
Department of Pharmaceutics and Pharmaceutical Technology
Faculty of Pharmacy, Yarmouk University
Irbid, Jordan

Nid'a Alshraiedeh
Department of Pharmaceutical Technology
Faculty of Pharmacy, Jordan University of Science and Technology
Irbid, Jordan

Lorca Alzoubi
Department of Medicinal Chemistry and Pharmacognosy
Faculty of Pharmacy, Yarmouk University
Irbid, Jordan

Amro A. Amara
Protein Research Department, Genetic Engineering and Biotechnology Research Institute (GEBRI)
City of Scientific Research and Technological Applications (SRTA-City), New Borg El-Arab City
Alexandria, Egypt

Nawal Abd El-Baky
Protein Research Department, Genetic Engineering and Biotechnology Research Institute (GEBRI)
City of Scientific Research and Technological Applications (SRTA-City), New Borg El-Arab City
Alexandria, Egypt

Mohamed El-Tanani
RAK College of Pharmacy
Ras Al Khaimah Medical and Health Sciences University
Ras Al Khaimah, United Arab Emirates

Omar Gammoh
Department of Clinical Pharmacy and Pharmacy Practice,
Faculty of Pharmacy, Yarmouk University
Irbid, Jordan

Maryam Golzardi
Department of Genetics and Bioengineering
Faculty of Engineering and Natural Sciences, International University of Sarajevo
Sarajevo, Bosnia and Herzegovina

Yassmen Hamzat
Department of Pharmaceutics and Pharmaceutical Technology
Faculty of Pharmacy, Yarmouk University
Irbid, Jordan

Altijana Hromić-Jahjefendić
Department of Genetics and Bioengineering
Faculty of Engineering and Natural Sciences
International University of Sarajevo
Sarajevo, Bosnia and Herzegovina

Aminata Lowe
School of Pharmaceutical Sciences
Lovely Professional University
Phagwara, India

Kenneth Lundstrom
PanTherapeutics
Lutry, Switzerland

Vijay Mishra
School of Pharmaceutical Sciences
Lovely Professional University
Phagwara, India

Yachana Mishra
School of Bioengineering and Biosciences
Lovely Professional University
Phagwara, India

Esam Qnais
Department of Biology and Biotechnology
Faculty of Science, The Hashemite University
Zarqa, Jordan

Elrashdy M. Redwan
Department of Protein Research
Genetic Engineering and Biotechnology Research Institute (GEBRI)
City of Scientific Research and Technological Applications (SRTA-City), New Borg El-Arab City
Alexandria, Egypt

Department of Biological Sciences
Faculty of Science, King Abdulaziz University
Jeddah, Saudi Arabia

Center of Excellence in Bionanoscience Research
King Abdulaziz University,
Jeddah, Saudi Arabia

Meriem Rezigue
Department of Pharmaceutics and Pharmaceutical Technology
Faculty of Pharmacy, Yarmouk University
Irbid, Jordan

Aman Shukla
School of Pharmaceutical Sciences
Lovely Professional University
Phagwara, India

Jasmin Šutković
Department of Genetics and Bioengineering
Faculty of Engineering and Natural Sciences, International University of Sarajevo
Sarajevo, Bosnia and Herzegovina

1 Unmasking the Silent Pandemic

Why Neglected Viruses Demand Our Attention

Alaa A. A. Aljabali and Meriem Rezigue

1.1 INTRODUCTION

Viral pathogens that cause infectious diseases have played a significant role in shaping human history, with many deaths and social structures changing over the centuries [1]. Much attention, for the most part, has focused on high-profile viral pathogens such as SARS-CoV-2, HIV, and influenza, but a variety of lesser-known but significant viruses threaten public health and have received little investigation [2]. Nevertheless, there is a lack of public awareness and competition with interventions, which is considered a high priority in global public health programs [3–5]. Nevertheless, such a lack of awareness is an increasingly important public health concern, with many underestimated viruses having the potential to cause epidemics and pandemics if intervention is not implemented [6, 7].

The Lassa, Nipah, and Chikungunya viruses disproportionately impact poor and marginalized communities living in the tropics and subtropics and, in doing so, widen current inequities in health [8, 9]. Despite high case fatality rates and long-term debilitating symptoms, these viruses receive little attention in surveillance, diagnostics, and vaccine development [10]. There is a lack of scientific data regarding such pathogens, representing not only an immediate danger to involved communities but also long-term danger through increased zoonotic spillover and ecological reorganization with urbanization, connectivity expansion, and climate change. There is a definite risk of such poorly defined pathogens spilling out of geographical regions and disseminating epidemics [11, 12].

Investigating underexplored viruses is critical for enhancing global health security [13, 14]. A greater understanding of their pathogenesis, routes of infection, and biological properties enables the development of effective preventive interventions, diagnostic tests, and infection controls [15]. In addition, supporting such work promotes fair access to care, reduces burden in economically disadvantaged countries, and strengthens preparedness for future pandemics. In an increasingly interconnected world, the investigation of understudied viruses is not simply a matter of scientific importance; it is a moral imperative to safeguard humanity from predictable and unpredictable threats [16, 17].

Lassa fever, Nipah virus infection, and Crimean-Congo hemorrhagic fever are examples of severe zoonotic diseases that continue to pose significant public-health risks in endemic regions [18, 19]. For instance, together, approximately one million cases per year, with less than 15% of the population in most endemic nations having access to effective diagnostic capacities for such pathogens as shown in Table 1.1 [20, 21]. Long-term underinvestment over a long period creates a disconnection between the disease burden and funding for research, compromising humanity's security for future catastrophes [22].

Conceptually, a "virosome" can be understood as referring to an invisible environment in which unregulated viruses multiply in unmonitored settings, taking advantage of

DOI: 10.1201/9781003607816-1

TABLE 1.1
Global Epidemiological Summary of Selected Neglected Viruses, Highlighting Disease Burden, Geographic Distribution, and Current Research Investment Levels

Virus	Natural Reservoir	Intermediate/ Amplifying Hosts	Transmission Route	Case Fatality Rate (CFR)	Major Clinical Manifestations	Vaccine/Drug Development Status
Lassa virus (Arenavirus)	Rodents (*Mastomys natalensis*)	Rarely humans via food contamination	Contact with rodent excreta, person-to-person	~1–2% overall, up to 15–20% in hospitalized cases	Hemorrhagic fever, neurological complications, long-term hearing loss	No licensed vaccine Ribavirin used experimentally
Nipah virus (Paramyxovirus)	Fruit bats (*Pteropus* spp.)	Pigs as amplifying hosts	Bat-to-human, pig-to-human, human-to-human	40–75%	Encephalitis, respiratory disease, high relapse risk	No licensed vaccine Monoclonal antibodies and vaccines in trials
Chikungunya virus (Alphavirus)	Primates, rodents	Mosquitoes (*Aedes aegypti*, *A. albopictus*)	Mosquito-borne	<1% (but high morbidity)	High fever, severe arthralgia, chronic joint pain	Vaccine candidates in advanced clinical trials Supportive care
Crimean-Congo hemorrhagic fever virus (CCHFV) (Nairovirus)	Wild and domestic animals (asymptomatic)	Ticks (*Hyalomma* spp.)	Tick bites, contact with animal blood	10–40%	Hemorrhagic fever, hepatic injury, shock	No licensed vaccine outside limited regional use Ribavirin under study
Oropouche virus (Orthobunyavirus)	Sloths, primates	Midges (*Culicoides paraensis*)	Vector-borne	<1%	Acute febrile illness, headache, meningitis in some cases	No vaccines or antivirals
Lymphocytic choriomeningitis virus (LCMV) (Arenavirus)	House mice (*Mus musculus*)	Rodents	Contact with rodent secretions	Low overall, severe in immunocompromised/ pregnant	Meningitis, encephalitis, congenital infection	No vaccines or antivirals

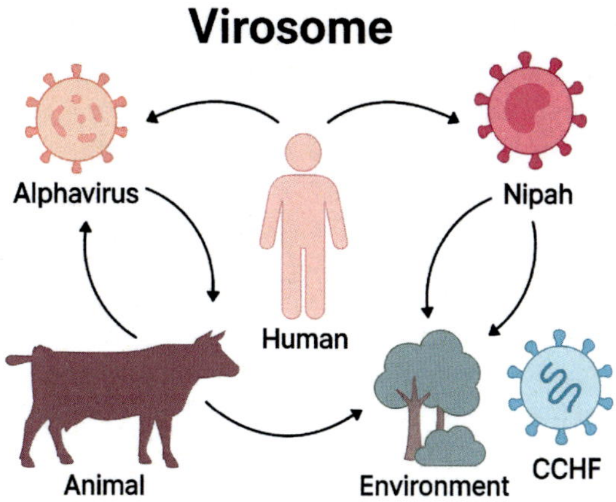

FIGURE 1.1 The virosome concept: ecological and epidemiological network of neglected viruses. The figure illustrates how viruses such as alphaviruses, Nipah virus, and Crimean-Congo hemorrhagic fever virus circulate between humans, animals, and the environment. This interconnected "virosome" reflects the complex and often unmonitored viral ecosystems that contribute to spillover events and emerging disease risks.

observational vulnerabilities such as predators in unpatrolled habitats as shown in Figure 1.1 [23]. Figuratively, it describes a tangled web of interrelated dangers posed by under-researched pathogens thriving at the nexus of poor living environments, environmental degradation, and ecological susceptibility, capable of untracked recombination, re-emerging subsequently with deadly efficiency, such as during the Ebola epidemic in Western Africa between 2013 and 2016 [24–27].

Neglected viruses constitute an important but poorly researched aspect of this disease worldwide [28]. Their high contribution to morbidity and mortality in specific at-risk groups is underappreciated [29]. Viruses such as those responsible for viral hemorrhagic fevers, such as Lassa fever and Nipah virus (NiV) disease, have a high endemicity but have received much less scientific investigation in comparison with high-profile disease entities such as influenza and HIV [30–32]. Consequently, key epidemiological and biological factors fall short of full comprehension and are often overlooked [33]. Unless such concerns are addressed, these viruses not only threaten the communities in which they have become endemic but can also have an impact on the worldwide community during a period of increased environmental uncertainty and increased international mobilization [11, 34].

1.1.1 Global Health Impact

This chapter systematically reviews the molecular biology, epidemiology, and public health importance of these neglected viruses. By shedding light on their modes of transmission, clinical presentation, and socioeconomic burdens, we hope to spur increased scientific interest and research investment in these areas [35]. Our in-depth review reveals how understanding these under-researched viral pathogens is key to the protection of susceptible populations and to enhancing global preparedness against infectious diseases [36, 37].

The effects of infectious diseases are particularly significant, extending not only to direct medical consequences but also to socioeconomic consequences [38, 39]. Economic disruption in local economies, increased medical expenses, and financial resources diverted from other important concerns cumulatively contribute to the worldwide importance of such infections [40, 41].

1.1.2 Current Research Landscape

Despite their apparent importance to public health, most under-recognized viruses have received relatively little funding for research in comparison to well-established disease entities [42]. Attempts at surveillance, laboratory development, and clinical studies regarding these pathogens have been intermittent and underfunded; therefore, a significant lack in our understanding of their biological profiles, transmission, and epidemic potential has ensued [43, 44].

Moreover, with changing epidemiologic trends concerning climate change, urbanization, and growing worldwide connectivity, the re-emergence, and emergence of pathogens in this under-recognized group represent an increasingly critical danger [45]. Because of disorganized scientific investigations and persistent insufficiencies in financial support, a heightened danger of widespread and catastrophic epidemic consequences prevails [46, 47].

1.1.3 Chapter Objectives

This chapter offers an evaluation of the molecular biology, epidemiological settings, and public health concerns related to viruses that are often overlooked. The goals include:

- *Elucidating Transmission Mechanisms*: By comparative evaluation of general and specific transmission routes (e.g., vectors, droplets, and reservoirs), an investigation can reveal the factors that enable the transmission of these viruses [48, 49].
- *Elucidating Clinical Manifestations*: A detailed analysis of different syndromes, including their severity, complications, and the possibility of challenging diagnostics in resource-poor settings, will yield important insights [50, 51].
- *Assessing Socioeconomic Implications*: An evaluation of wider social and economic consequences will foster increased demand for improved policies, disease surveillance, and healthcare programs [52, 53].
- *Encouraging Research Funding*: By revealing gaps in current knowledge, from studies of viral genomes to population-level epidemic dynamics, we hope to stimulate increased scientific investigation and mobilization of financial resources for these important but under-investigated pathogens [54, 55].

1.2 MOLECULAR BIOLOGY OF NEGLECTED VIRUSES

1.2.1 Entry and Replication Mechanisms

The processes through which neglected viruses enter and replicate in host organisms are complex events characterized by complex interactions between viral and host organismal cellular processes [56]. Identification of such processes is critical for successful antiviral development. The general mechanism of virus entry into target cells involves receptor-mediated endocytosis, with specific receptors in target cells allowing for virus internalization. For instance, pathogenic H5N1 avian influenza virus hemagglutinin proteins have been proven capable of overcoming initial barriers to replication and supporting productive replication in macrophages [57]. Identification of such a role for viral proteins in controlling both entry and secondary replication emphasizes their role in controlling both processes [57]. The involvement of phosphatidylserine receptors in the infection of enveloped viruses identifies a mechanism in which the expression of phosphatidylserine in viral envelopes can modulate replication efficacy, with several viruses utilizing such receptors for access in cells [58, 59].

Following the invasion of target cells, viruses must escape the endosomal compartment to gain access to the cytoplasm, which is a critical stage for successful infection and replication. Proteolytic enzymes, such as the escape mechanism of a porcine epidemic diarrhea virus from an endosome, play a key role in such processes, allowing for unimpeded discharge of viral genomes in the cytoplasm [60]. In addition, hijacking host reticulon proteins for membrane reorganization, which is critical for positive-strand RNA virus replication, identifies a mechanism in which viruses hijack cellular processes to generate environments conducive to efficient replication [61].

Upon entry into the host organism, viruses begin replication by hijacking the structures of the host cells. A relevant case in point is represented by murine noroviral NS1/2 proteins' interaction with host proteins, allowing for processes of replication post-entry but preceding RNA production, and therefore represents a case of the numerous roles played during replication by viral proteins [62, 63]. In addition, the development of viral replication organelles (VROs) is a key function of positive-strand RNA virus replication, with such structures combining both viral and key host factors involved in replication [64].

Coordination between viral and host factors is also seen through cytoskeletal manipulation, a mechanism adopted by a variety of viruses for increased mobility in the host cell and for virus egress [65]. Exploration of such coordination not only reveals processes driving viral pathogenesis but also identifies therapeutic candidates for intervention in such processes. In conclusion, processes involved in the entry and replication of lesser-characterized viruses exhibit both host-dependent machinery and hijacking of host factors for purposeful functioning. An understanding of such processes is important for the development of antiviral therapies to interfere with such processes, intending to curb infection with a virus.

1.2.1.1 Alphavirus Entry Pathways

Alphaviruses, such as the Chikungunya virus (CHIKV) and Semliki Forest virus (SFV), have complex processes for cell entry and are essential for pathogenicity and tissue tropism. Membrane clathrin-dependent endocytosis (CME) is an essential mechanism for viral entry into the cytoplasm of target cells [66, 67]. Conformational changes in envelope glycoproteins E1 and E2 in the acidic environment of an endosome are an essential phase in membrane fusion [67]. Membrane fusion facilitates the successive translocation of viral nucleocapsids into the cytoplasm, thereby initiating the viral life cycle [66, 67].

The E1 glycoprotein is an important contributor to fusion, and its conformational shape is determined by variation in the pH level [66, 67]. Some amino acid residues, such as E3 Y47, have been studied and found to contribute to E1/E2 heterodimer integrity in an acidic environment, and such integrity is important for effective viral entry as shown in Figure 1.2 [68]. Specific interactions between viruses and receptors are important for effective viral entry [66, 67]. For instance, the Sindbis virus (SINV) utilizes the NRAMP2 receptor, but an authentic receptor for SFV is yet to be identified [69]. Virus binding to these receptors facilitates membrane fusion and endocytosis [66, 67].

Recent studies have uncovered one of the most important characteristics of alphavirus infection: these pathogens possess a mechanism of bypassing the normal endocytic pathway under specific conditions. Alternative entry, involving a direct perturbative effect on the plasma membrane, involves the rapid delivery of viral RNA into the cytoplasm through a mechanism that does not rely on receptor-mediated endocytosis [70]. Not only does such adaptability complicate alphaviral pathogenesis, but it also reflects the presence of several routes for viral infection, possibly temperature- and specific-property-dependent [71].

Additionally, the role of host factors in alphavirus infection and replication cannot be exaggerated. It has been established that some host proteins, such as BST-2/tetherin, can restrict the budding and release of the virus, thereby regulating the overall virus lifecycle [69, 72]. Actin filaments and microtubules have also been implicated in the transport of viral replication complexes between the plasma membrane and intracellular compartments, strengthening the cooperative interaction between the cytoplasmic machinery of the host and virus [67, 73]. These factors must be identified to develop therapeutic interventions to inhibit viral replication and transmission.

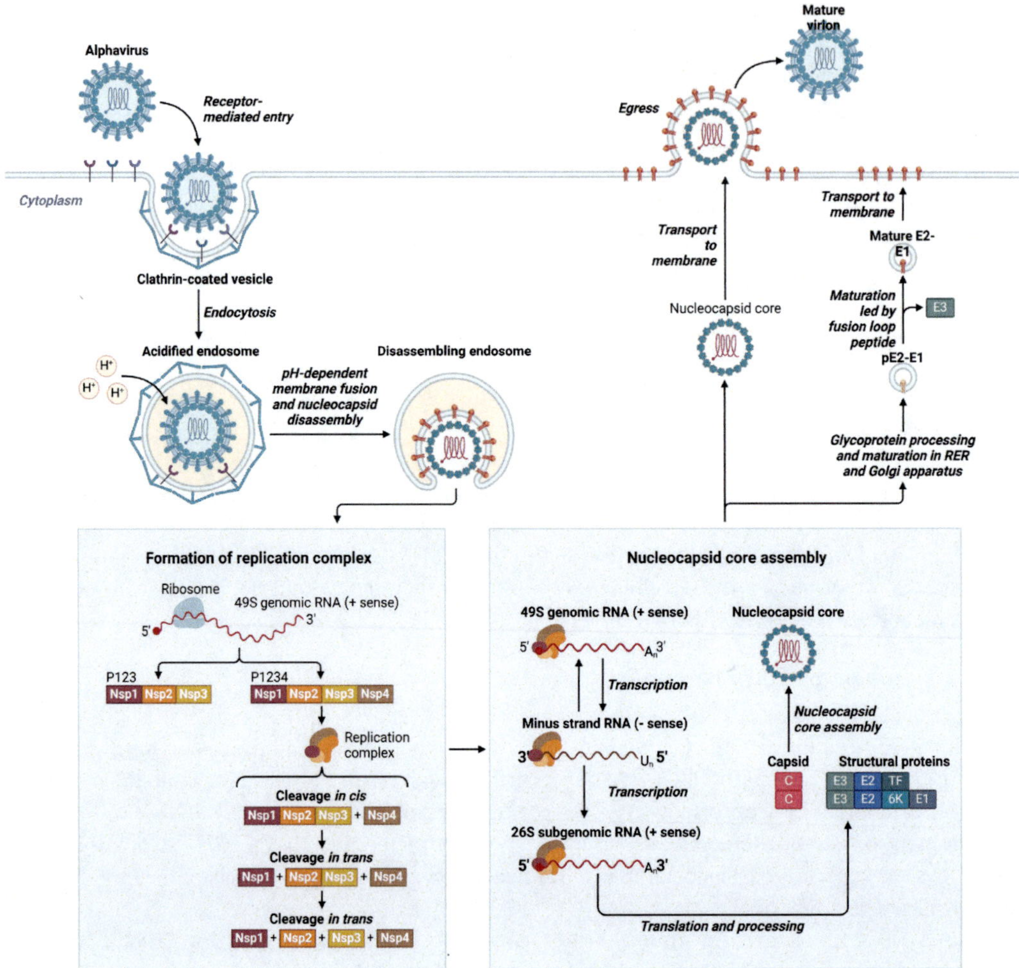

FIGURE 1.2 Alphavirus infection cycle and antibody-mediated inhibition mechanisms. The diagram illustrates the sequential stages of alphavirus entry into host cells: receptor binding at the plasma membrane, receptor-mediated endocytosis, and endosomal fusion leading to cytoplasmic release of the nucleocapsid. Once uncoated, viral genomic RNA undergoes translation, replication, and packaging of structural proteins in preparation for assembly and egress. The figure also depicts the roles of monoclonal antibodies DC2.112 and DC2.315 in modulating infection, showing their capacity for cell-specific neutralization (effective in N2a cells, but not Vero cells), Fc-mediated effector functions, and blocking of viral egress. Neither antibody inhibits receptor attachment or viral fusion directly, but both restrict egress of mature virions from infected cells. This highlights potential therapeutic (DC2.112) and prophylactic (DC2.315) antibody applications against alphaviral infections. (Created with BioRender.com.)

In addition, alphaviral biology can be distinguished by the development of VROs. VROs have been proposed to form through reorganization in cell membranes and act as key sites for alphaviral RNA synthesis and alphaviral component packaging [74, 75]. The integration of specific host proteins in such organelles could stimulate and inhibit viral replication, and molecular interactions in alphaviral infections must be understood [76]. In summary, alphaviral infection and replication involve complex multifaceted interactions between alphaviral and host cell proteins. The engagement and evasion of CME processes demonstrate the adaptability of these viruses to diverse cellular environments. Additional studies on such molecular complexity will provide insightful information about the pathogenicity of such underestimated pathogens and will lead to successful antiviral approaches.

1.2.1.2 Paramyxovirus Entry Strategies

Paramyxoviruses, including the NiV, utilize a sophisticated mechanism of entry, which is strikingly disparate when compared with alphaviruses and most virus groups [70]. Paramyxoviral entry involves two key glycoproteins: HN, an attachment glycoprotein, and F, a fusion glycoprotein [72, 77]. Conformational changes in the F protein in response to HN binding to target receptors induce its activation, membrane fusion with target cells, and subsequent delivery of the viral genome into the cytoplasm [78, 79].

One of the distinguishing features of paramyxoviruses is their cell-to-cell fusion capability, which produces multinucleated giant cells, a feature referred to as syncytia [68, 69]. This feature is one of the virulent factors for viruses, such as Nipah, in that it helps in the dissemination of the virus in tissue environments. Syncytia is created through a coordinated function of HN and F proteins, both of which play a role in the infection of cells with a virus and the infusion of adjacent cells with an infected one [80, 81]. Alphaviruses, however, predominantly use endocytic routes for infection and prefer not to cause syncytia development.

Studies have shown that the development of syncytia is correlated with increased dissemination and proliferation of the virus. For instance, studies on the NiV have proven that access to glycoproteins at the level of the cell interface is a requirement for initiation of the reaction of fusion and, therefore, for dissemination of the virus [82]. In addition, fusion dynamics can play a considerable role in both the intensity and velocity of syncytia development. In some cases, reduced velocities of fusion can contribute toward reduced development of syncytia, and therefore toward the overall pathogenicity of the virus [83]. The functions of host factors in controlling the paramyxovirus entry and fusion processes are extremely important. In a highly specific manner, viral glycoprotein and host receptor interactions can significantly affect both the efficiency of the fusion reaction and the ensuing post-fusion viral replication. In addition, actin cytoskeleton activity has been shown to contribute to increased fusion efficiency, specifically through its role in actin cytoskeleton reorganization involved in generating syncytia [84]. Greater familiarity with such processes provides valuable insights into the complex processes involved in paramyxovirus infection and identifies areas for therapeutic intervention to control viral dissemination.

The processes utilized for paramyxovirus entry, including that of the NiV, have a unifying feature of utilizing surface-dependent mechanisms for fusion facilitated through HN and F glycoproteins [71, 74]. Not only is the ability to form syncytia responsible for spreading infection, it also reflects a high level of pathogenicity when compared with an endosomal mechanism for virus entry [75, 85]. Continuing studies on the molecular dynamics involved in such processes will form a critical part of future antiviral therapy developments as shown in Figure 1.3 [85].

1.2.1.3 Role of Ionophores in Viral Entry

Ionophores have gained significant academic interest as therapeutic compounds for alphavirus infection therapy, partly because of their ability to modulate the pH in endosomal spaces, which is a key mechanism for alphavirus entry. Ionophores can be broadly classified into two categories: protons and weak bases [71, 74]. By facilitating proton movement across membranes, ionophores reduce the acidity of endosomes and counteract the acidic environment required for viral activation. Ionophores counteracting the acidic environment in an endosome are important because they are in an acidic environment where a conformational change in the alphavirus E1 glycoprotein takes place, allowing for membrane fusion and, consequently, for alphavirus entry [86].

In contrast, weak bases increase the pH in an endosome, inhibiting low-pH-dependent fusion, which is an important mechanism for alphavirus entry [86, 87]. Administration of a weak base, such

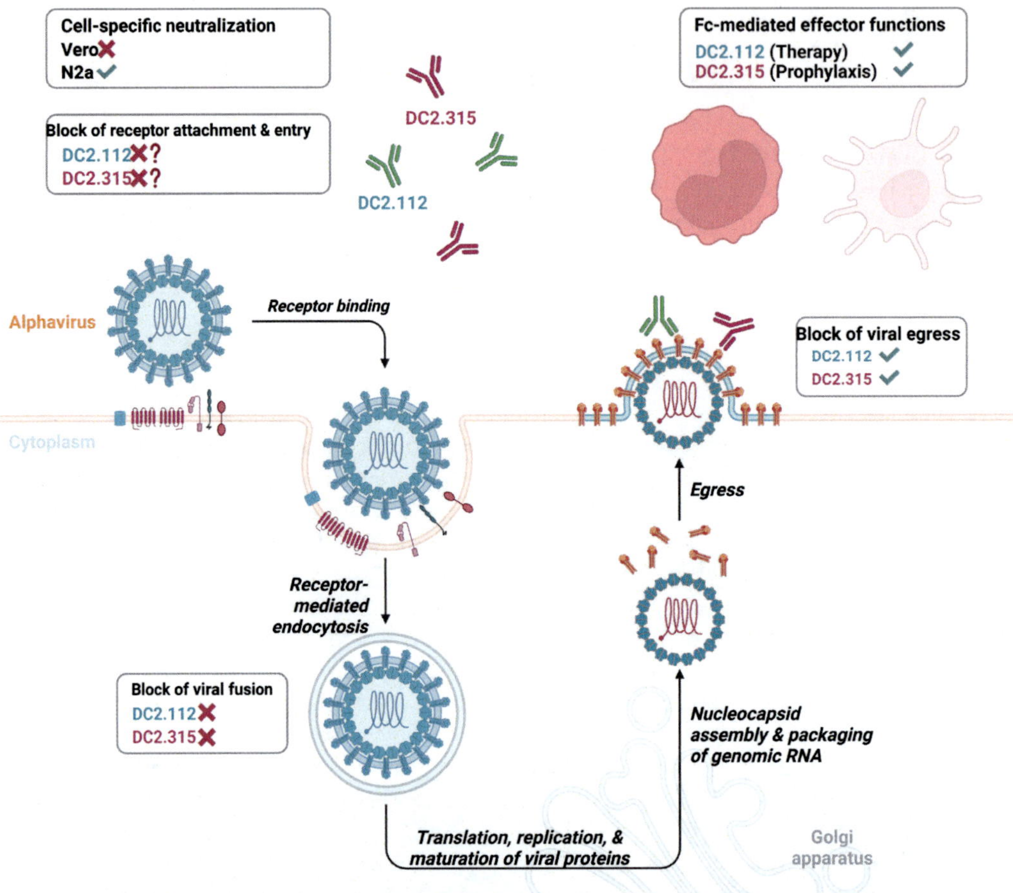

FIGURE 1.3 Molecular replication cycle of alphaviruses. The diagram outlines the stages of alphavirus replication following receptor-mediated endocytosis. After internalization via clathrin-coated vesicles, endosomal acidification triggers pH-dependent fusion of the viral envelope with the endosomal membrane, allowing nucleocapsid disassembly and genome release. The positive-sense 49S genomic RNA is translated into nonstructural polyproteins (P123, P1234), which are cleaved in cis and trans to form the replication complex. This complex synthesizes minus-strand RNA intermediates, which then serve as templates for the production of genomic (49S) and subgenomic (26S) RNAs. Subgenomic RNAs are translated into structural proteins (C, E3, E2, 6K, TF, and E1) that undergo glycoprotein processing in the rough endoplasmic reticulum (RER) and Golgi apparatus. Nucleocapsid cores assemble in the cytoplasm, migrate to the plasma membrane, and incorporate viral glycoproteins, producing mature virions that egress via budding. This coordinated process emphasizes the roles of nonstructural proteins (nsPs) in replication, structural proteins in virion assembly, and host organelles in facilitating viral maturation. (Created with BioRender.com.)

as ammonium chloride, has been proven effective in preventing membrane fusion of virions in the environment of an endosome and, therefore, in preventing cytoplasmic delivery of an encapsidated nucleocapsid of virions [86, 87]. Inhibition occurs at an early stage of alphavirus infection and is significant, suggesting that compounds with pH-modulating activity could have significant efficacy in controlling alphavirus infection by counteracting the early phases of infection [88].

Studies have shown that ionophores can substantially hinder the entry of alphaviruses into various cellular models. Targeted research on the SINV illustrated that ionophore treatment drastically suppressed alphaviral infectivity through effective inhibition of obligatory conformational E1 glycoprotein rearrangements [70]. In addition, treatment with a weak base can inhibit the acidification of endosomes, a requirement for virus fusion, and thus increase alphaviral entry [89]. Targeted ionophore treatment during alphaviral entry into its life cycle indicates its therapeutic use in alphaviral outbreaks.

Ionophore use can reach alphaviruses via an entry mechanism shared across other viral groups [89]. Ionophore action in endosomal pH modulation can provide novel therapeutic interventions for infections with viruses with the same entry mechanism [69, 77]. For instance, research on ionophore action in alphaviral infections has demonstrated that ionophores can block alphaviral entry via alterations in endosomal conditions [90]. The cross-applicability of therapeutic interventions across virus types plays a central role in pH modulation during antiviral therapies.

In summary, ionophores represent a family of compounds with therapeutic utility in the inhibition of alphaviral infection of target cells [91]. Ionophores inhibit acidification processes in endosomes and, in doing so, inhibit conformational changes in alphaviral fusion proteins, thereby inhibiting membrane fusion and infection initiation [92]. Ionophores have therapeutic utility for targeting early alphaviral infections and for controlling a variety of other types of infections, making ionophores a family of compounds of value for future development and investigation of antiviral therapeutic interventions [93].

1.2.2 Structural Insights

1.2.2.1 Key Viral Proteins

The intricate structures of alphaviruses and paramyxoviruses have proven to be important model systems for studying modern conceptions of viral entry processes and therapeutic targets. El and E2 glycoproteins play key roles in receptor modulation and membrane fusion [87, 88, 94]. Receptor contact induces significant conformational changes in integral membrane glycoproteins, which are critical for membrane fusion and the delivery of the viral nucleocapsid into target cells [95, 96]. Conformational shifts in these proteins face direct evolutionary selective pressures that have been optimized for contact with receptors in target cells.

The NiV is a model paramyxovirus in which HN and F proteins are specifically utilized for target recognition and processes involved in virion infection in target cells [97]. HN and F proteins enable target receptor contact and initiation of fusion between the virion envelope and target membrane, providing a model for successful membrane fusion between adjacent cells [98]. It creates multinucleated giant cells or syncytia and helps in the dissemination of the virus in tissues [99, 100]. Conformational shifts in such proteins reveal a level of specialization and a specific evolutionary path for such pathogens, and when placed in contrast to alphaviruses, it seems a discernible evolutionary path for such pathogens [101].

A detailed understanding of viral protein architecture is crucial for the development of antiviral strategies and vaccine platforms [6, 8]. Molecular interactions between different viral glycoproteins and their respective receptors are promising targets for inhibiting the mechanisms by which virions enter host cells [65, 83]. Strategies for inhibiting such molecular interactions using synthesized small-molecule drugs or antibody treatments could pave the way for the development of a strong strategy to counter such mechanisms, thus impeding the course of infection [58, 84, 90]. The conformational structures involved in fusion processes, resolved by cryo-electron microscopy and X-ray crystallography, can provide valuable information for structure-based and systematic therapeutic approaches [102].

In addition, the evolutionary strategies displayed by these viruses, especially for their ability to adapt to different environmental settings, highlight the significance of structural studies for understanding the pathogenic characteristics of virions [17, 34]. For instance, changes in certain glycoproteins may affect receptor-binding affinities and alter the dynamics of fusion events, potentially making such virions even more virulent [103, 104]. Evolved modifications illustrate the need for continuous observation and study of the structures of little-known virions, considering the emerging pathogenetically significant structures and characters in new strains and their potential for public health impact [7, 32, 54].

Accessory proteins play a significant role in virion integrity and function [95, 96]. In most cases, such proteins enable evasion of a host organism and even stimulate virion replication in the host environment [56, 83]. For example, specific accessory proteins modulate host cell signaling pathways, thereby creating a microenvironment that supports the maturation of virions [105, 106]. A thorough analysis of the structural features that enable virion-accessory protein interactions is critical for developing effective antiviral strategies. In conclusion, a comparative analysis of key viral proteins of relatively under-studied viruses, such as alphaviruses and paramyxoviruses, demonstrated a complex evolutionary landscape of adaptations responsible for infection and disease progression [87]. Not only can the individual structures of such proteins disclose evolutionary challenges encountered by such viruses, but such information can also contribute to developing target-specific antiviral interventions and immunizations [91]. As future studies increasingly expose the molecular complexity of such viruses, it will become increasingly important to apply this information to develop effective public health interventions [51, 90].

1.2.2.2 Proteomic Studies

Proteomic investigations have increasingly played a key role in shedding light on the relationship between underappreciated viruses and host cell functions [96]. By characterizing viral and host protein interactions, researchers can identify key interfaces in the life cycle of a virus, including processes such as entry into, replication, and exit from the host cell [94]. Gaining such information is important for developing targeted antiviral drugs and vaccines, specifically for viruses that have been underfunded in terms of financial investment and access to funding [107, 108].

A critical function of proteomic studies is their ability to reveal the processes through which viral proteins hijack host cellular processes to support their replication [109]. For example, it has been shown that the murine noroviral nonstructural protein NS1 engages with VAMP-associated protein A (VAPA), a host protein with a role in post-infection but pre-RNA synthesis [107]. Likewise, studies on CHIKV have shown that infection with CHIKV induces a change in the overall proteome in Aedes albopictus cells, underlining the complex role of host proteins in supporting genome replication and processing of proteins, while also revealing factors in a host that inhibit viral proliferation [108].

Despite the profound benefits imparted through cutting-edge proteomic techniques, its use in cases of regularly misdiagnosed viral infection has not yet been appropriately utilized [110, 111]. Exploitation stems, for the most part, from a lack of access to appropriately equipped laboratories following biosafety protocols in addition to the financial requirements for such studies. Such barriers are important because they can reveal valuable targets for the development of novel antiviral therapies and expand existing knowledge regarding viral pathogenesis [112, 113]. For example, the identification of host factors with which virions may come into contact can reveal valuable therapeutic targets. The interaction between enteroviral 3A and the discovered host factor ACBD3 can enable efficient recruitment of key factors in RNA replication [114].

In addition, proteomic analysis has identified that a virus can cause oxidative stress in the host organism, a characteristic that permits both capsid assembly and genomic RNA replication [115]. Viral protein and host cell function interactions describe the intricate processes involved in viral pathogenesis and determine the need for detailed investigation of such processes [116]. For instance, enteroviral protein interaction with phosphatidylinositol 4-kinase IIIβ at an RNA replication site specific for such an interaction is a relevant example in the context of a virus commandeering a host's processes because of its greater efficiency in replication [117].

The promise of proteomic analysis in revealing new antiviral targets can be seen in studies confirming the specific associations between viral proteins and host factors that are critical for viral entry and replication [64, 107]. For instance, hepatitis C virus protein and host protein associations have been documented to have an impact on both viral entry and replication kinetics, and their interference could potentially yield new targets for antiviral therapeutic development [118]. In addition, the determination of critical host proteins involved in herpes simplex virus replication underpinned the necessity for studying host-pathogen associations during a viral infection episode [119, 120].

In summary, proteomic analysis is a useful tool for analyzing poorly characterized viral interfaces with complex processes in host cells [110]. By disassembling key interfaces between proteins and recontextualizing their functional roles in the life cycle of a virus, antiviral intervention, and vaccine development targets can be defined [110]. Technical and infrastructure barriers in the analysis of poorly characterized viruses must first be overcome to fully realize the potential of high-tech proteomic approaches. Long-term studies and infrastructure investments are imperative to improve our current level of expertise in viral pathogenesis and therapeutic intervention development [121, 122].

1.2.3 Therapeutic Targets

Structural and proteomic information obtained from research on less-studied viruses, such as alphaviruses and paramyxoviruses, translates into a range of therapeutic targets. The three broad classes of targets are fusion proteins, ion channel modulators, and attachment proteins. There are certain intervention opportunities in each of the three classes with a significant potential to modulate viral pathogenesis and infection control as shown in Table 1.2.

TABLE 1.2
Molecular Therapeutic and Vaccine Targets in Selected Neglected Viruses

Virus	Key Viral Proteins Involved in Entry/Replication	Host Factors Involved	Potential Therapeutic Targets	Vaccine/Drug Development Status
Chikungunya virus (CHIKV) (Alphavirus)	El glycoprotein (fusion), E2 glycoprotein (receptor binding), nsP2 helicase	Actin filaments, BST-2/tetherin, reticulon proteins	Blocking El conformational changes E2 receptor-binding inhibition nsP2 helicase inhibitors	Several vaccine candidates (virus-like particles, live-attenuated) nsP2-based antivirals under study
Semliki Forest virus (SFV) (Alphavirus)	El/E2 glycoprotein heterodimer; viral replication organelles	PI3K–actin–microtubule network	Inhibiting clathrin-mediated endocytosis Targeting replication organelles	Used as model for alphavirus vaccine platforms
Nipah virus (NiV) (Paramyxovirus)	F protein (fusion), HN protein (attachment)	Actin cytoskeleton reorganization	Fusion inhibitors (F protein) Anti-HN antibodies	Human monoclonal antibodies (m102.4) in clinical evaluation Multiple vaccines in phase I/II
Lassa virus (Arenavirus)	Glycoprotein complex (GP1, GP2)	Dystroglycan receptor	GP-mediated entry inhibitors	DNA vaccines and recombinant vector vaccines in development
Crimean-Congo hemorrhagic fever virus (CCHFV) (Nairovirus)	Glycoproteins Gn/Gc	Tick–host interactions	Blocking glycoprotein processing RNA polymerase inhibitors	No licensed vaccine (outside limited regional use) Antivirals under development
Oropouche virus (OROV) (Orthobunyavirus)	Gc (fusion), Gn (attachment)	Midges as vectors	Glycoprotein inhibitors Host factor modulation	No vaccines/antivirals Experimental models under study

This table summarizes key viral proteins and host factors involved in the entry and replication of neglected viruses, highlighting their relevance as therapeutic targets. Fusion proteins (e.g., El in alphaviruses, F protein in paramyxoviruses), attachment proteins (e.g., E2, HN), and enzymatic regulators (e.g., nsP2 helicase in Chikungunya virus) represent critical antiviral intervention points. Host factors such as actin filaments, tetherin, and PI3K–microtubule networks illustrate additional mechanisms that can be targeted. The final column outlines available vaccines or candidate therapeutics, underscoring the persistent translational gap despite significant public-health need.

1.2.3.1 Fusion Proteins (F, E1)

Fusion proteins, such as alphaviral E1 glycoprotein and paramyxoviral F protein, play key roles in the infection of target cells by inducing specific conformational changes that allow fusion between the virus envelope membrane and target cell membranes. Inhibition of such conformational changes is an effective mechanism for preventing infection. For instance, antibodies binding to specific regions of the E2 protein in CHIKV block the disassembly of E1 and E2 proteins, subsequently impairing membrane fusion efficiency and, consequently, infection [123]. In addition, inhibition of paramyxoviral F proteins is a potential mechanism for inhibiting membrane fusion, an important infection and dissemination process [124].

1.2.3.2 Ion Channel Modulators

Ion channel modulators, including ionophores and agents that can modulate pH, block a variety of processes utilized by many viruses during their infectious life cycle, particularly endosomal fusion. By blocking the acidification process in endosomes, such agents disrupt conformational changes in viral fusion proteins, which is an essential step in infection initiation [125, 126]. For example, the addition of amines and alkaline substances such as ammonium chloride can prevent membrane fusion between virions and membranes found in endosomes, thus effectively preventing infection [86, 87]. Such a methodology has immense significance as a preventive strategy against alphaviruses and other alphaviral pathogen infections [73, 74].

1.2.3.3 Attachment Proteins (HN, E2)

The role of HN in alphaviruses and E2 in paramyxoviruses is critical, as both contribute to the interaction with receptors and subsequent invasion of host cells [127]. Interruption of the contact between specific receptors and their respective proteins can inhibit viral infection [128]. For example, the presence of anti-HN antibodies has the potential to inhibit receptor binding, thus prohibiting viral entry [129]. Targeting E2 in CHIKV with specific peptides is a therapeutic intervention with the potential for higher efficacy in virus neutralization, and this finding demonstrates that such a protein is a key target for vaccine development [128].

1.2.3.4 Progress in Vaccine Innovation and Drug Discovery

This has rekindled curiosity in fighting such underexploited viruses, specifically in terms of current and future infectious disease epidemics, spurring the development of drugs against specific viral pathogens and systemic vaccine approaches [130]. The integration of cutting-edge methodologies, such as cryo-electron microscopy, X-ray crystallography, and single-molecule proteomics, has facilitated the development of increasingly sophisticated and reliable viral protein structures, driving ongoing drug programs [35, 131]. CHIKV nsP2 helicase crystallization is a potential candidate for antiviral drug development, considering its critical role in viral replication [132].

Moreover, the recognition of host factors interacting with viral proteins can provide more details regarding therapeutic possibilities. For example, actin and Chikungunya viruses have been discovered to be essential for viral replication, and inhibition of such a complex could represent a novel antiviral possibility [129, 133]. In addition, the manipulation of host cell signaling processes through proteomic analysis has the potential to discover new intervention targets [134, 135]. In summary, proteomic and structural studies of underexploited viruses have revealed several potential therapeutic targets, including fusion proteins, ion channel regulators, and adhesion proteins. As technological progress and methodologies become increasingly sophisticated, future studies will surely contribute to a deeper understanding of viral pathogenesis, in addition to developing effective antiviral drugs and vaccines [6, 8]. By focusing on these critical areas, scientists can break new ground in fighting underdiagnosed and under-researched viral diseases and creating positive public health impacts [7, 39, 42]. Neglected viruses, such as alphaviruses and paramyxoviruses, present intricate entry mechanisms that remain underexplored owing to research prioritization and funding biases [55]. Nonetheless, recent technological advances in molecular biology and proteomics have begun to peel away the complex layers surrounding cell entry, structural proteins, and pathogenesis [121]. By mapping these entry pathways, particularly the roles of clathrin-mediated endocytosis, pH-dependent membrane fusion, and ionophore-based inhibition, researchers can target the critical steps in the viral life cycle [57, 71]. These findings underscore that, despite logistical and financial challenges, an enhanced understanding of neglected viruses holds immense potential for therapeutic innovation and the broader goal of mitigating global infectious disease threats [28].

1.3 IMMUNE EVASION STRATEGIES: CORNERSTONE IN NEGLECTED VIRUS PERSISTENCE AND PATHOGENESIS

The knowledge gained through studies of less-well-characterized viruses, such as alphaviruses and paramyxoviruses, converges to reveal many therapeutic targets. These targets can be grouped into three general categories: fusion proteins, ion-channel modulators, and attachment proteins. All three categories present new avenues for therapeutic intervention with considerable potential for both viral pathogenesis and infection management (Figure 1.4).

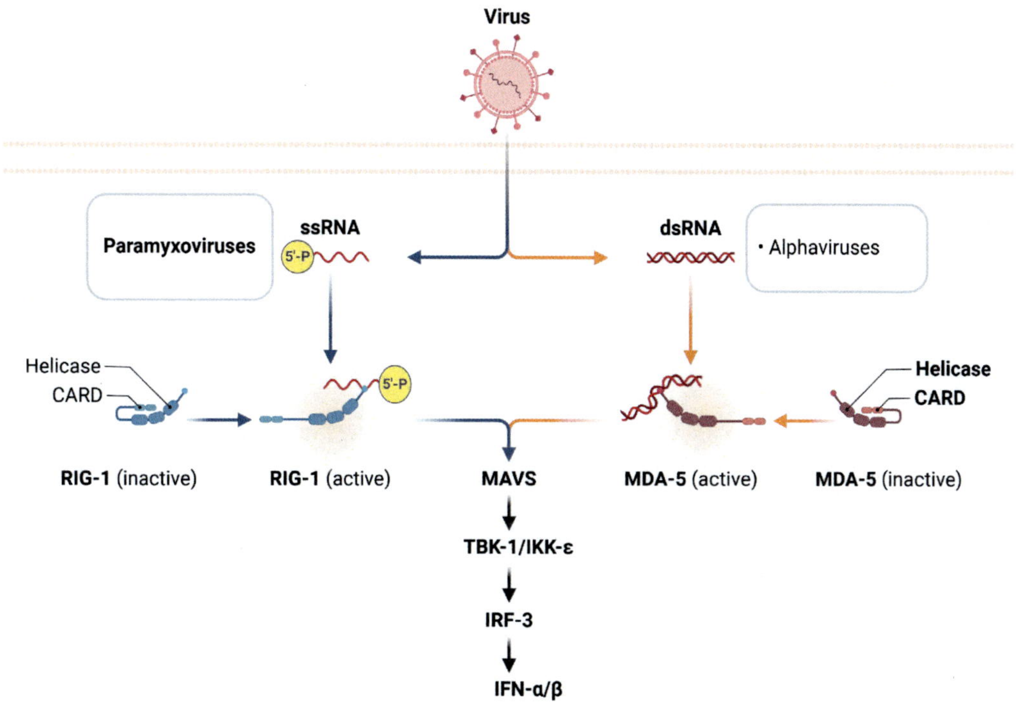

FIGURE 1.4 Innate immune sensing of alphaviruses and paramyxoviruses. Paramyxoviral single-stranded RNA (ssRNA, 5'-triphosphate) activates the cytosolic helicase RIG-I, whereas alphaviral double-stranded RNA (dsRNA) is recognized by MDA-5. Upon activation, both sensors signal via MAVS, leading to TBK1/IKKε activation, IRF3 phosphorylation, and induction of type I interferons (IFN-α/β). This pathway illustrates the first line of antiviral defense prior to viral immune evasion strategies. (Image was created by Biorender.com.)

1.3.1 Fusion Proteins (F, E1)

Fusion proteins, such as alphaviral El glycoprotein and paramyxoviral F protein, play critical roles in target cell infections. Specific conformational changes in proteins allow them to fuse with the target cell membranes. Suppression of conformational changes is a viable mechanism for interfering with viral infection processes. One such mechanism can be seen in CHIKV E2 protein-specific site inhibition through antibody complexation, effectively preventing E2 disassembly from El and its function in membrane fusion and, consequently, infection [136]. On the other hand, inhibition of the F protein in paramyxoviruses is an effective mechanism for interfering with membrane fusion, a critical activity in both the initiation and establishment of infection [137].

1.3.2 Ion Channel Modulators

Ion channel modulators, such as ionophores and acid-modifying drugs, can inhibit endosomal fusion, which is an important and targetable mechanism used by many pathogens in the infection process. Pharmacological agents prevent conformational changes in viral fusion proteins, an essential feature required for infection initiation, by preventing endosomal acidification. For instance, the use of ammonium chloride, a basic drug, has been proven to effectively impair infection by preventing the membrane fusion of both virions and endosomes, thus blocking infection [138]. In an ideal situation, such a mechanism would act as an efficient inhibitor of infection, specifically against alphaviruses and related pathogens.

1.3.3 Attachment Proteins (HN, E2)

The role played by such proteins, for instance, HN in alphaviruses and E2 in alphaviruses, is critical in terms of their interaction with receptors, allowing for viral entry into target cells and infection. Inhibition of such interactions between receptors and proteins can effectively hinder viral entry into target cells. For instance, the use of anti-HN antibodies has proven effective in interfering with virus-receptor contact, with significant inhibition of virus entry [139]. Furthermore, E2 inhibition in CHIKV by specific peptides has been reported to increase the effectiveness of the infection blockade through neutralization; thus, such a protein is targeted in vaccine development [140].

1.3.4 Vaccine Innovation and Drug Discovery Progress

The increased activity in working with relatively less-characterized viruses, and in terms of chronic and emerging disease epidemics, has generated significant breakthroughs in the systemic and systematic development of drugs and vaccine approaches. Application of sophisticated techniques, such as cryo-electron microscopy, X-ray crystallography, and single-molecule proteomics, have helped in high-resolution visualization of structures of viral proteins, and in the process, have guided the development of new antiviral drugs in a sophisticated model. For instance, crystallization of the CHIKV nsP2 helicase revealed candidates for antiviral drug development, highlighting the key role played by this protein in the mechanism of viral reproduction [141].

Moreover, the determination of interfaces between viral and host factors could reveal new therapeutic avenues. Investigation of the interaction of the CHIKV with actin is considered paramount for its reproduction, suggesting that inhibition of such interfaces could produce new antiviral therapeutic options [142]. In addition, the management of signaling processes in host cells through proteomic studies holds significant potential for new therapeutic target identification [143]. Altogether, studies of underexploited viral structures, in addition to proteomic studies, reveal a range of therapeutic options, such as fusion proteins, ion channel regulators, and adhesion proteins. With methodologies and tools in constant development, future studies have strong potential to enrich our current knowledge in terms of viral pathogenesis and enable the efficient development of antiviral drugs and vaccines. By focusing on such key areas, scientists can potentially make significant breakthroughs in controlling viral disease, many of them underdiagnosed and under-researched, and contribute toward a better public health scenario.

1.4 ZOONOTIC TRANSMISSION DYNAMICS

The mechanisms of zoonotic transmission play a critical role in explaining the emergence and dissemination of under-recognized viruses. Underrated viruses have complex ecological cycles in various animals before being transmitted to humans. Such dynamics must be understood to develop effective interventions and controls.

1.4.1 Reservoir Hosts

In most cases, reservoir hosts for zoonotic viruses have no disease symptoms, but they play a critical role in disease survival in the environment. A range of species, such as bats, rodents, and non-human primates, serve as reservoir hosts. Among other coronaviruses, the NiV has been seen as an asymptomatic reservoir among bat populations, exhibiting a significant level of resilience under various environmental conditions [65]. Lassa virus transmission has, in turn, been associated with specific rodent species, posing a significant public health issue to populations in Western Africa [19, 33]. A review of the unique physiological and immunological characteristics of these species indicates how they can carry such viral pathogens without showing clinical signs, thus facilitating their coexistence with the viruses [19].

1.4.2 Intermediate Hosts

Domestic animals act as intermediate hosts for zoonotic diseases, with the primary function of enabling interactions between humans and wildlife. For instance, studies have proven that pigs act as efficient amplifying hosts for the NiV in Bangladesh and Malaysia, with cases of transmission reported through direct contact between pigs and humans [84, 97]. In addition, it has been proven that livestock can be vectors for the Crimean-Congo hemorrhagic fever virus, thus greatly amplifying zoonotic transmission routes [107]. By acting as a bridge between humans and wildlife reservoirs, these intermediate hosts enable the spillover of viruses that might not naturally cause disease in humans, allowing pathogens to break free from their initial reservoirs and cause infections among human populations.

1.4.3 Environmental and Climate Factors

Environmental determinants, such as habitat degradation, fragmentation, and climate variability, play a crucial role in increasing zoonotic transmission dynamics. Environmental factors facilitate human-wildlife contact, alter vector distribution, and enhance virus adaptability to new hosts [17, 34]. Habitat degradation, for example, leads to increased contact between humans and wildlife and, subsequently, spillover events [17]. Climate change accelerates such risks even further through geographical range shifts in hosts and vectors and potentially extends the zoonotic disease range [45].

1.4.4 Influence on Vulnerable Groups

The intersection of social determinants of health and biomedical vulnerabilities intensifies the underdiagnosis of viral diseases in economically disadvantaged groups. About 1.7 billion individuals living in low- and middle-income countries (LMICs) live in high-transmission environments.

Rural and peri-urban communities, with limited access to care and poor diagnostics, are at the highest risk [144]. The underlying factors include the following.

- *Occupational Risk*: Agricultural workers, many of whom work in endemic areas, have increased contact with reservoirs and intermediate hosts. Poor occupational health regulations and poor access to protective gear (e.g., personal protective equipment [PPE]) further enhance vulnerability [145].
- *Gender Inequality*: Women have a higher infection rate than men, often taking on caregiver roles and working in occupations that increase their exposure to animals [146].

1.4.5 Socioeconomic Consequence

The economic burden of neglected viral diseases creates cycles of poverty. Affected family incomes go toward a significant portion of annual earnings for medical expenses, with many family incomes dipping below the poverty level in the following year [147]. Long-term productivity loss and impairment, in addition to ongoing medical expenses, cause an additional economic burden, particularly in pediatric infections, with remarkably high infection rates compared to adults [148]. Interruption of school studies can have long-term consequences for future generations and, therefore, cause cycles of poverty [149].

1.4.6 Mitigation of Vulnerabilities

For vulnerability mitigation in terms of the burden of neglected viral diseases, it is important to implement full-fledged interventions.

- *Surveillance Systems*: There is a dire necessity to strengthen molecular and epidemiological disease and spillover event surveillance and capabilities [150].
- *Healthcare Infrastructure*: Rural medical infrastructure and laboratory testing capabilities must be developed through investment for efficient disease management [15, 17].
- *Interdisciplinary Community Intervention*: Mobilization and awareness at the community level can slow transmission and promote positive disease outcomes [42].
- *Economic Policies*: Targeted interventions can effectively curtail disease burden, with positive economic implications in terms of increased productivity and reduced medical expenses [99]. In conclusion, the zoonotic disease transmission dynamics depend on complex ecological, environmental, and social factors. The identification of such dynamics is important for formulating efficient disease prevention and management strategies for neglected viral diseases. Mitigation of such diseases through vulnerabilities in such susceptible populations and strengthening disease and laboratory infrastructure can promote positive public health consequences.

1.5 PUBLIC HEALTH IMPACT

1.5.1 Epidemiological Burden

The repercussions of overlooked viruses on public health are significant but underacknowledged. Neglected viruses play a considerable role in the global disease burden, particularly in LMICs with poor medical infrastructure. This section discusses the epidemiological impact of these frequently neglected viruses, the difficulties involved in establishing effective intervention programs, and the socioeconomic and ethical factors that make public health efforts challenging [151].

Viruses that are often underdiagnosed present a significant challenge to public health, with high morbidity and mortality rates following infection. For example, an estimated 390 million cases of dengue infection occur annually, with significant consequences for both financial and economic security and health in terms of outcomes in the tropics. The most severe forms of dengue, such as dengue hemorrhagic fever, have alarmingly high case fatality and predominantly impact economically disadvantaged and poorly accessible communities with poor access to medical care [67].

Lassa fever, for its part, claims about 300,000 cases and over a thousand deaths a year in Western Africa alone. There is a high level of suspected underreporting of cases of Lassa fever, and therefore, a high chance that actual cases could be several times larger, yet unrecognized and unregistered, due to poor reporting and poor access to medical care [19, 33]. Ineffective therapeutic and diagnostic tools present a significant challenge, and complicated case management and delivery of proper care to impacted communities.

1.5.2 Challenges to Successful Responses

There are many challenges posed by overlooked viruses in terms of their control. One of these is a barrier to diagnosis. There is a lack of access to effective point-of-care diagnostics, with restricted access to timely reporting and management of cases. The inability to have effective diagnostics can result in underdiagnosis and delayed therapy, creating a barrier to effective public health interventions [68].

However, therapeutic constraints pose significant challenges. The unavailability of specific antiviral drugs mandates that therapy often involves supportive care, which may be ineffective in extreme cases [152]. In addition, concerning certain viruses, such as dengue, numerous lesser-known pathogens, such as Lassa and Nipah, have no effective vaccination options; therefore, communities are at increased risk [153].

1.5.3 Socioeconomic and Ethical Factors

Failure to prioritize these viruses leads to deeper systemic inequities. Socioeconomic barriers contribute to high infection rates of neglected viral diseases. Most affected individuals live in poor conditions and lack proper access to medical infrastructure; therefore, such populations become susceptible to infection [154].

Ethical concerns apply specifically about high-risk population studies. It is important to ensure fair access to interventions and benefits, with marginalized groups disproportionately burdened with disease, but with little care, support, or intervention [155]. In conclusion, the public health consequences of under-recognized viruses are significant, with significant epidemiologic barriers and several impediments to successful interventions. To address such concerns, a thorough and multidimensional approach with a focus on strengthening diagnostics, discovering targeted therapies, and offering fair access to care must be pursued. By focusing on such factors, public health programs can work to mitigate under-recognized virus impact and improve care for high-risk populations.

1.6 ONE HEALTH APPROACH

The "One Health" model presents a unifying model that recognizes the complex interdependence between human, animal, and environmental factors. For zoonotic viruses with an intertwined interrelationship in such a region, such a model is most relevant, with coordination of its constituent parts delivering effective planning. By integrating observations in a range of studies through its multidisciplinary model, One Health aims to amplify disease surveillance, coordinate a single, consolidated reaction, and, in the long-term, counteract zoonotic disease impact [156, 157].

1.6.1 Networks for Surveillance

An integral part of the One Health platform is the creation of multidisciplinary networks for surveillance programs. Networks enable the effective analysis of virus communities, which is a key consideration in spillover prediction between animals and humans [13, 150]. By combining veterinary and medical professionals with studies in the environment, such networks can conduct a thorough analysis of the factors responsible for zoonotic transmission [20]. For example, it has been noted that effective operations of the One Health network entail coordinated actions toward infectious disease emergence and curbing [16].

1.6.2 Synchronized Responses

The fundamental principle of coordinated intervention approaches in an integrated model is critical for intervention programs, specifically in terms of the availability of resources and complex dynamics in ecological systems [17, 34]. The model must not only promote proactive collaboration between healthcare professionals but also require integration of agricultural and environmental sectors in a manner that promotes thorough planning and intervention development to effectively reach all concerned communities [66]. This theoretical premise identifies a critical imperative for interventions against zoonotic viruses in the One Health model, which promotes community-led approaches for early detection, preparedness, and maintenance of operational capacity [66].

1.6.3 Dealing with Zoonotic Spillover

The One Health model identifies the need for intervention in factors driving zoonotic spillover in the environment. Deforestation, habitat fragmentation, and climate change cause increased contact between humans and wildlife, and subsequently, increased vulnerability to virus transmission [68]. Careful consideration of increased predictive capacities through heightened surveillance can allow for effective management of emerging mosquito-transmitted viruses and, subsequently, an imperative for proactive intervention in zoonotic diseases [68].

1.6.4 Relevance of Integrated Health Systems

The effectiveness of the One Health model necessitates the integration of health systems in all three sectors: humans, animals, and the environment. Integration promotes a single harmonized epidemic response and enables early zoonotic disease detection capabilities to demonstrate such integration through a case study of influenza transmission in pig farm settings. The effectiveness of such integration in enhancing one's understanding and controlling zoonoses is supported by considerable evidence [158].

1.6.5 Community Involvement and Ethical Factors

Ethical principles play a critical role in health, particularly in intervention and research in high-risk groups. Equitable access to such interventions and research benefits is critical for the development of trust and working with such groups [70]. Community asset mobilization is critical, with the integration of traditional, community-based information with participatory approaches strengthening intervention efficacy and allowing for the development of long-term zoonotic disease transmission approaches [71]. In conclusion, the One Health model is generally applicable for dealing with zoonotic spillover events, in addition to re-emerging and emerging infectious disease complications, which, in most instances, receive little consideration. By creating synergy between disease surveillance networks, harmonizing reaction protocols, and combining the health infrastructure between sectors, such a model can effectively boost zoonotic spillover incident prediction and prevention. Sustained funding for One Health programs, in conjunction with ethics and community integration, will become critical in enhancing public health outcomes and minimizing the burden of underemphasized viral disease.

1.7 IMPLICATIONS AND FUTURE DIRECTIONS

1.7.1 Future Directions in Mitigation of Neglected Viruses

The obstacles posed by neglected viruses are complex and multifaceted and include sophisticated immune evasion, structural complexity, and evolutionary processes. The challenges posed require an integrated strategy that includes disease investigation, disease surveillance, and public health interventions to decrease the disease burden. This section discusses the possible implications and potential avenues for research and interventions intended to mitigate the impact of underappreciated viruses [159].

Complex processes through which immune evasion viruses evade immune attacks in hosts are sophisticated and allow them to survive without eliciting immune defenses. Investigation of such processes has important therapeutic implications. For instance, approaches designed to counteract antigenic variations, interfere with interferon (IFN) signaling, and inhibit the complement cascade can enable the development of novel antiviral therapies. An in-depth investigation of the molecular basis of immune suppression can reveal the downregulation of major histocompatibility complex (MHC) molecules and cytokine modulation, providing a platform for vaccine and immunomodulatory therapy development [153, 160].

However, there is a striking lack of effective discovery of such immune evasion processes in these viruses in modern studies. There is a critical need for further investigation of such complex processes, and a deeper understanding of such pathogeneses could form a basis for enhancing diagnostics and reducing disease burden worldwide [161].

1.7.2 Proteomic and Structural Information

Recent developments in the field of structural biology have greatly improved our understanding of enigmatic viruses, particularly in terms of their functional mechanisms, as demonstrated by comparative investigations. The nsP2 protein of CHIKV is a prime example of its dual role in immune evasion and viral replication, making it a central target for therapeutic interventions [153]. The Z protein of the Lassa virus is also a central component in the budding and assembly processes, making it a key component in the formulation of antiviral therapy [162].

Proteomic studies immensely expand our knowledge regarding viral and host interactions and develop a complex platform for therapeutic intervention and personalized medical therapies. These studies emphasize the urgency of interventions for less-characterized viruses, with specific virion structures proving to be effective therapeutic options [143].

1.7.3 Evolutionary Dynamics and Risk for Spillover

Viruses that have not been well characterized present a unique view of evolutionary processes, specifically concerning zoonotic emergence and adaptation. To a certain extent, genetic variation in lesser-characterized viruses is determined by factors such as the inherently high polymerase error rate and high mutation rates, which have a significant impact on their potential for transmission between species [163]. Chances for spillover events with such viruses depend on complex interactions between selective pressures placed on reservoir animals, such as bats and rodents, which carry a range of zoonotic pathogens [164].

An in-depth examination of such processes is important to enhance predictive and preventive approaches regarding spillover events. Empirical studies comparing evolutionary patterns in lesser-characterized viruses can make significant additions to methodologies in tracking spillover events and in public health programs for minimizing spillover-related danger [165]. For instance, studies have shown that spillover events occur most often between closely related species of bats, with geographical distance and evolutionary ties being key factors [65]. Consequently, addressing the challenges posed by less-dominant viruses will require an overall multidisciplinary approach that integrates expertise in immunology, structural biology, and evolutionary biology. By prioritizing immune evasion mechanisms, leveraging structural information for therapeutic development, and understanding the evolutionary processes that allow spillover events, researchers can develop effective strategies to counter such oft-overlooked viral pathogens. Sustained funding for such studies, and collaboration between professionals in disparate disciplines, will be critical in countering the public health consequences linked with such underappreciated viruses and in strengthening worldwide health security [166].

1.8 CONCLUSIONS

Addressing the worldwide burden posed by these neglected viruses will require a deep understanding of their structural profiles, evolutionary processes, and public health implications. Such information will be critical in guiding focused intervention programs; driving vaccine development programs; and overcoming social, financial, and ethical barriers to disease management. The application of the One Health principle, with its integration of humans, animals, and the environment, will be critical for success in such ventures. Since most neglected viruses have zoonotic origins, investigating zoonotic disease emergence is critical. The consequences of such viruses are most profound in high-risk groups, whose medical inequity is compounded by social and financial obstacles to accessing care. For instance, neglected infection with a virus disproportionately affects poor communities, and such communities have increased morbidity and mortality in such groups. In such a scenario, an integral community-based intervention that integrates disease surveillance networks strengthens care infrastructure and actively engages communities in intervention and response programs will become critical. Overcoming the obstacles posed by such viruses will demand increased investments in disease studies and public health programs. The enhancement of financial investments in disease surveillance and care infrastructure, increased international collaborations, and proactive interventions through policies will go a long way to lessen the disease burden. In addition, developing climate resilience programs to overcome the changing environments that enable zoonotic virus circulation will become critical in suppressing disease prevalence. Targeted interventions for such high-risk groups, particularly in children and indigenous communities, will become critical for enhancing health equity and reducing emerging zoonotic disease threats. By focusing on interventions in such groups, public health programs can effectively channel such advances in research into real improvements in care for disadvantaged groups. In conclusion, the effective and timely management of overlooked viral ailments can be achieved through a thorough and coordinated strategy. With information gained through analysis in structural biology, evolutionary studies, and public health evaluations, stakeholders can develop multidimensional approaches that not only address immediate challenges posed by such viruses but also contribute toward long-term development in worldwide health equity.

REFERENCES

1. Okokhere, P.O.; Rafiu, M.O.; Bankole, I.A. Central nervous system disease associated with major viral haemorrhagic fever viruses of public health importance: Challenges in patient diagnosis, evaluation, and management in resource-limited settings. *EMJ Neurology* 2017, *5*, 73–81.
2. Hotez, P.J. *Forgotten People, Forgotten Diseases: The Neglected Tropical Diseases and Their Impact on Global Health and Development*; John Wiley & Sons: 2021.
3. Adetifa, I.; Muyembe, J.J.; Bausch, D.G.; Heymann, D.L. Mpox neglect and the smallpox niche: A problem for Africa, A problem for the world. *The Lancet* 2023, *401*, 1822–1824.
4. Oladoye, M.J. Monkeypox: A neglected viral zoonotic disease. *Electronic Journal of Medical and Educational Technologies* 2021, *14*, em2108.

5. Harapan, H.; Ophinni, Y.; Megawati, D.; Frediansyah, A.; Mamada, S.S.; Salampe, M.; Bin Emran, T.; Winardi, W.; Fathima, R.; Sirinam, S. Monkeypox: A comprehensive review. *Viruses* 2022, *14*, 2155.
6. Ferreira, L.L.G.; Andricopulo, A.D. Drugs and vaccines in the 21st century for neglected diseases. *The Lancet Infectious Diseases* 2019, *19*, 125–127.
7. Cupertino, M.C.; Resende, M.B.; Mayer, N.A.J.; Carvalho, L.M.; Siqueira-Batista, R. Emerging and re-emerging human infectious diseases: A systematic review of the role of wild animals with a focus on public health impact. *Asian Pacific Journal of Tropical Medicine* 2020, *13*, 99–106.
8. de Lima Cavalcanti, T.Y.V.; Pereira, M.R.; de Paula, S.O.; Franca, R.F.O. A review on chikungunya virus epidemiology, pathogenesis and current vaccine development. *Viruses* 2022, *14*, 969.
9. Mascarenhas, M.; Garasia, S.; Berthiaume, P.; Corrin, T.; Greig, J.; Ng, V.; Young, I.; Waddell, L. A scoping review of published literature on chikungunya virus. *PLOS One* 2018, *13*, e0207554.
10. Montalvo Zurbia-Flores, G.; Reyes-Sandoval, A.; Kim, Y.C. Chikungunya virus: Priority pathogen or passing trend? *Vaccines* 2023, *11*, 568.
11. Rougeron, V.; Sam, I.C.; Caron, M.; Nkoghe, D.; Leroy, E.; Roques, P. Chikungunya, a paradigm of neglected tropical disease that emerged to be a new health global risk. *Journal of Clinical Virology* 2015, *64*, 144–152.
12. Bartholomeeusen, K.; Daniel, M.; LaBeaud, D.A.; Gasque, P.; Peeling, R.W.; Stephenson, K.E.; Ng, L.F.P.; Ariën, K.K. Chikungunya fever. *Nature Reviews Disease Primers* 2023, *9*, 17.
13. Worsley-Tonks, K.E.L.; Bender, J.B.; Deem, S.L.; Ferguson, A.W.; Fèvre, E.M.; Martins, D.J.; Muloi, D.M.; Murray, S.; Mutinda, M.; Ogada, D. Strengthening global health security by improving disease surveillance in remote rural areas of low-income and middle-income countries. *The Lancet Global Health* 2022, *10*, e579–e584.
14. Pannu, J.; Barry, M. Global health security as it pertains to Zika, Ebola, and COVID-19. *Current Opinion in Infectious Diseases* 2021, *34*, 401–408.
15. Traore, T.; Shanks, S.; Haider, N.; Ahmed, K.; Jain, V.; Rüegg, S.R.; Razavi, A.; Kock, R.; Erondu, N.; Rahman-Shepherd, A. How prepared is the world? Identifying weaknesses in existing assessment frameworks for global health security through a one health approach. *The Lancet* 2023, *401*, 673–687.
16. Wolicki, S.B.; Nuzzo, J.B.; Blazes, D.L.; Pitts, D.L.; Iskander, J.K.; Tappero, J.W. Public health surveillance: At the core of the global health security agenda. *Health Security* 2016, *14*, 185–188.
17. Zinsstag, J.; Kaiser-Grolimund, A.; Heitz-Tokpa, K.; Sreedharan, R.; Lubroth, J.; Caya, F.; Stone, M.; Brown, H.; Bonfoh, B.; Dobell, E. Advancing one human–animal–environment health for global health security: What does the evidence say? *The Lancet* 2023, *401*, 591–604.
18. Ozdarendeli, A. Crimean–Congo hemorrhagic fever virus: Progress in vaccine development. *Diagnostics* 2023, *13*, 2708.
19. Aloke, C.; Obasi, N.A.; Aja, P.M.; Emelike, C.U.; Egwu, C.O.; Jeje, O.; Edeogu, C.O.; Onisuru, O.O.; Orji, O.U.; Achilonu, I. Combating Lassa fever in West African sub-region: Progress, challenges, and future perspectives. *Viruses* 2023, *15*, 146.
20. O'Hearn, A.E.; Voorhees, M.A.; Fetterer, D.P.; Wauquier, N.; Coomber, M.R.; Bangura, J.; Fair, J.N.; Gonzalez, J.P.; Schoepp, R.J. Serosurveillance of viral pathogens circulating in West Africa. *Virology Journal* 2016, *13*, 1–6.
21. Spengler, J.R.; Estrada-Peña, A.; Garrison, A.R.; Schmaljohn, C.; Spiropoulou, C.F.; Bergeron, É; Bente, D.A. A chronological review of experimental infection studies of the role of wild animals and livestock in the maintenance and transmission of Crimean-Congo hemorrhagic fever virus. *Antiviral Research* 2016, *135*, 31–47.
22. Nasirian, H. New aspects about Crimean-Congo hemorrhagic fever (CCHF) cases and associated fatality trends: A global systematic review and meta-analysis. *Comparative Immunology, Microbiology and Infectious Diseases* 2020, *69*, 101429.
23. Saga, K.; Kaneda, Y. Virosome presents multimodel cancer therapy without viral replication. *BioMed Research International* 2013, *2013*, 764706.
24. Kaneda, Y. Virosome: A novel vector to enable multi-modal strategies for cancer therapy. *Advanced Drug Delivery Reviews* 2012, *64*, 730–738.
25. Wilschut, J. Influenza vaccines: The virosome concept. *Immunology Letters* 2009, *122*, 118–121.
26. Huckriede, A.; Bungener, L.; Stegmann, T.; Daemen, T.; Medema, J.; Palache, A.M.; Wilschut, J. The virosome concept for influenza vaccines. *Vaccine* 2005, *23*, S26–S38.
27. de Jonge, J.; Holtrop, M.; Wilschut, J.; Huckriede, A. Reconstituted influenza virus envelopes as an efficient carrier system for cellular delivery of small-interfering RNAs. *Gene Therapy* 2006, *13*, 400–411.
28. Mackey, T.K.; Liang, B.A. Threats from emerging and re-emerging neglected tropical diseases (NTDs). *Infection Ecology & Epidemiology* 2012, *2*, 18667.

29. LaBeaud, A.D. Why arboviruses can be neglected tropical diseases. *PLOS Neglected Tropical Diseases* 2008, *2*, e247.
30. Utzinger, J.; Becker, S.L.; Knopp, S.; Blum, J.; Neumayr, A.L.; Keiser, J.; Hatz, C.F. Neglected tropical diseases: Diagnosis, clinical management, treatment and control. *Swiss Medical Weekly: Official Journal of the Swiss Society of Infectious Diseases, the Swiss Society of Internal Medicine, the Swiss Society of Pneumology* 2012, *142*.
31. Paessler, S.; Walker, D.H. Pathogenesis of the viral hemorrhagic fevers. *Annual Review of Pathology: Mechanisms of Disease* 2013, *8*, 411–440.
32. Flórez-Álvarez, L.; de Souza, E.E.; Botosso, V.F.; de Oliveira, D.B.L.; Ho, P.L.; Taborda, C.P.; Palmisano, G.; Capurro, M.L.; Pinho, J.R.R.; Ferreira, H.L. et al. Hemorrhagic fever viruses: Pathogenesis, therapeutics, and emerging and re-emerging potential. *Frontiers in Microbiology* 2022, *13*, 1040093. https://doi.org/10.3389/fmicb.2022.1040093.
33. Grove, J.N.; Branco, L.M.; Boisen, M.L.; Muncy, I.J.; Henderson, L.A.; Schieffellin, J.S.; Robinson, J.E.; Bangura, J.J.; Fonnie, M.; Schoepp, R.J. Capacity building permitting comprehensive monitoring of a severe case of Lassa hemorrhagic fever in Sierra Leone with a positive outcome: Case report. *Virology Journal* 2011, *8*, 1–14.
34. Mackey, T.K.; Liang, B.A.; Cuomo, R.; Hafen, R.; Brouwer, K.C.; Lee, D.E. Emerging and reemerging neglected tropical diseases: A review of key characteristics, risk factors, and the policy and innovation environment. *Clinical Microbiology Reviews* 2014, *27*, 949–979.
35. Baker, R.E.; Mahmud, A.S.; Miller, I.F.; Rajeev, M.; Rasambainarivo, F.; Rice, B.L.; Takahashi, S.; Tatem, A.J.; Wagner, C.E.; Wang, L.F. et al Infectious disease in an era of global change. *Nature Reviews Microbiology* 2022, *20*, 193–205. https://doi.org/10.1038/s41579-021-00639-z.
36. van Seventer, J.M.; Hochberg, N.S. *Principles of Infectious Diseases: Transmission, Diagnosis, Prevention, and Control*; International Encyclopedia of Public Health. 2017:22–39. https://doi.org/10.1016/B978-0-12-803678-5.00516-6.
37. Ryu, S.; Chun, J.Y.; Lee, S.; Yoo, D.; Kim, Y.; Ali, S.T.; Chun, B.C. Epidemiology and transmission dynamics of infectious diseases and control measures. *Viruses* 2022, *14*. https://doi.org/10.3390/v14112510.
38. Lindahl, J.F.; Grace, D. The consequences of human actions on risks for infectious diseases: A review. *Infection Ecology & Epidemiology* 2015, *5*, 30048. https://doi.org/10.3402/iee.v5.30048.
39. de Bengy Puyvallée, A.; Kittelsen, S. *"Disease Knows No Borders": Pandemics and the Politics of Global Health Security*; Pandemics, Publics, and Politics. 2018:59–73. https://doi.org/10.1007/978-981-13-2802-2_5.
40. Smith, K.M.; Machalaba, C.C.; Seifman, R.; Feferholtz, Y.; Karesh, W.B. Infectious disease and economics: The case for considering multi-sectoral impacts. *One Health* 2019, *7*, 100080. https://doi.org/10.1016/j.onehlt.2018.100080.
41. Branda, F.; Ali, A.Y.; Ceccarelli, G.; Albanese, M.; Binetti, E.; Giovanetti, M.; Ciccozzi, M.; Scarpa, F. Assessing the burden of neglected tropical diseases in low-income communities: Challenges and solutions. *Viruses* 2025, *17*. https://doi.org/10.3390/v17010029.
42. Morel, C.M. Neglected diseases: Under-funded research and inadequate health interventions. Can we change this reality? *EMBO Reports* 2003, *4*, S35–38. https://doi.org/10.1038/sj.embor.embor851.
43. Petti, S.; Andreasen, J.O.; Glendor, U.; Andersson, L. The fifth most prevalent disease is being neglected by public health organisations. *Lancet Global Health* 2018, *6*, e1070–e1071. https://doi.org/10.1016/s2214-109x(18)30380-2.
44. Naghavi, M.; Mestrovic, T.; Gray, A.; Gershberg Hayoon, A.; Swetschinski, L.R.; Robles Aguilar, G.; Davis Weaver, N.; Ikuta, K.S.; Chung, E.; Wool, E.E. et al. Global burden associated with 85 pathogens in 2019: A systematic analysis for the Global Burden of Disease Study 2019. *The Lancet Infectious Diseases* 2024, *24*, 868–895. https://doi.org/10.1016/S1473-3099(24)00158-0.
45. Anikeeva, O.; Hansen, A.; Varghese, B.; Borg, M.; Zhang, Y.; Xiang, J.; Bi, P. The impact of increasing temperatures due to climate change on infectious diseases. *BMJ* 2024, *387*.
46. Anderson, G.B.; Barnes, E.A.; Bell, M.L.; Dominici, F. The future of climate epidemiology: Opportunities for advancing health research in the context of climate change. *American Journal of Epidemiology* 2019, *188*, 866–872. https://doi.org/10.1093/aje/kwz034.
47. Zell, R. Global climate change and the emergence/re-emergence of infectious diseases. *International Journal of Medical Microbiology Supplements* 2004, *293*, 16–26. https://doi.org/10.1016/S1433-1128(04)80005-6.
48. Leung, N.H.L. Transmissibility and transmission of respiratory viruses. *Nature Reviews Microbiology* 2021, *19*, 528–545. https://doi.org/10.1038/s41579-021-00535-6.

49. Marr, L.C.; Tang, J.W. A paradigm shift to align transmission routes with mechanisms. *Clinical Infectious Diseases* 2021, *73*, 1747–1749. https://doi.org/10.1093/cid/ciab722.
50. Vilibic-Cavlek, T.; Savic, V.; Ferenc, T.; Mrzljak, A.; Barbic, L.; Bogdanic, M.; Stevanovic, V.; Tabain, I.; Ferencak, I.; Zidovec-Lepej, S. Lymphocytic choriomeningitis—emerging trends of a neglected virus: A narrative review. *Tropical Medicine and Infectious Disease* 2021, *6*, 88.
51. Da Rosa, J.F.T.; De Souza, W.M.; de Paula Pinheiro, F.; Figueiredo, M.L.; Cardoso, J.F.; Acrani, G.O.; Nunes, M.R.T. Oropouche Virus: Clinical, epidemiological, and molecular aspects of a neglected orthobunyavirus. *The American Journal of Tropical Medicine and Hygiene* 2017, *96*, 1019.
52. Conteh, L.; Engels, T.; Molyneux, D.H. Socioeconomic aspects of neglected tropical diseases. *The Lancet* 2010, *375*, 239–247.
53. Magalhães, A.R.; Codeço, C.T.; Svenning, J.-C.; Escobar, L.E.; Van de Vuurst, P.; Gonçalves-Souza, T. Neglected tropical diseases risk correlates with poverty and early ecosystem destruction. *Infectious Diseases of Poverty* 2023, *12*, 32.
54. Monrad, J.T.; Sandbrink, J.B.; Cherian, N.G. Promoting versatile vaccine development for emerging pandemics. *Npj Vaccines* 2021, *6*, 26.
55. Charani, E.; Abimbola, S.; Pai, M.; Adeyi, O.; Mendelson, M.; Laxminarayan, R.; Rasheed, M.A. Funders: The missing link in equitable global health research? *PLOS Global Public Health* 2022, *2*, e0000583.
56. Strumillo, S.T.; Kartavykh, D.; de Carvalho, F.F. Jr; Cruz, N.C.; de Souza Teodoro, A.C.; Sobhie Diaz, R.; Curcio, M.F. Host–virus interaction and viral evasion. *Cell Biology International* 2021, *45*, 1124–1147.
57. Cline, T.D.; Karlsson, E.A.; Seufzer, B.J.; Schultz-Cherry, S. The hemagglutinin protein of highly pathogenic H5N1 influenza viruses overcomes an early block in the replication cycle to promote productive replication in macrophages. *Journal of Virology* 2013, *87*, 1411–1419.
58. Morizono, K.; Chen, I.S.Y. Role of phosphatidylserine receptors in enveloped virus infection. *Journal of Virology* 2014, *88*, 4275–4290.
59. Jemielity, S.; Wang, J.J.; Chan, Y.K.; Ahmed, A.A.; Li, W.; Monahan, S.; Bu, X.; Farzan, M.; Freeman, G.J.; Umetsu, D.T. TIM-family proteins promote infection of multiple enveloped viruses through virion-associated phosphatidylserine. *PLOS Pathogens* 2013, *9*, e1003232.
60. Oh, C.; Kim, Y.; Chang, K. O. Proteases facilitate the endosomal escape of porcine epidemic diarrhea virus during entry into host cells. *Virus Research* 2019, *272*, 197730.
61. Diaz, A.; Ahlquist, P. Role of host reticulon proteins in rearranging membranes for positive-strand RNA virus replication. *Current Opinion in Microbiology* 2012, *15*, 519–524.
62. Budicini, M.R.; Rodriguez-Irizarry, V.J.; Maples, R.W.; Pfeiffer, J.K. Murine norovirus mutants adapted to replicate in human cells reveal a post-entry restriction. *bioRxiv* 2024. https://doi.org/10.1101/2024.01.11.575274.
63. Nice, T.J.; Strong, D.W.; McCune, B.T.; Pohl, C.S.; Virgin, H.W. A single-amino-acid change in murine norovirus NS1/2 is sufficient for colonic tropism and persistence. *Journal of Virology* 2013, *87*, 327–334.
64. He, R.; Li, Y.; Bernards, M.A.; Wang, A. Manipulation of the cellular membrane-cytoskeleton network for RNA virus replication and movement in plants. *Viruses* 2023, *15*, 744.
65. Seo, D.; Gammon, D.B. Manipulation of host microtubule networks by viral microtubule-associated proteins. *Viruses* 2022, *14*, 979.
66. Weston, S.; Czieso, S.; White, I.J.; Smith, S.E.; Wash, R.S.; Diaz-Soria, C.; Kellam, P.; Marsh, M. Alphavirus restriction by IFITM proteins. *Traffic* 2016, *17*, 997–1013.
67. Spuul, P.; Balistreri, G.; Kääriäinen, L.; Ahola, T. Phosphatidylinositol 3-kinase-, actin-, and microtubule-dependent transport of Semliki Forest virus replication complexes from the plasma membrane to modified lysosomes. *Journal of Virology* 2010, *84*, 7543–7557.
68. Fields, W.; Kielian, M. Interactions involved in pH protection of the alphavirus fusion protein. *Virology* 2015, *486*, 173–179.
69. Ooi, Y.S.; Dubé, M.; Kielian, M. BST2/tetherin inhibition of alphavirus exit. *Viruses* 2015, *7*, 2147–2167.
70. Vancini, R.; Wang, G.; Ferreira, D.; Hernandez, R.; Brown, D.T. Alphavirus genome delivery occurs directly at the plasma membrane in a time-and temperature-dependent process. *Journal of Virology* 2013, *87*, 4352–4359.
71. Schuchman, R.M.; Vancini, R.; Piper, A.; Breuer, D.; Ribeiro, M.; Ferreira, D.; Magliocca, J.; Emmerich, V.; Hernandez, R.; Brown, D.T. Role of the vacuolar ATPase in the alphavirus replication cycle. *Heliyon* 2018, *4*.

72. Jones, P.H.; Maric, M.; Madison, M.N.; Maury, W.; Roller, R.J.; Okeoma, C.M. BST-2/tetherin-mediated restriction of chikungunya (CHIKV) VLP budding is counteracted by CHIKV non-structural protein 1 (nsP1). *Virology* 2013, *438*, 37–49.
73. Radoshitzky, S.R.; Pegoraro, G.; Chī, X.; Dǒng, L.; Chiang, C.Y.; Jozwick, L.; Clester, J.C.; Cooper, C.L.; Courier, D.; Langan, D.P. siRNA screen identifies trafficking host factors that modulate alphavirus infection. *PLOS Pathogens* 2016, *12*, e1005466.
74. Pietilä, M.K.; van Hemert, M.J.; Ahola, T. Purification of highly active alphavirus replication complexes demonstrates altered fractionation of multiple cellular membranes. *Journal of Virology* 2018, *92*, 10–1128.
75. Embarc-Buh, A.; Francisco-Velilla, R.; Martinez-Salas, E. RNA-binding proteins at the host-pathogen interface targeting viral regulatory elements. *Viruses* 2021, *13*, 952.
76. Varjak, M.; Saul, S.; Arike, L.; Lulla, A.; Peil, L.; Merits, A. Magnetic fractionation and proteomic dissection of cellular organelles occupied by the late replication complexes of Semliki Forest virus. *Journal of Virology* 2013, *87*, 10295–10312.
77. Ooi, Y.S.; Stiles, K.M.; Liu, C.Y.; Taylor, G.M.; Kielian, M. Genome-wide RNAi screen identifies novel host proteins required for alphavirus entry. *PLOS Pathogens* 2013, *9*, e1003835.
78. Plattet, P.; Plemper, R.K. Envelope protein dynamics in paramyxovirus entry. *MBio* 2013, *4*, 10–1128.
79. Bose, S.; Song, A.S.; Jardetzky, T.S.; Lamb, R.A. Fusion activation through attachment protein stalk domains indicates a conserved core mechanism of paramyxovirus entry into cells. *Journal of Virology* 2014, *88*, 3925–3941.
80. Sun, Y.; Yu, S.; Ding, N.; Meng, C.; Meng, S.; Zhang, S.; Zhan, Y.; Qiu, X.; Tan, L.; Chen, H. Autophagy benefits the replication of Newcastle disease virus in chicken cells and tissues. *Journal of Virology* 2014, *88*, 525–537.
81. Garner, O.B.; Aguilar, H.C.; Fulcher, J.A.; Levroney, E.L.; Harrison, R.; Wright, L.; Robinson, L.R.; Aspericueta, V.; Panico, M.; Haslam, S.M. Endothelial galectin-1 binds to specific glycans on Nipah virus fusion protein and inhibits maturation, mobility, and function to block syncytia formation. *PLOS Pathogens* 2010, *6*, e1000993.
82. DeBuysscher, B.L.; de Wit, E.; Munster, V.J.; Scott, D.; Feldmann, H.; Prescott, J. Comparison of the pathogenicity of Nipah virus isolates from Bangladesh and Malaysia in the Syrian hamster. *PLOS Neglected Tropical Diseases* 2013, *7*, e2024.
83. Nikolin, V.; Hatsue Sobreda Doi, L.; Sieg, M.; Busch, J.; Böttcher, D.; Tedeschi, L.; Poulard, A.; Staszewski, V.; Vahlenkamp, T.; Poulet, H. In vitro growth, receptor usage and pathogenesis of feline morbillivirus in the natural host. *Viruses* 2022, *14*, 1503.
84. Vázquez, C.A.; Widerspick, L.; Thuenauer, R.; Schneider, C.; Reimer, R.; Neira, P.; Olal, C.; Heung, M.; Niemetz, L.; Lawrence, P. Nipah virus infection generates ordered structures in cellulo. *Viruses* 2022, *14*, 1523.
85. Dutch, R.E. Entry and fusion of emerging paramyxoviruses. *PLOS Pathogens* 2010, *6*, e1000881.
86. Tsetsarkin, K.A.; McGee, C.E.; Higgs, S. Chikungunya virus adaptation to aedes albopictus mosquitoes does not correlate with acquisition of cholesterol dependence or decreased pH threshold for fusion reaction. *Virology Journal* 2011, *8*, 1–12.
87. Yeh, J.X.; Park, E.; Schultz, K.L.W.; Griffin, D.E. NF-κB activation promotes alphavirus replication in mature neurons. *Journal of Virology* 2019, *93*, 10–1128.
88. Qian, Q.; Zhou, H.; Shu, T.; Mu, J.; Fang, Y.; Xu, J.; Li, T.; Kong, J.; Qiu, Y.; Zhou, X. The capsid protein of Semliki Forest virus antagonizes RNA interference in mammalian cells. *Journal of Virology* 2020, *94*, 10–1128.
89. Bernard, E.; Solignat, M.; Gay, B.; Chazal, N.; Higgs, S.; Devaux, C.; Briant, L. Endocytosis of chikungunya virus into mammalian cells: Role of clathrin and early endosomal compartments. *PLOS One* 2010, *5*, e11479.
90. Saxena, T.; Tandon, B.; Sharma, S.; Chameettachal, S.; Ray, P.; Ray, A.R.; Kulshreshtha, R. Combined miRNA and mRNA signature identifies key molecular players and pathways involved in chikungunya virus infection in human cells. *PLOS One* 2013, *8*, e79886.
91. Li, G.; De Oliveira, D.M.P.; Walker, M.J. The antimicrobial and immunomodulatory effects of ionophores for the treatment of human infection. *Journal of Inorganic Biochemistry* 2022, *227*, 111661. https://doi.org/10.1016/j.jinorgbio.2021.111661.
92. Chadwick, S.R.; Grinstein, S.; Freeman, S.A. From the inside out: Ion fluxes at the centre of endocytic traffic. *Current Opinion in Cell Biology* 2021, *71*, 77–86. https://doi.org/10.1016/j.ceb.2021.02.006.

93. Kaushik, V.; Yakisich, J.S.; Kumar, A.; Azad, N.; Iyer, A.K.V. Ionophores: Potential use as anticancer drugs and chemosensitizers. *Cancers (Basel)* 2018, *10*. https:/doi.org/10.3390/cancers10100360.
94. Pound, H.L.; Gann, E.R.; Tang, X.; Krausfeldt, L.E.; Huff, M.; Staton, M.E.; Talmy, D.; Wilhelm, S.W. The "neglected viruses" of Taihu: Abundant transcripts for viruses infecting eukaryotes and their potential role in phytoplankton succession. *Frontiers in Microbiology* 2020, *11*, 338.
95. Kalynych, S.; Přidal, A.; Pálková, L.; Levdansky, Y.; de Miranda, J.R.; Plevka, P. Virion structure of iflavirus slow bee paralysis virus at 2.6-angstrom resolution. *Journal of virology* 2016, *90*, 7444–7455.
96. Procházková, M.; Füzik, T.; Škubník, K.; Moravcová, J.; Ubiparip, Z.; Přidal, A.; Plevka, P. Virion structure and genome delivery mechanism of sacbrood honeybee virus. *Proceedings of the National Academy of Sciences* 2018, *115*, 7759–7764.
97. Sun, B.; Jia, L.; Liang, B.; Chen, Q.; Liu, D. Phylogeography, transmission, and viral proteins of Nipah virus. *Virologica Sinica* 2018, *33*, 385–393. https://doi.org/10.1007/s12250-018-0050-1.
98. Skehel, J.J.; Wiley, D.C. Receptor binding and membrane fusion in virus entry: The influenza hemagglutinin. *Annual Review of Biochemistry* 2000, *69*, 531–569. https://doi.org/10.1146/annurev.biochem.69.1.531.
99. Thomas, V.; Bertelli, C.; Collyn, F.; Casson, N.; Telenti, A.; Goesmann, A.; Croxatto, A.; Greub, G. Lausannevirus, a giant amoebal virus encoding histone doublets. *Environmental Microbiology* 2011, *13*, 1454–1466.
100. Gao, Y.; Liu, S.; Huang, J.; Wang, Q.; Li, K.; He, J.; He, J.; Weng, S.; Zhang, Q. Cryo-electron microscopy structures of novel viruses from mud crab *Scylla paramamosain* with multiple infections. *Journal of Virology* 2019, *93*, 10–1128.
101. Subhadarshini, S.; Tandon, H.; Srinivasan, N.; Sowdhamini, R. Normal mode analysis elicits conformational shifts in proteins at both proximal and distal regions to the phosphosite stemming from single-site phosphorylation. *ACS Omega* 2024, *9*, 24520–24537. https://doi.org/10.1021/acsomega.4c00523.
102. Kimura, K.; Tomaru, Y. Discovery of two novel viruses expands the diversity of single-stranded DNA and single-stranded RNA viruses infecting a cosmopolitan marine diatom. *Applied and Environmental Microbiology* 2015, *81*, 1120–1131.
103. Hsu, N.Y.; Ilnytska, O.; Belov, G.; Santiana, M.; Chen, Y.H.; Takvorian, P.M.; Pau, C.; van der Schaar, H.; Kaushik-Basu, N.; Balla, T. Viral reorganization of the secretory pathway generates distinct organelles for RNA replication. *Cell* 2010, *141*, 799–811.
104. Watanabe, T.; Kawaoka, Y. Villains or heroes? The raison d'être of viruses. *Clinical & Translational Immunology* 2020, *9*, e01114.
105. Spurny, R.; Přidal, A.; Pálková, L.; Kiem, H.K.T.; de Miranda, J.R.; Plevka, P. Virion structure of black queen cell virus, a common honeybee pathogen. *Journal of Virology* 2017, *91*, 10–1128.
106. Zhu, L.; Wang, X.; Ren, J.; Porta, C.; Wenham, H.; Ekström, J.O.; Panjwani, A.; Knowles, N.J.; Kotecha, A.; Siebert, C.A. Structure of Ljungan virus provides insight into genome packaging of this picornavirus. *Nature Communications* 2015, *6*, 8316.
107. Budicini, M.R.; Rodriguez-Irizarry, V.J.; Maples, R.W.; Pfeiffer, J.K. Murine norovirus mutants adapted to replicate in human cells reveal a post-entry restriction. *Journal of Virology* 2024, *98*, e00047–00024.
108. Kumar, R.; Mehta, D.; Chaudhary, S.; Nayak, D.; Sunil, S. Impact of CHIKV replication on the global proteome of Aedes albopictus cells. *Proteomes* 2022, *10*, 38.
109. Ravenhill, B.J.; Oliveira, M.; Wood, G.; Di, Y.; Kite, J.; Wang, X.; Davies, C.T.R.; Lu, Y.; Antrobus, R.; Elliott, G. et al. Spatial proteomics identifies a CRTC-dependent viral signaling pathway that stimulates production of interleukin-11. *Cell Reports* 2025, *44*. https://doi.org/10.1016/j.celrep.2025.115263.
110. Sathe, G.; Sapkota, G.P. Proteomic approaches advancing targeted protein degradation. *Trends in Pharmacological Sciences* 2023, *44*, 786–801. https://doi.org/10.1016/j.tips.2023.08.007.
111. Sun, B.B.; Suhre, K.; Gibson, B.W. Promises and challenges of populational proteomics in health and disease. *Molecular & Cellular Proteomics* 2024, *23*, 100786. https://doi.org/10.1016/j.mcpro.2024.100786.
112. Kesic, M.J.; Simmons, S.O.; Bauer, R.; Jaspers, I. Nrf2 expression modifies influenza A entry and replication in nasal epithelial cells. *Free Radical Biology and Medicine* 2011, *51*, 444–453.
113. Reyes, E.D.; Kulej, K.; Pancholi, N.J.; Akhtar, L.N.; Avgousti, D.C.; Kim, E.T.; Bricker, D.K.; Spruce, L.A.; Koniski, S.A.; Seeholzer, S.H. Identifying host factors associated with DNA replicated during virus infection. *Molecular & Cellular Proteomics* 2017, *16*, 2079–2097.
114. Xiao, X.; Lei, X.; Zhang, Z.; Ma, Y.; Qi, J.; Wu, C.; Xiao, Y.; Li, L.; He, B.; Wang, J. Enterovirus 3A facilitates viral replication by promoting phosphatidylinositol 4-kinase IIIβ–ACBD3 interaction. *Journal of Virology* 2017, *91*, 10–1128.
115. Gullberg, R.C.; Steel, J.J.; Moon, S.L.; Soltani, E.; Geiss, B.J. Oxidative stress influences positive strand RNA virus genome synthesis and capping. *Virology* 2015, *475*, 219–229.

116. Brito, A.F.; Pinney, J.W. protein-protein interactions in virus-host systems. *Front Microbiol* 2017, *8*, 1557. https://doi.org/10.3389/fmicb.2017.01557.
117. Sasaki, J.; Ishikawa, K.; Arita, M.; Taniguchi, K. ACBD3-mediated recruitment of PI4KB to picornavirus RNA replication sites. *The EMBO Journal* 2012, *31*, 754–766.
118. Bush, C.O.; Greenstein, A.E.; Delaney Iv, W.E.; Beran, R.K.F. Hepatitis C viral entry inhibitors prolong viral suppression by replication inhibitors in persistently-infected Huh7 cultures. *PLOS One* 2013, *8*, e65273.
119. Mohni, K.N.; Mastrocola, A.S.; Bai, P.; Weller, S.K.; Heinen, C.D. DNA mismatch repair proteins are required for efficient herpes simplex virus 1 replication. *Journal of Virology* 2011, *85*, 12241–12253.
120. Dembowski, J.A.; DeLuca, N.A. Temporal viral genome-protein interactions define distinct stages of productive herpesviral infection. *MBio* 2018, *9*, 10–1128.
121. Khan Mirzaei, M.; Xue, J.; Costa, R.; Ru, J.; Schulz, S.; Taranu, Z.E.; Deng, L. Challenges of studying the human virome – Relevant emerging technologies. *Trends in Microbiology* 2021, *29*, 171–181. https://doi.org/10.1016/j.tim.2020.05.021.
122. Lauber, C.; Seitz, S. Opportunities and challenges of data-driven virus discovery. *Biomolecules* 2022, *12*. https://doi.org/10.3390/biom12081073.
123. Weber, C.; Büchner, S.M.; Schnierle, B.S. A small antigenic determinant of the Chikungunya virus E2 protein is sufficient to induce neutralizing antibodies which are partially protective in mice. *PLOS Neglected Tropical Diseases* 2015, *9*, e0003684.
124. Gil, P.I.; Albrieu-Llinás, G.; Mlewski, E.C.; Monetti, M.; Fozzatti, L.; Cuffini, C.; Fernández Romero, J.; Kunda, P.; Paglini, M.G. Pixuna virus modifies host cell cytoskeleton to secure infection. *Scientific Reports* 2017, *7*, 5757.
125. Dai, G. Signaling by ion channels: Pathways, dynamics and channelopathies. *Missouri Medicine* 2023, *120*, 367–373.
126. Subramanyam, P.; Colecraft, H.M. Ion channel engineering: Perspectives and strategies. *Journal of Molecular Biology* 2015, *427*, 190–204. https://doi.org/10.1016/j.jmb.2014.09.001.
127. Holmes, A.C.; Basore, K.; Fremont, D.H.; Diamond, M.S. A molecular understanding of alphavirus entry. *PLOS Pathogens* 2020, *16*, e1008876. https://doi.org/10.1371/journal.ppat.1008876.
128. Schein, C.H.; Rafael, G.; Baker, W.S.; Anaya, E.S.; Schmidt, J.G.; Weaver, S.C.; Negi, S.; Braun, W. Peptide and protein alphavirus antigens for broad spectrum vaccine design. *bioRxiv* 2022, 2022–2005.
129. Wintachai, P.; Kaur, P.; Lee, R.C.H.; Ramphan, S.; Kuadkitkan, A.; Wikan, N.; Ubol, S.; Roytrakul, S.; Chu, J.J.H.; Smith, D.R. Activity of andrographolide against chikungunya virus infection. *Scientific Reports* 2015, *5*, 14179.
130. Graham, B.S.; Sullivan, N.J. Emerging viral diseases from a vaccinology perspective: Preparing for the next pandemic. *Nature Immunology* 2018, *19*, 20–28. https://doi.org/10.1038/s41590-017-0007-9.
131. Muthukutty, P.; MacDonald, J.; Yoo, S.Y. Combating emerging respiratory viruses: Lessons and future antiviral strategies. *Vaccines (Basel)* 2024, *12*. https://doi.org/10.3390/vaccines12111220.
132. Law, Y.S.; Utt, A.; Tan, Y.B.; Zheng, J.; Wang, S.; Chen, M.W.; Griffin, P.R.; Merits, A.; Luo, D. Structural insights into RNA recognition by the Chikungunya virus nsP2 helicase. *Proceedings of the National Academy of Sciences* 2019, *116*, 9558–9567.
133. Lee, R.C.H.; Chu, J.J.H. Proteomics profiling of chikungunya-infected Aedes albopictus C6/36 cells reveal important mosquito cell factors in virus replication. *PLOS Neglected Tropical Diseases* 2015, *9*, e0003544.
134. Srivastava, P.; Chaudhary, S.; Malhotra, S.; Varma, B.; Sunil, S. Transcriptome analysis of human macrophages upon chikungunya virus (CHIKV) infection reveals regulation of distinct signaling and metabolic pathways during the early and late stages of infection. *Heliyon* 2023, *9*.
135. Vasconcellos, A.F.; Melo, R.M.; Mandacaru, S.C.; de Oliveira, L.S.; de Oliveira, A.S.; Moraes, E.C.S.; Trugilho, M.R.O.; Ricart, C.A.O.; Báo, S.N.; Resende, R.O. Aedes aegypti Aag-2 cell proteome modulation in response to chikungunya virus infection. *Frontiers in Cellular and Infection Microbiology* 2022, *12*, 920425.
136. Marcink, T.C.; Kicmal, T.; Armbruster, E.; Zhang, Z.; Zipursky, G.; Golub, K.L.; Idris, M.; Khao, J.; Drew-Bear, J.; McGill, G. Intermediates in SARS-CoV-2 spike-mediated cell entry. *Science Advances* 2022, *8*, eabo3153.
137. Kim, Y.H.; Donald, J.E.; Grigoryan, G.; Leser, G.P.; Fadeev, A.Y.; Lamb, R.A.; DeGrado, W.F. Capture and imaging of a prehairpin fusion intermediate of the paramyxovirus PIV5. *Proceedings of the National Academy of Sciences* 2011, *108*, 20992–20997.

138. Dahmani, I.; Ludwig, K.; Chiantia, S. Influenza a matrix protein M1 induces lipid membrane deformation via protein multimerization. *Bioscience Reports* 2019, *39*, BSR20191024.
139. Klein, S.; Cortese, M.; Winter, S.L.; Wachsmuth-Melm, M.; Neufeldt, C.J.; Cerikan, B.; Stanifer, M.L.; Boulant, S.; Bartenschlager, R.; Chlanda, P. SARS-CoV-2 structure and replication characterized by in situ cryo-electron tomography. *Nature Communications* 2020, *11*, 5885.
140. Dent, K.C.; Thompson, R.; Barker, A.M.; Hiscox, J.A.; Barr, J.N.; Stockley, P.G.; Ranson, N.A. The asymmetric structure of an icosahedral virus bound to its receptor suggests a mechanism for genome release. *Structure* 2013, *21*, 1225–1234.
141. Liu, X.; Zhang, Q.; Murata, K.; Baker, M.L.; Sullivan, M.B.; Fu, C.; Dougherty, M.T.; Schmid, M.F.; Osburne, M.S.; Chisholm, S.W. Structural changes in a marine podovirus associated with release of its genome into prochlorococcus. *Nature Structural & Molecular Biology* 2010, *17*, 830–836.
142. Hryc, C.F.; Chen, D.H.; Chiu, W. Near-atomic resolution cryo-EM for molecular virology. *Current Opinion in Virology* 2011, *1*, 110–117.
143. Wang, H.; Salaipeth, L.; Miyazaki, N.; Suzuki, N.; Okamoto, K. Capsid structure of a metazoan fungal dsRNA megabirnavirus reveals its uniquely acquired structures. *bioRxiv* 2022, 2022–2009.
144. Marmot, M.; Bell, R. Social determinants and non-communicable diseases: Time for integrated action. *BMJ* 2019, *364*.
145. Nordgren, T.M.; Charavaryamath, C. Agriculture occupational exposures and factors affecting health effects. *Current Allergy and Asthma Reports* 2018, *18*, 65. https://doi.org/10.1007/s11882-018-0820-8.
146. Gay, L.; Melenotte, C.; Lakbar, I.; Mezouar, S.; Devaux, C.; Raoult, D.; Bendiane, M.K.; Leone, M.; Mège, J.L. Sexual dimorphism and gender in infectious diseases. *Frontiers in Immunology* 2021, *12*, 698121. https://doi.org/10.3389/fimmu.2021.698121.
147. Li, X.C.; Zhang, Y.Y.; Zhang, Q.Y.; Liu, J.S.; Ran, J.J.; Han, L.F.; Zhang, X.X. global burden of viral infectious diseases of poverty based on global burden of diseases study 2021. *Infectious Diseases of Poverty* 2024, *13*, 71. https://doi.org/10.1186/s40249-024-01234-z.
148. Niewiadomski, P.; Ortega-Ortega, M.; Łyszczarz, B. Productivity losses due to health problems arising from COVID-19 pandemic: A systematic review of population-level studies worldwide. *Applied Health Economics and Health Policy* 2025, *23*, 231–251. https://doi.org/10.1007/s40258-024-00935-8.
149. Richards, F.; Kodjamanova, P.; Chen, X.; Li, N.; Atanasov, P.; Bennetts, L.; Patterson, B.J.; Yektashenas, B.; Mesa-Frias, M.; Tronczynski, K. et al. Economic burden of COVID-19: A systematic review. *ClinicoEconomics and Outcomes Research* 2022, *14*, 293–307. https://doi.org/10.2147/ceor.S338225.
150. Murray, J.; Cohen, A.L. *Infectious Disease Surveillance*; International Encyclopedia of Public Health. 2017:222–229. https://doi.org/10.1016/B978-0-12-803678-5.00517-8.
151. Murray, C.J.L. Findings from the Global Burden of Disease Study 2021. *The Lancet* 2024, *403*, 2259–2262. https://doi.org/10.1016/S0140-6736(24)00769-4.
152. Rosa, A.C.; Mio, M.; Andreadou, I.; Sumbayev, V.V. Editorial: The challenge of new therapeutic approaches for unmet therapeutic needs. *Frontiers in Pharmacology* 2020, *11*, 01341. https://doi.org/10.3389/fphar.2020.01341.
153. Wang, L.; Huang, A.T.; Katzelnick, L.C.; Lefrancq, N.; Escoto, A.C.; Duret, L.; Chowdhury, N.; Jarman, R.; Conte, M.A.; Berry, I.M. et al. Antigenic distance between primary and secondary dengue infections correlates with disease risk. *Science Translational Medicine* 2024, *16*, eadk3259. https://doi.org/10.1126/scitranslmed.adk3259.
154. Quinn, S.C.; Kumar, S. Health inequalities and infectious disease epidemics: A challenge for global health security. *Biosecur Bioterror* 2014, *12*, 263–273. https://doi.org/10.1089/bsp.2014.0032.
155. Tan, C.; Ofner, M.; Candon, H.L.; Reel, K.; Bean, S.; Chan, A.K.; Leis, J.A. An ethical framework adapted for infection prevention and control. *Infection Control & Hospital Epidemiology* 2023, *44*, 2044–2049. https://doi.org/10.1017/ice.2023.121.
156. Danasekaran, R. One health: A holistic approach to tackling global health issues. *Indian Journal of Community Medicine* 2024, *49*, 260–263. https://doi.org/10.4103/ijcm.ijcm_521_23.
157. World Health Organization. One Health. https://www.who.int/health-topics/one-health#tab=tab_1 (accessed on 19/02/2025).
158. Piquer-Martinez, C.; Urionagüena, A.; Benrimoj, S.I.; Calvo, B.; Dineen-Griffin, S.; Garcia-Cardenas, V.; Fernandez-Llimos, F.; Martinez-Martinez, F.; Gastelurrutia, M.A. Theories, models and frameworks for health systems integration. A scoping review. *Health Policy* 2024, *141*, 104997. https://doi.org/10.1016/j.healthpol.2024.104997.
159. Lewnard, J.A.; Reingold, A.L. Emerging challenges and opportunities in infectious disease epidemiology. *American Journal of Epidemiology* 2019, *188*, 873–882. https://doi.org/10.1093/aje/kwy264.

160. World Health Organization. *Global Report on Neglected Tropical Diseases 2023*; World Health Organization: 2023.
161. Tay, D.J.W.; Lew, Z.Z.R.; Chu, J.J.H.; Tan, K.S. Uncovering novel viral innate immune evasion strategies: What has SARS-CoV-2 taught us? *Frontiers in Microbiology* 2022, *13*, 844447. https://doi.org/10.3389/fmicb.2022.844447.
162. Strecker, T.; Eichler, R.; Meulen, J.; Weissenhorn, W.; Dieter Klenk, H.; Garten, W.; Lenz, O. Lassa virus Z protein is a matrix protein and sufficient for the release of virus-like particles [corrected]. *Journal of Virology* 2003, *77*, 10700–10705. https://doi.org/10.1128/jvi.77.19.10700-10705.2003.
163. Letko, M.; Seifert, S.N.; Olival, K.J.; Plowright, R.K.; Munster, V.J. Bat-borne virus diversity, spillover and emergence. *Nature Reviews Microbiology* 2020, *18*, 461–471. https://doi.org/10.1038/s41579-020-0394-z.
164. Payne, S. *Virus Evolution and Genetics*; Viruses. 2017:81–6. https://doi.org/10.1016/B978-0-12-803109-4.00008-8.
165. Pearson, T.A.; Califf, R.M.; Roper, R.; Engelgau, M.M.; Khoury, M.J.; Alcantara, C.; Blakely, C.; Boyce, C.A.; Brown, M.; Croxton, T.L. et al. Precision health analytics with predictive analytics and implementation research: JACC state-of-the-art review. *Journal of the American College of Cardiology* 2020, *76*, 306–320. https://doi.org/10.1016/j.jacc.2020.05.043.
166. Singh, B.K.; Delgado-Baquerizo, M.; Egidi, E.; Guirado, E.; Leach, J.E.; Liu, H.; Trivedi, P. Climate change impacts on plant pathogens, food security and paths forward. *Nature Reviews Microbiology* 2023, *21*, 640–656. https://doi.org/10.1038/s41579-023-00900-7.

2 Alphaviruses
Neglected Viruses

Kenneth Lundstrom

2.1 INTRODUCTION

Alphaviruses belonging to the Togaviridae genus carry a positive-sense single-stranded RNA (ssRNA) surrounded by an envelope structure [1]. Their capability to infect both vertebrates and invertebrates has demonstrated their broad host range. For example, mammals, birds, amphibians, fish, and insects have demonstrated susceptibility to alphavirus infections. Due to their transmission by mosquitoes between species, alphaviruses are listed among arthropod-borne viruses. For this reason, alphaviruses have shown pathogenicity in domestic animals and humans and have been identified as the cause of epidemics in different parts of the world. Alphaviruses have commonly been used in cell and molecular biology for studies on virus-host interactions, viral lifecycles investigations and post-translational modifications and transport of proteins [2]. They have also been subjected to structural studies, and their replication has been outlined [1]. The broad knowledge of alphavirus biology and their use as expression systems have made them attractive vehicles for therapeutic applications [3–5]. Moreover, as they act as pathogens they have been targets for therapeutic interventions and vaccine development [6].

In this chapter, an overview of alphaviruses is presented. Furthermore, the biology, immunogenicity, pathogenicity, epidemiology, clinical manifestations, and therapeutic strategies for alphaviruses are discussed.

2.2 HISTORICAL CONTEXT

Historically, alphaviruses have been divided into Old-World and New-World species, although Sindbis virus (SIN) for example, has a very broad distribution [7]. Typically, Semliki Forest virus (SFV) and SIN have been classified as Old-World viruses with geographical locations in Europe, Asia, Australia, and Africa. Moreover, Chikungunya (CHIKV), Whataroa, AURA virus, O'nyong-nyong (ONN), Ockelbo, and Ross River (RRV) viruses are Old-World alphaviruses [8, 9]. On the other hand, Venezuelan equine encephalitis virus (VEE), Eastern equine encephalitis virus (EEE), Western equine encephalitis virus WEE), and Mayaro virus (MAYV) belong to New World viruses [10, 11]. Moreover, the Highland J virus (HJ) isolated in the eastern United States shows serological resemblance to WEE although oligonucleotide fingerprinting suggests that it is unrelated to WEE [12]. However, sequence data indicated that HJ is a recombinant-like WEE and therefore should be classified as WEE virus.

2.3 EPIDEMIOLOGY AND GLOBAL SPREAD

Alphaviruses have been widely studied due to their potential threat to human health in different areas of the world [13–16]. For example, EEE and WEE have been identified as causing restricted cases of fatal encephalitis in humans and domestic animals in North and South America [17]. Likewise, VEE has been identified in connection with encephalitis in horses and humans in South America [18]. Additionally, CHIKV and the closely related ONN have been responsible for millions of infected cases of humans [19]. For example, CHIKV outbreaks in La Reunion [20] and South

DOI: 10.1201/9781003607816-2

America [21] caused extensive spread of diseases and deaths. Although both SFV [22] and SIN [23] have been associated with outbreaks in Africa and Asia resulting in fever, rash, arthralgia, laboratory strains have been considered avirulent.

2.4 CLINICAL MANIFESTATIONS

The clinical outcome of alphaviruses varies quite drastically from one type of virus to another. It is well documented that alphaviruses cause febrile illness, arthritis and in severe cases encephalitis [1]. Although CHIKV has been determined to be responsible for millions of cases of infections in humans, they are characterized by fever, nausea, rash, polyarthralgia, myalgia and skin rash, which are not life-threatening [24]. ONN, responsible for sporadic outbreaks in Africa are characterized by acute febrile symptoms and polyarthralgia [25]. Moreover, RR infection leads to persistent epidemic polyarthritis, which can last for months to years [26]. On the other hand, VEE, EEE, and WEE can cause fatal encephalitis with severe consequences [17]. Despite SFV and SIN laboratory strains being considered avirulent, one fatal case of SFV-based encephalitis has been documented in a laboratory worker [27]. However, in this case was develop more virulent strains.

On the cellular level, the pathogenicity of alphaviruses has been associated with induced apoptosis leading to killing of host cells [28]. The apoptotic mechanism has been linked to loss of membrane potential in mitochondria leading to activation of caspase-3, -8, and -9 [29]. In addition, alphaviruses have been demonstrated to dramatically shut down the host cell protein synthesis by hijacking the host cell protein synthesis machinery [30]. Moreover, the alphaviral nsP2 and nsP3 proteins are involved in the shut-off process, specifically suppressing antiviral pathways [30].

2.5 VIRAL STRUCTURE AND REPLICATION

Structurally, Alphaviruses are composed of an icosahedral nucleocapsid enveloped in an icosahedral lattice composed of envelope glycoproteins [1]. The nucleocapsid structure has been solved for SIN, SFV, and RRV by cryoelectron microscopy [31–33]. X-ray diffraction studies confirmed that the nucleocapsid was compatible with a T = 4 particle symmetry containing 240 copies of the capsid protein [34, 35]. Moreover, cryoelectron microscopy has demonstrated the T = 4 based on one-to-one relationship between capsid and envelope proteins with 240 copies [36]. The viral envelope consists of the glycoproteins El and E2 embedded in a host cell-derived lipid bilayer [37]. The stable El and E2 envelope proteins form heterodimers and three heterodimers assemble into spikes on the surface of viral particles [38]. The icosahedral lattice of 240 heterodimers assembles into 80 spikes per virus particle [33]. Moreover, the third alphavirus envelope protein E3 remains associated with viral particles in the case of SFV [39], whereas it is cleaved off for SIN [40]. The small hydrophobic 6K membrane protein is involved in the processing of viral glycoproteins, membrane permeabilization, and budding of viral particles [41].

Obviously, the surface structure plays an important role related to infectivity and host tropism of alphaviruses. Although alphaviruses can infect both arthropod and vertebrate hosts the infection is persistent and lifelong in arthropods [42] whereas it is acute and generally short-lived in vertebrates leading to cytotoxic effect [30]. The first step of alphavirus infection involves the recognition of host cell receptors [43]. The broad range of alphavirus hosts including fish, insects, amphibians, birds and mammals relates to different receptors targeted. For example, laminin [44], heparan sulfate [45], NRAMP [46], PHB1 [47], CD147 complex [48], MXRA8 [49], LDLRAD3 [50], and VLDR and ApoER2 [51] have been identified as entry receptors for various alphaviruses.

After receptor recognition, alphavirus particles enter the host cells by fusion to the cell membrane [52, 20] or through endocytosis in clathrin-coated vesicles followed by the transfer to endosomes [53, 54]. Next, the positive-sense 11–12 kb ssRNA alphavirus genome, containing two open reading frames coding for structural and non-structural proteins and flanked by a 5′ end cap and a

3′ poly(A) tail, is released into the cytoplasm [55]. There, the non-structural proteins 1–4 (nsP1-nsP4) are produced as the P1234 polyprotein, which is assembled into the replicase complex (RC) in modified endosomal and lysosomal membranes [56]. The early RC composed of the P123 polyprotein and nsP4 synthesizes minus strand ssRNA, whereas late RC based on the fully processed nsP1–4 generates extensive replication of genomic and subgenomic ssRNA molecules [57]. The strong 26S subgenomic promoter is responsible for high-level expression of alphavirus capsid and envelope proteins. The capsid protein forms nucleocapsids by encapsulation of viral RNA [58]. In contrast, the viral envelope proteins are glycosylated, transported through the endoplasmic reticulum and the Golgi subjected to post-translational modification events [59] before being delivered to the cell membrane [60]. Subsequently, nucleocapsids are surrounded by envelope proteins embedded in a lipid bilayer and mature viral particles are released by budding [61]. The budding of viral particles is rapid and approximately 2000 plaque forming units are released within 4–6 hours post-infection [1]. The broad tropism including insects, mammals and birds mean that alphaviruses grow over a wide range of temperatures from 25°C to 40°C.

2.6 IMMUNE RESPONSE AND PATHOGENESIS

Immune responses against alphaviruses have been studied intensively and reviewed [62]. They can be divided in innate immune and adaptive responses against alphaviruses. Within innate immune responses, pattern recognition receptors such as Toll-like receptors TL3, TL7, and TLR8 recognize alphavirus RNA, which triggers downstream signaling molecules [63]. Moreover, the RIG-I and MDA5 receptors recognize dsRNA, activate downstream adaptor and signaling molecules resulting in translocation of interferon regulatory factors to the nucleus and induction of type I interferons and proinflammatory cytokines and chemokines [64, 65]. It has been discovered that interferon regulatory factor (IRF) dimers bind to DNA in the nucleus contributing to the regulation of type I interferons and interferon-stimulated genes (ISGs) [66]. Other studies have demonstrated that the lack of type I interferon receptor IFNAR caused increased replication and rapid death in mice [67]. Moreover, binding of interferon to IFNAR1/IFNAR2 heterodimer receptors activate the Janus kinase signal transducer and activator of transcription (JAK/STAT) pathway and induction of the transcription of ISGs controlling alphavirus infections [68]. As ISGs can inhibit translation of RNA in alphavirus-infected cells, reduced replication of CHIKV, SIN, and VEE was observed in cells expressing ISG20 [69, 70]. Moreover, in SIN and CHIKV infected cells type I interferons induce ISG15 resulting in reduced viral replication [71, 72]. It has also been demonstrated that alphavirus infections can activate the NLRP3 inflammasome leading to IL-1β and IL-18 production [73, 74].

Cellular immunity, being part of the innate immune response, contributes to the control of alphavirus replication, clearance of infected cells and stimulation of adaptive immune responses [62]. For example, myeloid cells such as dendritic cells (DCs), macrophages, and neutrophils are mobilized to the site of alphavirus infection, which will contribute to the protection [75] or pathogenesis [76]. In the context of plasmacytoid dendritic cells (pDCs), their production of type I interferon controlled viral replication and provided protection of Irf3/Irf7–/– mice from lethal challenges with CHIKV [77]. Furthermore, pDC-based type I interferon production limited CHIKV and MAYV infections of monocytes [78]. In contrast, reduced production of type I interferon in pDCs increased CHIK and MAY infections of monocytes. In another study, it was demonstrated that pDCs were able to protect mice against challenges with an interferon-sensitive RRV mutant [79]. It has also been postulated that pDC-based type I interferon production can control encephalitic VEE infections [80]. In the context of arthritogenic alphavirus infections, monocytes and macrophages have been found to be accumulated in musculoskeletal tissues, which can result in both protective [81] and pathogenic functions [82]. For example, depletion of macrophages during CHIKV and RRV infections decreased skeletal muscle damage and reduced clinical manifestation of disease, whereas no impact on viral burden was observed [83, 84]. Moreover, deficiency in expression of the MIF and CXCL10 chemo-attractants resulted in limited monocyte and macrophage recruitment, which decrease clinical symptoms and immune cell infiltration after RRV, CHIKV, and ONN infections [85, 86]. In contrast, CCL2 receptor-deficient Ccr2–/– mice developed foot swelling after CHIKV infection [87]. Neutrophils have demonstrated contrasting protective and pathological functions during encephalitic alphavirus infections [88]. For example, EEE was significantly more pathogenic than VEE in in cynomolgus macaques. There was no difference in neuropathogenicity between subcutaneous and aerosol administration of EEV [88]. Moreover, subcutaneous exposure of WEE caused only limited disease in contrast to previous studies on aerosol administration resulting in fever and encephalitis in macaques [89].

Activated natural killer (NK) cells have been detected in blood and musculoskeletal tissues after CHIKV [90] and RRV [91] infections and in the brain after SFV [92] and SIN [93] infections. It has been demonstrated that NK cell depletion protected mice from challenges with an attenuated VEE strain, whereas transfer of NK cells to depleted mice enhanced virus pathogenicity [94]. In another study, the lack of NK cells during SFV infections extended the lifespan in mice although the infection rate was higher than in control mice [92].

Skin-resident T lymphocytes, γδ T cells, play a critical role in cytotoxicity, cytokine secretion, and promotion of DC maturation [95, 96]. For example, the increase in γδ T cells has demonstrated a protective role after CHIKV infections in mouse footpads [97]. In contrast, absence of γδ T cells showed increased monocyte infiltration and enhanced levels of proinflammatory mediators such as IFN-γ, CCL2, and CXCL9, which led to more extensive tissue damage and severity of disease (98).

In contrast to innate immune responses, adaptive immunity has been less explored. In this context, studies in mice have revealed that CD8+ T cells can mediate cell damage in the CNS during SFV infections, where the upregulation of MHC-1 antigens contributed to the cytotoxic killing of oligodendrocytes [99]. Moreover, depletion of CD8+ T cells resulted in reduced inflammation and elimination of demyelinating lesions in SFV-infected mice [100, 101]. T helper cell subsets, Th1, Th2, and Th17, play a crucial role in immune responses against viral infections [102–188]. As Th1 cells mediate the control of intracellular pathogens through activation of myeloid cells at infection sites, they can contribute to the pathogenesis of arthritis during CHIKV infections [103, 104]. In contrast, mice with deficiencies of MHC-II or CD4+ T cells showed only minor or no joint pathology after CHIKV infections [83, 105]. Moreover, when CD4+ T cells from CHIKV-infected individuals were transferred to TCRαβ–/– mice, the joint swelling was restored [104]. In another study, mice deficient in the serine protease granzyme A did not show joint swelling after CHIKV infections [106]. In the case of neurotrophic alphaviruses, conflicting results have been reported for their pathogenic roles. For example, CD4+ T cells expressing IFN-γ promoted disease resolution in VEE-infected mice [107]. Moreover, transfer of CD4+ T cells to TCRαβ–/– mice provided protection against intranasal challenges with VEE [108]. In the context of neurovirulent SIN strains, CD4+ T cells have been associated with both pathogenic and protective effects in infected mice [109, 110].

Related to the Th2 subset of CD4+ T cells, their production of such cytokines as IL-4, IL-5, IL-10, and IL-13 modulated humoral immunity in response against viral infections [111]. For example, higher levels of IL-13 have been associated with a delayed disease resolution and prolonged musculoskeletal symptoms [112]. Moreover, in the absence of IFN-γ expression, Th17 cells have been associated with encephalomyelitis after infection with neurotrophic SINs [113]. TGF-β can promote the differentiation of naive CD4+ T cells into Tregs, which are responsible for the production of anti-inflammatory cytokines such as IL-10 and TGF-β for suppression of immune cell activity and protection of arthritogenic and encephalitic alphavirus infections [114, 115]. Regulatory immune responses after infection of neurotrophic SINs have been related to Tregs-based IL-10 [116]. Moreover, it was shown that IL-10 deficiency accelerated fatal disease development [110].

In another context, B cells have been demonstrated to produce virus-specific antibodies after alphavirus infections, which clears virus from the bloodstream [117]. Moreover, clearance of arthritogenic and encephalitic alphaviruses target tissues and modulation of disease progression are related to B cells and antibodies [117, 118]. In B cell-deficient mice CHIKV [118] and VEE [119] caused persistent infections in joint tissues and the brain, respectively. Moreover, monoclonal antibodies (mAbs) and immune serum demonstrated protection against arthritogenic and encephalitic alphavirus infections and disease development [120, 121]. Protective antibodies can interfere with virus attachment and entry [122], endosomal fusion, virus assembly and release [123]. For example, neutralizing IgM antibodies reduced viral load in mice after CHIKV and SIN infections [124]. Moreover, it has been established that high-level CHIKV replication during early part of the infection induced production of anti-CHIKV IgG3 antibodies in humans, which reduced the progress of disease [125]. Infiltration and retention of B cells secreting virus-specific antibodies have been associated with long-term control of viral replication in the CNS [126]. However, high levels of neutralizing antibodies were not able to eliminate neurotrophic SIN and SFV from the brain during acute infections [127]. In another study, pathogenic CHIKV strains were able to delay the induction of humoral immune responses by triggering neutrophil and monocyte influx into the draining lymph nodes, which impairs B cells responses against CHIKV [128].

The pathogenesis of CHIKV has been reviewed [129]. In this context, it has been demonstrated that CHIKV infection of human osteoblasts causes cytopathic effects [130]. Moreover, persistent arthritis has been detected in subsets of CHIKV infected patients [131]. One possibility is that persistent infections occur demonstrated by the detection of CHIKV proteins in macrophages in macaques long after the acute phase of infection has passed [132].

2.7 THERAPEUTIC STRATEGIES

Alphaviruses have been used as delivery vectors for therapeutics and vaccines [6, 133]. Moreover, due to their pathogenicity and association with epidemics, they have also been targets for prophylactic and therapeutic interventions. In this chapter, the focus is on alphavirus infections, whereas alphavirus vector-based therapeutics and vaccines are described in Chapter 9 on Immunotherapies and Vaccines.

In the case of alphaviruses, therapeutic and prophylactic strategies based on antiviral drugs, monoclonal antibodies and vaccines have been investigated as described below and listed in Tables 2.1–2.3.

2.7.1 Antiviral Drugs against Alphaviruses

Antiviral drugs against CHIKV have been evaluated in both in vitro and in vivo studies as extensively reviewed [134] (Table 2.1). For example, arbidol (ARB) has previously shown antiviral activity against influenza virus and other respiratory viral infections originally licensed in Russia [135]. In the case of CHIKV, ARB showed potent virus inhibition in Vero cells and primary MRC-5 lung fibroblasts [135]. In another study, suramin treatment of C57BL/6 mice decreased CHIKV viral loads and significantly reduced acute foot lesions [136]. Moreover, favipiravir showed therapeutic efficacy in C57BL/6 mice infected with CHIKV resulting in complete inhibition of viral spread and no detection of viral RNA in non-inoculated feet when favipiravir was administered at day 4 post-infection [137]. However, when the treatment started on day 49 post-injection no reduction of viral RNA was seen in the joints. Sofosbuvir has previously shown efficacy against flaviviruses and compared to the pan-antiviral drug ribavirin (RIBV) demonstrated three times more selective inhibition of CHIKV production in human hepatoma cells [138]. Moreover, sofosbuvir prevented CHIKV-induced paw edema in adult mice and prevented mortality in a neonatal mouse model [138]. In another approach, treatment with the sphingosine-1-phosphate receptor agonist fingolimod resulted in abrogated joint pathology in CHIKV infected mice by blocking the migration of CD4+ cells into joints [139].

Several antiviral drugs have been evaluated against MAYV [140]. Typically, plant extracts such as the flavonoid proanthocyanidin isolated from the root of *Maytenus imbricata* showed 100% inhibition of MAYV proliferation [141]. Moreover, Ginkgolic acid has been isolated from

TABLE 2.1
Antiviral Drugs against Alphaviruses

Strategy	Alphavirus	Findings	References
Arbidol	CHIKV	Inhibition of CHIK in Vero and MRC-5 cells	[135]
Suramin	CHIKV	Decrease in viral load and foot lesions in mice	[136]
Favipiravir	CHIKV	Complete inhibition of viral spread, no detection of viral RNA in mice	[137]
Sofosbuvir	CHIKV	Prevention of paw edema in adult mice and prevention of mortality in neonatal mice	[138]
Fingolimod	CHIKV	Abrogated joint pain in CHIK infected mice	[139]
Proanthocyanidin	MAYV	100% inhibition of MAYV proliferation	[141]
Ginkgolic acid	CHIKV, MAYV	Potential antiviral activity against CHIKV and MAYV	[142]
Co-protoporphyrin IX, Sn-protoporphyrin IV	CHIKV, MAYV, SIN	Antiviral activity against CHIKV, MAYV, and SIN	[143]
Tetrahydroamentoflavone Agathisflavone	MAYV	Strong antiviral activity, reduced MAYV infection	[144]
Punicalagin	MAYV	Antiviral activity against MAYV	[145]
Ribavirin	CHIKV	Efficacy against CHKIV in mice with doxycycline	[148]
Ribavirin	CHIKV	Reduced joint pain in CHIKV-infected patients	[149]
Quercetin	MAYV	Superior antiviral activity compared to RIBV	[150]
Epicatechin	MAYV	Antiviral activity against MAYV	[151]
Thienopyridine derivative	MAYV	Antiviral activity against MAYV	[152]
Favipiravir	MAYV, CHIKV, WEE	Reduced infection and viral production in pre-treated mice	[137, 153, 154]
Brefeldin A	MAYV	Antiviral activity against MAYV	[155]
Prostaglandin A1	SIN	Antiviral activity against SIN	[156]
Cerulenin	MAYV	Reduced MAYV replication	[157]
MG132, lactacystin	CHIKV	Inhibition of CHIKV replication	[158]
Bortozomib	VEE	Inhibition of VEE replication	[159]
	EEE, WEE	Potent inhibition of EEE and WEE	
Silymarin	MAYV	Significantly reduced MAYV infection in mice	[160]

Abbreviations: CHIKV, Chikungunya virus; EEE, Eastern equine encephalitis virus; MAYV, Mayaro virus; RIBV, ribavirin; SIN, Sindbis virus; VEE, Venezuelan equine encephalitis virus; WEE, Western equine encephalitis virus.

the *Gingko biloba* plant and has shown potential antiviral activity against CHIKV and MAYV [142]. Also, porphyrins such as Co-protoporphyrin IX and Sn-protoporphyrin IX have demonstrated antiviral activity against CHIKV, MAYV, and SIN by causing viral envelope protein loss [143]. In another study, the extract from *Schinus terebinthifolius* fruit showed antiviral activity against MAYV in vitro [144]. For example, the bioflavonoids tetrahydroamentoflavone and agathisflavone showed strong antiviral activity and reduced MAYV infections. Moreover, punicalagin isolated from pomegranates has demonstrated antiviral activity against MAYV [145].

RIBV, a compound used in treatment of other viral infections such as hepatitis C virus [146] and hemorrhagic fever virus [147], has also been repurposed for alphaviruses. For example, RIBV has demonstrated efficacy against CHIKV in mice in combination with doxycycline [148]. Furthermore, reduction in joint pains was observed in all 10 CHIKV-infected patients after treatment with RIBV [149]. Quercetin, isolated from the *Bauhinia longifolia* plant, showed superior anti-MAYV activity compared to RIBV [150]. Moreover, epicatechin extracted from *Salacia crassifolia* has been demonstrated to contribute to the inhibition of viral replication resulting in antiviral activity in cases of MAYV infection [151]. Synthetic molecules such as the thienopyridine derivative compound 104 can inhibit MAYV replication by interference in the nsP2 transcription leading to deficient assembly of MAYV particles [152]. Reduced infection and decrease in RNA copy number and viral production were observed in C57BL/6 mice pre-treated with favipiravir prior to MAYV [153], CHIKV [137], and WEE [154] infections. However, favipiravir treatment post-infection showed only a decrease in CHIKV infection in the quadriceps muscles and no other tissue.

The fungal metabolite brefeldin A interferes with protein transport in the endoplasmic reticulum resulting in the inhibition of MAYV replication [155]. Moreover, prostaglandin A1 has demonstrated antiviral efficacy against SIN [156]. In the context of inhibitors of lipid synthesis, cerulenin has been demonstrated to reduce MAYV replication, which was more pronounced at the beginning of the viral infection [157]. Another class affecting the replication of CHIKV comprises proteasome inhibitors [158]. For example, MG132 and lactacystin inhibited CHIKV replication. Moreover, another proteasome inhibitor, Bortzomib, turned out to be a potent inhibitor of VEE multiplication [159]. Additionally, studies on EEE and WEE indicated that Bortozomib was a broad-spectrum inhibitor in alphaviruses. Also, silymarin, a compound derived from the Silybum marianum plant demonstrated a significant reduction of MAYV infection in mice [160].

2.7.2 Antibodies against Alphaviruses

Various antibodies have been evaluated for therapeutic and prophylactic use against alphaviruses (Table 2.2). For example, anti-cytotoxic T-lymphocyte-associated protein 4 (CTLA-4) blocking antibodies showed certain clinical improvement in mice, the combination of the antirheumatic drug CLTA4-Ig and CTLA-4 resulted in superior therapeutic effects, which controlled acute CHIKV infection and arthritis in CHIKV-infected mice [161]. Moreover, neutralizing antibodies against the MAYV E2 envelope protein demonstrated inhibition of not only MAYV but also cross-reactivity

TABLE 2.2
Antibodies against Alphaviruses

Strategy	Alphavirus	Findings	References
Anti-CTLA-4	CHIKV	Control of arthritis in mice	[161]
Anti-E2 MAYV	MAYV, CHIKV, UNAV	Inhibition of MAYV, cross-reactivity against CHIKV and UNAV	[162]
Anti-E2 CHIKV	CHIKV	Neutralizing antibodies against CHIKV	[122]
Anti-E2 EEE	EEE	Neutralizing antibodies against EEE	[120]
Anti-E2 CHIKV/MAYV mAbs DC2. M16, DC.M357	CHIKV, MAYV	Protection against CHIKV/MAYV in mice	[163]
Anti-CHIKV 5F10, 8B10 mAbs	CHIKV	Neutralization of CHIKV/ONNV in mice	[166]
		Delayed pathogenesis of CHIKV, still lethal	[168]
Anti-CHIKV C9 mAb	CHIKV	Protection against CHIKV in mice	[170]
Anti-CHIKV 4J21, 5M16 mAbs	CHIKV	Protection against CHIKV in mice	[171]
Anti-CHIKV 102, 152, 166, 263 mAbs	CHIKV	Complete protection against CHIKV in mice	[165]
Anti-CHIKV 152, 166 mAbs	CHIKV	Reduced viral spread and infection in primates	[172]
Anti-CHIKV 187, 265 mAbs in conserved E2 region	CHIKV	Protection against CHIKV, MAYV, and ONNV	[122]
Anti-EEE 138 E1-specific mAb	EEE	Moderate neutralization of SIN	[173]
		Cross-protection of CHIKV	
Anti-E2 VEE mVEEV-8, -36, 68 mAbs	VEE	Prophylactic and therapeutic protection in mice	[174]

Abbreviations: CHIKV, Chikungunya virus; EEE, Eastern equine encephalitis virus; MAYV, Mayaro virus; ONNV, O'nyong-nyong virus; SIN, Sindbis virus; UNAV, Una virus.

against Una virus (UNAV) and CHIKV [162]. Likewise, neutralizing antibodies against CHIKV [122] and EEE [120] were generated. Additionally, mAbs have been engineered by expression of the alphavirus E3-E2-E1 proteins [163]. For example, the DC2.M16 and DC.M357 mAbs generated against the B domain of the E2 protein, which protected against CHIKV and MAYV challenges in mice [163].

Several studies on mAbs particularly related to CHIKV inhibition and protection have been reviewed [164]. For example, the human mAbs 5F10 and 8B10 isolated from CHIKV-infected patients, showed potent neutralization of African CHIKV strains but not Asian strains [165, 166]. Typically, 5F10 and 8B10 specifically neutralized CHIKV and ONNV but no other alphaviruses [166]. Combination treatment with 5F10 and 8B10 delayed CHIKV pathogenesis and prevented lethality in mice [167, 168]. Structural studies revealed that 5F10 binds to the B domain of the CHIKV E2 protein preventing the exposure of the E1 fusion loop, whereas 8B10 blocks receptor binding to the E2 protein [169]. Moreover, the human mAb C9 protected mice against CHIKV viremia and arthritis [170]. Protection was still obtained when the C9 mAb was administered up to 18 hours after CHIKV infection [170]. Mice were also protected against CHIKV after administration of the human 4J21 and 5M16 mAbs [171]. Combination therapy with different mAbs has been suggested to limit the viral resistance seen against mAb treatment [165]. In this context, the combination of CHIKV mouse 102, 152, 166, and 263 mAbs provided complete protection against CHIKV challenges. Moreover, the combination therapy prolonged the treatment window to 60 h post-infection and was superior to treatment with the 152 mAb alone [165]. Administration of the 152 and 166 mAbs to rhesus macaques reduced viral spread and CHIKV infection although viral RNA was still detectable [172]. In another study, the 187 and 265 neutralizing mAbs for a highly conserved epitope of the E2 protein across diverse alphaviruses provided protection against CHIKV, MAYV, and ONNV in mice [122].

In the context of EEE, E1-specific human mAbs were isolated from survivors of EEE infections [173]. For example, the EEE-138 mAb moderately neutralized SIN and reduced joint swelling and CHIKV RNA levels showing cross-protection in mice [173]. In another study, mouse and human VEE E2-specific mAbs were generated in mice immunized with an attenuated VEE strain [174]. For example, the mVEEV-8, mVEEV-36 and mVEEVE-68 mAbs provided both prophylactic and therapeutic protection against lethal aerosol challenges with VEE in mice.

2.7.3 Vaccines against Alphaviruses

The development of vaccines against alphaviruses have played an essential role in both prophylactic and therapeutic applications [175]. Examples are presented below and summarized in Table 2.3. For example, the CHIKV envelope protein genes E3-E2-6K-E1 replaced the G protein in vesicular stomatitis virus (VSV) G protein in viral particles, which elicited neutralizing antibodies and provided protection against CHIKV challenges in mice [176]. Moreover, an attenuated measles virus (MV) strain expressing the CHIKV capsid and envelope proteins generated strong neutralizing antibody responses and protected mice against CHIKV challenges after the administration of a single vaccine dose [177]. Additionally, neutralizing antibodies and CHIKV protection were achieved in macaques immunized with MV-CHIKV [178]. In a phase I clinical trial the MV-CHIKV vaccine generated 44–92% seroconversion depending on the dose applied and 100% seroconversion after a second immunization independent of the dose used [179]. In a phase II study, no adverse events were recorded, strong neutralizing antibody responses were detected and the seroconversion rate varied between 50% and 95.9% dependent on the vaccine dose [180]. In another phase II trial, a prime-boost vaccination strategy with MV-CHIKV elicited superior neutralizing antibody responses compared to prime vaccinations [181]. In another phase II study, the safety of CHIKV virus-like particles (VLPs) was evaluated for safety and tolerability [182]. The CHIKV was well tolerated, and no serious adverse events were registered. Vaccine-induced CHIKV responses were

TABLE 2.3
Vaccines against Alphaviruses

Vector	Alphavirus construct	Findings	References
VSV	CHIKV E3-E2-6K-E1	nAbs, protection against CHIKV in mice	[176]
MV	CHIKV C-E3-E2-6K-E1	nAbs, protection against CHIKV in mice	[177]
MV	CHIKV C-E3-E2-6K-E1	nAbs, protection against CHIKV in macaques	[178]
MV	CHIKV C-E3-E2-6K-E1	44–92% seroconversion after first dose, 100% seroconversion after second dose in Phase I	[179]
MV	CHIKV C-E3-E2-6K-E1	No adverse events, strong nAb responses, seroconversion 50–95.9%	[180]
MV	CHIKV C-E3-E2-6K-E1	Superior nAb responses after a prime-boost strategy compared to prime vaccination in Phase II	[181]
CHIKV	CHIKV VLPs	Good safety and tolerability in Phase II	[182]
CHIKV	CHIKV live attenuated	Strong immunogenicity for 1 year in Phase I	[183]
EEE, VEE, WEE	EEE, VEE, WEE	Protection after combination therapy in mice, primates Complete (EEE, VEE), partial (WEE) protection in primates	[184]
VEE	VEE DNA replicon TC-83	Strong nAbs, protection against VEE in mice	[185]
VEE	VEE-DNA replicon TC-83	Strong nAbs, protection against VEE in primates	[186]
VEE	VEE DNA replicon E3-E2-6K-E1	No serious adverse events, strong nAb responses in Phase I	[187]
VEE	VEE-RNA replicon-CNE	Protection against VEE in mice	[188]
MVA-BN	Individual EEE, VEE, WEE Trivalent EEE/VEE/WEE	Protection against challenges with correspondent viruses, 90–100% against WEE/VEE, 60–90% for EEE with trivalent vaccine	[189]

Abbreviations: CHIKV, Chikungunya virus; CNE, cationic nanoemulsion; EEE, Eastern equine encephalitis virus; MAYV, Mayaro virus; MV, measles virus; MVA-BN, modified vaccinia virus Ankara-Bavarian Nordic strain; nAbs, neutralization antibodies; ONNV, O'nyong-nyong virus; TC-83, attenuated VEE strain; UNAV, Una virus; VEE. Venezuelan equine encephalitis virus; VLPs, virus-like particles; VSV, vesicular stomatitis virus; WEE, Western equine encephalitis virus.

detected, which needs further confirmation in phase III. In another approach, a live attenuated CHIKV was well tolerated, elicited strong immunogenicity with 100% seroconversion sustained for one year in phase I, which could contribute to the prophylactic application against CHIKV worldwide [183].

Viral vectors have been engineered for EEE, VEE and WEE by removing the furin cleavage site between the corresponding E2 and E3 envelope proteins [184]. Immunization of mice with individual EEV, VEE, and WEE or their combination elicited strong immune responses and generated protection against challenges with lethal doses of the individual viruses. In primates, protection was obtained after immunization individually with VEE and EEE particles or the combination of all three alphaviruses, whereas WEE administration to macaques resulted in only partial protection [184]. A VEE DNA replicon vector based on the TC-83 strain carrying the VEE capsid gene showed no adverse events, induced strong neutralizing antibody responses, and protected mice against challenges with VEE [185]. Furthermore, intramuscular administration of VEE DNA replicons induced strong immune responses and protected macaques against aerosol challenges with lethal VEE strains [186]. In a phase I trial, intramuscular or intradermal administration of VEE DNA replicons expressing the E3-E2-6K-E1 polyprotein caused no vaccine-related serious adverse events and neutralizing antibody responses were elicited [187].

In another approach, RNA-based VEE replicons have been formulated in synthetic cationic nanoemulsion (CNE) as live-attenuated VEE particles (VEE-LAV) based on TC-83 genome [188].

Immunization with VEE-LAV elicited strong immune response and protected mice against challenges in 100% of mice. Deletion of the capsid gene generated the replication-deficient VEE-IAV vector, which provided partial protection in mice [188].

The vaccine platform based on modified vaccinia virus Ankara-Bavarian Nordic (MVA-BN) strain was applied for the engineering of individual WEE (MVA-BN-W), EEE (MVA-BN-E), and VEE (MVA-BN-V) vaccines and the trivalent EEE/VEE/WEE (MVA-BN-WEV) vaccine [189]. All individual vaccines provided full protection for each correspondent virus type. The trivalent vaccine showed 90–100% protection against challenges with WEE and VEE, and 60–90% protection against EEE.

2.8 CONCLUSIONS

The family of ssRNA alphaviruses was originally considered as important models in molecular biology and virology to study the viral lifecycle and cellular protein transport in general. Although alphaviruses have been long known to cause disease, ranging from mild fever and rash to arthritis and encephalitis even leading to fatal outcome, recent climate change has facilitated the global spread of alphaviruses. For this reason, monitoring of alphaviruses such as CHIKV have become of high priority. Likewise, as described above, much attention has been dedicated to the development of antiviral drugs. In this context, antiviral drugs developed for other types of viral infections have been monitored and also novel drugs have been introduced. An area which has received plenty of attention is the development of antibodies against alphavirus infections, particularly monoclonal antibodies. Vaccine development is another area, which has been widely investigated. However, further research and development is mandatory to better understand the alphavirus biology and ensure the best possible readiness for prophylactic and therapeutic interventions.

REFERENCES

1. Strauss JH, Strauss EG. The alphaviruses: Gene expression, replication and evolution. Microbiol Rev 1994; 58: 491–562.
2. Atkins GJ, Sheahan BJ, Liljestrom P. The molecular pathogenesis of Semliki Forest virus: A model virus made useful? J Gen Virol 1999; 80: 2287–2297.
3. Liljestrom P, Garoff H. A new generation of animal cell expression vectors based on the Semliki Forest virus replicon. Biotechnology 1991; 9: 1356–1361.
4. Xiong C, Levis R, Shen P, Schlesinger S, Rice CM, Huang HV. Sindbis virus: An efficient, broad host range vector for gene expression in animal cells. Science 1989; 243: 1188–1191.
5. Davis NL, Willis LV, Smith JF, Johnston RF. In vitro synthesis of infectious Venezuelan equine encephalitis virus RNA from a cDNAclone: Analysis of a viable deletion mutant. Virology 1989; 171: 189–204.
6. Ljungberg K, Liljeström P. Self-replicating alphavirus RNA vaccines. Expert Rev Vaccines 2015; 14: 177–194.
7. Calisher CH, Karabatsos N. Arbovirus serogroups: Definition and geographic distribution. In The Arboviruses: Epidemiology and Ecology. Monath TP (Ed.). 1988; 19–57. CRC Press, Inc., Baca Raton, USA.
8. Shirako Y, Niklasson B, Dalrymple JM, Strauss EG, Strauss JH. Structure of the Ockelbo virus genome and its relationship to other Sindbis viruses. Virology 1991; 182: 753–764.
9. Yuen KY, Bielefeldt-Ohmann H. Ross River virus infection: A cross-disciplinary review with a veterinary perspective. Pathogens 2021; 10: 357. https://doi.org/10.3390/pathogens10030357.
10. Zacks MA, Paessler S. Encephalitic alphaviruses. Vet Microbiol 2010; 140: 281–286. https://doi.org/10.1016/j.vetmic.2009.08.023.
11. Wei LLL, Tom R, Kim YC. An emerging alphavirus in the Americas. Viruses 2024; 16: 1297. https://doi.org/10.3390/v16081297.

12. Karabatsos N, Lewis AL, Calisher CH, Hunt AR, Roehrig JT. Identification of Highlands J virus from a Florida horse. Am J Trop Med Hyg 1988; 39: 603–606.
13. Chamberlain RW. Epidemiology of arthropod-borne Togaviruses: The role of arthropods as hosts and vectors and of vertebrate hosts in natural transmission cycles. In The Togaviruses: Biology, Structure, Replication. Schlesinger RW (Ed.). 1980; 175–227. Academic Press, Inc, New York.
14. Peters CJ, Dalrymple JM. Alphaviruses. In Virology. Fields BN, Knipe DM (Eds.). 1990; 713–761. Raven Press, New York.
15. Phillips DA, Murray JR, Asakov JG, Wiemers MA. Clinical and subclinical Barmah Forest virus infection in Queensland. Med J Aust 1990; 152: 463–466.
16. Rennels MB. Arthropod-borne virus infections of the central nervous system. Neurol Clin 1984; 2: 241–254.
17. Petrov AA, Lebedev VN, Kulish VS, Pyshnaya NS, Stovba LF, Borisevich SV. Epidemiological analysis of outbreaks of diseases caused by American equine encephalitis causative agents in endemic regions. Zh Mikrobiol Epidemiol Immunobiol 2015; 5: 103–110.
18. Weaver S, Salas R, Rico-Hesse R, Ludwig GV, Oberste MS, Boshell J, Tesh JB, VEE Study Group. Re-emergence of epidemic Venezuelan equine encephalomyelitis in South America. Lancet 1996; 348: 43640.
19. Kelvin A. Outbreak of Chikungunya in the Republic of Congo and the global picture. J. Infect. Dev. Ctries 2011; 5: 4414.
20. Jansen K. The 20A5-2N7 Chikungunya epidemic in reunion: Ambiguous etiologies, memories, and meaning-making. Med Anthropol 2013; 32: 174–189.
21. Rodrigues Faria N, Lourenco J, Marques de Cerqueira E, Maia de Lima M, Pybus O, Alcantara LCJ. Epidemiology of Chikungunya virus in Bahia, Brazil, 2014–2015. PLOS Curr 2016; 8.
22. Mathiot C, Grimaud G, Garry P, Bouquety JC, Mada A, Daguisy AM, Georges AJ. An outbreak of human Semliki Forest virus infection in Central African Republic. Am J Trop Med Hyg 1990; 42: 386–393.
23. Niklasson B. Sindbis and Sindbis-like viruses. In The Arboviruses: Epidemiology and Ecology. Monath TP (Ed.). 1988; 167–176. CRC Press Inc, Boca Raton, FL, USA.
24. Vairo F, Halder N, Kock R, Ntoumi F, Ippolito G, Zumla A. Chikungunya: Epidemiology, pathogenesis, clinical features, management, and prevention. Infect Dis Clin North Am 2019; 33: 1003–1025.
25. Tong Jia Ming S, Tan Yi Jun K, Carissimo G. Pathogenicity and virulence of O'nyong-nyong virus: A less studied Togaviridae with pandemic potential. Virulence 2024; 15: 2355201.
26. Fraser JR. Epidemic polyarthritis and Ross River virus disease. Clin Rheum Dis 1986; 12: 369–388.
27. Willems WR, Kaluza G, Boschek CB, Barrier H, Hager H, Schutz HJ, Feistner H. 1979. Semliki Forest virus: Cause of a fatal case of human encephalitis. Science 1979; 203: 1127–1129.
28. Levine B, Huang Q, Isaacs JT, Reed JC, Griffin DE, Hardwick JM. Conversion of lytic to persistent alphavirus infection by the bcl-2 cellular oncogene. Nature 1993; 361: 739–742.
29. Barry G, Fragkoudis R, Ferguson MC, Lulla A, Merits A, Kohl A, Fazakerley JKJ. Semliki Forest virus-induced endoplasmic reticulum stress accelerates apoptotic death of mammalian cells. J Virol 2010; 84: 7369–7377.
30. Fros JJ, Pijlman GP. Alphavirus infection: Host cell shut-off and inhibition of antiviral responses. Viruses 2016; 8: 166.
31. Fuller SD. The T=4 envelope of Sindbis virus is organized by interactions with a complementary T=3 capsid. Cell 1987; 48: 923–934.
32. Paredes AM, Brown DT, Rothnagel RB, Chiu W, Schoepp RJ, Johnston RE, Prasad BVV. Three dimensional structure of a membrane-containing virus. Proc Natl Acad Sci USA 1993; 90: 9095–9099.
33. Vogel RH, Provencher SW, von Bonsdorff CH, Adrian M, Dubochet J. Envelope structure of Semliki Forest virus reconstructed from cryoelectron micrographs. Nature 1986; 320: 533–535.
34. Choi H-K, Tong L, Minor W, Dumas P, Boege U, Rossmann MG, Wengler G. Structure of Sindbis virus core protein reveals a chymotrypsin-like serine proteinase and the organization of the virion. Nature 1991; 354: 37–43.
35. Tong L, Wengler G, Rossmann MG. Refined structure of Sindbis virus core protein and comparison with other chymotrypsin-like serine proteinase structures. J Mol Biol 1993; 230: 228–247.
36. Cheng RH, Kuhn RJ, Olson NH, Rossmann MG, Choi H-K, Baker TS. Nucleocapsid and glycoprotein organization in an enveloped virus. Cell 1995; 80: 621–630.
37. Gahmberg CG, Utermann G, Simons K. The membrane proteins of Semliki Forest virus have a hydrophobic part attached to the viral membrane. FEBS Lett 1972; 28: 179–182.

38. Ziemiecki A, Garoff H, Simons K. Formation of the Semliki Forest virus membrane glycoprotein complexes in the infected cell. J. Gen. Virol 1980; 50: 111–123.
39. Jain SK, DeCandido S, Kielian M. Processing of the p62 envelope precursor protein of Semliki Forest virus. J Biol Chem 1991; 266: 5756–5761.
40. Mayne JT, Rice CM, Strauss EG, Hunkapiller MW, Strauss JH. Biochemical studies of the maturation of the small Sindbis virus glycoprotein E3. Virology 1984; 134: 338–357.
41. Sanz MA, Madan V, Carrasco L, Nieva JL. Interfacial domains in Sindbis virus 6K protein. Detection and functional characterization. J Biol Chem 2003; 278: 2051–2057.
42. Bowers DF, Abell BA, Brown DT. Replication and tissue tropism of the alphavirus Sindbis in the mosquito *Aedes albopictus*. Virology 1995; 212: 1–12.
43. Zimmerman O, Holmes AC, Kafai NM, Adams LJ, Diamond MS. Entry receptors – The gateway to alphavirus infection. J Clin Invest 2023; 133: e165307.
44. Wang K-S, Kuhn RJ, Strauss EG, Ou S, Strauss JH. High-affinity laminin receptor is a receptor for Sindbis virus in mammalian cells. J Virology 1992; 66: 4992–5001.
45. ZhuW, Wang L, Yang Y, Jia J, Fu S, Feng Y, He Y, Li JP, Liang G. Interaction of E2 glycoprotein with heparan sulfate is crucial for cellular infection of Sindbis virus. PLOS One 2010; 5: e9656.
46. Rose PP, Hanna SL, Spiridigliozzi A, Wannissorn N, Belting DP, Ross SR et al. Natural resistance-associated macrophage protein is a cellular receptor for Sindbis virus in both insect and mammalian hosts. Cell Host Microbe 2011; 10: 97–104.
47. Wintachai P, Wikan N, Kuadtkitkan A, Jaimipuk T, Ubol S, Pulmanausahakul R et al. Identification of prohibitin as a Chikungunya virus receptor protein. J Med Virol 2012; 84: 1757–1770.
48. De Caluwe L, Coppens S, Vereecken K, Daled S, Dhaenens M, Van Ostade X et al. The CD147 protein complex is involved in entry of Chikungunya virus and related alphaviruses in human cells. Front Microbiol 2021; 12: 615165.
49. Zhang R, Kim AS, Fox JM, Nair S, Basore K, Klimstra WB et al. Mxra8 is a receptor for multiple arthritogenic alphaviruses. Nature 2018; 557: 570–574.
50. Ma H, Kim AS, Kafai NM, Earnest JT, Shah AP, Case JB et al. LDLRAD3 is a receptor for Venezuelan equine encephalitis virus. Nature 2020; 588: 308–314.
51. Clark LE, Clark SA, Lin C, Liu J, Coscia A, Nabel KG et al. VLDLR and ApoER2 are receptors for multiple alphaviruses. Nature 2022; 602: 475–480.
52. Hoekstra D, Kok JW. Entry mechanisms of enveloped viruses. Implications for fusion of intracellular membranes. Biosci Rep 1989; 9: 273–305.
53. Marsh M. The entry of enveloped viruses into cells by endocytosis. Biochem J 1984; 218: 1–10.
54. Stegmann TR, Doms RW, Helenius A. Protein mediated membrane fusion. Annu Rev Biophys Biophys Chem 1989; 18: 187–211.
55. Strauss EG, Strauss JH. Structure and replication of the alphavirus genome. In The Togaviridae and Flaviviridae. Schlesinger S, Schlesinger MJ (Eds.). 1986; 35–90. Plenum Publishing Corp, New York.
56. Ahola T, McInerney G, Merits A. Alphavirus replication in vertebrate cells. Adv Virus Res 2021; 111: 111–156.
57. Pietilä MK, Hellström K, Ahola T. Alphavirus polymerase and RNA replication. Virus Res 2017; 234: 44–57.
58. Mendes A, Kuhn RJ. Alphavirus nucleocapsid packaging and assembly. Viruses 2018; 10: 138.
59. Doms RW, Lamb RA, Rose JK, Helenius A. Folding and assembly of viral membrane glycoproteins. Virology 1993; 193: 545–562.
60. deCurtis I, Simons K. Dissection of Semliki Forest virus glycoprotein delivery from the trans-golgi network to the cell surface in permeabilized BHK cells. Proc Natl Acad Sci USA 1988; 85: 8052–8056.
61. Cadd TL, Skoging U, Liljeström P. Budding of enveloped viruses from the plasma membrane. Bioessays 1997; 19: 993–1000.
62. Kafai NM, Diamond MS, Fox JM. Distinct cellular tropism and immune responses to alphavirus infection. Annu Rev Immunol 2022; 40: 615–649.
63. Kawasaki T, Kawai T. Toll-like receptor signaling pathways. Front Immunol 2014; 5: 461.
64. Hopfner K-P, Hornung V. Molecular mechanisms and cellular functions of cGAS–STING signalling. Nat Rev Mol Cell Biol 2020; 21: 501–521.
65. Webb LG, Veloz J, Pintado-Silva J, Zhu T, Rangel MV, Mutetwa T et al. Chikungunya virus antagonizes cGAS-STING mediated type-I interferon responses by degrading cGAS. PLOS Pathog 2020; 16: e1008999.

66. Thompson CD, Matta B, Barnes BJ. Therapeutic targeting of IRFs: Pathway-dependence or structure-based? Front Immunol 2018; 9: 2622.
67. Grieder FB, Vogel SN. Role of interferon and interferon regulatory factors in early protection against Venezuelan equine encephalitis virus infection. Virology 1999; 257: 106–118.
68. Sadler AJ, Williams BR. Interferon-inducible antiviral effectors. Nat Rev Immunol 2008; 8: 559–568.
69. Weiss CM, Trobaugh DW, Sun C, Lucas TM, Diamond MS, Ryman KD et al. The interferon-induced exonuclease ISG20 exerts antiviral activity through upregulation of type I interferon response proteins. mSphere 2018; 3: e00209–18.
70. Zhang Y, Burke CW, Ryman KD, Klimstra WB. Identification and characterization of interferon-induced proteins that inhibit alphavirus replication. J Virol 2007; 81: 11246–11255.
71. Werneke SW, Schilte C, Rohatgi A, Monte KJ, Michault A, Arenzana-Seisdedos F et al. ISG15 is critical in the control of Chikungunya virus infection independent of UbE1L mediated conjugation. PLOS Pathog 2011; 7: e1002322.
72. Lenschow DJ, Giannakopoulos NV, Gunn LJ, Johnston C, O'Guin AK, Schmidt RE et al. 2005. Identification of interferon-stimulated gene 15 as an antiviral molecule during Sindbis virus infection in vivo. J Virol 2005; 79: 13974–13983.
73. de Castro-Jorge LA, de Carvalho RVH, Klein TM, Hiroki CH, Lopes AH, Guimaraes RM et al. The NLRP3 inflammasome is involved with the pathogenesis of Mayaro virus. PLOS Pathog 2019; 15: e1007934.
74. Chen W, Foo SS, Zaid A, Teng TS, Herrero LJ, Wolf S et al. Specific inhibition of NLRP3 in Chikungunya disease reveals a role for inflammasomes in alphavirus-induced inflammation. Nat Microbiol 2017; 2: 1435–1445.
75. Atella MO, Carvalho AS, Da Poian AT. Role of macrophages in the onset, maintenance, or control of arthritis caused by alphaviruses. Exp Biol Med 2023; 248: 2039–2044.
76. MacDonald GH, Johnston RE. 2000. Role of dendritic cell targeting in Venezuelan equine encephalitis virus pathogenesis. J Virol 2000; 74: 914–922.
77. Webster B, Werneke SW, Zafirova B, This S, Coléon S, Dcembre E et al. Plasmacytoid dendritic cells control dengue and Chikungunya virus infections via IRF7-regulated interferon responses. eLife 2018; 7: e34273.
78. Winkler ES, Shrihari S, Hykes BL Jr., Handley SA, Andhey PS, Huang YS et al. The intestinal microbiome restricts alphavirus infection and dissemination through a bile acid-type i IFN signaling axis. Cell 2020; 182: 901–18.e18.
79. Haist KC, Carpentier KS, Davenport BJ, Morrison TE. Plasmacytoid dendritic cells mediate control of Ross River virus infection via a type i interferon-dependent, MAVS independent mechanism. J Virol 2021; 95: e01538–20.
80. Bhalla N, Gardner CL, Downs SN, Dunn M, Sun C, Klimstra WB. 2019. Macromolecular synthesis shutoff resistance by myeloid cells is critical to IRF7-dependent systemic interferon alpha/beta induction after alphavirus infection. J Virol 2019; 93: e00872–19.
81. Haist KC, Burrack KS, Davenport BJ, Morrison TE. Inflammatory monocytes mediate control of acute alphavirus infection in mice. PLOS Pathog 2017; 13: e1006748.
82. Rulli NE, Guglielmotti A, Mangano G, Rolph MS, Apicella C, Zaid A et al. Amelioration of alphavirus induced arthritis and myositis in a mouse model by treatment with bindarit, an inhibitor of monocyte chemotactic proteins. Arthritis Rheum 2009; 60: 2513–2523.
83. Gardner J, Anraku I, Le TT, Larcher T, Major L, Roques P et al. Chikungunya virus arthritis in adult wild-type mice. J Virol 2010; 84: 8021–8032.
84. Lidbury BA, Rulli NE, Suhrbier A, Smith PN, McColl SR, Cunningham AL et al. Macrophage-derived proinflammatory factors contribute to the development of arthritis and myositis after infection with an arthrogenic alphavirus. J Infect Dis 2008; 197: 1585–1593.
85. Herrero LJ, Sheng KC, Jian P, Taylor A, Her Z, Herring BL et al. 2013. Macrophage migration inhibitory factor receptor CD74 mediates alphavirus-induced arthritis and myositis in murine models of alphavirus infection. Arthritis Rheum 2013; 65: 2724–2736.
86. Lin T, Geng T, Harrison AG, Yang D, Vella AT, Fikrig E et al. CXCL10 signaling contributes to the pathogenesis of arthritogenic alphaviruses. Viruses 2020; 12: 1252.
87. Poo YS, Nakaya H, Gardner J, Larcher T, Schroder WA, Le TT et al. CCR2 deficiency promotes exacerbated chronic erosive neutrophil-dominated Chikungunya virus arthritis. J Virol 2014; 88: 6862–6872.

88. Smith DR, Schmaljohn CS, Badger C, Ostrowski K, Zeng X, Grimes SD et al. Comparative pathology study of Venezuelan, eastern, and western equine encephalitis viruses in non-human primates. Antivir Res 2020; 182: 104875.
89. Reed DS, Larsen T, Sullivan LJ, Lind CM, Lackemeyer MG, Pratt WD et al. Aerosol exposure to western equine encephalitis virus causes fever and encephalitis in cynomolgus macaques. J Infect Dis 2005; 192: 1173–1182.
90. Petitdemange C, Becquart P, Wauquier N, Béziat V, Debré P, Leroy EM et al. Unconventional repertoire profile is imprinted during acute Chikungunya infection for natural killer cells polarization toward cytotoxicity. PLOS Pathog 2011; 7: e1002268.
91. Hazelton RA, Hughes C, Aaskov JG. The inflammatory response in the synovium of a patient with Ross River arbovirus infection. Aust N Z J Med 1985; 15: 336–339.
92. Alsharifi M, Lobigs M, Simon MM, Kersten A, Müller K, Koskinen A et al. NK cell-mediated immunopathology during an acute viral infection of the CNS. Eur J Immunol 2006; 36: 887–896.
93. Griffin DE, Hess JL. Cells with natural killer activity in the cerebrospinal fluid of normal mice and athymic nude mice with acute Sindbis virus encephalitis. J Immunol 1986; 136: 1841–1845.
94. Taylor K, Kolokoltsova O, Patterson M, Poussard A, Smith J, Esters DM et al. Natural killer cell mediated pathogenesis determines outcome of central nervous system infection with Venezuelan equine encephalitis virus in C3H/HeN mice. Vaccine 2012; 30: 4095–4105.
95. Hayday AC. $\gamma\delta$ Cells: A right time and a right place for a conserved third way of protection. Annu Rev Immunol 2000; 18: 975–1026.
96. Born W, Cady C, Jones-Carson J, Mukasa A, Lahn M, O'Brien R. Immunoregulatory functions of gamma delta T cells. Adv Immunol 1999; 71: 77–144.
97. Long KM, Ferris MT, Whitmore AC, Montgomery SA, Thurlow LR, McGee CE et al. Gamma delta T cells play a protective role in Chikungunya virus-induced disease. J Virol 2016; 90: 433–443.
98. Fu YX, Roark CE, Kelly K, Drevets D, Campbell P, O'Brien R et al. Immune protection and control of inflammatory tissue necrosis by gamma delta T cells. J Immunol 1994; 153: 3101.
99. Fazakerley JK, Pathak S, Scallan M, Amor S, Dyson H. Replication of the A7(74) strain of Semliki Forest virus is restricted in neurons. Virology 1993; 195: 627–637.
100. Subak-Sharpe I, Dyson H, Fazakerley J. In vivo depletion of CD8+ T cells prevents lesions of demyelination in Semliki Forest virus infection. J Virol 1993; 67: 7629–7633.
101. Fazakerley JK, Webb HE. Semliki Forest virus-induced, immune-mediated demyelination: Adoptive transfer studies and viral persistence in nude mice. J Gen Virol 1987; 68: 377–385.
102. Rosendahl Huber S, van Beek J, de Jonge J, Luytjes W, van Baarle D. T cell responses to viral infections—Opportunities for peptide vaccination. Front Immunol 2014; 5: 171.
103. Hoarau J-J, Jaffar Bandjee M-C, Krejbich Trotot P, Das T, Lee-Pat-Yuen G, Sahoo R et al. Persistent chronic inflammation and infection by Chikungunya arthritogenic alphavirus in spite of a robust host immune response. J Immunol 2010; 184: 5914.
104. Teo T-H, Lum F-M, Claser C, Lulla V, Lulla A, Merits A et al. A pathogenic role for CD4+ T cells during Chikungunya virus infection in mice. J Immunol 2013; 190: 259.
105. Poo YS, Rudd PA, Gardner J, Wilson JA, Larcher T, Colle M-A et al. Multiple immune factors are involved in controlling acute and chronic Chikungunya virus infection. PLOS Negl Trop Dis 2014; 8: e3354.
106. Wilson JA, Prow NA, Schroder WA, Ellis JJ, Cumming HE, Gearing LJ et al. RNA-seq analysis of Chikungunya virus infection and identification of granzyme a as a major promoter of arthritic inflammation. PLOS Pathog 2017; 13: e1006155.
107. Paessler S, Yun NE, Judy BM, Dziuba N, Zacks MA, Grund HA et al. Alpha-beta T cells provide protection against lethal encephalitis in the murine model of VEEV infection. Virology 2007; 367: 307–323.
108. Yun NE, Peng BH, Bertke AS, Borisevich V, Smith JK, Smith JN et al. CD4+ T cells provide protection against acute lethal encephalitis caused by Venezuelan equine encephalitis virus. Vaccine 2009; 27: 4064–4073.
109. Rowell JF, Griffin DE. Contribution of T cells to mortality in neurovirulent Sindbis virus encephalomyelitis. J Neuroimmunol 2002; 127: 106–114.
110. Kulcsar KA, Baxter VK, Abraham R, Nelson A, Griffin DE, Dermody TS. Distinct immune responses in resistant and susceptible strains of mice during neurovirulent alphavirus encephalomyelitis. J Virol 2015; 89: 8280–8291.
111. Spellberg B, Edwards JE Jr. Type 1/type 2 immunity in infectious diseases. Clin Infect Dis 2001; 32: 76–102.

112. Venugopalan A, Ghorpade RP, Chopra A. Cytokines in acute Chikungunya. PLOS One 2014; 9: e111305.
113. Baxter VK, Griffin DE. Interferon-gamma modulation of the local T cell response to alphavirus encephalomyelitis. Viruses 2020; 12: 113.
114. Lee GR. 2018. The balance of Th17 versus Treg cells in autoimmunity. Int J Mol Sci 2018; 19: 730.
115. Lee WW, Teo T-H, Her Z, Lum F-M, Kam Y-W, Haase D et al. Expanding regulatory T cells alleviates Chikungunya virus-induced pathology in mice. J Virol 2015; 89: 7893–7904.
116. Kulcsar KA, Griffin DE. T cell-derived interleukin-10 is an important regulator of the Th17 response during lethal alphavirus encephalomyelitis. J Neuroimmunol 2016; 295–296: 60–67.
117. Metcalf TU, Griffin DE. Alphavirus-induced encephalomyelitis: Antibody-secreting cells and viral clearance from the nervous system. J Virol 2011; 85: 11490–11501.
118. Hawman DW, Fox JM, Ashbrook AW, May NA, Schroeder KMS, Torres RM et al. Pathogenic Chikungunya virus evades B cell responses to establish persistence. Cell Rep 2016; 16: 1326–1338.
119. Brooke CB, Deming DJ, Whitmore AC, White LJ, Johnston RE. T cells facilitate recovery from Venezuelan equine encephalitis virus-induced encephalomyelitis in the absence of antibody. J Virol 2010; 84: 4556–4568.
120. Kim AS, Austin SK, Gardner CL, Zuiani A, Reed DS, Trobaugh DW et al. Protective antibodies against Eastern equine encephalitis virus bind to epitopes in domains A and B of the E2 glycoprotein. Nat Microbiol 2019; 4: 187–197.
121. Fox JM, Diamond MS. Immune-mediated protection and pathogenesis of Chikungunya virus. J Immunol 2016; 197: 4210–4218.
122. Fox JM, Long F, Edeling MA, Lin H, Van Duijl-Richter MKS, Fong RH et al. Broadly neutralizing alphavirus antibodies bind an epitope on E2 and inhibit entry and egress. Cell 2015; 163: 1095–1107.
123. Jin J, Liss NM, Chen D-H, Liao M, Fox JM, Shimak RM et al. Neutralizing monoclonal antibodies block Chikungunya virus entry and release by targeting an epitope critical to viral pathogenesis. Cell Rep 2015; 13: 2553–2564.
124. Petitdemange C, Wauquier N, Vieillard V. Control of immunopathology during Chikungunya virus infection. J Allergy Clin Immunol 2015; 135: 846–855.
125. Kam Y-W, Simarmata D, Chow A, Her Z, Teng T-S, Ong EKS et al. 2012. Early appearance of neutralizing immunoglobulin G3 antibodies is associated with Chikungunya virus clearance and long-term clinical protection. J Infect Dis 2012; 205: 1147–1154.
126. Fragkoudis R, Dixon-Ballany CM, Zagrajek AK, Kedzierski L, Fazakerley JK. Following acute encephalitis, Semliki Forest virus is undetectable in the brain by infectivity assays but functional virus RNA capable of generating infectious virus persists for life. Viruses 2018; 10: 273.
127. Griffin D, Levine B, Tyor W, Ubol S, Desprès P. 1997. The role of antibody in recovery from alphavirus encephalitis. Immunol Rev 1997; 159: 155–161.
128. McCarthy MK, Reynoso GV, Winkler ES, Mack M, Diamond MS, Hickman HD et al. MyD88-dependent influx of monocytes and neutrophils impairs lymph node B cell responses to Chikungunya virus infection via Irf5, Nos2 and Nox2. PLOS Pathog 2020; 16: e1008292.
129. Burt FJ, Chen W, Miner JJ, Lenschow DJ, Merits A, Schnettler E et al. Chikungunya virus: An update on the biology and pathogenesis of this emerging pathogen. Lancet Infect Dis 2017; 17: e107–e117.
130. Noret M, Herrero L, Rulli N, Rolph M, Smith PN, Li RW et al. Interleukin 6, RANKL, and osteoprotegerin expression by Chikungunya virus-infected human osteoblasts. J Infect Dis 2012; 206: 455–459.
131. Miner JJ, Aw-Yeang HX, Fox JM, Taffner S, Malkova ON, Oh ST et al. Brief report: Chikungunya viral arthritis in the United States: A mimic of seronegative rheumatoid arthritis. Arthritis Rheumatol 2015; 67: 1214–1220.
132. Labadie K, Larcher T, Joubert C, Mannioui A, Delache B, Brochard P et al. Chikungunya disease in nonhuman primates involves long-term viral persistence in macrophages. J Clin Invest 2010; 120: 894–906.
133. Lundstrom K. Alphaviruses in immunotherapy and anticancer therapy. Biomedicines 2022; 10: 2263.
134. Battisti V, Urban E, Langer T. Antivirals against the Chikungunya virus. Viruses 2021; 13: 1307.
135. Delogu I, Pastorino B, Baronti C, Nougairède A, Bonnet E, de Lamballerie X. In vitro antiviral activity of arbidol against Chikungunya virus and characteristics of a selected resistant mutant. Antivir Res 2011; 90: 99–107.
136. Kuo S-C, Wang Y-M, Ho Y-J, Chang T-Y, Lai Z-Z, Tsui P-Y et al. Suramin treatment reduces Chikungunya pathogenesis in mice. Antivir Res 2016; 134: 89–96.

137. Abdelnabi R, Jochmans D, Verbeken E, Neyts J, Delang L. Antiviral treatment efficiently inhibits Chikungunya virus infection in the joints of mice during the acute but not during the chronic phase of the infection. Antivir Res 2017; 149: 113–117.
138. Ferreira AC, Reis PA, de Freitas CS, Sacramento CQ, Hoelz LVB, Bastos MM et al. Beyond members of the Flaviviridae family, sofosbuvir also inhibits Chikungunya virus replication. Antimicrob Agents Chemother 2019; 63: e01389–18.
139. Teo TH, Chan YH, Lee WW, Lum FM, Amrun SN, Her Z et al. Fingolimod treatment abrogates Chikungunya virus-induced arthralgia. Sci Transl Med 2017; 9: eaal1333.
140. Andreolla AP, Borges AA, Bordignon J, Nunes Duarte dos Santos C. Mayaro Virus: The state-of-the-art for antiviral drug development. Viruses 2022; 14: 1787.
141. Coelho Ferraz A, Silva Moraes TF, da Cruz Nizer WS, dos Santos M, Helvécio Totola A, Siqueira Ferreira JM et al. Virucidal activity of proanthocyanidin against Mayaro virus. Antivir Res 2019; 168: 76–81.
142. Campos D, Navarro S, Llamas-González YY, Sugasti M, González-Santamaría J. Broad antiviral activity of ginkgolic acid against Chikungunya, Mayaro, Una, and Zika viruses. Viruses 2020; 12: 449.
143. Neris RLS, Figueiredo CM, Higa LM, Araujo DF, Carvalho CAM, Verçoza BRF et al. Co-protoporphyrin IX and Sn-protoporphyrin IX inactivate Zika, Chikungunya and other arboviruses by targeting the viral envelope. Sci Rep 2018; 8: 9805.
144. Salles TS, Meneses MDF, Yamamoto KA, Sá-Guimarães TE, Caldas LA, Silva JHS et al. Chemical composition and anti-Mayaro virus activity of *Schinus terebinthifolius* fruits. Virus Disease 2021; 32: 526–534.
145. Salles TS, Meneses MDF, Caldas LA, Sá-Guimarães TE, de Oliveira DM, Ventura JA et al. Virucidal and antiviral activities of pomegranate (*Punica granatum*) extract against the mosquito Borne Mayaro virus. Parasites Vectors 2021; 14: 443.
146. Cummings KJ, Lee SM, West ES, Cid-Ruzafa J, Fein SG, Aoki Y et al. interferon and ribavirin vs interferon alone in the re-treatment of chronic hepatitis C previously nonresponsive to interferon a meta-analysis of randomized trials. J Am Med Assoc 2001; 285: 193–199.
147. Andrei G, De Clercq E. Molecular approaches for the treatment of hemorrhagic fever virus infections. Antivir Res 1993; 22: 45–75.
148. Rothan HA, Bahrani H, Mohamed Z, Teoh TC, Shankar EM, Rahman NA et al. Combination of doxycycline and ribavirin alleviated Chikungunya infection. PLOS One 2015; 10: e0126360.
149. Ravichandran R, Manian M. Ribavirin therapy for Chikungunya arthritis. J Infect Dev Ctries 2008; 2: 140–142.
150. Dos Santos AE, Kuster RM, Yamamoto KA, Salles TS, Campos R, De Meneses MDF et al. Quercetin and quercetin 3-O-glycosides from *Bauhinia longifolia* (Bong.) Steud. Show anti-Mayaro virus activity. Parasites Vectors 2014; 7: 130.
151. Ferreira PG, Ferraz AC, Figueiredo JE, Lima CF, Rodrigues VG, Taranto AG et al. Detection of the antiviral activity of epicatechin isolated from *Salacia crassifolia* (Celas traceae) against Mayaro virus based on protein C homology modelling and virtual screening. Arch Virol 2018; 163: 1567–1576.
152. Amorim R, de Meneses MDF, Borges JC, da Silva Pinheiro LC, Caldas LA, Cirne-Santos CC et al. Thieno[2,3-b]pyridine derivatives: A new class of antiviral drugs against Mayaro virus. Arch Virol 2017; 162: 1577–1587.
153. Bengue M, Pintong AR, Liegeois F, Nougairède A, Hamel R, Pompon J et al. Favipiravir inhibits Mayaro virus infection in mice. Viruses 2021; 13: 2213.
154. Julander JG, Smee DF, Morrey JD, Furuta Y. Effect of T-705 treatment on Western equine encephalitis in a mouse model. Antivir Res 2009; 82: 169–171.
155. Da Costa LJ, Rebello MA. Effect of brefeldin A on Mayaro virus replication in *Aedes albopictus* and Vero cells. Acta Virol 1999; 43: 357–360.
156. Burlandy FM, Meneses MDF. Prostaglandin A inhibits the replication of Sindbis virus in monkey kidney and mosquito cells. Rev Ciênc Médicas E Biológicas 2008; 7: 169–174.
157. Pereira HS, Rebello MA. Inhibition of Mayaro virus replication by cerulenin in *Aedes albopictus* cells. Acta Virol 1998; 42: 383–388.
158. Karpe YA, Pingale KD, Kanade GD. Activities of proteasome and M-calpain are essential for Chikungunya virus replication. Virus Genes 2016; 52: 716–721.
159. Amaya M, Keck F, Lindquist M, Voss K, Scavone L, Kehn-Hall K, et al. The ubiquitin proteasome system plays a role in Venezuelan Equine Encephalitis Virus infection. PLOS One 2015; 10: e0124792.

160. Ferraz AC, Almeida LT, da Silva Caetano CC, da Silva Menegatto MB, Souza Lima RL, de Senna JPN et al. Hepatoprotective, antioxidant, anti inflammatory, and antiviral activities of silymarin against Mayaro virus infection. Antivir Res 2021; 194: 105168.
161. Miner JJ, Cook LE, Hong JP, Smith AM, Richner JM, Shimak RM et al. 2017. Therapy with CTLA4-Ig and an antiviral monoclonal antibody controls Chikungunya virus arthritis. Sc Transl Med 2017; 9: eaah3438.
162. Earnest JT, Basore K, Roy V, Bailey AL, Wang D, Alter G et al. Neutralizing antibodies against Mayaro virus require Fc effector functions for protective activity. J Exp Med 2019; 216: 2282–2301.
163. Malonis RJ, Earnest JT, Kim AS, Angeliadis M, Holtsberg FW, Javad Aman M et al. Near-germline human monoclonal antibodies neutralize and protect against multiple arthritogenic alphaviruses. Proc Natl Acad Sci USA 2021; 118: e2100104118.
164. Clayton AM. Monoclonal antibodies as prophylactic and therapeutic agents against Chikungunya virus. J Infect Dis 2016; 214: S506–S509.
165. Pal P, Dowd KA, Brien JD, Edeling MA, Gorlatov S, Johnson S et al. Development of a highly protective combination monoclonal antibody therapy against Chikungunya virus. PLOS Pathog 2013; 9: e1003312. 21.
166. Warter L, Lee CY, Thiagarajan R, Grandadam M, Lebecque S, Lin RTP et al. Chikungunya virus envelope-specific human monoclonal antibodies with broad neutralization potency. J Immunol 2011; 186: 3258–3264.
167. Fric J, Bertin-Maghit S, Wang C-I, Nardin A, Warter L. Use of human monoclonal antibodies to treat Chikungunya virus infection. J Infect Dis 2013; 207: 319–322.
168. Weger-Lucarelli J, Aliota MT, Kamlangdee A, Osorio JE. Identifying the role of E2 domains on alphavirus neutralization and protective immune responses. PLOS Negl Trop Dis 2015; 9: e0004163.
169. Porta J, Mangala Prasad V, Wang C-I, Akahata W, Ng LFP, Rossmann MG. Structural studies of Chikungunya virus-like particles complexed with human antibodies: Neutralization and cell-to-cell transmission. J Virol 2016; 90: 1169–77.
170. Selvarajah S, Sexton NR, Kahle KM, Fong RH, Mattia K-A, Gardner J et al. A neutralizing monoclonal antibody targeting the acid-sensitive region in Chikungunya virus E2 protects from disease. PLOS Negl Trop Dis 2013; 7: e2423.
171. Smith SA, Silva LA, Fox JM, Flyak AI, Cose N, Sapparapu G et al. Isolation and characterization of broad and ultrapotent human monoclonal antibodies with therapeutic activity against Chikungunya virus. Cell Host Microbe 2015; 18: 86–95.
172. Pal P, Fox JM, Hawman DW, Huang Y-JS, Messaoudi I, Kreklywich C et al. Chikungunya viruses that escape monoclonal antibody therapy are clinically attenuated, stable, and not purified in mosquitoes. J Virol 2014; 88: 8213–8226.
173. Williamson LE, Reeder KM, Bailey K, Tran MH, Roy V, Fouch ME et al. Therapeutic alphavirus cross-reactive E1 human antibodies inhibit viral egress. Cell 2021; 184: 4430–4446.e22.
174. Kafai NM, Williamson LE, Binshtein E, Sukupolvi-Petty S, Gardner CL, Liu J et al. Neutralizing antibodies protect mice against Venezuelan equine encephalitis virus aerosol challenge. Exp Med 2022; 219: e20212532.
175. Lundstrom K. Self-replicating RNA viruses for vaccine development. In Trends in mRNA Vaccine Research. Szabo GT, Pardi N (Eds.). 2025; 109–160. Wiley Online Library.
176. Chattopadhyay A, Aquilar PV, Bopp NE, Yarovinsky TO, Weaver SC, Rose JK. A recombinant virus vaccine that protects both against Chikungunya and Zika virus infections. Vaccine 2018; 36: 3894–3900.
177. Brandler S, Ruffié C, Combredet C, Brault JB, Najburg V, Prevost MC et al. A recombinant measles vaccine expressing Chikungunya virus-like particles is strongly immunogenic and protects mice from lethal challenge with Chikungunya virus. Vaccine 2013; 31: 3718–3725.
178. Rossi SL, Comer JE, Wang E, Azar SR, Lawrence WS, Plante JA et al. Immunogenicity and efficacy of a measles virus-vectored Chikungunya vaccine in nonhuman primates. J Infect Dis 2019; 220: 735–742.
179. Ramsauer K, Schwameis M, Firbas C, Müllner M, Putnak RJ, Thomas SJ et al. Immunogenicity, safety, and tolerability of a recombinant measles virus-based Chikungunya vaccine: A randomised, double blind, placebo controlled, active-comparator, first-in-man trial. Lancet Infect Dis 2015; 15: 519–527.
180. Reisinger EC, Tschismarov R, Beubler E, Wiedermann U, Firbas C, Loebermann M et al. (2018). Immunogenicity, safety, and tolerability of the measles-vectored Chikungunya virus vaccine MV-CHIK: A double-blind, randomised, placebo-controlled and active-controlled phase 2 trial. Lancet 2018; 392: 2718–2727.

181. Tschismarov R, Zellweger RM, Koh MJ, Leong YS, Low JG, Eong Ooi E et al. Antibody effector analysis of prime versus prime-boost immunizations with a recombinant measles-vectored Chikungunya virus vaccine. JCI Insight 2021; 6: e151095.
182. Chen GL, Coates EE, Plummer SH, Carter CA, Berkowitz N, Conan-Cibotti M et al. 2020. Effect of a Chikungunya virus-like particle vaccine on safety and tolerability outcomes: A randomized clinical trial. JAMA 2020; 323: 1369–1377.
183. Wressnigg N, Hochreiter R, Zoihsl O, Fritzer A, Bézay N, Klingler A et al. Single-shot live-attenuated Chikungunya vaccine in healthy adults: A phase 1, randomised controlled trial. Lancet Infect Dis 2020; 20: 1193–1203.
184. Reed DS, Glass PJ, Bakken RR, Barth JF, Lind CM, da Silva L et al. Combined alphavirus replicon particle vaccine induces durable and cross-protective immune responses against equine encephalitis virus. J Virol 2014; 88: 12077–12086.
185. Tretyakova I, Tibbens A, Jokinen JD, Johnson DM, Lukashevich IS, Pushko P et al. Novel DNA-launched Venezuelan equine encephalitis virus vaccine with rearranged genome. Vaccine 2019; 37: 3317–3325.
186. Tretyakova I, Plante KS, Rossi SL, Lawrence WS, Peel JE, Gudjohnsen S et al. Venezuelan equine encephalitis vaccine with rearranged genome resists reversion and protects non-human primates from viremia after aerosol challenge. Vaccine 2020; 38: 3378–3386.
187. Hannaman D, Dupuy LC, Ellefsen B, Schmaljohn CS. A phase 1 clinical trial of a DNA vaccine for Venezuelan equine encephalitis delivered by intramuscular or intradermal electroporation. Vaccine 2016; 34: 3607–3612.
188. Samsa MM, Dupuy LC, Beard CW, Six CM, Schmaljohn CS, Mason PW et al. Self-amplifying RNA vaccines for Venezuelan equine encephalitis virus induce robust protective immunogenicity in mice. Mol Ther 2019; 27: 850–865.
189. Hu W-G, Steigerwald R, Kalla M, Volkmann A, Noll D, Nagata LP. Protective efficacy of monovalent and trivalent recombinant MVA-based vaccines against three encephalitic alphaviruses. Vaccine 2018; 36: 5194–5203.

3 Lassa Virus
Hidden Threats and Unseen Dangers

Vijay Mishra, Aminata Lowe, Alaa A. A. Aljabali, and Yachana Mishra

3.1 INTRODUCTION

Lassa fever, an acute viral haemorrhagic disease that is native to West Africa, is caused by the enveloped, single-stranded RNA Lassa virus (LASV), which is a member of the *Arenaviridae* family [1]. Following an epidemic in Lassa, Nigeria, where missionary nurses experienced a severe feverish illness, the virus was initially discovered in 1969 [2]. With sporadic outbreaks and an increasing number of cases recorded each year, LASV has since been identified as a serious public health concern, especially in Nigeria, Sierra Leone, Liberia, and Guinea [3].

Contact with *Mastomys natalensis*, a rodent species that acts as the virus's natural reservoir, is the main way that Lassa fever is spread [4]. Human infections can be contracted by direct human-to-human transmission, especially in medical facilities, or by inhaling or consuming food tainted with rat urine or faeces [5]. From asymptomatic or mild fever to serious consequences such as bleeding, multi-organ failure, and death, the disease can present with a wide range of clinical symptoms [6]. Significantly, there is an increased risk of serious illness and death for immunocompromised people and pregnant women [7].

Due to its low diagnostic capacity, vague symptoms that resemble those of other feverish diseases like malaria, and a lack of surveillance data, Lassa fever continues to be an underreported and neglected disease despite its substantial impact [8]. There is presently no approved vaccine, and although the antiviral medication ribavirin has been used as a treatment, it works best when given early in the course of infection [7]. To lessen the impact of Lassa fever in endemic areas, improved surveillance, quick diagnosis, and vaccine research are urgently needed due to the disease's rising incidence and potential for large-scale outbreaks [3].

3.2 LASSA VIRUS: AN OVERVIEW

In West Africa, especially in nations like Nigeria, Sierra Leone, Guinea, and Liberia, the LASV, an acute viral haemorrhagic fever (VHF) pathogen that belongs to the *Adenoviridae* family, is endemic [5]. The *M. natalensis* rodent, also referred to as the multimammate rat, is the main reservoir of the zoonotic virus [4]. The main ways that humans become sick are by direct contact with infected people, contaminated food, or exposure to rodent excrement [1].

The symptoms of Lassa fever, which is caused by LASV, range from a low fever to severe haemorrhagic complications and multi-organ failure, and the illness has an incubation period of 6–21 days [6]. Although the majority of infections are mild or asymptomatic, severe instances can cause neurological abnormalities, deafness, and high mortality rates, especially in pregnant women [9].

Due to its vague early symptoms that resemble those of other feverish diseases, including typhoid fever and malaria, LASV is difficult to diagnose [8]. Reverse transcription polymerase chain reaction (RT-PCR), enzyme-linked immunosorbent assay (ELISA), or viral isolation are the methods used in laboratories for confirmation [2]. Although there is currently no approved

DOI: 10.1201/9781003607816-3

vaccine, treatment options include ribavirin, an antiviral that has demonstrated effectiveness when given early in infection [7]. LASV is regarded as a public health problem because of its high morbidity and mortality, which has led to efforts to produce a vaccine, conduct monitoring, and diagnose cases early [3].

3.2.1 Historical Background of Lassa Virus

Following a major febrile disease outbreak among missionary nurses in 1969, the LASV was initially discovered in the town of Lassa in Borno State, Nigeria [2]. At first, the illness was confused with other endemic illnesses like typhoid fever or malaria but after more research, virologists discovered a new virus in patient samples, which led to the first Lassa fever case ever recorded [10].

Three American missionary nurses were involved in the first outbreak in Nigeria; two of them died from the illness, and the third lived after being sent to the United States for treatment [2]. Because of its structural resemblance to other arenaviruses, the isolated virus was later examined at the Yale Arbovirus Research Unit and assigned to the *Arenaviridae* family [10]. The discovery of LASV was noteworthy since it added more haemorrhagic fever viruses to the list of known ones than those previously identified, including the filoviruses that cause Marburg virus and Ebola.

LASV is endemic in West Africa, specifically in Nigeria, Sierra Leone, Liberia, and Guinea, according to later studies conducted in the 1970s and 1980s [11]. Lassa fever is zoonotic, as evidenced by epidemiological research that found *M. natalensis*, a common rodent species, to be the main reservoir [12]. Researchers were able to examine LASV genetic diversity, pathogenesis, and transmission dynamics over time thanks to developments in molecular and serological tools, which are still essential for creating preventative and treatment plans [9].

3.2.1.1 Geographic Prevalence and Endemic Regions

Lassa fever is an acute viral haemorrhagic disease that is endemic to many West African countries. The disease is mainly found in areas where the LASV's natural reservoir, the *M. natalensis* rodent, is common. The World Health Organisation (WHO) lists the following countries as endemic: Benin, Ghana, Guinea, Liberia, Mali, Nigeria, Sierra Leone, and Togo, though the virus is probably also present in other nearby countries [13].

In Nigeria, Lassa fever is reported all year long, with over half of the cases occurring in the first quarter. Recent data shows that the number of cases and the geographic spread of the virus are increasing in Nigeria [14]. Ecological modelling was used to forecast Lassa fever risk areas in a study published in *PLOS Neglected Tropical Diseases*. According to the findings, roughly 30% of Côte d'Ivoire, Togo, and Benin, 50% of Guinea, 40% of Nigeria, and 80% of Sierra Leone and Liberia are in danger. Roughly 10% of Ghana is thought to be at risk [4].

The habitat of the *M. natalensis* rodent, which flourishes in regions with particular ecological circumstances, is intimately related to the distribution of Lassa fever. Implementing efficient monitoring and control mechanisms requires an understanding of these patterns. The WHO has comprehensive maps showing the location of Lassa fever cases throughout West Africa for a visual depiction of the endemic areas (Figure 3.1) [3]. Even though certain locations have been designated as endemic, Lassa fever's true distribution can differ, and isolated cases may arise outside of these boundaries. Our knowledge of the disease's geographic prevalence must be updated and improved by ongoing research and monitoring.

3.2.2 Epidemiology and Public Health Importance

The LASV, which is widespread throughout West Africa, is a serious public health hazard. Between 100,000 and 300,000 infections and over 5,000 fatalities are attributed to LASV each year in the area [13]. Contact with the urine or faeces of infected *M. natalensis* rodents, also referred to as multimammate rats, is the main way that the virus is spread to people. Additionally, human-to-human

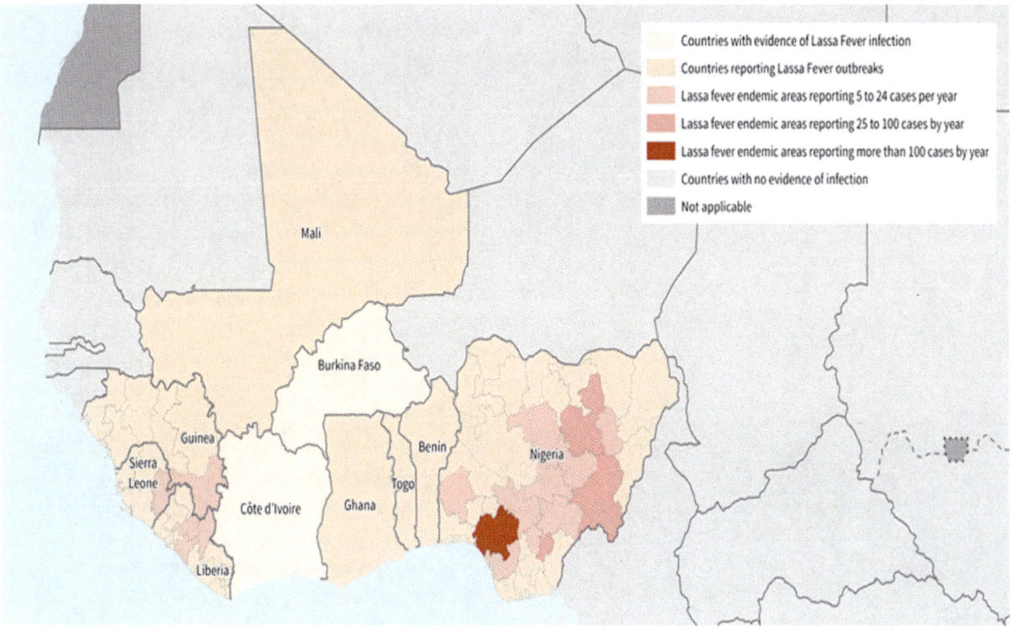

FIGURE 3.1 Geographic distribution of Lassa virus in West Africa (1969–2023). (From Ref. [3].)

transmission is possible, particularly in medical facilities with inadequate infection prevention and control protocols [13]. Approximately 80% of LASV infections are moderate or asymptomatic in clinical settings. The remaining 20%, however, may develop a serious illness that affects several organs, including the kidneys, liver, and spleen. Although the average case-fatality rate (CFR) is under 1%, it can be as high as 15% among patients who are hospitalised with severe Lassa fever [13].

The economic and health costs of Lassa fever in West Africa have been brought to light by recent studies. One study, for example, anticipated the prospective advantages of vaccination programs in the area and calculated the health-economic cost of Lassa fever, both of which were reported in *Nature Medicine* [15]. LASV's potential for epidemics, difficulties in early detection because of its vague symptoms, and lack of an approved vaccine highlight the virus's significance for public health. To lessen the effects of Lassa fever in areas where it is endemic, efforts must be made to increase monitoring, strengthen diagnostic skills, and create efficient vaccinations (Figure 3.2) [13].

3.2.3 Infection Rates in Endemic Regions

With infection rates differing throughout endemic locales, Lassa fever is a serious public health concern in West Africa. Between 100,000 and 300,000 LASV infections are estimated to occur each year, with 5,000 deaths [13]. These numbers are estimates, though, because the region's reporting and monitoring systems are not applied equally. The Nigeria Centre for Disease Control (NCDC) reported 5,057 suspected Lassa fever cases in 2019, of which 833 were confirmed, including 174 deaths among the confirmed cases, in Nigeria, one of the most impacted countries [16]. Compared to the prior year, when there were 3,498 suspected cases and 633 confirmed cases, this represented an increase.

According to research, around 2.1–3.4 million LASV infections occur in West Africa each year, resulting in 15,000–35,000 hospitalisations and 6,000–12,000 deaths [15]. These increased estimates highlight the need for better surveillance in the area and the possibility of underreporting. It's crucial to remember that underreporting, difficult diagnosis, and restricted access to healthcare

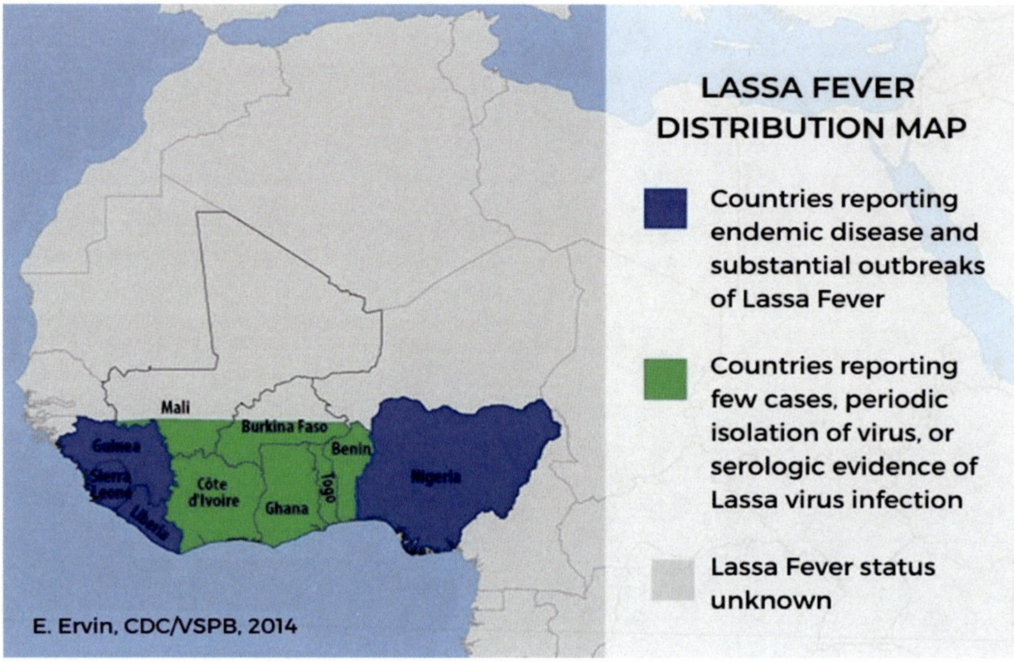

FIGURE 3.2 Lassa distribution map. (From Ref. [13].)

mean that the actual impact of Lassa fever is probably underestimated. To get more precise information on infection rates and carry out successful public health actions, surveillance systems must be strengthened, and diagnostic capabilities must be enhanced.

3.2.4 Mortality Rates

The LASV poses serious risks to the lives of those who come in contact to it as well as to their long-term health.

a. **Mortality Rates**
   - *General Population*: About 1% of people die from Lassa fever [17].
   - *Hospitalised Patients*: The CFR rises to about 15% when patients are admitted with severe symptoms [17].
   - *Pregnant Women*: Maternal death rates from infections during pregnancy can exceed 80%, with the third trimester being particularly severe [18].

The necessity for thorough post-recovery treatment, including auditory rehabilitation and support services, is highlighted by the high prevalence of hearing loss among LASV survivors. Improving the quality of life for Lassa fever patients requires addressing these long-term effects [19, 20].

3.3 VIROLOGY OF LASSA VIRUS

Although Lassa fever is now limited to West Africa, there is a chance that it could spread elsewhere, especially as international travel increases. The need for worldwide awareness and readiness is highlighted by recent events, such as the death of an Iowa resident in the United States after visiting

West Africa and suspected Lassa fever [21]. The main goal is to thoroughly evaluate the risks that the LASV poses and its substantial influence on world health.

Since there is presently no vaccine to prevent rodents, prevention strategies concentrate on encouraging communal hygiene. Although ribavirin, an antiviral medication, has been used for treatment, there are still questions about how effective it is and what dosages are best [13]. The possible advantages of vaccination campaigns have been emphasised by recent studies. Preventive vaccination campaigns could greatly lessen the disease's negative effects on health and the economy, according to a study that was published in *Nature Medicine* and calculated the health-economic burden of Lassa fever in West Africa [15].

Specifically categorised as an Old-World arenavirus, LASV is an enclosed, single-stranded RNA virus that is a member of the *Arenaviridae* family. It is the etiological agent of Lassa fever, a West African haemorrhagic disease.

3.3.1 Taxonomy and Classification

The complex family of enclosed, single-stranded RNA viruses known as the *Arenaviridae* is capable of infecting both humans and rodents. Based on their geographic distribution and evolutionary traits, these viruses are mainly separated into two complexes: Old World and New World arenaviruses [22].

3.3.1.1 *Arenaviridae* Family: Overview

a. *Old World Arenaviruses*: Old World arenaviruses are primarily found in the Eastern Hemisphere and include well-known species like (i) LASV, causing Lassa fever, a haemorrhagic disease with a high morbidity and fatality rate. It is endemic to West Africa. (ii) Lymphocytic choriomeningitis virus (LCMV) is found all over the world and can lead to neurological disorders in people [8].
b. *New World Arenaviruses*: Clades A, B, C, and D are further classifications for New World arenaviruses, which are mainly found in the Western Hemisphere. Notable species consist of: (i) Junín virus (JUNV), which causes haemorrhagic fever in Argentina, (ii) Machupo virus (MACV), which causes Bolivian haemorrhagic fever, (iii) Guanarito virus (GTOV) causes Venezuelan haemorrhagic fever [23].

3.3.1.2 Similarities between New World and Old World Arenaviruses

Despite their disparate geographic locations, Old and New World arenaviruses have several traits in common:

a. *Genomic structure*: They have an ambisense, bisegmented RNA genome that codes for four proteins: zinc-binding matrix protein (Z), RNA-dependent RNA polymerase (RdRp; L), glycoprotein precursor (GPC), and nucleoprotein (NP) [24].
b. *Natural reservoirs*: Both species mostly infect rodents as their natural hosts, causing chronic infections in these creatures with little to no clinical manifestation.
c. *Human transmission*: Infections in humans usually happen when contaminated rat excrement comes into direct contact with humans or when aerosolised particles are inhaled [25].

Old and New World arenaviruses also show some differences, which are listed in Table 3.1.

Understanding these similarities and differences is crucial for developing targeted therapeutic strategies and vaccines against arenavirus-induced diseases.

TABLE 3.1
Differences between Old and New World Arenaviruses

Features	Old World Arenaviruses	New World Arenaviruses	References
Geographical distribution	Africa, Asia, and Europe	North and South America	25
Examples	LASV, LCMV	Junín virus (JUNV), MACV, GTOV, Sabiá virus (SBAV), Chapare virus (CHAPV)	25
Primary host reservoirs	Mastomys rodents (e.g., *M. natalensis*)	Sigmodontine rodents (e.g., *Calomys* spp.)	24
Receptor for cell entry	α-dystroglycan	Transferrin receptor 1 (TfR1)	24
Pathogenesis	Generalised immune suppression, persistent viremia (e.g., Lassa fever)	Strong immune activation, cytokine storm (e.g., Argentine haemorrhagic fever)	25
Human-to-human transmission	Common (e.g., LASV spreads in hospitals)	Less common but possible (e.g., Machupo and Chapare viruses)	25
Disease penetration	Haemorrhagic fever (Lassa fever) or neurological disease (LCMV)	Haemorrhagic fevers with high fatality rates (e.g., Argentine haemorrhagic fever)	23
Vaccine availability	No licensed vaccine for Lassa fever	Junín virus (JUNV) vaccine available (Candid #1)	25
Endemic regions	West Africa (Nigeria, Sierra Leone, Guinea, Liberia)	South America (Argentina, Bolivia, Venezuela, Brazil)	25

3.3.2 Specific Characteristics of Lassa Virus

LASV exhibits several unique characteristics that distinguish it from other arenaviruses:

3.3.2.1 Geographic Distribution and Genetic Diversity

- *Endemic region*: LASV is primarily found in West Africa, notably in countries such as Nigeria, Sierra Leone, Liberia, and Guinea [9].
- *Genetic lineages*: Research has identified multiple lineages of LASV, with three distinct lineages in Nigeria and a fourth encompassing strains from Guinea, Liberia, and Sierra Leone. This genetic diversity is more pronounced in LASV compared to other arenaviruses [9].

3.3.2.2 Transmission Dynamics

- *Zoonotic reservoir*: The primary reservoir for LASV is the *M. natalensis* rodent, commonly known as the Natal multimammate mouse. Human infection typically occurs through exposure to the urine or faeces of these rodents [26].
- *Human-to-human transmission*: Unlike many other arenaviruses, LASV can spread between humans, particularly in healthcare settings, making nosocomial outbreaks a significant concern [4].

3.3.2.3 Clinical Manifestations

- *Asymptomatic cases*: Approximately 80% of LASV infections are asymptomatic, a higher proportion than observed with other pathogenic arenaviruses [4].
- *Hearing loss*: A notable complication of Lassa fever is sensorineural hearing loss, occurring in about one-third of cases, which is less commonly reported with other arenavirus infections.

3.3.2.4 Immune Evasion Mechanisms

- *Nucleoprotein Function*: LASV's NP possesses exonuclease activity that degrades double-stranded RNA, thereby suppressing the host's interferon (IFN) response. This immune evasion strategy is particularly pronounced in LASV compared to other arenaviruses [27].

3.3.2.5 Receptor Utilisation

- *Cell Entry*: LASV utilises α-dystroglycan (α-DG) as its cellular receptor for entry. This feature is shown by some but not all arenaviruses [27].

These distinct features of LASV have significant implications for its transmission dynamics, clinical management, and the development of targeted therapeutic and preventive measures.

3.3.3 Genetic Structure and Genomic Organisation

3.3.3.1 RNA Structure and Segments

The single-stranded RNA genome of LASV is bisegmented and uses an ambisense coding approach [28]. There are two sections to this genome:

a. *Small (S) Segment*: It is about 3.4 kilobases (kb) in size [28]. Viral RNA is encapsulated by NP, which forms the ribonucleoprotein (RNP) complex necessary for transcription and replication [29]. GPC produces two envelope glycoproteins (GPs) through post-translational cleavage. Receptor binding is mediated by GP1 [29]. When a host cell enters, GP2 promotes membrane fusion [29].
b. *Large (L) Segment*: It is roughly 7.2 kb in size [28]. The L protein catalyses the transcription and replication of the viral RNA by acting as an RdRp [29]. Z Protein, A little protein with a RING domain that controls transcription and replication and is essential for virion assembly and budding [29].

A stable, non-coding intergenic region separates each RNA segment's two open reading frames in opposite directions, giving each segment an ambisense orientation. During the viral life cycle, timing and gene expression can be controlled thanks to this special structure [28]. A double-stranded RNA panhandle structure that is essential for the start of viral RNA production is formed by the highly conserved terminal 19 nucleotides at both the 5′ and 3′ ends of each segment, which create complementary sequences [29].

3.3.3.2 Viral Proteins and Their Roles

LASV encodes several critical proteins that aid in its replication, immune evasion, and pathogenesis. These proteins include the zinc-binding protein (Z), NP, L polymerase (L), and GPC. Each of these proteins has a unique function in the viral life cycle.

a. *GPC and Surface Glycoproteins (GP1 and GP2)*: The GPC is produced from the small (S) segment of the LASV genome and is cleaved by host subtilisin kexin isozyme-1/site-1 protease to produce two mature GPs: (i) GP1: Mediates virus attachment by binding to the host receptor α-DG, starting viral entry [30], (ii) GP2: Release viral genome into the host cytoplasm, which promotes membrane fusion between the viral and endosomal membranes [31]. These GPs are essential for immune system evasion and host cell tropism, which enable LASV to effectively establish infection.
b. *NP*: The most prevalent LASV protein, NP, has several functions in both immune evasion and viral replication. It forms the RNP complex, which is necessary for transcription

and replication, by encasing the viral RNA genome [32]. Additionally, by breaking down double-stranded RNA, NP suppresses antiviral immune responses by preventing the formation of type I IFN.

c. *L Polymerase (L Protein)*: The L protein, which is encoded by the large (L) segment of the genome, is an RdRp that catalyses the viral genome's transcription and replication. It is essential to LASV's ambisense gene expression strategy because it makes viral mRNA and antigenomic RNA production possible [33].
d. *Zinc-Binding (Z) Protein*: A tiny matrix protein called the Z protein regulates viral assembly and budding. Through the Endosomal Sorting Complex Required for Transport (ESCRT) pathway, its RING domain facilitates virus egress by interacting with host cellular machinery [34]. Furthermore, Z protein functions as a crucial regulatory factor by interacting with L polymerase to prevent viral transcription in the latter phases of infection [35] (Figure 3.3).

3.3.3.3 Replication Cycle and Mechanisms

Entry into host cells, genome replication, transcription, and viral assembly are all steps in the unique replication cycle of the enveloped, ambisense RNA LASV. The host cellular machinery and particular receptor contacts are crucial to the process.

a. *Mechanisms of Entry into Host Cells*: Through receptor-mediated endocytosis, LASV enters host cells, and its main cellular receptor is α-DG, which is a broadly expressed extracellular matrix receptor that is selectively bound by the viral GP1 [30]. Internalisation into the endosomal compartment results from clathrin-mediated endocytosis, which is facilitated by this interaction [36]. Acidification causes GP2, the fusion subunit of the viral GP complex, to undergo a conformational shift once it enters the endosome. This step facilitates the release of the viral RNP complex into the cytoplasm by promoting membrane fusion between the endosomal membrane and the viral envelope [37].

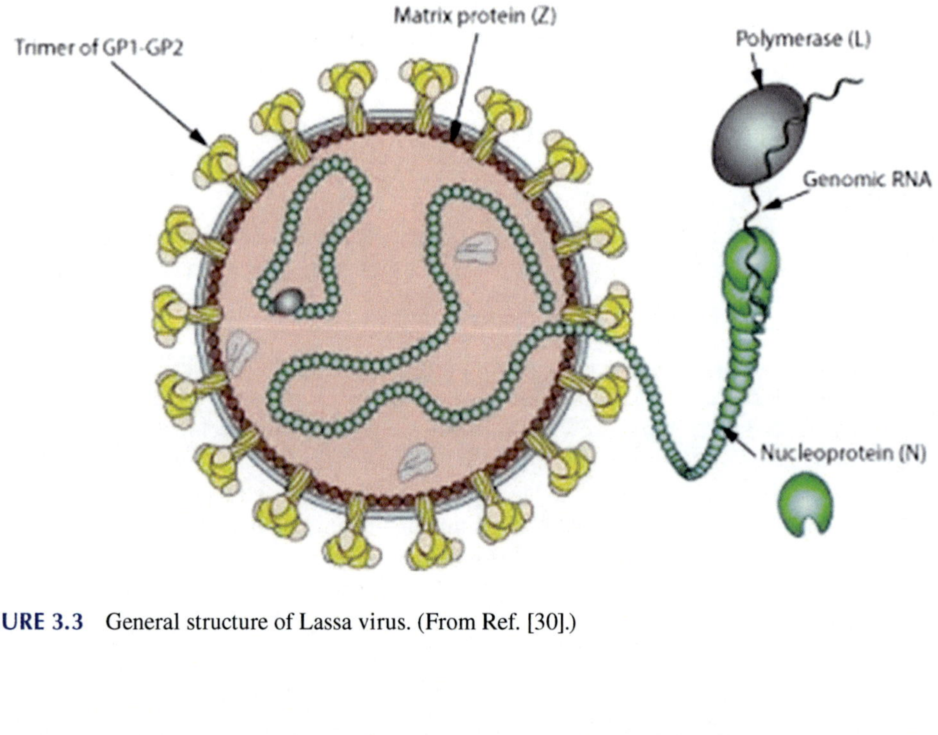

FIGURE 3.3 General structure of Lassa virus. (From Ref. [30].)

b. *Translation and Transcription*: After entering the cytoplasm, the viral L polymerase (RdRp) transcribes and replicates the LASV genome, which is made up of two ambisense RNA segments (S and L segments) [33]. The viral RNA is encapsulated by the NP, which shields it from the host's defences [32].
c. *Transcription*: Using host ribosomes, the viral L polymerase converts both genome segments into subgenomic mRNAs, which are subsequently translated into viral proteins [38].
d. *Replication*: To guarantee the creation of offspring virions, the whole viral genome and antigenome are synthesised. NP and L polymerase aid in the complementary RNA strand synthesis that takes place during the process.
e. *Assembly and Release*: Mature RNP complexes are created in the cytoplasm by the assembly of freshly synthesised viral proteins and genomic RNA. By interacting with the host cell's ESCRT machinery, the matrix protein Z promotes virus budding [39]. The lipid envelope is acquired by the fully constructed virions as they bud from the plasma membrane. Following their release into the extracellular space, the mature LASV particles are prepared to infect fresh host cells and carry on the replication cycle [35].

3.4 TRANSMISSION PATHWAYS OF LASSA VIRUS

Mastomys natalensis, a multimammate rodent species that is common in West Africa, serves as the natural reservoir for LASV. However, human-to-human transmission plays a major role in outbreaks, especially in healthcare settings. Understanding disease epidemiology and putting control measures in place depend on knowing the LASV transmission pathways.

a. *Animal reservoirs*: The multimammate rat (*M. natalensis*) is the main reservoir host for LASV. The risk of transmission is increased by the fact that these rodents are common throughout West Africa and frequently live in human homes. Throughout their lives, infected *M. natalensis* can release the virus in their urine and faeces without showing any symptoms, which makes it easier for the environment to become contaminated. The virus usually infects humans when they come into close contact with these excreta or when they eat contaminated food. Transmission between rodents occurs via: (i) Direct touch with infected rats' urine, faeces, and saliva. Intrauterine transmission results in children who are infected [40], (ii) Inhaling aerosolised rodent excrement in residences, food storage facilities, and agricultural fields is one way that humans contract the disease, and (iii) Eating contaminated rodents [41]. *Mastomys natalensis* reproduces quickly and is widely distributed; rodent control is still a major problem in endemic areas.
b. *Human-to-Human Transmission and Bodily Fluids*: Although zoonotic transmission is the main method, direct contact with an infected person's blood, urine, faeces, or other body secretions can result in human-to-human transmission of LASV. In hospital settings, where insufficient infection control procedures might result in the virus spreading among patients and healthcare personnel, this route of transmission is particularly important. Exposure to infected individuals' blood, saliva, urine, vomit, or faeces is among the modes of transmission. Sexual transmission; months after infection, LASV RNA was found in semen [18]. High maternal and foetal mortality results from vertical transmission [42]. When family carers and close contacts handle contaminated objects without taking the necessary precautions, their risk of contracting LASV increases.
c. *Nosocomial Infections*: There are significant dangers associated with nosocomial transmission of LASV, particularly in areas where the virus is prevalent. Outbreaks related to healthcare have been reported; these are frequently caused by conventional precautions being neglected and infectious items being handled without the proper protective gear. These environments have the potential to increase the virus's spread, which emphasises the necessity of strict infection control procedures to safeguard patients and healthcare workers.

Needlestick injuries and contact with tainted medical equipment are the two ways that the disease is spread. Transfer of aerosols during invasive procedures like suctioning and intubation [43]. The high death rate among healthcare personnel emphasises the importance of early detection, stringent barrier nursing measures, and isolating suspected cases.

3.5 CLINICAL PRESENTATION AND SYMPTOMS OF LASSA FEVER

Lassa fever is an acute viral haemorrhagic disease that can cause a wide range of symptoms, from severe multisystem disease to asymptomatic instances.

3.5.1 Incubation Period and Early Symptoms

The majority of Lassa fever cases show symptoms 1–3 weeks after exposure, with an incubation period ranging from 2 to 21 days [13]. Lassa fever usually develops slowly, starting with vague symptoms like fever, malaise, and general weakness. Headache, sore throat, chest discomfort, muscle soreness, nausea, vomiting, diarrhoea, cough, and stomach pain are frequently present together with these early symptoms [13]. Due to their lack of specificity, these early symptoms are often confused with those of other diseases, such as typhoid fever, malaria, and other VHFs [6].

About 20% of infected people may experience more serious symptoms as the disease worsens, such as bleeding, vomiting, breathing difficulties, facial swelling, chest, back, and abdominal pain, and shock [13]. Lassa fever, an acute viral haemorrhagic illness endemic to West Africa, presents a wide spectrum of clinical manifestations, ranging from mild symptoms to severe, life-threatening conditions.

3.5.2 Progression to Severe Disease

About 20% of LASV infections develop into severe illness that affects several organ systems, whilst the remaining 80% of cases are asymptomatic or moderate [17]. Patients may have haemorrhagic symptoms, such as bleeding from the nose, eyes, or gums, in extreme situations [44]. These signs point to increased permeability and vascular involvement, which could result in hypovolemic shock.

In the later stages of Lassa fever, neurological problems are possible, though they are less frequent. Tremors, convulsions, disorientation, and coma are among them; they indicate involvement of the central nervous system and are associated with a higher risk of death [45]. With the virus damaging organs like the liver, spleen, and kidneys, multi-organ failure is a serious worry that increases the risk of death [17].

3.5.3 Mortality Rates and Long-Term Complications

Lassa fever, a West African acute viral haemorrhagic disease, poses serious problems because of its high death rates and persistent consequences for survivors. Lassa fever has an approximate 1% CFR overall. However, it is estimated that the CFR among hospitalised patients with severe clinical presentations is approximately 15% [17].

Sensorineural hearing loss is the most prominent long-term consequence among survivors, involving up to one-third of patients [13]. Other side effects that could affect quality of life after recovery include ataxia, iridocyclitis, and temporary blindness [46].

The severity of Lassa fever and the population it affects determine the overall CFR. The CFR is around 1% in the general population. Nonetheless, the CFR can rise to 15% in hospitalised patients with severe symptoms [13]. Notably, greater CFRs have been reported during particular outbreaks. Nigeria, for example, reported CFRs of 13.4% in 2020, 19.6% in 2019, 23.7% in 2018, and 26.5% in 2017 [47].

Lassa fever survivors frequently experience serious long-term issues that lower their quality of life. About one-third of people who recover from the disease experience sensorineural hearing loss, one of the most common consequences [48]. Significant difficulties in everyday communication and social interactions result from this hearing loss, which can range in severity from modest to profound and is usually permanent [13].

Hearing loss usually starts 5–22 days after a severe sickness, during the recovery phase. Although the precise mechanism is unknown, it is thought to entail either an immune-mediated reaction or direct viral damage to the inner ear tissues [48]. Although they are less frequent than hearing loss, neurological symptoms, vision impairment, joint discomfort, temporary hair loss, and psychological issues are other persistent sequelae that have been described among Lassa fever survivors [13]. The necessity for thorough post-recovery treatment, including audiological evaluations and rehabilitation services, to enhance the quality of life for those impacted by LASV is highlighted by the high prevalence of hearing loss among survivors.

3.6 DIAGNOSIS OF LASSA VIRUS INFECTION

Due to its vague symptoms that mimic those of other endemic infectious diseases like typhoid fever, malaria, and other VHF, such as Ebola and Marburg virus infections, diagnosing LASV infection is difficult [1]. Laboratory testing is essential for diagnosis accuracy because clinical presentation is frequently insufficient for distinction [13].

3.6.1 Challenges in Diagnosis

Clinical diagnosis of early-stage LASV infection is challenging because of flu-like symptoms [6]. Moreover, prompt detection is hampered in many endemic locations by the absence of extensive diagnostic infrastructure [17]. Identification of cases is made more difficult by overlapping symptoms with other VHFs, which necessitate laboratory confirmation [9].

3.6.2 Laboratory Methods

The detection of IgM and IgG antibodies specific to LASV is frequently accomplished using serological techniques such as ELISA. In contrast to IgG persistence, which shows prior exposure, IgM detection indicated current infection [4]. However, delayed antibody generation may make serological assays unreliable in the early stages of infection [6].

The gold standard for LASV detection is RT-PCR, which has excellent sensitivity and specificity [8]. This technique is helpful in early-stage infections prior to the development of antibodies and can identify viral RNA in blood, urine, or tissue samples [49]. In environments with limited resources, recent developments in point-of-care diagnostic technologies, like lateral flow tests, are intended to allow faster LASV detection [50]. Even though these tests show promise, more verification is necessary to guarantee accuracy [44].

3.6.2.1 Differential Diagnosis

Lassa fever and other VHFs overlap clinical characteristics; hence, laboratory confirmation is necessary for precise classification. One characteristic that sets Lassa fever apart from Ebola and Marburg is hearing loss, which is uncommon in cases of severe bleeding [2]. The diagnosis of LASV may be made more difficult by co-infections with bacterial infections, malaria, or other tropical illnesses. In endemic areas, multi-pathogen testing techniques are advised to increase diagnostic precision [51].

3.6.3 Immune Response and Pathophysiology

A complicated immunological response is triggered by LASV infection, and this response is crucial to the course and outcome of the disease. Some people have a strong immune response that clears the virus, whereas others have immunological dysregulation that causes serious illness [52].

3.6.3.1 Host Immune Response to LASV Infection

The initial line of protection against LASV infection is innate immunity. LASV primarily targets dendritic cells and macrophages, where it inhibits type I IFN responses to avoid early immunological detection [29]. Unchecked viral spread results from LASV infection, which inhibits the activation of natural killer cells, which are essential for regulating viral replication [53].

Clearance of LASV is largely dependent on the adaptive immunological response, especially T-cell activation. Although CD8+ T-cells are necessary for the removal of viruses, T-cell fatigue is frequently seen in severe cases [54]. In contrast to other VHFs, LASV infection typically results in weak and delayed neutralising antibodies, although antibody synthesis does occur [55].

3.6.3.2 Mechanisms of Immune Evasion

LASV uses its NP and GPC to avoid immune detection. Through breaking down double-stranded RNA, the NP prevents IFN signalling [56], but the GPC facilitates viral entry and protects viral antigens from immune recognition [32]. The efficiency of the adaptive immune response is decreased by LASV infection, which inhibits T-cell activity and proliferation [57]. Furthermore, the virus inhibits the production of the major histocompatibility complex, which restricts the presentation of antigens and delays the activation of B cells [58].

3.6.3.3 Pathophysiological Mechanisms of Disease Progression

A cytokine storm, which is defined by an overabundance of pro-inflammatory cytokines including interleukin-6, tumour necrosis factor-alpha, and IFN-gamma (IFN-γ), is caused by dysregulated immune activation in severe LASV infections. Shock, multi-organ failure, and vascular leakage are all exacerbated by this severe inflammation. LASV can result in vascular instability and endothelial dysfunction, even though it usually does not produce significant haemorrhaging like Ebola [4]. The virus's effects on the central nervous system and immune-mediated neuronal destruction are associated with neurological symptoms, such as encephalopathy and seizures [59].

3.7 CURRENT TREATMENT STRATEGIES

Antiviral medication and supportive care are the mainstays of Lassa fever treatment, which aims to lower mortality and control complications. The cornerstone of antiviral therapy is still ribavirin, but new therapies and better supportive strategies are being investigated to enhance patient outcomes [60].

3.7.1 Antiviral Therapies

The main medication for Lassa fever is ribavirin, a broad-spectrum antiviral. According to studies, mortality is considerably decreased by early intravenous injection during the first 6 days after the onset of symptoms [60]. However, ribavirin's use is limited by side effects such as haemolytic anaemia, and its efficacy decreases if administered later in the course of the disease [61]. There are several novel antiviral candidates under investigation. An RNA polymerase inhibitor called favipiravir has shown effectiveness against LASV in vitro [62]. Preclinical research on monoclonal antibody treatments, including Arevirumab-3, has yielded encouraging outcomes [63]. The potential efficacy of broad-spectrum antivirals, such as remdesivir, is also being assessed [64].

3.7.2 Supportive Care

Lassa fever patients often suffer from dehydration as a result of their fever, diarrhoea, and vomiting. To avoid hypovolemic shock and organ failure, aggressive fluid resuscitation using balanced electrolyte solutions is essential [5]. Support from an intensive care unit (ICU) may be required in extreme situations. Patients with LASV who have respiratory distress syndrome may need mechanical ventilation or oxygen therapy [17]. Acute renal injury, another consequence, typically requires dialysis [65].

3.7.3 Limitations and Challenges in Treatment

In many areas where LASV is endemic, ribavirin is costly and not easily accessible. Concern over possible resistance as a result of extensive use is also rising [66]. Effective supportive care is hampered in resource-constrained situations by inadequate ICU facilities, a lack of qualified staff, and a poor hospital infrastructure [19]. Economic limitations raise death rates in impacted areas and further restrict access to therapies.

3.8 VACCINES AND PREVENTION STRATEGIES

Despite advancements, obstacles like cultural barriers and resource constraints impede effective implementation of prevention strategies that centre on vaccine development, vector control, and public health initiatives. Lassa fever is still a major public health concern in West Africa, where there is currently no widely available licensed vaccine.

3.8.1 Current Status of Vaccine Development

There are numerous vaccine candidates in development that focus on various immune response mechanisms. Clinical trials for the recombinant vesicular stomatitis virus-based vaccination (rVSV-LASV) are currently underway, and preclinical research has yielded encouraging findings [67]. Other candidates have shown safety and immunogenicity by entering Phase I trials, such as the DNA-based INO-4500 vaccine [68]. To offer long-term protection, vaccines based on live-attenuated and virus-like particle are also being developed [69]. The LASV GP and NP, which are essential for viral entry and replication, are the main targets of vaccine efforts. While NP-based strategies concentrate on cellular immunity through T-cell activation, GP-based vaccines produce potent neutralising antibody responses [70].

3.8.2 Preventive Measures in Endemic Regions

The main rodent reservoir for LASV, *M. natalensis*, flourishes in populated areas, raising the possibility of zoonotic transmission. According to Fivtet-Calvet and Rogers [4], community-wide mouse traps, food source sealing, and improved sanitation are all effective rodent control measures. Rodent populations are further reduced by environmental modification initiatives, such as minimising shrub encroachment around residences [71].

In areas where LASV is endemic, educational initiatives that encourage cleanliness, safe food storage, and avoiding rat ingestion have proven extremely important. According to Olayemi et al. [70], community engagement, especially through local leaders, improves adherence to preventive interventions. In healthcare settings, early detection and isolation procedures have also contributed to a decrease in nosocomial transmission [17].

3.8.3 Challenges to Effective Prevention

Misinformation, mistrust of healthcare systems, and cultural attitudes are the main causes of vaccine reluctance, which is a serious problem. Uptake rates are decreased in certain groups where traditional medicine is valued more than vaccination [72]. Programs for community-based sensitisation have been successful in removing these obstacles and boosting vaccination uptake [73].

Financial limitations restrict access to vaccines and preventive measures in many LASV-endemic areas. In rural areas, cold chain logistics, which are essential for the delivery and storage of vaccines, continue to be a major concern [8]. Widespread vaccination campaigns are made more difficult by a lack of adequate healthcare infrastructure, underscoring the necessity of international financing and cooperation.

3.9 GLOBAL PUBLIC HEALTH IMPACT AND EMERGING CONCERNS

Due to its endemicity in West Africa, Lassa fever presents a serious public health concern. The illness has a significant negative influence on economies and healthcare systems in addition to its effects on afflicted individuals. Urgent attention is required due to the possibility of regional and worldwide spread, which is exacerbated by urbanisation and climate change. Lessons from past outbreaks of comparable VHFs, such as Ebola and Marburg, can also help guide future outbreak management plans.

3.9.1 Impact on Endemic Communities

In endemic areas, Lassa fever has a major impact on the socioeconomic structure. High morbidity and mortality rates caused by the illness put a tremendous burden on healthcare systems and lower worker productivity [70]. Affected families frequently experience financial instability as a result of income loss and the need for ongoing medical care; therefore, the financial impact goes beyond just medical expenses [74]. Additionally, in areas with limited resources, outbreaks worsen poverty levels by interfering with local enterprises, trade, and agriculture [75].

Flawless healthcare systems are further taxed by Lassa fever epidemics, especially in remote regions with few medical resources. The disease can spread through bodily fluids and cause nosocomial infections, which puts healthcare personnel at significant risk. Healthcare services are further burdened by medical staff absenteeism due to fear of catching LASV [17]. Programs for mother and child health are among the regular medical services that are impacted by the allocation of resources to outbreak control [76].

3.9.2 Risks of Regional and Global Spread

Although LASV is mainly endemic in West Africa, there are worries that it may spread outside of this region due to increased international travel and migration. The risk of worldwide transmission has been highlighted by the reports of imported cases in nations such as the United States, the United Kingdom, and Germany [77]. In non-endemic areas, air travel increases the chance of undetected spread since it drastically reduces the period between exposure and diagnosis [78]. To identify and stop instances early, surveillance systems need to be improved, especially at international points of entry.

Changes in the environment are having an impact on Lassa fever epidemiology. *Mastomys natalensis* is the main rodent reservoir for LASV, and its habitat has changed due to rising temperatures, deforestation, and changing rainfall patterns [79]. Variability in the climate may cause rodent populations to spread geographically, increasing the risk of virus transmission for humans. In a similar vein, overcrowding and poor sanitation brought about by fast urbanisation have made it easier for zoonotic diseases to spread [71]. Developing predictive models to foresee future epidemics requires an understanding of these ecological alterations.

3.9.3 LASV in the Context of Other Emerging Infectious Diseases

Similar to other VHFs like Ebola and Marburg, Lassa fever also poses certain difficulties. LASV has a lower CFR, ranging from 1% to 15%, compared to Ebola, which can have CFRs of up to 90%, even though all three viruses induce severe sickness and haemorrhagic symptoms [70]. LASV is more difficult to diagnose and contain early than Ebola because of its lengthy incubation period and large percentage of asymptomatic individuals [8]. Furthermore, as LASV is mostly zoonotic, rodent management is an essential component of its preventative strategy, whereas Ebola and Marburg are primarily spread through human-to-human contact [7].

Insights on outbreak response tactics have been gained from the handling of previous VHF epidemics, especially the West African Ebola epidemic that lasted from 2014 to 2016. Lassa fever can benefit from the same rapid diagnostic tests, community involvement, and enhanced infection control strategies that were essential in the fight against Ebola [80]. For LASV to be detected and contained early, laboratory capability and genomic surveillance must be strengthened [17]. Furthermore, encouraging global cooperation and financing programs will guarantee that endemic areas have access to the tools required for efficient outbreak control.

3.10 RECENT ADVANCES IN RESEARCH

Thanks to advancements in molecular biology, genetics, and diagnostics, research on LASV has made considerable strides. Researchers are learning new things about the genetic diversity, evolution, and pathogenicity of the virus due to the growing availability of proteomics and next-generation sequencing technology. Furthermore, new antiviral treatments and quick diagnostic techniques are enhancing early detection and treatment methods. These advancements are essential for reducing LASV's effects in endemic areas and averting further outbreaks.

3.10.1 Advances in Molecular Research and Genetics

The evolutionary dynamics of LASV have been better understood because of developments in genome sequencing. According to whole-genome sequencing, LASV has a significant level of genetic heterogeneity that may affect its immunological evasion, transmissibility, and virulence [81]. Several different LASV lineages are circulating in West Africa by phylogenetic analysis, indicating that the virus is constantly undergoing genetic drift and recombination [82]. Given that genetic abnormalities may affect immune recognition and medication resistance, this information is crucial for the development of vaccines and antivirals [83].

Variations in disease severity and immunological response are caused by the substantial genetic variety of LASV strains across various geographic regions. The viral genome has recently been mapped to pinpoint certain alterations that support pathogenicity and immune evasion. Targeted surveillance and the creation of region-specific diagnoses and treatments are made possible by an understanding of these variances. Furthermore, studies on viral quasi-species have shown that LASV can quickly adapt inside hosts, making treatment plans more difficult.

3.10.2 Novel Diagnostic and Therapeutic Approaches

Early treatment and outbreak containment depend on a timely and precise LASV diagnosis. The speed and sensitivity of LASV detection have increased because of recent developments in point-of-care diagnostics, including loop-mediated isothermal amplification (LAMP) testing and CRISPR-based assays [84]. These resources lessen reliance on centralised labs, which are frequently unavailable in rural endemic areas, and allow for earlier intervention [85]. Furthermore, real-time surveillance is being conducted using nanopore sequencing technologies, which enable the prompt identification of LASV strains and alterations.

Although ribavirin is still the go-to antiviral medication for Lassa fever, its effectiveness is reduced when given towards the end of the illness. New antiviral drugs that have shown promise in preclinical models, like favipiravir and remdesivir, are being investigated by researchers. Another promising therapy strategy is the use of monoclonal antibody treatments that target LASV GPs [86]. Furthermore, research is being done on immunomodulatory strategies to reduce the excessive inflammatory response linked to severe Lassa fever [32].

3.10.3 Role of Genomics and Proteomics in Understanding LASV

A deeper understanding of LASV-host interactions is being made possible by omics technologies, including transcriptomics and proteomics. Important host variables implicated in immune response and virus replication have been found by proteomic investigations; these factors may be targets for future medications. Differential gene expression patterns in LASV-infected patients have been identified by transcriptomic analysis, providing insight into immune evasion strategies and the course of the disease [81]. These investigations are crucial for figuring out new treatment targets and comprehending the molecular underpinnings of LASV pathogenesis.

3.10.4 Biomarkers for Diagnostics and Prognosis

A primary goal of current research is to find trustworthy biomarkers for early LASV diagnosis and disease prognosis. To forecast the severity of the disease, several possible biomarkers have been proposed, such as viral load thresholds and cytokine profiles [85]. Targeted diagnostics may be possible because of proteomic investigations that have revealed host proteins that are expressed differently in severe vs moderate cases [79]. To improve early diagnosis and risk classification in LASV-infected patients, future research attempts to validate these biomarkers for clinical use.

3.11 FUTURE DIRECTIONS AND CHALLENGES

A comprehensive strategy that incorporates research, public health campaigns, and international cooperation is required to address the persistent LASV problem. There are still many important gaps in our knowledge of LASV pathogenesis, transmission dynamics, and long-term immunity, despite developments in molecular biology, vaccine development, and diagnostics. The possibilities for LASV pandemic preparedness, strategic control techniques, and future research needs are examined in this section.

3.11.1 Research Gaps and Needs

Many facets of LASV pathophysiology are still poorly understood despite decades of investigation. The role that host genetic variables play in determining the severity of the disease is one significant gap. Individual genetic variants may impact immunological responses to LASV, according to studies; however, further study is required to clarify these processes [81]. Furthermore, more research is needed to determine the virus's potential for chronic infection and its long-term persistence in survivors [87].

From an epidemiological perspective, nothing is known about the specific dynamics of rodent-human transmission. Other rodent species might contribute to the spread of LASV, even though *M. natalensis* is the main reservoir [4]. A more comprehensive understanding of LASV's ecology and transmission hazards may be possible by extending surveillance investigations to encompass more rodent species and geographical areas. Furthermore, even while there is evidence that shifting environmental conditions may lead to an increase in rodent-human interactions, the effect of urbanisation and climate change on LASV outbreaks is still poorly understood [83].

3.11.2 Strategic Approaches for Control and Eradication

Controlling LASV in endemic areas requires an integrated "One Health" strategy that takes into account environmental and human factors [71]. Reducing human exposure to infectious rodents requires rodent control efforts, which include better sanitation and habitat management. Furthermore, improving laboratory-based surveillance can aid in the early detection of epidemics and the prevention of their extensive spread [85].

One of the most promising approaches to LASV control is still vaccination. Clinical studies are being conducted on existing vaccine candidates, including the Lassa vaccine (rVSV-LASV), which is based on the recombinant vesicular stomatitis virus [86]. However, logistical and cultural obstacles, including vaccination hesitancy and limits in healthcare infrastructure, must be overcome for vaccine deployment to be successful in endemic countries [88].

Governments, academic institutions, and groups like the WHO and the Coalition for Epidemic Preparedness Innovations must work together internationally to effectively combat LASV. In its R&D Blueprint for Emerging Diseases, WHO designated Lassa fever as a priority pathogen, highlighting the necessity of expedited research and response tactics [17]. International collaborations can boost laboratory capacity in endemic areas, promote epidemiological research, and make it easier to finance vaccine development. Furthermore, international LASV surveillance and epidemic response can be enhanced by data-sharing programs among nations [89–91].

3.12 CONCLUSION

Even though LASV is mostly endemic in West Africa, there is still a chance that it could spread worldwide. The necessity for strong pandemic preparedness frameworks is highlighted by the fact that prior Lassa fever cases were brought into non-endemic nations through international travel. LASV's potential to produce widespread outbreaks if not properly controlled is highlighted by its placement in the WHO Blueprint for Priority Diseases.

The creation of scalable diagnostic and treatment methods is a crucial part of pandemic preparedness. Early detection and outbreak containment can be facilitated by investments in quick diagnostic technologies, such as CRISPR-based tests. Furthermore, creating monoclonal antibody therapeutics and accumulating antiviral medications like ribavirin can enhance therapy alternatives during outbreaks. The necessity of quick vaccine development and distribution has been highlighted by the international reaction to COVID-19. The combination of mRNA and viral vector vaccine platforms to speed up vaccine production is one example of how COVID-19 lessons might be used for LASV. Additionally, enhancing the healthcare system in areas where LASV is endemic can enhance the ability to respond to outbreaks and lessen the possibility of uncontrolled viral transmission.

REFERENCES

1. Richmond JK, Baglole DJ. Lassa fever: Epidemiology, clinical features, and social consequences. BMJ 2003;327(7426):1271–1275.
2. Frame JD, Baldwin JM, Gocke DJ, Troup JM. Lassa fever, a new virus disease of man from West Africa. Am J Trop Med Hyg 1970;19(4):670–676.
3. World Health Organization (WHO). Lassa Fever – West Africa. Geneva: WHO; 2022. Available from: https://www.who.int (Accessed on 2025 Mar 17).
4. Fichet-Calvet E, Rogers DJ. Risk maps of Lassa fever in West Africa. PLOS Negl Trop Dis 2009;3(3):e388.
5. Ogbu O, Ajuluchukwu E, Uneke CJ. Lassa fever in West African sub-region: An overview. J Vector Borne Dis 2007;44(1):1–11.
6. McCormick JB, Fisher-Hoch SP. Lassa fever. Curr Top Microbial Immunol 2002;262:75–109.
7. Bausch DG, Hadi CM, Khan SH, Lertora JJ. Review of the literature and proposed guidelines for the use of ribavirin as postexposure prophylaxis for Lassa fever. Clin Infect Dis 2001;33(5):723–730.
8. Gunther S, Lenz O. Lassa virus. Crit Rev Clin Lab Sci 2022;41(4):339–390.
9. Bowen MD, Rollin PE, Ksiazek TG, Hustad HL, Bausch DG, Demby AH et al. Genetic diversity among Lassa virus strains. J Virol 2000;74(15):6992–7004.
10. Buckley SM, Casals J. Lassa fever, a new virus disease of man from West Africa. Isolation and characterization of the virus. Bull World Health Organ 1970;42(4):592–598.
11. Monath TP, Mertens PE, Patton R, Moser CR, Baum JJ, Pinneo L et al. Arenaviruses in West Africa: IV. Virological and serological studies in Sierra Leone, Nigeria, and Zaire. Bull World Health Organ 1974;51(1):113–120.

12. Walker DH, Wulff H, Lange JV, Murphy FA. Comparative pathology of Lassa virus infection in monkeys, Guinea pigs, and *Mastomys natalensis*. Bull World Health Organ 1975;52(4–6):523–534.
13. Centres for Disease Control and Prevention (CDC). About Lassa Fever Atlanta: CDC.
14. Dalhat MM, Olayinka A, Meremikwu MM, Dan-Nwafor C, Iniobong A, Ntoimo LF et al. Epidemiological trends of Lassa fever in Nigeria, 2018–2021. PLOS One 2022;17(12):e0279467.
15. Smith DR, Turner J, Fahr P, Attfield LA, Bessell PR, Donnelly CA et al. Health and economic impacts of Lassa vaccination campaigns in West Africa. Nat Med 2024;28:1–10.
16. European Centre for Disease Prevention and Control (ECDC). Lassa Fever Annual Epidemiological Report 2019 [Internet]. Stockholm: ECDC; 2019 [cited 2025 Mar 17]. Available from: https://www.ecdc.europa.eu/sites/default/files/documents/AER-Lassa-2019
17. World Health Organization (WHO). Lassa Fever Fact Sheet, 2023. WHO, Geneva.
18. Mateer EJ, Huang C, Shehu NY, Paessler S, Huang Y. Lassa fever-induced hearing loss: The neglected disability of haemorrhagic fever. Front Neurol 2021;12:626764.
19. Yun NE, Walker DH. Lassa fever-induced sensorineural hearing loss: A neglected public health and social burden. PLOS Negl Trop Dis 2012;6(11):e1906.
20. Africa Centres for Disease Control and Prevention (Africa CDC). Lassa Fever. Africa CDC; 2025.
21. Aleccia J. Lassa fever suspected in death of Iowa resident who travelled to West Africa. AP News; 2024.
22. Buchmeier MJ, de la Torre JC, Peters CJ. *Arenaviridae*. In Fields Virology. 6th ed. Lippincott Williams & Wilkins; 2013. pp. 1283–1303.
23. Charrel RN, de Lamballerie X. Arenaviruses other than Lassa virus. Antiviral Res 2010;90(2):89–100.
24. Eyzaguirre EJ, Marta RF. The ever-expanding diversity and complexity of the *Arenaviridae* family. Virulence 2023;14(1):21505594.
25. McLay L, Liang Y, Ly H. Comparative analysis of disease pathogenesis and molecular mechanisms of New World and Old-World arenavirus infections. Future Virol 2014;9(1):57–74.
26. Goeijenbier M, Wagenaar J, Goris M, Martina B, Henttonen H, Vaheri A et al. Rodent-borne haemorrhagic fevers: Under-recognized, widely spread and preventable – Epidemiology, diagnostics and treatment. Crit Rev Microbiol 2013;39(1):26–42.
27. Donaldson RI. The Lassa Ward: One Man's Fight against One of the World's Deadliest Diseases, 2009. St. Martin's Press.
28. Lukashevich IS, Djavani M, Shapiro K, Sanchez A, Ravkov EV, Nichol ST et al. The Lassa fever virus L gene: Nucleotide sequence, comparison, and expression of a minigenome. Virus Res 1997;48(1):1–15.
29. Kranzusch PJ, Whelan SP. Arenavirus Z protein controls viral RNA synthesis by locking a polymerase-promoter complex. Proc Natl Acad Sci USA 2011;108(49):19743–19748.
30. Kunz S, Sevilla N, McGavern DB, Campbell KP, Oldstone MB. Molecular analysis of the interaction of Lassa virus with its cellular receptor, α-dystroglycan. J Virol 2005;79(10):5979–5987.
31. Tian Y, Bassit B, Chau D, Li YM. An APP inhibitory domain containing the Flemish mutation residue modulates γ-secretase activity for Aβ production. Nat Struct Mol Biol 2010;17(2):151–158.
32. Hastie KM, Kimberlin CR, Zandonatti MA, MacRae IJ, Saphire EO. Structure of the Lassa virus nucleoprotein reveals a dsRNA-specific 3′ to 5′ exonuclease activity essential for immune suppression. Proc Natl Acad Sci USA 2011;108(6):2396–2401.
33. Morin B, Coutard B, Lelke M, Ferron F, Kerber R, Jamal S et al. The N-terminal domain of the arenavirus L protein binds to the viral RNA and activates the polymerase. PLOS Pathog 2010;6(9):e1001038.
34. Cai Y, Liu Y, Zhang X. Induction of transcription factor Egr-1 gene expression in astrocytoma cells by Murine coronavirus infection. Virology 2006;355(2):152–163.
35. Aryal S. Lassa Virus – An Overview. Microbe Notes. Swiss Institute of Bioinformatics; 2021.
36. Abraham J, Kwong JA, Albarino CG, Lu JG, Radoshitzky SR, Salazar-Bravo J et al. Host–species barriers to Lassa virus entry identified by structural differences in GP1 receptor binding. Nat Struct Mol Biol 2010;17(1):127–134.
37. Cohen-Dvashi H, Israeli H, Shani O, Katz A, Diskin R. Role of Lassa virus surface glycoprotein in cell entry. J Virol 2016;90(11):5287–5297.
38. García CC, Scolaro LA, Damonte EB. Molecular mechanisms of arenavirus transcription and replication. Viruses 2022;14(7):1425.
39. Urata S, Noda T, Kawaoka Y, Morikawa S, Yokosawa H, Yasuda J. Interaction of Tsg101 with Lassa virus Z protein. Virology 2006;358(1):84–91.
40. Lecompte E, Fichet-Calvet E, Daffis S, Granjon L. *Mastomys natalensis* and Lassa fever: Ecological and epidemiological factors influencing transmission in West Africa. Vector Borne Zoonotic Dis 2006;6(3):213–220.

41. Olayemi A, Cadar D, Magassouba N, Obadare A, Kourouma F, Löscher T. Human-mastomys interactions in Lassa fever-endemic areas of Guinea: Implications for disease transmission. PLOS Negl Trop Dis 2016;10(11):e0005215.
42. Price ME, Bruenn JA, Berry AC. Lassa fever in pregnancy: Epidemiological, clinical, and virological aspects. Lancet Infect Dis 2020;20(3):e92–e99.
43. Andersen KG, Shapiro BJ, Matranga CB, Sealfon RSG, Lin AE, Moses LM et al. Clinical and epidemiological aspects of Lassa fever outbreaks. Lancet Infect Dis 2008;8(8):487–496.
44. National Emerging Special Pathogens Training and Education Center. Lassa fever. NETEC Resource Library.
45. Okokhere PO, Bankole IA, Akpede GO. Central Nervous system manifestations of Lassa fever in Nigeria and the effect on mortality. J Neurol Sci 2013;333:e604.
46. Yuill TM. Lassa Fever: Overview and Treatment, 2023. Merck Manuals.
47. Yaro CA, Kogi E, Opara KN, Batiha GE, Baty RS, Albrakati A et al. Infection pattern, case fatality rate and spread of Lassa virus in Nigeria. BMC Infect Dis 2021;21:1–9.
48. Ficenec SC, Percak J, Arguello S, Bays A, Goba A, Gbakie M et al. Lassa fever induced hearing loss: The neglected disability of haemorrhagic fever. Int J Infect Dis 2020;100:82–87.
49. Klitting R, Mehta SB, Oguzie JU, Oluniyi PE, Pauthner MG, Siddle KJ et al. Lassa virus genetics. In Lassa Fever: Epidemiology, Immunology, Diagnostics, and Therapeutics, 2020; 23–65. Springer International Publishing, Cham.
50. Besson ME, Pépin M, Metral PA. Lassa fever: Critical review and prospects for control. Trop Med Infect Dis 2024;9(8):178.
51. Demby AH, Chamberlain J, Brown DW, Clegg CS. Early diagnosis of Lassa fever by reverse transcription-PCR. J Clin Microbiol 1994;32(12):2898–2903.
52. Fisher-Hoch SP, Tomori O, Nasidi A, Perez-Oronoz GI, Fakile Y, Hutwagner L et al. Review of cases of nosocomial Lassa fever in Nigeria: The high price of poor medical practice. BMJ 1995;311(7009):857–859.
53. Russier M, Reynard S, Carnec X, Baize S. The exonuclease domain of Lassa virus nucleoprotein is involved in antigen-presenting-cell-mediated NK cell responses. J Virol 2014;88(23):13811–13820.
54. McElroy AK, Akondy RS, Harmon JR, Ellebedy AH, Cannon D, Klena JD et al. A case of human Lassa virus infection with robust acute T-cell activation and long-term virus-specific T-cell responses. J Infect Dis 2017;215(12):1862–1872.
55. Baize S, Pannetier D, Faure C, Marianneau P, Marendat I, Georges-Courbot MC et al. Role of interferons in the control of Lassa virus replication in human dendritic cells and macrophages. Microbes Infect 2006;8(5):1194–1202.
56. Qi X, Lan S, Wang W, Schelde LM, Dong H, Wallat GD et al. Cap binding and immune evasion revealed by Lassa nucleoprotein structure. Nature 2010;468(7325):779–783.
57. Warner BM, Safronetz D, Stein DR. Current perspectives on vaccines and therapeutics for Lassa fever. Virol J 2024;21(1):320.
58. Kühl A, Pöhlmann S. How Ebola virus counters the interferon system. Zoonoses Public Health 2012;59:116–131.
59. Cummins D, McCormick JB, Bennett D, Samba JA, Farrar B, Machin SJ et al. Acute sensorineural deafness in Lassa fever. JAMA 1990;264(16):2093–2096.
60. McCormick JB, King IJ, Webb PA, Johnson KM, O'Sullivan R, Smith ES et al. A case-control study of the clinical diagnosis and course of Lassa fever. J Infect Dis 1987;155(3):445–455.
61. Huggins JW. Prospects for treatment of viral haemorrhagic fevers with ribavirin, a broad-spectrum antiviral drug. Rev Infect Dis 1989;11(Suppl_4):S750–S761.
62. Safronetz D, Rosenke K, Westover JB, Martellaro C, Okumura A, Furuta Y et al. The broad-spectrum antiviral favipiravir protects guinea pigs from lethal Lassa virus infection post-disease onset. Sci Rep 2015;5:14775.
63. Cross RW, Mire CE, Branco LM, Geisbert JB, Rowland MM, Heinrich ML et al. Treatment of Lassa virus infection in outbred Guinea pigs with first-in-class human monoclonal antibodies. Antiviral Res 2016;133:218–222.
64. Orosco FL. Breaking the chains: Advancements in antiviral strategies to combat Nipah virus infections. Int J One Health 2023;9(2):2455–8931.
65. Ibekwe TS, Okokhere PO, Asogun D, Blackie FF, Nwegbu MM, Wahab KW et al. Early-onset sensorineural hearing loss in Lassa fever. Eur Arch Otorhinolaryngol 2011;268:197–201.
66. Mari Saez A, Cherif Haidara M, Camara A, Kourouma F, Sage M, Magassouba NF et al. Rodent control to fight Lassa fever: Evaluation and lessons learned from a 4-year study in upper Guinea. PLOS Negl Trop Dis 2018;12(11):e0006829.

67. Cross RW, Woolsey C, Prasad AN, Borisevich V, Agans KN, Deer DJ, et al. A recombinant VSV-vectored vaccine rapidly protects nonhuman primates against heterologous lethal Lassa fever. Cell Rep. 2022;40(3)
68. Newswire PR. INOVIO completes enrolment of Phase 1B clinical trial for its DNA vaccine candidate against Lassa Fever, INO-4500, in West Africa.
69. Aloke C, Obasi NA, Aja PM, Emelike CU, Egwu CO, Jeje O et al. Combating Lassa fever in West African sub-region: Progress, challenges, and future perspectives. Viruses 2023;15(1):146.
70. Murphy H, Ly H. Understanding immune responses to Lassa virus infection and to its candidate vaccines. Vaccines 2022;10(10):1668.
71. Bonwitt J, Kelly AH, Ansumana R, Agbla S, Sahr F, Saez AM et al. Rat-atouille: A mixed method study to characterize rodent hunting and consumption in the context of Lassa fever. Ecohealth 2016;13:234–247.
72. Abaya ST, Ogoina D, Stow J, Abaye BB, Emeka C, Raimi MO. Beyond the epidemic: Effective public health strategies in response to Nigeria's first Lassa fever outbreak in a non-endemic region. JMIR Preprints 2024;19(8):2024.
73. Vanderslott S, Van Ryneveld M, Marchant M, Lees S, Nolna SK, Marsh V. How can community engagement in health research be strengthened for infectious disease outbreaks in Sub-Saharan Africa? A scoping review of the literature. BMC Public Health 2021;21:1–6.
74. Ilori EA, Furuse Y, Ipadeola OB, Dan-Nwafor CC, Abubakar A, Womi-Eteng OE et al. Epidemiologic and clinical features of Lassa fever outbreak in Nigeria. Emerg Infect Dis 2019;25(6):1066.
75. Izah SC, Ovuru KF, Ogwu MC. Lassa fever in Nigeria: Social and ecological risk factors exacerbating transmission and sustainable management strategies. Int J Trop Dis 2022;5(2):065.
76. Ezechukwu C. Abstracts of proceedings: 50th annual general and scientific conference of the paediatrics association of Nigeria (PANCONF). Niger J Paediatric 2019;46(2):73–121.
77. Kitching A, Addiman S, Cathcart S, Bischop L, Krahé D, Nicholas M et al. A fatal case of Lassa fever in London, 2009. Euro Surveill 2009;14(6):19117.
78. Ostroff SM, Kozarsky P. Emerging infectious diseases and travel medicine. Infect Dis Clin North Am 1998;12(1):231–241.
79. Hay SI, Battle KE, Pigott DM, Smith DL, Moyes CL, Bhatt S et al. Global mapping of infectious disease. Philos Trans R Soc Lond B Biol Sci 2013;368(1614):20120250.
80. Kuppalli K. Ebola: Ten years later—Lessons learned and future pandemic preparedness. PLOS Glob Public Health 2024;4(9):e0003662.
81. Andersen KG, Shapiro BJ, Matranga CB, Sealfon R, Lin AE, Moses LM et al. Clinical sequencing uncovers origins and evolution of Lassa virus. Cell 2015;162(4):738–750.
82. Kafetzopoulou LE, Pullan ST, Lemey P, Suchard MA, Ehichioya DU, Pahlmann M et al. Metagenomic sequencing at the epicentre of the Nigeria 2018 Lassa fever outbreak. Science 2019;363(6422):74–77.
83. Redding DW, Moses LM, Cunningham AA, Wood J, Jones KE. Environmental-mechanistic modelling of the impact of global change on human zoonotic disease emergence: A case study of Lassa fever. Methods Ecol Evol 2016;7(6):646–655.
84. Bao M, Waitkus J, Liu L, Chang Y, Xu Z, Qin P et al. Micro-and nanosystems for the detection of haemorrhagic fever viruses. Lab Chip 2023;23(19):4173–4200.
85. Siddle KJ, Eromon P, Barnes KG, Mehta S, Oguzie JU, Odia I et al. Genomic analysis of Lassa virus during an increase in cases in Nigeria in 2018. N Engl J Med 2018;379(18):1745–1753.
86. Cross RW, Hastie KM, Mire CE, Robinson JE, Geisbert TW, Branco LM et al. Antibody therapy for Lassa fever. Curr Opin Virol 2019;37:97–104.
87. Malhotra S, Yen JY, Honko AN, Garamszegi S, Caballero IS, Johnson JC et al. Transcriptional profiling of the circulating immune response to Lassa virus in an aerosol model of exposure. PLOS Negl Trop Dis 2013;7(4):e2171.
88. Gibb R, Moses LM, Redding DW, Jones KE. Understanding the cryptic nature of Lassa fever in West Africa. Pathog Glob Health 2017;111(6):276–288.
89. Iannetta M, Di Caro A, Nicastri E, Vairo F, Masanja H, Kobinger G et al. Viral haemorrhagic fevers other than Ebola and Lassa. Infect Dis Clin North Am 2019;33(4):977–1002.
90. Raabe V, Koehler J. Laboratory diagnosis of Lassa fever. J Clin Microbiol 2017;55(6):1629–1637.
91. Oestereich L, Rieger T, Lüdtke A, Ruibal P, Wurr S, Pallasch E et al. Efficacy of favipiravir alone and in combination with ribavirin in a lethal, immunocompetent mouse model of Lassa fever. J Infect Dis 2016;213(6):934–938.

4 Nipah Virus

Nid'a Alshraiedeh

4.1 INTRODUCTION

Nipah virus (NiV) is a deadly zoonotic pathogen that can cause neurological and respiratory diseases in humans. NiV belongs to a relatively new genus *Henipavirus in Paramyxoviridae* family. NiV has been categorized as a biosafety level-4 (BSL4) pathogen, which means that this virus imposes high risk to the health and security of the public [1]. When infecting humans, this highly fatal NiV usually would typically have 4–14 days of incubation period, and thus far, there is no effective prophylactic or therapeutic interventions available [2]. NiV is most commonly hosted by the *Pteropus* genus of fruit bats from the *Pteropodidae* family and is transmitted through the saliva and urine of these bats [3]. These bats, which can be found in Australia, South Asia and Southeast Asia, carry the virus with no sign of illness. They can fly very far, up to 600 km, increasing the risk of spread and emergence of outbreak in different places [4, 5]. There have been a number of reported outbreaks of NiV infections in regions where these bats are commonly found, especially in South Asia (India and Bangladesh) and Southeast Asia (Malaysia, Singapore, and the Philippines) [6]. The first documented outbreak was in 1998 in Malaysia, caused by infected pigs – in that case, pigs were the main amplifying hosts, but dogs and horses can also carry the virus [7]. The outbreak was then spread to Singapore, through importation of these infected pigs [8].

At present, two variants of NiV have been identified, namely NiV-M and NiV-B. The former was isolated from the outbreaks that occurred in Malaysia and Singapore, while the latter was isolated from the outbreaks that occurred in India and Bangladesh. Notably, each variant has different fatality rate, symptoms and transmission pathways. However, both carry a high genetic identity and amino acid homology [9, 10].

4.2 OUTBREAKS

The NiV derived its name from Sungai Nipah village in Malaysia. The village was the place where the virus was first discovered, in an outbreak which occurred from 1998 to 1999 [11]. This first outbreak involved 283 cases, resulting in more than 100 deaths [12]. The victims, mainly pig farmers and those in close contact with pigs, developed encephalitis and respiratory illness [11]. The virus spread across peninsular Malaysia through shipments of infected animals. Through these shipments, the virus entered Singapore, infecting 11 individuals in 1999, causing 1 death [13]. The virus was then dormant for two years, before remerging in India and Bangladesh – which are in different region – in 2001. India recorded its first NiV outbreak in 2001 in Siliguri, West Bengal, involving 66 reported cases with 45 deaths, and in 2007, another outbreak (a minor one) occurred also in West Bengal claiming the lives of 5 individuals (100% death rate) [14, 15]. These outbreaks occurred across the border from Bangladesh's Nipah belt. A brutal outbreak reoccurred in India in May 2018, in the districts of Kozhikode and Malappuram in Kerala, involving 19 reported cases that resulted in 17 deaths [16]. Kerala is geographically distinct from previous region (West Bengal) as it is located in the South of India, and this outbreak was the first one to have occurred in the south region of the country [15]. Then, in 2019 and 2021, two more outbreaks occurred in Kerala, with one reported death each [15]. Between August and September 2023, 6 cases were reported in Kerala also, and 2 deaths reportedly occurred [2].

In Bangladesh, the first outbreak was reported in 2001, and since then, NiV outbreaks kept occurring in the country. The outbreaks in Bangladesh seemed to follow a seasonal pattern, and

DOI: 10.1201/9781003607816-4

the peak season would be the cold months [17]. From 2001 to 2024, there were 341 reported NiV infection cases in Bangladesh, with 242 fatalities [6]. In contrast, there were no new reported cases in Malaysia and Singapore after the initial outbreak in 1998. In the Philippines, the first case of NiV was reported in 2014 involving 17 cases. The virus, which was spread by infected horses, caused 9 deaths (53% fatality rate) [18].

During the past 25 years – for the region of Southeast Asia, represented by three countries namely Malaysia, Singapore and the Philippines – 311 NiV cases have been reported with 119 fatalities (38%). For the region of South Asia, represented by two countries namely India and Bangladesh, 443 NiV cases have been reported with 316 fatalities (71%) [6].

At present, cases of NiV infections to humans are mainly reported in Southeast Asia and South Asia. This seems to correspond with the geographic distribution of *Pteropus* bats. It can thus be deduced that fruit bats from the genus *Pteropus* is the main risk factor for the disease spread. Nonetheless, extensive observations have resulted in the discovery of Anti-NiV, anti-NiV antibodies and NiV RNA in other countries, including Brazil, Ghana, Cameroon, Indonesia, Papua New Guinea, Cambodia, China, Madagascar, New Caledonia, Thailand, Vietnam, and Timor-Leste [6, 19–27]. Clearly, there has been a global spread of the disease and this has raised global concerns on the potential future global NiV spread.

4.3 ROUTES OF TRANSMISSION

The transmission of NiV from bats to humans follows three major pathways as follows: ingestion of tainted food, contact with domestic animals, and direct exposure to excretions from NiV-infected bats (Figure 4.1) [28], Ingestion of tainted food was the most common way of getting infected by NiV, and many people in Bangladesh were infected after consuming fresh date palm sap tainted with the urine and saliva of the infected bats [29]. In fact, NiV outbreaks in Bangladesh have been mainly linked to the consumption of tainted raw date palm sap – in Bangladesh, date palm sap is collected during the winter months [17]. Hence, the recurring NiV outbreaks in Bangladesh during the winter months is justifiable. During harvest, bats would

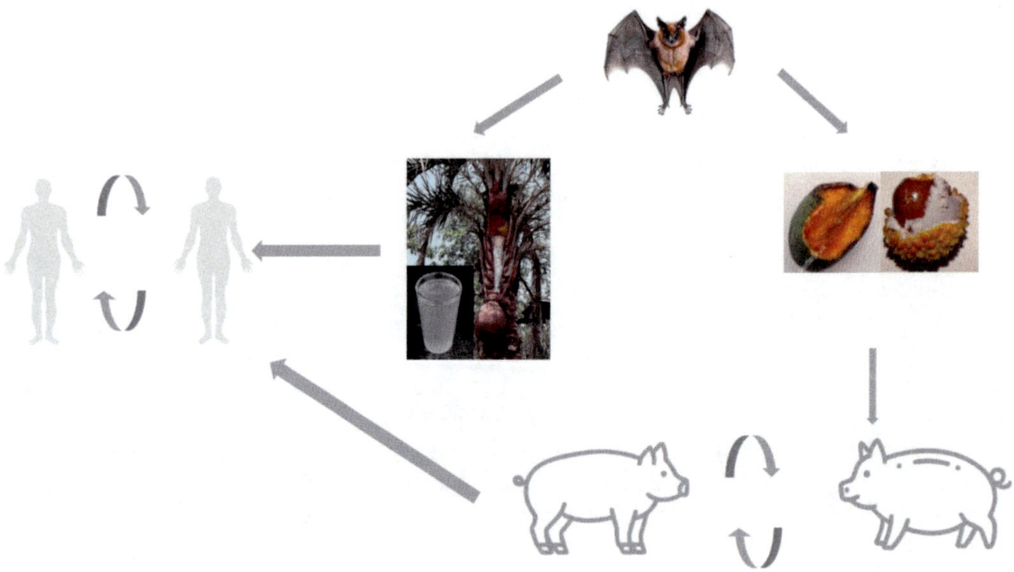

FIGURE 4.1 Nipah virus routes of transmission.

come and lick the sap, contaminating it with its NiV infested saliva [30]. Additionally, NiV has the capacity to live for days in liquids that contain high level of sugar, like date palm sap, lychee juice and mango juice [31].

Contact with domestic animals, especially pigs, was the primary way NiV was transmitted to humans during its initial outbreak in Malaysia in 1998. The outbreak had cause death to 109 people, mainly pig farmers and those in close contact with the pigs [12]. Consequently, about 1 million pigs were culled, to curb the disease [32, 33]. The following year, an outbreak occurred in Singapore, and the outbreak was also linked to the importation of infected pigs from Malaysia [13]. Usually, domestic animals acquire the virus from the *Pteropus* fruit bats (flying foxes) indirectly. Domestic animal usually catch the virus from eating pasture or feed tainted by excretions of the bats such as the saliva, urine, feces, semen [29]. Although the initial outbreak was primarily associated with infected pigs, the NiV can infect different animal species that have cellular receptors known as ephrin-B2 and ephrin-B3, which facilitate the virus's entry into the host [21, 34–36].

It should be noted that the NiV spread is not merely limited to bat-to-human transmission. Human-to-human transmission can also occur when an infected person transmits the virus to other person through his bodily fluids, mainly respiratory secretions [15, 37–39]. The recurring NiV outbreaks in Bangladesh are an example of outbreaks caused by human-to-human transmission. In Bangladesh, the outbreaks that occurred from 2001 to 2007 involved 122 cases, and 62 of these cases (51%) were caused by close contact with infected person. Additionally, 9 of the infected individuals (7% from the reported 122 cases), transmitted NiV to 62 other individuals [37, 40]. Recurring NiV outbreaks in Bangladesh that are mainly caused by human-to-human transmission, demonstrate the significant role of social norms and caregiving practices such as close physical contact, sharing of personal items, and sleeping with infected persons during outbreaks [41].

Nosocomial transmission, which is a real threat, also reported with NiV, as evident with the NiV outbreak in Siliguri, India, where 33 healthcare workers and hospital visitors were infected by the virus after being exposed to the infected patients [35, 42]. A person also can become infected when exposed to NiV RNA on hospital surfaces – this further affirms the shedding of the virus from the infected patients into their surroundings, which facilitates the transmission further [43]. However, in the outbreaks in Malaysia and Singapore, studies on healthcare workers showed low potential of nosocomial transmission [44].

4.4 VIROLOGY

NiV is a pleomorphic, enveloped virus. The outer membrane of this virus known as the envelope, enfolds a viral capsid containing a single-stranded negative-sense non-segmented RNA genome of roughly 18.2 kb with six successive genes arranged in the sequence 3′-N-P-M-F-G-L-5′ encoding several structural proteins, including the nucleocapsid (N), phosphoprotein (P), matrix (M), fusion glycoprotein (F), attachment glycoprotein (G), and long polymerase (L) as well as non-structural proteins, namely V, W, and C proteins (Figure 4.2) [36, 45].

The F and G glycoproteins are imbedded in the envelope of the virus. Both glycoproteins allow the host cell attachment. The attachment protein (G) of NiV in viral envelope recognizes and binds to cell surface receptors of host, specifically ephrin B2/3. Meanwhile, paramyxovirus fusion (F) proteins encompass type I transmembrane proteins, with a signal peptide at their N-terminus and a transmembrane anchor at their C-terminus. Initially, these proteins are an inactive precursor (F0) to be cleaved by a host cell enzyme, so that it becomes active biologically. When G protein is bound with ephrin B2/3 on host cells, a conformational change is triggered in G, causing the F protein to be activated. Through activated fusion (F) protein, the viral and host cell membranes can fuse, leading to viral entry. The viral genome is released upon entry, and replication begins with the generation of full-length anti-genome that functions as

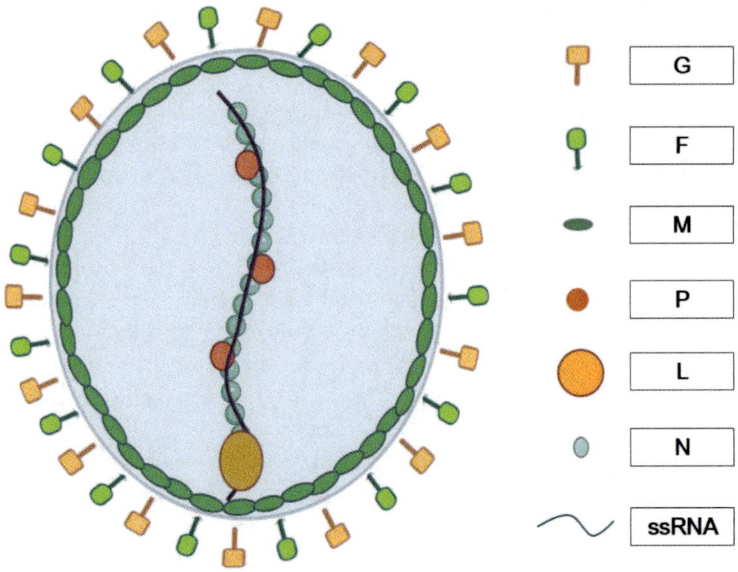

FIGURE 4.2 Schematic representation of the viral structure.

a template for the generation of numerous copies of viral RNA through the ribonucleoprotein (RNP) complex. RNP complex is a resultant of N, P, and L proteins attachment to the NiV RNA. RNP complex carries a transcriptase and replicase activity which are vital for viral transcription and replication. As for the capsid, it is generated by the M protein. In viral assembly, capsid provides linkage between the nucleocapsid and surface glycoproteins, and release of virus, through budding [46, 47].

The P gene has a complex nature, because it encodes both the structural P protein and the three non-structural proteins, namely C, V, and W, through RNA editing and alternative translation. Protein C is generated by alternative reading frame in the P gene, while the V and W proteins are resultant of mRNA editing process through the inclusion of one (for V) or two (for W) guanosine residues. The non-structural proteins regulate the life cycle of the virus. Together, these proteins obstruct the interferon response of the host [48].

NiV is a fairly recently discovered single-stranded RNA virus. It belongs in the order of Mononegavirales, *Paramyxoviridae* family, *Paramyxovirinae* subfamily (others include *Respirovirus*, *Morbillovirus*, *Avulavirus*, and *Rubulavirus*) and the relatively new genus *Henipavirus*. Classed into the Paramyxovirinae subfamily, the *Henipavirus* genus is a taxonomic classification emerged following the discovery of Hendra virus (HeV) [45, 49]. This genus includes two recently identified viruses; the first one was HeV which was first discovered in Australia in 1994, while the other was NiV which was first discovered in Malaysia in 1998. Both HeV and NiV are pathogenic and are classified as BSL4 pathogens. Additionally, they have zoonotic potential; carried by fruit bats (*Pteropus* species). HeV was first spread to humans through infected horses while NiV was first spread to humans through infected pigs [36]. The morphological characteristics of both HeV and NiV are akin to those viruses from the *Paramyxovirinae* subfamily. However, both have distinct biological, serological, and molecular properties that resulted in the emergence of the *Henipavirus* genus. HeV and NiV have much larger genomes (18.2 kb) than other members of the *Paramyxovirinae* subfamily whose genomes are much smaller (15.1–15.9 kb) [50].

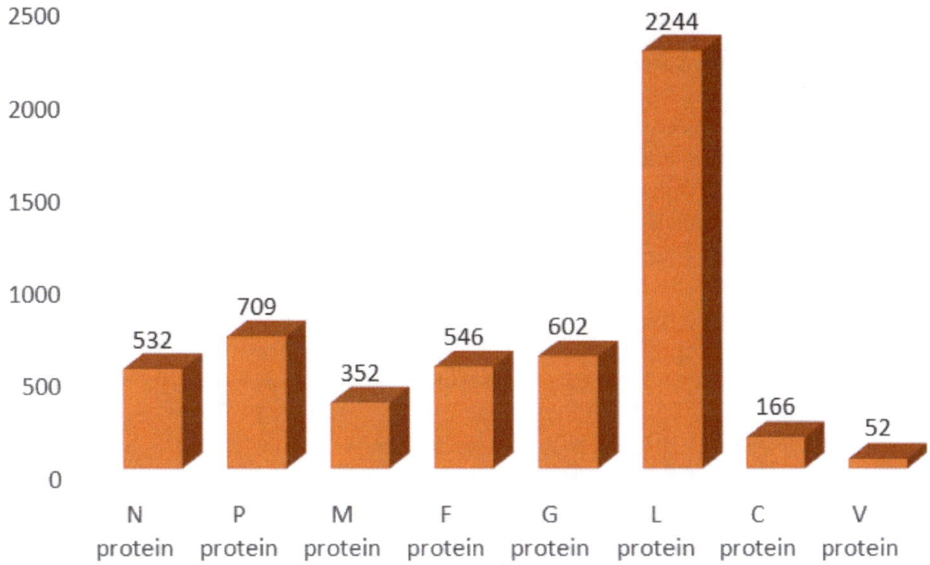

FIGURE 4.3 The length in amino acids of the N, P, M, F, G, and L proteins of NiV.

4.5 DIFFERENCE BETWEEN NIPAH VIRUS AND HENDRA VIRUS

The genetic sequences of both HeV and NiV have been ascertained, and from phylogenetic analysis, both viruses are closely linked. However, based on gene sequencing, there seem to be dissimilarities in individual gene sequences and amino acid compositions for each virus. The regions of both HeV and NiV genomes that code for the N, P, C, V, M, F, and G proteins share 70%–90% genetic resemblance. At the amino acid level, the homology in open reading frame (ORF) is 92.1% for N protein, 86.6% for L protein, 83.3% for G protein, and 67.6% for P protein, 81.1% for V protein, 83.2% for C protein and 89% for the M (Figures 4.3 and 4.4) [45, 50].

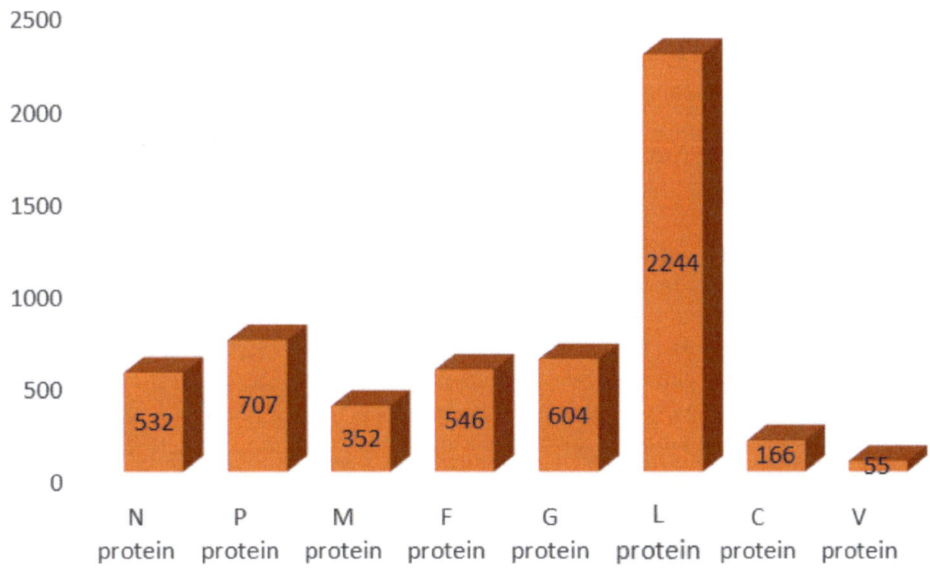

FIGURE 4.4 The length in amino acids of the N, P, M, F, G, and L proteins of HeV.

4.6 VARIANTS OF NIPAH VIRUS

Genetic and phylogenetic analyses showed two genetically distinct lineages of NiV namely the Malaysia lineage (NiV-M) and the Bangladesh lineage (NiV-B) with roughly 92% nucleotide sequence homology [9, 10]. Genomes sequencing of NiV-B and NiV-M showed high homology. However, the NiV-B genome is a bit lengthy, owing to the 6 extra nucleotides located on the 5′ non-translated region of the F protein gene, signifying growing virus evolution. The main difference is in the F protein at position 42. Here, NiV-M carries valine, as opposed to isoleucine that can be found in NiV-B – this affects the exposure of the fusion peptide. Other variations can be observed at three positions, namely 384, 386, and, particularly, position 498 of G protein. These variations alter the receptor-binding affinity. Variations can also be found in the untranslated regions of the P gene in non-structural proteins between these two strains, where, the V protein of NiV-M is three amino acids shorter in comparison to that of NiV-B. Also, there is a variation at site 1645 of L protein. This may affect the viral activity, when methyl groups are added to the 5′ RNA cap structure [10, 51].

NiV-B and NiV-M have distinct transmission dynamics, where, NiV-B infection is majorly caused by human-to-human transmission and bat-to-human transmission (eating bat-contaminated date palm sap), while NiV-M generally needs intermediate amplifying mammalian host to facilitate the transmission of the virus to the human. A bat-to-pig-to-human is the common route for NiV-M infection [17, 40, 52]

Compared to NiV-M, NiV-B seems more deadly with a much greater mortality rate. Additionally, NiV-B causes more recurring and serious respiratory symptoms like acute respiratory distress syndrome, in addition to neurological problems, like encephalitis and altered mental status. On the other hand, NiV-M would generally cause encephalitis with negligible respiratory symptoms. The period of incubation for both variants also varies; between 6 and 10 days for NiV-B, and between 14 days and 2 months for NiV-M [9, 53, 54].

4.7 VIRAL REPLICATION AND SPREAD

NiV pathogenesis begins with G protein attachment on NiV exterior with ephrin-B2 and ephrin-B3 expressed on the exterior of host respiratory epithelial cells which is NiV's initial target. Here, fusion with the host cell membrane is mediated by activated F protein. Upon entry into the host, NiV would multiply, mainly in the respiratory epithelial cells. These newly formed NiVs are then dispersed to the entire body of the host, mainly into the central nervous system (CNS) by crossing the blood-brain barrier (BBB) after multiplication in endothelial cells that made up the BBB. The virus can also travel along the olfactory nerve after multiplying in olfactory epithelial cells in the respiratory tract. In CNS, NiV targets the microvascular endothelial cells and this causes encephalitis.

Through the bloodstream, NiV also can reach various organs like the heart, kidneys, digestive system, and excretory systems. The entry of NiV into these organs results in multi-organ vasculitis, ischemia, syncytia formation, thrombosis, as well as parenchymal inflammation [55]. The systemic spread of NiV is highly influenced by the dissemination of numerous cell types expressing the ephrin-B2 and ephrin-B3 receptors. These include neurons, immune cells, endothelial cells and pneumocytes [56]. The ensuing intracellular multiplication inside the epithelial cells of the respiratory and urinary tracts aids the viral shedding and intensifies transmission [57].

It is possible for NiV to persist without multiplying in the immune cells such as in lymphocytes and monocytes – this situation could further facilitate the virus dissemination [58].

4.8 CLINICAL PRESENTATION

The infection of NiV typically emerges as respiratory distress or encephalitis. The infection of this virus can be asymptomatic at early stage [59] or begins with general clinical symptoms that are akin to the flu; patient would experience headache, dizziness, muscle pain, nausea, fever, general discomfort, cough, vomiting, as well as confusion [60, 61]. Brain involvement may present as myoclonus, areflexia, altered mental status, hypotonia, altered sensorium, hypertension, tachycardia and convulsions. Respiratory symptoms may include coughing, tachypnea, difficulty breathing, atypical pneumonia, and acute respiratory distress syndrome [62].

Patients who suffer from encephalitis mostly would develop vasculitis and necrosis in the CNS. Encephalitis is clinically regarded as a poor prognostic factor because it amplifies the fatality rate. Typically, a patient would die within a median of 6 days after the development of the symptoms. Patients suffering severe encephalitis with drowsiness and disorientation are likely to progress into seizures and fall into a coma within 48 hours. Also, many people that survive encephalitis would have long-term neurological dysfunction, such as enduring seizures, draining fatigue, and behavioral irregularities [63]. Patient may not develop encephalitis until later (late onset) or patient may experience a relapse encephalitis months after recovery. Individuals with delayed onset encephalitis have a reduced mortality rate compared to those who suffer from acute onset encephalitis. [60, 64].

4.9 DIAGNOSIS

Infection of NiV shows symptoms comparable to those of other feverish illnesses. In order to contain the outbreaks, it is important to diagnose NiV early and accurately. Not only that, early diagnosis could increase the chance of survival, with proper medical care. Various direct or indirect laboratory tests can be performed for diagnosis, such as nucleic acid testing, serological tests, histopathology with immunohistochemistry, and virus isolation and neutralization assay [65–67]. It is important to collect specimens during the early stages of the disease, and the most commonly collected specimens include urine, blood, throat, and nasal swabs [65]. Some of the commonly used tests are as explained in the following.

4.9.1 Nucleic Acid Tests

PCR and sequencing are reliable assays for diagnosis of active NiV infection because of their high specificity and high sensitivity in detecting even a tiny amount of viral genetic material [68]. Somehow, in endemic areas with poor infrastructure, performing these tests can be challenging. Not only that, the presence of genetic variants can reduce the accurateness of these tests. Notably, probes employed in this assay are usually designed for use in highly conserved sequences of NiV RNA genome existing in all the identified pathogen strains [69]. Several nucleic acid test can be employed including [65, 70–72]:

1. Reverse transcription real-time PCR (RT-PCR) has been recommended as the main method to diagnose acute NiV infection. As opposed to conventional PCR, RT-PCR has higher sensitivity.
2. Multiplex PCR assays which allowing for the concurrent detection and differentiation of various pathogens or viral strains with identical manifestations, facilitating a quick diagnosis that could lower mortality rates and minimize treatment delays.
3. The Truenat™ point-of-care test utilizing a portable and battery-powered PCR system allow for quick identification of NiV. By using this technology, there is no longer need for the extensive infrastructure, materials, and equipment generally needed for viral detection through real-time reverse transcription polymerase chain reaction, a method known for its sensitivity, specifiicty in diagnosing NiV.
4. Recombinase polymerase amplification assay using CRISPR/Cas12a protein or CRISPR/Cas13a protein provide a safe and cost-effective techniques requiring BSL-2 environment [73, 74].

5. DNA sequencing such as next generation sequencing provides user with complete genomic data for viral identification and evolutionary analysis. This method is employed mainly for research and phylogenetic studies, not for routine diagnostics, owing to its cost and complexity.

4.9.2 Serological Tests

Antigen and antibody (IgM and IgG) ELISAs are serological tests used for the detection viral antigens or the antibodies of the host (IgM and IgG). Serological tests typically exhibit lower sensitivity compared to PCR in detecting minor genetic variations and tend to have greater cross-reactivity among subtypes and strains. NiV-specific IgM ELISA is used in detecting active infection when PCR is unavailable. NiV-specific IgM enzyme-linked immunosorbent assay (ELISA) is used to detect active infection when polymerase chain reaction (PCR) testing is unavailable. In contrast, NiV-specific IgG ELISA is suitable for epidemiological studies and surveillance because IgG antibodies can persist for several months [65].

The development of rapid diagnostic tests (RDTs) grounded upon ELISA technology is ongoing. The purpose is to facilitate application, that is, the tests could be performed within a non-laboratory environment, with minimal specimen processing. At present, NiV (or HeV) RDTs for humans or animals are still unavailable [65, 74].

4.9.3 Histopathology with Immunohistochemistry

This method is usually carried out for post-mortem confirmation. It utilizes formalin fixed tissue that can be obtained from different NiV-infected organs [65, 75].

4.9.4 Virus Isolation and Neutralization Assays

These tests are usually carried out for definitive confirmatory purposes. Confirmatory Sero-neutralization assays employ either live NiV under a high-containment BSL-4 environment or non-infectious pseudo-typed virus particles obtained from vesicular stomatitis virus or lentivirus (HIV), which require a BSL-2 environment [65, 76].

4.10 THERAPEUTIC INTERVENTIONS

At present, approved accurate antiviral medication for NiV infection treatment is still unavailable, and clinical management is mainly supportive care based on the specific condition and the symptoms manifested by each patient. Among the commonly used intervention methods include nutritional support, hydration, oxygen therapy, antipyretics to manage fever, anticonvulsants to curb seizures, and empirical treatment to manage secondary infections. Intensive care is provided to patients with severe respiratory and/or neurological problems. Numerous potential therapeutic strategies are currently being scrutinized for treating NiV infections (Table 4.1). However, the efficacy of these strategies are very much based on preclinical animal studies because of the unfeasibility of human challenge trials and the limited number of cases found during sporadic outbreaks.

4.11 PREVENTION AND CONTROL MEASURES

At present, the outbreaks of NiV are likely to occur in regions with the carrier bat populations. Still, outbreaks in new locations can occur – this has become a concern because of climate change and anthropogenic disruption of pteropodid habitats. It is important to have a complete understanding of the transmission dynamics of bat-borne virus and disease biogeography so that targeted prevention and control strategies could be developed. It is necessary to implement diverse, comprehensive and adaptable preventive and control measures at homes and the healthcare facility to safeguard both

TABLE 4.1
Potential Therapeutic Strategies against Nipah Virus

Therapeutic strategy	Drug	Mechanism of action	References
Nucleoside analogue antiviral agents	Ribavirin	Inhibiting RNA-dependent RNA polymerase (RdRp)	[77–79]
	Remdesivir (GS-5734)	Inhibiting RNA-dependent RNA polymerase (RdRp)	[80, 81]
	Favipiravir (T-705)	Inhibiting RNA-dependent RNA polymerase (RdRp) inhibitors	[82]
Monoclonal antibodies (mAb)	Human mAb m102.4	Interfering with the attachment of viral glycoprotein G of both NiV and HeV to the ephrin B2 and B3 receptors on the prone host cells	[49, 76, 83, 84]
	Human mAb NiV41 and its mature forn NiV41-6	Targeting the NiV G protein	[85, 86]
	Humanized mAb h5B3	Targeting the F protein, by interfering with fusion protein-mediated viral entry	[87]
	Murine mAb IF 5 and 12B2 and their humanized counnterpart h1F5 and h12B2	Targeting the F protein, by interfering with fusion protein-mediated viral entry	[77]
Antimalarial agent	Chloroquine	Targeting cathepsin L that is vital for F protein processing	[88–90]
Immunomodulator	Poly(I)-poly(C12U)	Stimulating the production of IFN-α due to Its immunomodulatory powerfully	[91]

humans and animals against NiV infection. Zoonotic transmissions to livestock should be prevented – this is a very important task, especially in regions in which livestock becomes intermediate hosts to NiV. Furthermore, the roosting sites of carrier fruit bats should be far away from agricultural facilities and grazing lands, to break the transmission cycle, as demonstrated in Malaysia. Considering that NiV could be transmitted to humans through ingestion of tainted food, such as the consumption of raw contaminated date palm sap in Bangladesh which contributed to the NiV infection outbreaks, behavioral modifications are necessary. People should be made aware of the potential risks of consuming raw palm sap. Additionally, some measures should be taken to prevent bats from accessing the date palm sap collection sites, and collection vessels should be kept clean and protected at all times to prevent contamination. As part of the efforts in curbing the spread of the virus; it is important to practice fundamental hygiene, such as thorough washing of fruits and vegetables to remove any potential contamination from bat excreta. This will reduce the risk of being in contact with and ingesting the virus. Also, considering that NiV can infect human through human-to-human transmission, it is necessary to protect oneself, by avoiding direct contact with infected individuals. The use of personal protective equipment like gloves, masks and eyewear, is necessary when around the infected individuals.

Vigorous public health interventions are necessary in assuring effective outbreak control. Among the important components in outbreak management are active surveillance to detect cases early, thorough contact tracing to identify and monitor exposed individuals, and quarantining the high-risk individuals and healthcare professionals. The development of effective and rapid diagnostic tools, therapeutic modalities, and vaccines is urgent [92].

4.12 VACCINATION

The absence of effective therapies, especially for those at risk, highlight the urgent need for preventive vaccines to lessen the impact of NiV, a disease with a broad geographic spread. Comprehensive preclinical investigations in various animal models have facilitated the identification and development of multiple immunization strategies against both Hendra and Nipah viruses summarized in Table 4.2.

TABLE 4.2
Potential Vaccination Approaches against Nipah Virus

Immunization approaches	Vaccine	Description	References
DNA vaccines	DNA-G	DNA expressing NiV-G	[93]
Virus-like particles	Nipah virus-like particles (NiV-VLPs)	Self-assembled G, F, and M viral proteins forming particles mimicking the size, shape and surface structure of NiV with no genetic material	[94, 95]
Viral vectors	Recombinant measles virus-based NiV vaccine	Recombinant engineered measles virus expressing NiV-G)	[96]
	Recombinant vaccinia virus-based vaccine	Recombinant modified vaccinia Ankara (MVA) strain expressing NiV-F or NiV-G)	[97]
	Recombinant vesicular stomatitis virus (rVSV)	rVSV expressing NiV G or NiV F or NiV G along with the envelope glycoprotein of Ebola virus	[89, 90]
	Rabies virus-based vaccine candidates	Live attenuated rERAG333E strain of rabies virus expressing NiV G or NiV F	[98–100]
	Adeno-associated virus (AAV) vectors	AAV expressing the NiV G	[101]
	ChAdOx1 vector vaccine	Chimpanzee adenovirus vector (recombinant simian adenovirus vector) expressing the NiV G	[93, 102]
	Newcastle disease virus (NDV)-based Vaccine	NDV expressing NiV G or F	[103]
	Canarypox Virus-based Vaccine	live, attenuated canarypox virus vector (ALVAC) expressing the NiV F or NiV-G or both	[103]
mRNA Vaccine	mRNA-1215 (Moderna)	mRNA encodes the secreted stable pre-fusion NiV F protein covalently coupled with its G monomer (pre-F/G)	[104]

REFERENCES

1. Branda, F., Ceccarelli, G., Giovanetti, M., Albanese, M., Binetti, E., Ciccozzi, M., & Scarpa, F. (2025). Nipah virus: A zoonotic threat re-emerging in the wake of global public health challenges. *Microorganisms*, *13*(1), 124. https://doi.org/10.3390/microorganisms13010124
2. Verma, A., Jain, H., Sulaiman, S. A., Pokhrel, P., Goyal, A., & Dave, T. (2024). An impending public health threat: Analysis of the recent Nipah virus outbreak and future recommendations – An editorial. *Annals of Medicine & Surgery*, *86*(2), 638–642. https://doi.org/10.1097/MS9.0000000000001627
3. Li, H., Kim, J.-Y. V., & Pickering, B. S. (2023). *Henipavirus* zoonosis: Outbreaks, animal hosts and potential new emergence. *Frontiers in Microbiology*, *14*, 1167085. https://doi.org/10.3389/fmicb.2023.1167085
4. Deka, M. A., & Morshed, N. (2018). Mapping disease transmission risk of Nipah virus in South and Southeast Asia. *Tropical Medicine and Infectious Disease*, *3*(2), 57. https://doi.org/10.3390/tropicalmed3020057
5. Chua, K. B., Lek Koh, C., Hooi, P. S., Wee, K. F., Khong, J. H., Chua, B. H., Chan, Y. P., Lim, M. E., & Lam, S. K. (2002). Isolation of Nipah virus from Malaysian Island flying-foxes. *Microbes and Infection*, *4*(2), 145–151. https://doi.org/10.1016/S1286-4579(01)01522-2
6. Khan, S., Akbar, S. M. F., Mahtab, M. A., Uddin, M. N., Rashid, M. M., Yahiro, T., Hashimoto, T., Kimitsuki, K., & Nishizono, A. (2024). Twenty-five years of Nipah outbreaks in Southeast Asia: A persistent threat to global health. *IJID Regions*, *13*, 100434. https://doi.org/10.1016/j.ijregi.2024.100434
7. Bruno, L., Nappo, M. A., Ferrari, L., Di Lecce, R., Guarnieri, C., Cantoni, A. M., & Corradi, A. (2022). Nipah virus disease: Epidemiological, clinical, diagnostic and legislative aspects of this unpredictable emerging zoonosis. *Animals*, *13*(1), 159. https://doi.org/10.3390/ani13010159
8. Field, H., Young, P., Yob, J. M., Mills, J., Hall, L., & Mackenzie, J. (2001). The natural history of Hendra and Nipah viruses. *Microbes and Infection*, *3*(4), 307–314. https://doi.org/10.1016/S1286-4579(01)01384-3

9. Garbuglia, A. R., Lapa, D., Pauciullo, S., Raoul, H., & Pannetier, D. (2023). Nipah virus: An overview of the current status of diagnostics and their role in preparedness in endemic countries. *Viruses*, *15*(10), 2062. https://doi.org/10.3390/v15102062
10. Harcourt, B. H., Lowe, L., Tamin, A., Liu, X., Bankamp, B., Bowden, N., Rollin, P. E., Comer, J. A., Ksiazek, T. G., Hossain, M. J., Gurley, E. S., Breiman, R. F., Bellini, W. J., & Rota, P. A. (2005). Genetic characterization of Nipah virus, Bangladesh, 2004. *Emerging Infectious Diseases*, *11*(10), 1594–1597. https://doi.org/10.3201/eid1110.050513
11. Chua, K. B., Goh, K. J., Wong, K. T., Kamarulzaman, A., Tan, P. S. K., Ksiazek, T. G., Zaki, S. R., Paul, G., Lam, S. K., & Tan, C. T. (1999). Fatal encephalitis due to Nipah virus among pig-farmers in Malaysia. *The Lancet*, *354*(9186), 1257–1259. https://doi.org/10.1016/S0140-6736(99)04299-3
12. Chua, K. B. (2003). Nipah virus outbreak in Malaysia. *Journal of Clinical Virology*, *26*(3), 265–275. https://doi.org/10.1016/S1386-6532(02)00268-8
13. Chew, M. H. L., Arguin, P. M., Shay, D. K., Goh, K., Rollin, P. E., Shieh, W., Zaki, S. R., Rota, P. A., Ling, A., Ksiazek, T. G., Chew, S., & Anderson, L. J. (2000). Risk factors for Nipah virus infection among abattoir workers in Singapore. *The Journal of Infectious Diseases*, *181*(5), 1760–1763. https://doi.org/10.1086/315443
14. Harit, A. K., Ichhpujani, R. L., Gupta, S., Gill, K. S., Lal, S., Ganguly, N. K., & Agarwal, S. P. (2006). Nipah/Hendra virus outbreak in Siliguri, West Bengal, India in 2001. *The Indian Journal of Medical Research*, *123*(4), 553–560.
15. Uwishema, O., Wellington, J., Berjaoui, C., Muoka, K. O., Onyeaka, C. V. P., & Onyeaka, H. (2022). A short communication of Nipah virus outbreak in India: An urgent rising concern. *Annals of Medicine & Surgery*, *82*. https://doi.org/10.1016/j.amsu.2022.104599
16. Anoop A. S., A. (2018). Deadly Nipah outbreak in Kerala: Lessons learned for the future. *Indian Journal of Critical Care Medicine*, *22*(7), 475–476. https://doi.org/10.4103/ijccm.IJCCM_282_18
17. Nahar, N., Sultana, R., Gurley, E. S., Hossain, M. J., & Luby, S. P. (2008). Date palm sap collection: Exploring opportunities to prevent Nipah transmission. *International Journal of Infectious Diseases*, *12*, e135–e136. https://doi.org/10.1016/j.ijid.2008.05.337
18. Ching, P. K. G., De Los Reyes, V. C., Sucaldito, M. N., Tayag, E., Columna-Vingno, A. B., Malbas, F. F., Bolo, G. C., Sejvar, J. J., Eagles, D., Playford, G., Dueger, E., Kaku, Y., Morikawa, S., Kuroda, M., Marsh, G. A., McCullough, S., & Foxwell, A. R. (2015). Outbreak of *Henipavirus* infection, Philippines, 2014. *Emerging Infectious Diseases*, *21*(2), 328–331. https://doi.org/10.3201/eid2102.141433
19. Sun, Y.-Q., Zhang, Y.-Y., Liu, M.-C., Chen, J.-J., Li, T.-T., Liu, Y.-N., Zhang, L.-Y., Wang, T., Yu, L.-J., Che, T.-L., Tang, T., Xu, Q., Lv, C.-L., Jiang, B.-G., Golding, N., Mehlman, M. L., Hay, S. I., Fang, L.-Q., & Liu, W. (2024). Mapping the distribution of Nipah virus infections: A geospatial modelling analysis. *The Lancet Planetary Health*, *8*(7), e463–e475. https://doi.org/10.1016/S2542-5196(24)00119-0
20. Breed, A. C., Yu, M., Barr, J. A., Crameri, G., Thalmann, C. M., & Wang, L.-F. (2010). Prevalence of *Henipavirus* and *Rubulavirus* antibodies in Pteropid bats, Papua New Guinea. *Emerging Infectious Diseases*, *16*(12), 1997–1999. https://doi.org/10.3201/eid1612.100879
21. Hasebe, F., Thuy, N. T. T., Inoue, S., Yu, F., Kaku, Y., Watanabe, S., Akashi, H., Dat, D. T., Mai, L. T. Q., & Morita, K. (2012). Serologic evidence of Nipah virus infection in bats, Vietnam. *Emerging Infectious Diseases*, *18*(3), 536–537. https://doi.org/10.3201/eid1803.111121
22. Wacharapluesadee, S., Boongird, K., Wanghongsa, S., Ratanasetyuth, N., Supavonwong, P., Saengsen, D., Gongal, G. N., & Hemachudha, T. (2010). A longitudinal study of the prevalence of Nipah virus in *Pteropus lylei* bats in Thailand: Evidence for seasonal preference in disease transmission. *Vector-Borne and Zoonotic Diseases*, *10*(2), 183–190. https://doi.org/10.1089/vbz.2008.0105
23. Sendow, I., Field, H. E., Adjid, A., Ratnawati, A., Breed, A. C., Darminto, Morrissy, C., & Daniels, P. (2010). Screening for Nipah virus infection in West Kalimantan province, Indonesia. *Zoonoses and Public Health*, *57*(7–8), 499–503. https://doi.org/10.1111/j.1863-2378.2009.01252.x
24. Hayman, D. T. S., Suu-Ire, R., Breed, A. C., McEachern, J. A., Wang, L., Wood, J. L. N., & Cunningham, A. A. (2008). Evidence of *Henipavirus* infection in West African fruit bats. *PLOS One*, *3*(7), e2739. https://doi.org/10.1371/journal.pone.0002739
25. Iehlé, C., Razafitrimo, G., Razainirina, J., Andriaholinirina, N., Goodman, S. M., Faure, C., Georges-Courbot, M.-C., Rousset, D., & Reynes, J.-M. (2007). *Henipavirus* and Tioman virus antibodies in Pteropodid bats, Madagascar. *Emerging Infectious Diseases*, *13*(1), 159–161. https://doi.org/10.3201/eid1301.060791
26. Li, Y., Wang, J., Hickey, A. C., Zhang, Y., Li, Y., Wu, Y., Zhang, H., Yuan, J., Han, Z., McEachern, J., Broder, C. C., Wang, L.-F., & Shi, Z. (2008). Antibodies to Nipah or Nipah-like viruses in bats, China. *Emerging Infectious Diseases*, *14*(12), 1974–1976. https://doi.org/10.3201/eid1412.080359

27. Cappelle, J., Hoem, T., Hul, V., Furey, N., Nguon, K., Prigent, S., Dupon, L., Ken, S., Neung, C., Hok, V., Pring, L., Lim, T., Bumrungsri, S., Duboz, R., Buchy, P., Ly, S., Duong, V., Tarantola, A., Binot, A., & Dussart, P. (2020). Nipah virus circulation at human–bat interfaces, Cambodia. *Bulletin of the World Health Organization*, *98*(8), 539–547. https://doi.org/10.2471/BLT.20.254227
28. Luby, S., Rahman, M., Hossain, M., Blum, L., Husain, M., Gurley, E., Khan, R., Ahmed, B.-N., Rahman, S., Nahar, N., Kenah, E., Comer, J., & Ksiazek, T. (2006). Foodborne transmission of Nipah virus, Bangladesh. *Emerging Infectious Diseases*, *12*(12), 1888–1894. https://doi.org/10.3201/eid1212.060732
29. Halpin, K., Hyatt, A. D., Fogarty, R., Middleton, D., Bingham, J., Epstein, J. H., Rahman, S. A., Hughes, T., Smith, C., Field, H. E., & Daszak, P. (2011). Pteropid bats are confirmed as the reservoir hosts of Henipaviruses: A comprehensive experimental study of virus transmission. *The American Society of Tropical Medicine and Hygiene*, *85*(5), 946–951. https://doi.org/10.4269/ajtmh.2011.10-0567
30. Nahar, N., Paul, R. C., Sultana, R., Gurley, E. S., Garcia, F., Abedin, J., Sumon, S. A., Banik, K. C., Asaduzzaman, M., Rimi, N. A., Rahman, M., & Luby, S. P. (2015). Raw sap consumption habits and its association with knowledge of Nipah virus in two endemic districts in Bangladesh. *PLOS One*, *10*(11), e0142292. https://doi.org/10.1371/journal.pone.0142292
31. Fogarty, R., Halpin, K., Hyatt, A. D., Daszak, P., & Mungall, B. A. (2008). *Henipavirus* susceptibility to environmental variables. *Virus Research*, *132*(1–2), 140–144. https://doi.org/10.1016/j.virusres.2007.11.010
32. Mhod Nor, M. N., Gan, C. H., & Ong, B. L. (2000). Nipah virus infection of pigs in peninsular Malaysia: -EN- -FR- -ES-. *Revue Scientifique et Technique de l'OIE*, *19*(1), 160–165. https://doi.org/10.20506/rst.19.1.1202
33. Chua, K. B., Bellini, W. J., Rota, P. A., Harcourt, B. H., Tamin, A., Lam, S. K., Ksiazek, T. G., Rollin, P. E., Zaki, S. R., Shieh, W.-J., Goldsmith, C. S., Gubler, D. J., Roehrig, J. T., Eaton, B., Gould, A. R., Olson, J., Field, H., Daniels, P., Ling, A. E., … Mahy, B. W. J. (2000). Nipah virus: A recently emergent deadly paramyxovirus. *Science*, *288*(5470), 1432–1435. https://doi.org/10.1126/science.288.5470.1432
34. Hsu, V. P., Hossain, M. J., Parashar, U. D., Ali, M. M., Ksiazek, T. G., Kuzmin, I., Niezgoda, M., Rupprecht, C., Bresee, J., & Breiman, R. F. (2004). Nipah virus encephalitis reemergence, Bangladesh. *Emerging Infectious Diseases*, *10*(12), 2082–2087. https://doi.org/10.3201/eid1012.040701
35. Luby, S. P., Gurley, E. S., & Hossain, M. J. (2009). Transmission of human infection with Nipah virus. *Clinical Infectious Diseases*, *49*(11), 1743–1748. https://doi.org/10.1086/647951
36. Eaton, B. T., Broder, C. C., Middleton, D., & Wang, L.-F. (2006). Hendra and Nipah viruses: Different and dangerous. *Nature Reviews Microbiology*, *4*(1), 23–35. https://doi.org/10.1038/nrmicro1323
37. Luby, S. P., Hossain, M. J., Gurley, E. S., Ahmed, B.-N., Banu, S., Khan, S. U., Homaira, N., Rota, P. A., Rollin, P. E., Comer, J. A., Kenah, E., Ksiazek, T. G., & Rahman, M. (2009). Recurrent zoonotic transmission of Nipah virus into humans, Bangladesh, 2001–2007. *Emerging Infectious Diseases*, *15*(8), 1229–1235. https://doi.org/10.3201/eid1508.081237
38. Hahn, M. B., Epstein, J. H., Gurley, E. S., Islam, M. S., Luby, S. P., Daszak, P., & Patz, J. A. (2014). Roosting behaviour and habitat selection of *Pteropus giganteus* reveal potential links to Nipah virus epidemiology. *Journal of Applied Ecology*, *51*(2), 376–387. https://doi.org/10.1111/1365-2664.12212
39. Chua, K. B., Lam, S. K., Goh, K. J., Hooi, P. S., Ksiazek, T. G., Kamarulzaman, A., Olson, J., & Tan, C. T. (2001). The presence of Nipah virus in respiratory secretions and urine of patients during an outbreak of Nipah virus encephalitis in Malaysia. *Journal of Infection*, *42*(1), 40–43. https://doi.org/10.1053/jinf.2000.0782
40. Homaira, N., Rahman, M., Hossain, M. J., Epstein, J. H., Sultana, R., Khan, M. S. U., Podder, G., Nahar, K., Ahmed, B., Gurley, E. S., Daszak, P., Lipkin, W. I., Rollin, P. E., Comer, J. A., Ksiazek, T. G., & Luby, S. P. (2010). Nipah virus outbreak with person-to-person transmission in a district of Bangladesh, 2007. *Epidemiology and Infection*, *138*(11), 1630–1636. https://doi.org/10.1017/S0950268810000695
41. Blum, L. S., Khan, R., Nahar, N., & Breiman, R. F. (2009). In-depth assessment of an outbreak of Nipah encephalitis with person-to-person transmission in Bangladesh: Implications for prevention and control strategies. *The American Journal of Tropical Medicine and Hygiene*, *80*(1), 96–102.
42. Chadha, M. S., Comer, J. A., Lowe, L., Rota, P. A., Rollin, P. E., Bellini, W. J., Ksiazek, T. G., & Mishra, A. C. (2006). Nipah virus-associated encephalitis outbreak, Siliguri, India. *Emerging Infectious Diseases*, *12*(2), 235–240. https://doi.org/10.3201/eid1202.051247
43. Hassan, M. Z., Sazzad, H. M. S., Luby, S. P., Sturm-Ramirez, K., Bhuiyan, M. U., Rahman, M. Z., Islam, M. M., Ströher, U., Sultana, S., Kafi, M. A. H., Daszak, P., Rahman, M., & Gurley, E. S. (2018). Nipah virus contamination of hospital surfaces during outbreaks, Bangladesh, 2013–2014. *Emerging Infectious Diseases*, *24*(1), 15–21. https://doi.org/10.3201/eid2401.161758

44. Mounts, A. W., Kaur, H., Parashar, U. D., Ksiazek, T. G., Cannon, D., Arokiasamy, J. T., Anderson, L. J., & Lye, M. S. (2001). A cohort study of health care workers to assess nosocomial transmissibility of Nipah virus, Malaysia, 1999. *The Journal of Infectious Diseases*, *183*(5), 810–813. https://doi.org/10.1086/318822
45. Harcourt, B. H., Tamin, A., Ksiazek, T. G., Rollin, P. E., Anderson, L. J., Bellini, W. J., & Rota, P. A. (2000). Molecular characterization of Nipah virus, a newly emergent paramyxovirus. *Virology*, *271*(2), 334–349. https://doi.org/10.1006/viro.2000.0340
46. Yang, G., Wang, D., & Liu, B. (2024). Structure of the Nipah virus polymerase phosphoprotein complex. *Nature Communications*, *15*(1), 8673. https://doi.org/10.1038/s41467-024-52701-y
47. Madhukalya, R., Yadav, U., Parray, H. A., Raj, N., Lupitha, S. S., Kumar, V., Saroj, A., Agarwal, V., Kumar, D., Das, S., & Kumar, R. (2025). Nipah virus: Pathogenesis, genome, diagnosis, and treatment. *Applied Microbiology and Biotechnology*, *109*(1), 158. https://doi.org/10.1007/s00253-025-13474-6
48. Yoneda, M., Guillaume, V., Sato, H., Fujita, K., Georges-Courbot, M.-C., Ikeda, F., Omi, M., Muto-Terao, Y., Wild, T. F., & Kai, C. (2010). The nonstructural proteins of Nipah virus play a key role in pathogenicity in experimentally infected animals. *PLOS One*, *5*(9), e12709. https://doi.org/10.1371/journal.pone.0012709
49. Wang, L.-F., Yu, M., Hansson, E., Pritchard, L. I., Shiell, B., Michalski, W. P., & Eaton, B. T. (2000). The exceptionally large genome of Hendra virus: Support for creation of a new genus within the family *Paramyxoviridae*. *Journal of Virology*, *74*(21), 9972–9979. https://doi.org/10.1128/JVI.74.21.9972-9979.2000
50. Wang, L.-F., Harcourt, B. H., Yu, M., Tamin, A., Rota, P. A., Bellini, W. J., & Eaton, B. T. (2001). Molecular biology of Hendra and Nipah viruses. *Microbes and Infection*, *3*(4), 279–287. https://doi.org/10.1016/S1286-4579(01)01381-8
51. Hauser, N., Gushiken, A. C., Narayanan, S., Kottilil, S., & Chua, J. V. (2021). Evolution of Nipah virus infection: Past, present, and future considerations. *Tropical Medicine and Infectious Disease*, *6*(1), 24. https://doi.org/10.3390/tropicalmed6010024
52. Gurley, E. S., Montgomery, J. M., Hossain, M. J., Bell, M., Azad, A. K., Islam, M. R., Molla, M. A. R., Carroll, D. S., Ksiazek, T. G., Rota, P. A., Lowe, L., Comer, J. A., Rollin, P., Czub, M., Grolla, A., Feldmann, H., Luby, S. P., Woodward, J. L., & Breiman, R. F. (2007). person-to-person transmission of Nipah virus in a Bangladeshi community. *Emerging Infectious Diseases*, *13*(7), 1031–1037. https://doi.org/10.3201/eid1307.061128
53. Hossain, M. J., Gurley, E. S., Montgomery, J. M., Bell, M., Carroll, D. S., Hsu, V. P., Formenty, P., Croisier, A., Bertherat, E., Faiz, M. A., Azad, A. K., Islam, R., Molla, M. A. R., Ksiazek, T. G., Rota, P. A., Comer, J. A., Rollin, P. E., Luby, S. P., & Breiman, R. F. (2008). Clinical presentation of Nipah virus infection in Bangladesh. *Clinical Infectious Diseases*, *46*(7), 977–984. https://doi.org/10.1086/529147
54. Lo, M. K., & Rota, P. A. (2008). The emergence of Nipah virus, a highly pathogenic paramyxovirus. *Journal of Clinical Virology*, *43*(4), 396–400. https://doi.org/10.1016/j.jcv.2008.08.007
55. Wong, K. T., Shieh, W.-J., Kumar, S., Norain, K., Abdullah, W., Guarner, J., Goldsmith, C. S., Chua, K. B., Lam, S. K., Tan, C. T., Goh, K. J., Chong, H. T., Jusoh, R., Rollin, P. E., Ksiazek, T. G., & Zaki, S. R. (2002). Nipah virus infection. *The American Journal of Pathology*, *161*(6), 2153–2167. https://doi.org/10.1016/S0002-9440(10)64493-8
56. Hafner, C., Schmitz, G., Meyer, S., Bataille, F., Hau, P., Langmann, T., Dietmaier, W., Landthaler, M., & Vogt, T. (2004). Differential gene expression of Eph receptors and ephrins in benign human tissues and cancers. *Clinical Chemistry*, *50*(3), 490–499. https://doi.org/10.1373/clinchem.2003.026849
57. Lamp, B., Dietzel, E., Kolesnikova, L., Sauerhering, L., Erbar, S., Weingartl, H., & Maisner, A. (2013). Nipah virus entry and egress from polarized epithelial cells. *Journal of Virology*, *87*(6), 3143–3154. https://doi.org/10.1128/JVI.02696-12
58. Mathieu, C., Pohl, C., Szecsi, J., Trajkovic-Bodennec, S., Devergnas, S., Raoul, H., Cosset, F.-L., Gerlier, D., Wild, T. F., & Horvat, B. (2011). Nipah virus uses leukocytes for efficient dissemination within a host. *Journal of Virology*, *85*(15), 7863–7871. https://doi.org/10.1128/JVI.00549-11
59. Banerjee, S., Gupta, N., Kodan, P., Mittal, A., Ray, Y., Nischal, N., Soneja, M., Biswas, A., & Wig, N. (2019). Nipah virus disease: A rare and intractable disease. *Intractable & Rare Diseases Research*, *8*(1), 1–8. https://doi.org/10.5582/irdr.2018.01130
60. Arif, S. M., Basher, A., Quddus, M. R., & Faiz, M. A. (2012). Re-emergence Nipah—A review. *Mymensingh Medical Journal*, *21*(4), 772–779.

61. Chong, H. T., Kunjapan, S. R., Thayaparan, T., Geok Tong, J. M., Petharunam, V., Jusoh, M. R., & Tan, C. T. (2002). Nipah encephalitis outbreak in Malaysia, clinical features in patients from Seremban. *Canadian Journal of Neurological Sciences/Journal Canadien Des Sciences Neurologiques*, *29*(1), 83–87. https://doi.org/10.1017/S0317167100001785
62. Chandni, R., Renjith, T. P., Fazal, A., Yoosef, N., Ashhar, C., Thulaseedharan, N. K., Suraj, K. P., Sreejith, M. K., Sajeeth Kumar, K. G., Rajendran, V. R., Remla Beevi, A., Sarita, R. L., Sugunan, A. P., Arunkumar, G., Mourya, D. T., & Murhekar, M. (2020). Clinical manifestations of Nipah virus–infected patients who presented to the emergency department during an outbreak in Kerala state in India, May 2018. *Clinical Infectious Diseases*, *71*(1), 152–157. https://doi.org/10.1093/cid/ciz789
63. Sejvar, J. J., Hossain, J., Saha, S. K., Gurley, E. S., Banu, S., Hamadani, J. D., Faiz, M. A., Siddiqui, F. M., Mohammad, Q. D., Mollah, A. H., Uddin, R., Alam, R., Rahman, R., Tan, C. T., Bellini, W., Rota, P., Breiman, R. F., & Luby, S. P. (2007). Long-term neurological and functional outcome in Nipah virus infection. *Annals of Neurology*, *62*(3), 235–242. https://doi.org/10.1002/ana.21178
64. Tan, C. T., Goh, K. J., Wong, K. T., Sarji, S. A., Chua, K. B., Chew, N. K., Murugasu, P., Loh, Y. L., Chong, H. T., Tan, K. S., Thayaparan, T., Kumar, S., & Jusoh, M. R. (2002). Relapsed and late-onset Nipah encephalitis. *Annals of Neurology*, *51*(6), 703–708. https://doi.org/10.1002/ana.10212
65. Mazzola, L. T., & Kelly-Cirino, C. (2019). Diagnostics for Nipah virus: A zoonotic pathogen endemic to Southeast Asia. *BMJ Global Health*, *4*(Suppl 2), e001118. https://doi.org/10.1136/bmjgh-2018-001118
66. Wang, L.-F., & Daniels, P. (2012). Diagnosis of *Henipavirus* infection: Current capabilities and future directions. In B. Lee & P. A. Rota (Eds.), *Henipavirus* (Vol. *359*, pp. 179–196). Springer Berlin Heidelberg. https://doi.org/10.1007/82_2012_215
67. Daniels, P., Ksiazek, T., & Eaton, B. T. (2001). Laboratory diagnosis of Nipah and Hendra virus infections. *Microbes and Infection*, *3*(4), 289–295. https://doi.org/10.1016/S1286-4579(01)01382-X
68. Feldman, K. S., Foord, A., Heine, H. G., Smith, I. L., Boyd, V., Marsh, G. A., Wood, J. L. N., Cunningham, A. A., & Wang, L.-F. (2009). Design and evaluation of consensus PCR assays for henipaviruses. *Journal of Virological Methods*, *161*(1), 52–57. https://doi.org/10.1016/j.jviromet.2009.05.014
69. Tong, S., Chern, S.-W. W., Li, Y., Pallansch, M. A., & Anderson, L. J. (2008). Sensitive and broadly reactive reverse transcription-PCR assays to detect novel paramyxoviruses. *Journal of Clinical Microbiology*, *46*(8), 2652–2658. https://doi.org/10.1128/JCM.00192-08
70. Van Dessel, G., Lagrou, A., Hilderson, H. J., Dierick, W., & Dacremont, G. (1975). [Proceedings: Lipid composition of bovine thyroid gland. IV. Glycolipids]. *Archives Internationales De Physiologie Et De Biochimie*, *83*(2), 412–414.
71. Guillaume, V., Lefeuvre, A., Faure, C., Marianneau, P., Buckland, R., Lam, S. K., Wild, T. F., & Deubel, V. (2004). Specific detection of Nipah virus using real-time RT-PCR (TaqMan). *Journal of Virological Methods*, *120*(2), 229–237. https://doi.org/10.1016/j.jviromet.2004.05.018
72. D. Yadav, P., Majumdar, T., Gupta, N., Kumar, M. A., Shete, A., Pardeshi, P., Sultana, S., R. Sahay, R., Manoj, M. N., Patil, S., Floura, S., Gangakhedkar, R., & T. Mourya, D. (2021). Standardization & validation of Truenat™ point-of-care test for rapid diagnosis of Nipah. *Indian Journal of Medical Research*, *154*(4), 645–649. https://doi.org/10.4103/ijmr.IJMR_4717_20
73. Yang, X., Xu, K., Li, S., Zhang, J., Xie, Y., Lou, Y., & Xiao, X. (2024). Novel methods for the rapid and sensitive detection of Nipah virus based on a CRISPR/Cas12a system. *The Analyst*, *149*(9), 2586–2593. https://doi.org/10.1039/D4AN00027G
74. Miao, J., Zuo, L., He, D., Fang, Z., Berthet, N., Yu, C., & Wong, G. (2023). Rapid detection of Nipah virus using the one-pot RPA-CRISPR/Cas13a assay. *Virus Research*, *332*, 199130. https://doi.org/10.1016/j.virusres.2023.199130
75. Tanimura, N., Imada, T., Kashiwazaki, Y., Shamusudin, S., Syed Hassan, S., Jamaluddin, A., Russell, G., & White, J. (2004). Reactivity of anti-Nipah virus monoclonal antibodies to formalin-fixed, paraffin-embedded lung tissues from experimental Nipah and Hendra virus infections. *Journal of Veterinary Medical Science*, *66*(10), 1263–1266. https://doi.org/10.1292/jvms.66.1263
76. Kaku, Y., Noguchi, A., Marsh, G. A., McEachern, J. A., Okutani, A., Hotta, K., Bazartseren, B., Fukushi, S., Broder, C. C., Yamada, A., Inoue, S., & Wang, L.-F. (2009). A neutralization test for specific detection of Nipah virus antibodies using pseudotyped vesicular stomatitis virus expressing green fluorescent protein. *Journal of Virological Methods*, *160*(1–2), 7–13. https://doi.org/10.1016/j.jviromet.2009.04.037
77. Kallon, M. K., Mami, D. M., Mami, E. T., Romba, M., Andrew, M. P., & Martain, M. S. (2024). Therapeutic advancement in treatment and prevention of Nipah viral infection: A review. *Asian Journal of Research in Infectious Diseases*, *15*(4), 51–65. https://doi.org/10.9734/ajrid/2024/v15i4343

78. Chong, H.-T., Kamarulzaman, A., Tan, C.-T., Goh, K.-J., Thayaparan, T., Kunjapan, S. R., Chew, N.-K., Chua, K.-B., & Lam, S.-K. (2001). Treatment of acute Nipah encephalitis with ribavirin. *Annals of Neurology*, *49*(6), 810–813. https://doi.org/10.1002/ana.1062
79. Rockx, B., Bossart, K. N., Feldmann, F., Geisbert, J. B., Hickey, A. C., Brining, D., Callison, J., Safronetz, D., Marzi, A., Kercher, L., Long, D., Broder, C. C., Feldmann, H., & Geisbert, T. W. (2010). A novel model of lethal Hendra virus infection in African green monkeys and the effectiveness of ribavirin treatment. *Journal of Virology*, *84*(19), 9831–9839. https://doi.org/10.1128/JVI.01163-10
80. De Wit, E., Williamson, B. N., Feldmann, F., Goldin, K., Lo, M. K., Okumura, A., Lovaglio, J., Bunyan, E., Porter, D. P., Cihlar, T., Saturday, G., Spiropoulou, C. F., & Feldmann, H. (2023). Late remdesivir treatment initiation partially protects African green monkeys from lethal Nipah virus infection. *Antiviral Research*, *216*, 105658. https://doi.org/10.1016/j.antiviral.2023.105658
81. Lo, M. K., Feldmann, F., Gary, J. M., Jordan, R., Bannister, R., Cronin, J., Patel, N. R., Klena, J. D., Nichol, S. T., Cihlar, T., Zaki, S. R., Feldmann, H., Spiropoulou, C. F., & De Wit, E. (2019). Remdesivir (GS-5734) protects African green monkeys from Nipah virus challenge. *Science Translational Medicine*, *11*(494), eaau9242. https://doi.org/10.1126/scitranslmed.aau9242
82. Dawes, B. E., Kalveram, B., Ikegami, T., Juelich, T., Smith, J. K., Zhang, L., Park, A., Lee, B., Komeno, T., Furuta, Y., & Freiberg, A. N. (2018). Favipiravir (T-705) protects against Nipah virus infection in the hamster model. *Scientific Reports*, *8*(1), 7604. https://doi.org/10.1038/s41598-018-25780-3
83. Playford, E. G., Munro, T., Mahler, S. M., Elliott, S., Gerometta, M., Hoger, K. L., Jones, M. L., Griffin, P., Lynch, K. D., Carroll, H., El Saadi, D., Gilmour, M. E., Hughes, B., Hughes, K., Huang, E., De Bakker, C., Klein, R., Scher, M. G., Smith, I. L., … Broder, C. C. (2020). Safety, tolerability, pharmacokinetics, and immunogenicity of a human monoclonal antibody targeting the G glycoprotein of henipaviruses in healthy adults: A first-in-human, randomised, controlled, phase 1 study. *The Lancet Infectious Diseases*, *20*(4), 445–454. https://doi.org/10.1016/S1473-3099(19)30634-6
84. Zhu, Z., Bossart, K. N., Bishop, K. A., Crameri, G., Dimitrov, A. S., McEachern, J. A., Feng, Y., Middleton, D., Wang, L.-F., Broder, C. C., & Dimitrov, D. S. (2008). Exceptionally potent cross-reactive neutralization of Nipah and Hendra viruses by a human monoclonal antibody. *The Journal of Infectious Diseases*, *197*(6), 846–853. https://doi.org/10.1086/528801
85. Doyle, M. P., Kose, N., Borisevich, V., Binshtein, E., Amaya, M., Nagel, M., Annand, E. J., Armstrong, E., Bombardi, R., Dong, J., Schey, K. L., Broder, C. C., Zeitlin, L., Kuang, E. A., Bornholdt, Z. A., West, B. R., Geisbert, T. W., Cross, R. W., & Crowe, J. E. (2021). Cooperativity mediated by rationally selected combinations of human monoclonal antibodies targeting the *Henipavirus* receptor binding protein. *Cell Reports*, *36*(9), 109628. https://doi.org/10.1016/j.celrep.2021.109628
86. Dong, J., Cross, R. W., Doyle, M. P., Kose, N., Mousa, J. J., Annand, E. J., Borisevich, V., Agans, K. N., Sutton, R., Nargi, R., Majedi, M., Fenton, K. A., Reichard, W., Bombardi, R. G., Geisbert, T. W., & Crowe, J. E. (2020). Potent *Henipavirus* neutralization by antibodies recognizing diverse sites on Hendra and Nipah virus receptor binding protein. *Cell*, *183*(6), 1536–1550.e17. https://doi.org/10.1016/j.cell.2020.11.023
87. Mire, C. E., Chan, Y.-P., Borisevich, V., Cross, R. W., Yan, L., Agans, K. N., Dang, H. V., Veesler, D., Fenton, K. A., Geisbert, T. W., & Broder, C. C. (2020). A cross-reactive humanized monoclonal antibody targeting fusion glycoprotein function protects ferrets against lethal Nipah virus and Hendra virus infection. *The Journal of Infectious Diseases*, *221*(Supplement_4), S471–S479. https://doi.org/10.1093/infdis/jiz515
88. Al-Bari, M. A. A. (2017). Targeting endosomal acidification by chloroquine analogs as a promising strategy for the treatment of emerging viral diseases. *Pharmacology Research & Perspectives*, *5*(1), e00293. https://doi.org/10.1002/prp2.293
89. Freiberg, A. N., Worthy, M. N., Lee, B., & Holbrook, M. R. (2010). Combined chloroquine and ribavirin treatment does not prevent death in a hamster model of Nipah and Hendra virus infection. *Journal of General Virology*, *91*(3), 765–772. https://doi.org/10.1099/vir.0.017269-0
90. Pallister, J., Middleton, D., Crameri, G., Yamada, M., Klein, R., Hancock, T. J., Foord, A., Shiell, B., Michalski, W., Broder, C. C., & Wang, L.-F. (2009). Chloroquine administration does not prevent Nipah virus infection and disease in ferrets. *Journal of Virology*, *83*(22), 11979–11982. https://doi.org/10.1128/JVI.01847-09
91. Georges-Courbot, M. C., Contamin, H., Faure, C., Loth, P., Baize, S., Leyssen, P., Neyts, J., & Deubel, V. (2006). Poly(I)-poly(C12 U) but not ribavirin prevents death in a hamster model of Nipah virus infection. *Antimicrobial Agents and Chemotherapy*, *50*(5), 1768–1772. https://doi.org/10.1128/AAC.50.5.1768-1772.2006

92. Chattu, V., Kumar, R., Kumary, S., Kajal, F., & David, J. (2018). Nipah virus epidemic in southern India and emphasizing "One health" approach to ensure global health security. *Journal of Family Medicine and Primary Care*, *7*(2), 275. https://doi.org/10.4103/jfmpc.jfmpc_137_18
93. Lu, M., Yao, Y., Zhang, X., Liu, H., Gao, G., Peng, Y., Chen, M., Zhao, J., Zhang, X., Yin, C., Guo, W., Yang, P., Hu, X., Rao, J., Li, E., Chen, T., Chiu, S., Wong, G., Yuan, Z., … Shan, C. (2023). Both chimpanzee adenovirus-vectored and DNA vaccines induced long-term immunity against Nipah virus infection. *Npj Vaccines*, *8*(1), 170. https://doi.org/10.1038/s41541-023-00762-3
94. Walpita, P., Cong, Y., Jahrling, P. B., Rojas, O., Postnikova, E., Yu, S., Johns, L., & Holbrook, M. R. (2017). A VLP-based vaccine provides complete protection against Nipah virus challenge following multiple-dose or single-dose vaccination schedules in a hamster model. *Npj Vaccines*, *2*(1), 21. https://doi.org/10.1038/s41541-017-0023-7
95. Walpita, P., Barr, J., Sherman, M., Basler, C. F., & Wang, L. (2011). Vaccine potential of Nipah virus-like particles. *PLOS One*, *6*(4), e18437. https://doi.org/10.1371/journal.pone.0018437
96. Pedrera, M., Macchi, F., McLean, R. K., Franceschi, V., Thakur, N., Russo, L., Medfai, L., Todd, S., Tchilian, E. Z., Audonnet, J.-C., Chappell, K., Isaacs, A., Watterson, D., Young, P. R., Marsh, G. A., Bailey, D., Graham, S. P., & Donofrio, G. (2020). Bovine herpesvirus-4-vectored delivery of Nipah virus glycoproteins enhances T cell immunogenicity in pigs. *Vaccines*, *8*(1), 115. https://doi.org/10.3390/vaccines8010115
97. Volz, A., & Sutter, G. (2017). Modified Vaccinia Virus Ankara. In *Advances in Virus Research* (Vol. 97, pp. 187–243). Elsevier. https://doi.org/10.1016/bs.aivir.2016.07.001
98. Shuai, L., Ge, J., Wen, Z., Wang, J., Wang, X., & Bu, Z. (2020). Immune responses in mice and pigs after oral vaccination with rabies virus vectored Nipah disease vaccines. *Veterinary Microbiology*, *241*, 108549. https://doi.org/10.1016/j.vetmic.2019.108549
99. Keshwara, R., Shiels, T., Postnikova, E., Kurup, D., Wirblich, C., Johnson, R. F., & Schnell, M. J. (2019). Rabies-based vaccine induces potent immune responses against Nipah virus. *Npj Vaccines*, *4*(1), 15. https://doi.org/10.1038/s41541-019-0109-5
100. Li, Z., Zhu, Y., Yan, F., Jin, H., Wang, Q., Zhao, Y., Feng, N., Wang, T., Li, N., Yang, S., Xia, X., & Cong, Y. (2023). Inactivated recombinant rabies virus displaying the Nipah virus envelope glycoproteins induces systemic immune responses in mice. *Vaccines*, *11*(12), 1758. https://doi.org/10.3390/vaccines11121758
101. Ploquin, A., Szécsi, J., Mathieu, C., Guillaume, V., Barateau, V., Ong, K. C., Wong, K. T., Cosset, F.-L., Horvat, B., & Salvetti, A. (2013). Protection against *Henipavirus* infection by use of recombinant adeno-associated virus–vector vaccines. *The Journal of Infectious Diseases*, *207*(3), 469–478. https://doi.org/10.1093/infdis/jis699
102. Van Doremalen, N., Lambe, T., Sebastian, S., Bushmaker, T., Fischer, R., Feldmann, F., Haddock, E., Letko, M., Avanzato, V. A., Rissanen, I., LaCasse, R., Scott, D., Bowden, T. A., Gilbert, S., & Munster, V. (2019). A single-dose ChAdOx1-vectored vaccine provides complete protection against Nipah Bangladesh and Malaysia in Syrian golden hamsters. *PLOS Neglected Tropical Diseases*, *13*(6), e0007462. https://doi.org/10.1371/journal.pntd.0007462
103. Kong, D., Wen, Z., Su, H., Ge, J., Chen, W., Wang, X., Wu, C., Yang, C., Chen, H., & Bu, Z. (2012). Newcastle disease virus-vectored Nipah encephalitis vaccines induce B and T cell responses in mice and long-lasting neutralizing antibodies in pigs. *Virology*, *432*(2), 327–335. https://doi.org/10.1016/j.virol.2012.06.001
104. Loomis, R. J., DiPiazza, A. T., Falcone, S., Ruckwardt, T. J., Morabito, K. M., Abiona, O. M., Chang, L. A., Caringal, R. T., Presnyak, V., Narayanan, E., Tsybovsky, Y., Nair, D., Hutchinson, G. B., Stewart-Jones, G. B. E., Kueltzo, L. A., Himansu, S., Mascola, J. R., Carfi, A., & Graham, B. S. (2021). Chimeric fusion (F) and attachment (G) glycoprotein antigen delivery by mRNA as a candidate Nipah vaccine. *Frontiers in Immunology*, *12*, 772864. https://doi.org/10.3389/fimmu.2021.772864

5 Hantaviruses
Bridging Species, Crossing Borders

Rawan H. Alsharedeh

5.1 INTRODUCTION

Hantaviruses encompass enveloped single-stranded ribonucleic acid virus (ssRNA), from the *Bunyavirales* order. Rodents (the main Hantavirus carrier) primarily transmit this virus, which is also a form of zoonotic pathogen. Worldwide, about 200000 cases of Hantavirus would be reported each year [1, 2].

Hantavirus has been classed as the A category pathogen by the National Institute of Allergy and Infectious Diseases (NIAID). This virus, according to NIAID, in addition to being easily transmitted, increases morbidity and mortality rates, and no known approved therapeutic treatment for it, is indeed a threat to the human health. Hantavirus causes the development of two main hyperinflammatory syndromes: hemorrhagic fever renal syndrome (HFRS) and nephropathia epidemica (NE) as its milder form, and Hantavirus pulmonary syndrome/Hantavirus cardiopulmonary syndrome (HPS/HCPS). HFRS patients are likely to experience hemorrhagic fever, and renal complications, while HPS patients are mainly experience severe respiratory and cardiovascular complications. In diagnosing the Hantavirus infection, serology test is often performed. At present, specific Food and Drug Administration (FDA) approved treatment or preventive vaccine against Hantavirus diseases are still unavailable. Thus far, those infected are treated based on symptoms [2–8].

5.2 HANTAVIRUS ETIOLOGY

Hantavirus falls into two classes: the old world Hantavirus (OWH) and the new world Hantavirus (NWH). OWH was first discovered during the Korean War between 1951 and 1953, when the virus infected more than 3000 Korean soldiers. These soldiers were showing a syndrome observed with hemorrhagic fever and kidney failure. Then, in 1978, the etiologic agent for the syndrome was isolated form the striped field mouse as the source of this virus. Most of the reported Hantavirus cases were near Hantaan River in South Korea, hence the name Hantavirus and HFRS was the described condition. In 1980s, a milder HFRS known as NE was reported. Bank voles, which are a type of mouse, have been identified as the carrier of the disease. This case was actually reported earlier (in 1930s) in Scandinavia. Another Hantavirus cases were reported in 1990s in Slovenia and Dobrava–Belgrade virus was identified as the causative agent and the field mouse was identified as the carrier of the virus [3–6].

Concerning the NWH, the first reported case was in 1993 in the Four Corners region in the United States. A number of persons from the Navajo nation were suffered from an acute respiratory disease, currently known as HPS and Sin Nombre virus carried by deer mouse was the identified etiological causative agent [6, 7].

5.3 HANTAVIRUS PHYLOGENY

Given the persistent obtainability of the new sequenced viral genomic data, the taxonomy classifications would often be revised and updated. For instance, the viral family *Bunyaviridae* – the early known Hantavirus family, the largest family of ssRNA viruses with more than 300 species – was modified to be the viral order *Bunyavirales* in 2017. Additionally, the viral genus *Hantavirus* was

DOI: 10.1201/9781003607816-5

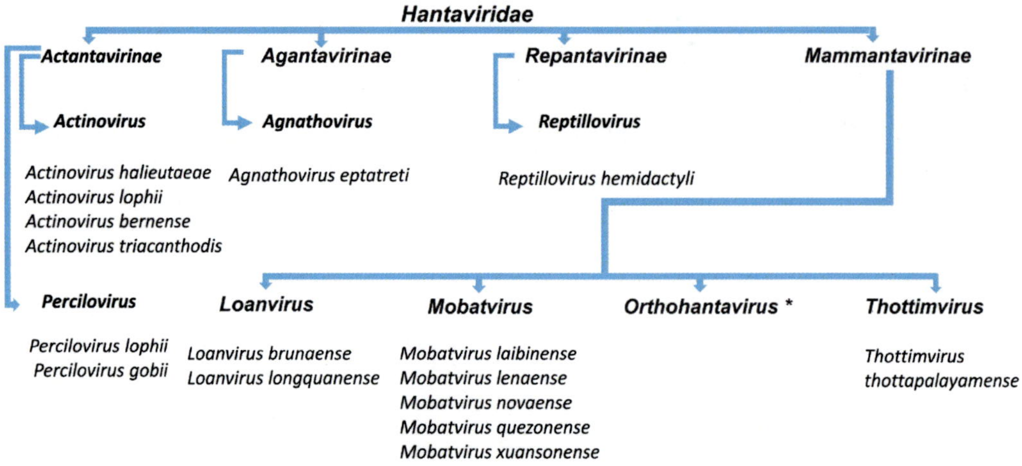

FIGURE 5.1 The current taxonomy of *Hantaviridae* family. The flowchart shows *Hantaviridae* classification into 4 subfamilies, 7 genera, and 53 species.

modified to be the viral family *Hantaviridae*. In 2018, the *Hantaviridae* family only encompassed the *Orthohantavirus* genus, but the following year (2019), three new genera were added. They are *Loanvirus*, *Mobatvirus*, and *Thottimvirus*. At present, the *Hantaviridae* family comprises 4 subfamilies, 7 genera, and 53 species (Figure 5.1). Multiple Hantaviruses are infective to humans and can lead to HFRS, NE, and HPS/HCPS. For example, Hantaan virus (*Orthohantavirus hantanense*), Bayou virus (*Orthohantavirus bayoui*), Black Creek Canal virus (*Orthohantavirus nigrorivense*), Choclo virus (*Orthohantavirus chocloense*), Amur virus (*Orthohantavirus hantanense*), Dobrava virus (*Orthohantavirus dobravaense*), Laguna Negra virus (*Orthohantavirus negraense*), Hokkaido virus *(Orthohantavirus puumalaense*), Gōu virus (*Orthohantavirus seoulense*), New York virus (*Orthohantavirus sinnombreense*), Adler virus (*Orthohantavirus tulaense*), Seoul virus (*Orthohantavirus seoulense*), Sin Nombre virus (*Orthohantavirus sinnombreense*), and Andes virus (*Orthohantavirus andesense*) [6–11].

5.4 HANTAVIRUS MORPHOLOGY

Hantaviruses encompass an enveloped tri-segmented negative ssRNA virus, normally with spherical and spiked virions with a quasi-symmetric icosahedral lattice of 80–160 nm in diameter. The viral envelope has lipid bilayer with approximately 5 nm in thickness. The surface of the viral envelope is spiky, and the spikes comprise transmembrane glycoproteins Gn and Gc in the form of heterodimers, which is assemble into tetrameric spikes with length of 10 nm away from the lipid bilayer. The genomes of Hantavirus comprise the designated segments: L (large), M (medium), and S (small), and in total, there are approximately 10000–15000 nucleotides packed into helical nucleocapsids (see Figure 5.2). These genomic segments are encoded for RNA-dependent RNA polymerase (RdRp), nucleoprotein N, and transmembrane glycoproteins Gn and Gc [1, 4, 11–14].

The three segments are as described below:

1. The L-segment, which comprises 6500 nucleotides, encoded for RdRp biosynthesis.
2. The M-segment, which comprises 3613–3707 nucleotides, encoded for the transmembrane glycoprotein precursor, that is co-translate to generate Gn and Gc glycoprotein.
3. The S-segment, which comprises 1700–2100 nucleotides, encoded for the N protein that synthesizes the nucleocapsid of the virion. However, there are some Hantaviruses whose S segment also encodes a nonstructural protein.

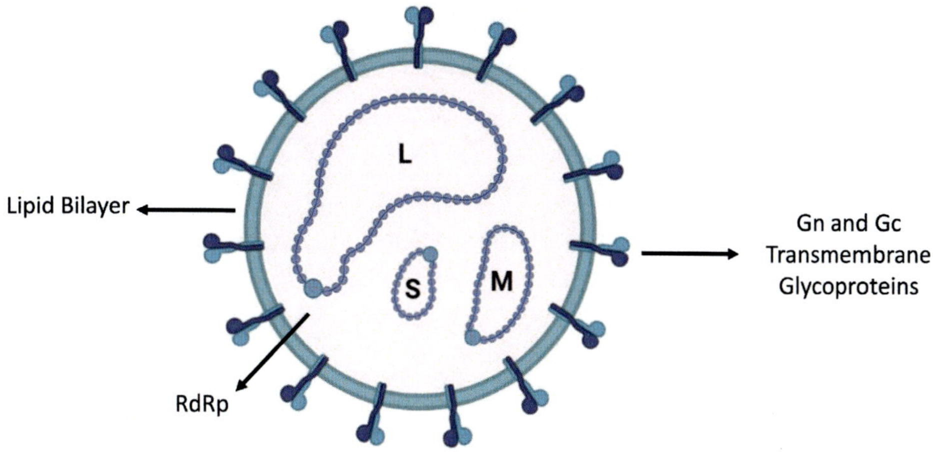

FIGURE 5.2 Schematic representation of Hantavirus morphology. The ssRNA segments S, M, and L and their encoding proteins. (Image generated by www.biorender.com.)

5.5 HANTAVIRUS EPIDEMIOLOGY

Since the last several decades, knowledge on Hantavirus infection has globally improved. Somehow, the outbreaks of Hantavirus have been on the rise, both in intensity and scale. Hantaviruses are now a global endemic. As previously mentioned, there are OWH and NWH. Hantaviruses are classed according to the global geographical spreading of the viruses (Figure 5.3). The most important OWH include Hantaan virus, Puumala virus, Seoul virus, and Dobrava virus, and are primarily found infecting people in Europe and Asia. These viruses are mainly circulated by rodent and insectivorous species. The major syndrome caused by these viruses is HFRS. Meanwhile, Hantaan virus mostly infect people in the Far East countries, like China,

FIGURE 5.3 Geographical distribution of Hantavirus etiological agents. Each *Hantavirus* etiological agents are marked in different color.

TABLE 5.1
Geographic Distribution of Important Pathogenic OWH

OWH genotypes	Geographic distribution	Associated disease	Comments
Hantaan virus	China, Russia, Korea, Central Europe	HFRS	Main Asian virus *Mortality rate*: 10–15%
Seoul virus	Globally distributed	HFRS	*Mortality rate*: 1–5%
Puumala virus	Europe, Russia, Americas	HFRS/NE/HPS	Main Hantavirus across Europe *Mortality rate*: 0.1–0.4%
Dobrava virus	Europe	HFRS	Main Hantavirus in Greece *Mortality rates*: 7–12%
Tula virus	Europe, Russia	HFRS	
Amur/Soochong virus	Far East Russia, China, Korea	HFRS	
Luxi virus	China	HFRS	
Thailand virus	Southeast Asia	HFRS	
Saaremaa virus	Europe	HFRS/NE	
Sangassou virus	Africa	HFRS	First African virus, spillover infections to bats

South Korea, and Russia. Puumala is the main type of Hantavirus infecting the people in Europe, France, Germany, Sweden, and the Balkans, with principal outbreak in Germany in 2012 involving approximately 3000 people and with the highest incidence in Finland with approximately 1000–3000 serological diagnoses annually. Infection by Seoul virus which is carried by the brown rat (*Rattus norvegicus*) is the only worldwide spread Hantavirus. This owes to the migratory nature of these rats, as well the brown rats can be found worldwide. As for Dobrava–Belgrade virus, the majorly reported cases are in the southeastern Europe and in Greece. This Hantavirus considered lethal with 12% mortality rate. The main OWH genotypes and their geographic distribution can be viewed in Table 5.1 [4, 5, 8, 12, 14–23].

The NWH are primarily found infecting people in Americas and often result in the development of HPS/HCPS with mortality rates reaching to 60% based on the outbreaks. Each year, these viruses would infect a number of hundred people. Sin Nombre virus was first discovered in 1993 in the Four Corners region in the southwestern United States. Those infected by this virus would develop an acute pulmonary distress syndrome. At first, the syndrome was called the "unexplained adult respiratory syndrome." During an outbreak with 23 infected individuals, 12 did not survive. In another outbreak in 2017 at the west of the Mississippi River mostly, 697 individuals were reported being infected. The main NWH genotypes and their geographic distribution can be viewed in Table 5.2 [5, 6, 8, 14, 24–28].

5.6 HANTAVIRUS HOSTS AND RESERVOIRS

Hantavirus infects various hosts based on their subfamilies. For instance, fish and reptiles are hosts for the three subfamilies of Hantavirus, namely *Acanthavirinae*, *Agantavirinae*, and *Repantavirinae*. Meanwhile, one subfamily of Hantavirus namely *Mammantavirinae* can be hosted by rodents, moles, shrews and bats. Worldwide, 53 Hantavirus types have been identified thus far, and many of these viruses cause HFRS, NE, and HPS/HCPS when they infect humans, as they are the dead-end hosts for the virus. The main carrier of Hantavirus is rodents. However, there are instances where the virus is spread by insectivore and bats. The natural reservoir and ecology of Hantavirus are connected to the geographical distribution. For the OWH, the main reservoirs are four rodent genera namely *Myodes*, *Microtus*, *Apodemus*, and *Rattus*, alongside two insectivore families namely *Soridae* and *Talpidae*. For the NWH, the main reservoir is the rodent subfamily S*igmodontines*. However, in the Mediterranean area, the main reservoir for the NWH is the bank vole [4–6, 9, 29, 30].

TABLE 5.2
Geographic Distribution of Important Pathogenic NWH

NWH genotypes	Geographic distribution	Associated disease	Comments
Sin Nombre Virus	Canada, United States	HPS/HCPS	Main North American virus *Mortality rate*: 40–50%
Andes Virus	Argentina, Chile, Uruguay	HPS/HCPS	Main South American virus Only Hantavirus with a person-to-person transmission *Mortality rate*: 43–54%
Black Creek Canal Virus	Florida	HPS/HCPS	*Mortality rate*: >40%
Araraquara Virus	Brazil	HPS/HCPS	Mortality rate: 50%
Laguna Negra Virus	Paraguay, Bolivia	HPS/HCPS	First South American Hantavirus *Mortality rate*: 12–15%
Choclo Virus	Panama	HPS/HCPS	*Mortality rate*: 12–15%
New York Virus	Eastern United States	HPS/HCPS	
Monongahela Virus	Eastern United States	HPS/HCPS	
Bayou Virus	Southeastern United States	HPS/HCPS	
Oran Virus	Argentina	HPS/HCPS	
Rio Mamore Virus	Peru	HPS/HCPS	
Lechiguanas Virus	Argentina	HPS/HCPS	
Anajatuba virus	South America	HPS/HCPS	
Castelo dos Sonhos virus	South America	HPS/HCPS	
Bermejo virus	Argentina, Bolivia	HPS/HCPS	

Since the time of Hantavirus discovery until now, substantial improvement has been made on the knowledge about the potential mechanisms of Hantavirus transmission. Form the Hantavirus seroprevalence studies, results highlight that viral complex transmission dynamics is highly associated with the exposure and the geographical distribution. Now it is known that Hantaviruses are generally transmitted horizontally, although it can also be transmitted vertically (Figure 5.4). Hantaviruses spread between chronically infected host reservoirs, namely rodents, insectivore and bats. However, most times, the virus does not cause any disease to these hosts, and so, these hosts become asymptomatic carriers of the virus. There are a number of ways that prone animals could contract Hantavirus disease: 1) through the aerosols and droplets from the saliva, urine and feces of the infected animals, 2) through scratches and bites of the infected animals, 3) through fecal–oral route (consumption of contaminated food), 4) from the mother

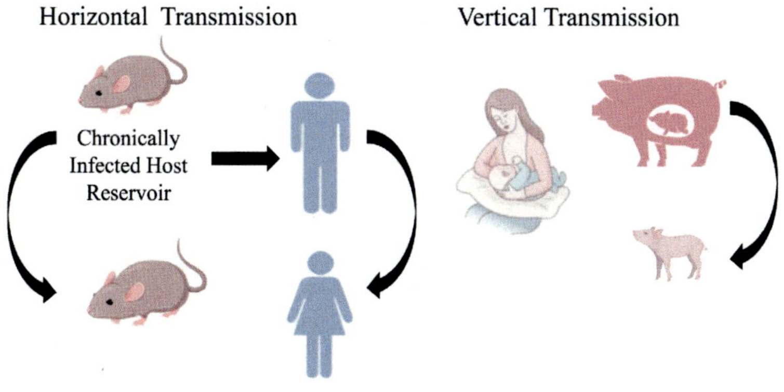

FIGURE 5.4 Illustration of the typical horizontal and vertical *Hantavirus* transmission.

of the animal (e.g., pregnant sow) to offspring during pregnancy, and 5) through mating. For humans, Hantavirus infection can occur through inhalation of aerosols or droplets from infected carrier, consumption of contaminated food, and through scratches and bites from the carrier. It is noteworthy to indicate that Hantavirus infection is regarded as orthozoonosis, means a direct zoonotic disease in which the recipients infected by the direct exposure to the aerosols or droplets of the carrier, or infrequently by the bites and scratches of the infected carrier [5, 8, 31–35].

Generally, humans contract Hantavirus disease from rodents, but Andes virus – which is also a type of Hantavirus – can also infect humans from another human. In other words, Andes virus can be spread from human-to-human, usually through close contact with an infected individual. The first case involving Andes virus transmission from human to human was first documented in 1996 (September 22 and December 5). It was an outbreak involving 19 cases of HPS in El Bolsón, Bariloche, and Esquel in southern Argentina. Five of the infected individuals were physicians, three of which were directly involved in the direct health care of HPS cases. This outbreak strongly suggested person-to-person transmission of HPS, considering the low population of rodent in these towns [36, 37]. Another example of human-to-human transmission was during the outbreak 2018–2019 that took place in Chubut Province, Argentina, where one infected individual had led to 33 other individuals to be infected as chains of transmissions. Additionally, several reports from Chile were claiming the ability of Andes virus to be transmitted from one individual to the other. The virus may be secreted into human saliva and spread through the respiratory tract through airborne droplets and also by sexual contact. Studies have also shown the transmission of Andes virus to infants via breastfeeding [38–40].

The main Hantaviruses and their specific rodent vector are listed below [6, 15, 18, 24, 41–50].

- *Puumala virus*: Bank vole (*Myodes glareolus*)
- *Dobrava–Belgrade virus*: Field mouse (*Apodemus flavicollis*)
- *Kurkino*: Striped field mouse (*Apodemus agrarius agrarius*)
- *Seoul virus*: Rats; mainly the brown rat (*Rattus norvegicus*) and the black rat (*Rattus rattus*)
- *Luxi virus*: Yunnan red-backed vole (*Eothenomys miletus*)
- *Thailand virus*: Greater bandicoot rat (*Bandicota indica*)
- *Saaremaa virus*: Striped field mouse (*Apodemus agrarius*)
- *Sangassou virus*: African Wood Mouse (*Hylomyscus simus*)
- *Amur/Soochong virus*: Korean field mouse (*Apodemus peninsulae*)
- *Tula virus*: Common vole (*Microtus arvalis*)
- *Hantaan virus*: Striped field mouse (*Apodemus agrarius*)
- *Sin Nombre virus*: Deer mouse (*Peromyscus maniculatus*)
- *New York virus*: White-footed mice (*Peromyscus leucopus*)
- *Monongahela virus*: Deer mouse (*Peromyscus maniculatus*)
- *Bayou virus*: Marsh rice rat (*Oryzomys palustris*)
- *Black Creek Canal virus*: Hispid cotton rats (*Sigmodon hispidus*)
- *Laguna Negra virus*: The small vesper mouse (*Calomys laucha*) and the large vesper mouse (*Calomys callosus*)
- *Andes virus*: Long-tailed colilargo (*Oligoryzomys longicaudatus*)
- *Oran virus*: Oran longicaudatus
- *Choclo virus*: Oligoryzomys fulvescens (*Pygmy rice rat*)
- *Rio Mamore virus*: Bristly mouse (*Neacomys spinosus*)
- *Lechiguanas virus*: Oligoryzomys fulvescens (*Pygmy rice rat*)
- *Araraquara virus*: Hairy-tailed bolo mouse (*Necromys lasiurus*)

5.7 CLINICAL ASPECTS OF HANTAVIRUS

Hantaviruses have been recognized as etiologic agents of a number of life threatening syndromes in humans, with a highly variable mortality rates subjected to the considered Hantavirus type, but it can reach up to 60% as in the case of Chile outbreak in 1997. For HFRS in the old-world category;

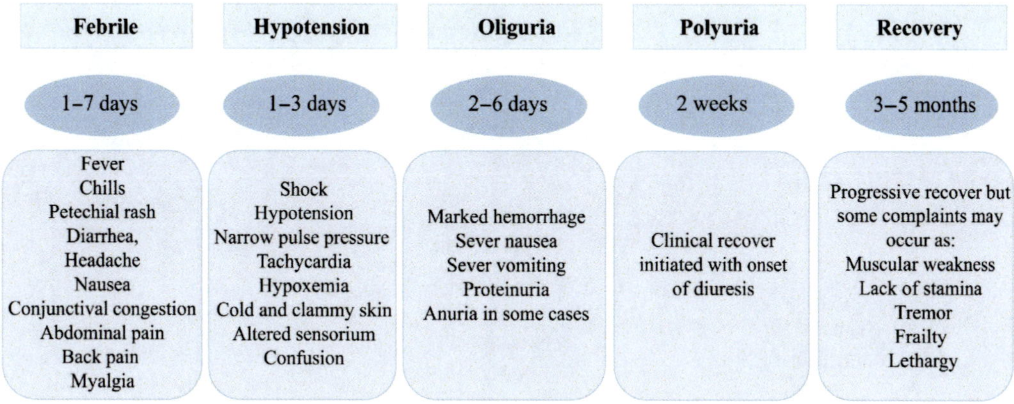

FIGURE 5.5 Clinical course and symptoms of the five phases of HFRS.

Hantaan virus, Dobrava virus, Seoul virus, and Puumala are the primary causative agents. These viruses are mainly found in Asia and Europe. In Europe, these viruses can result in the development of a milder form the NE – Puumala virus is the main culprit. This infection is usually acute and mild with a low mortality rate 0.1–0.4%; however, some patients may suffer from a more severe cases and required intensive care unit admission. For HPS/HCPS in the NWH category, Sin Nombre virus and Andes virus are the main causative agents, and are primarily occurred in the Americas (North, South and Central America) [8, 27, 51–58].

HFRS and NE conditions mostly presents with non-specific symptoms similar to flu-like symptoms and the illness can be progress in five phases: febrile, hypotension and shock, oliguria, polyuria, and recovery (Figure 5.5). The affliction usually happens after a period of incubation – the incubation period takes from 2 to 4 weeks. Acute kidney injury is the hallmark feature. In severe cases the phases are demarcated; however they may be overlapped or not presented in mild to moderate cases. The five phases of HFRS are as described below [1, 8, 59–63]:

1. *Febrile phase*: generally characterized by chills, high fever, headache, malaise, acute myopia, low back pain, gastrointestinal symptoms, hemorrhagic symptoms, hematuria, hemoptysis, conjunctival congestion, as well intracranial hemorrhage. In severe cases, disseminated intravascular coagulation (DIC) may occur.
2. *Hypotension phase*: during this phase, the patient experiences sudden hypotension and shock. The hypotension and shock are caused by microvascular leakage and thrombocytopenia, which may lead to tachycardia and hypoxemia, and sudden death.
3. *Oliguria phase*: this phase results from renal failure. The two major symptoms are oliguria (typically <400 mL/day) and proteinuria, for 50% of all deaths from HFRS are due to hypervolemia and renal impairment in this phase.
4. *Polyuria phase*: in this phase the increase in the urine output from 3 to 6 L/day usually occurs and the gradual recovery of the kidneys take place in this phase.
5. *Recovery phase*: during this phase, the symptoms begin to improve. After the recovery phase patients may present with chronic renal failure, hypertension or normal functioning kidneys. Aside from the common HFRS related renal disease, some patients may develop extra renal symptoms (cholecystitis, pericarditis, encephalitis), in addition to acute respiratory distress syndrome.

Affliction of HPS/HCPS involves three major phases namely the prodromal phase, the cardiopulmonary phase and the recovery phase. Affliction of HPS/HCPS usually occurs after the incubation period which would take between 6 and 39 days (Figure 5.6). As the virus primarily affects the lungs, patient would develop acute respiratory distress syndrome in addition to microvascular leakage [1, 4, 6–8, 64, 65].

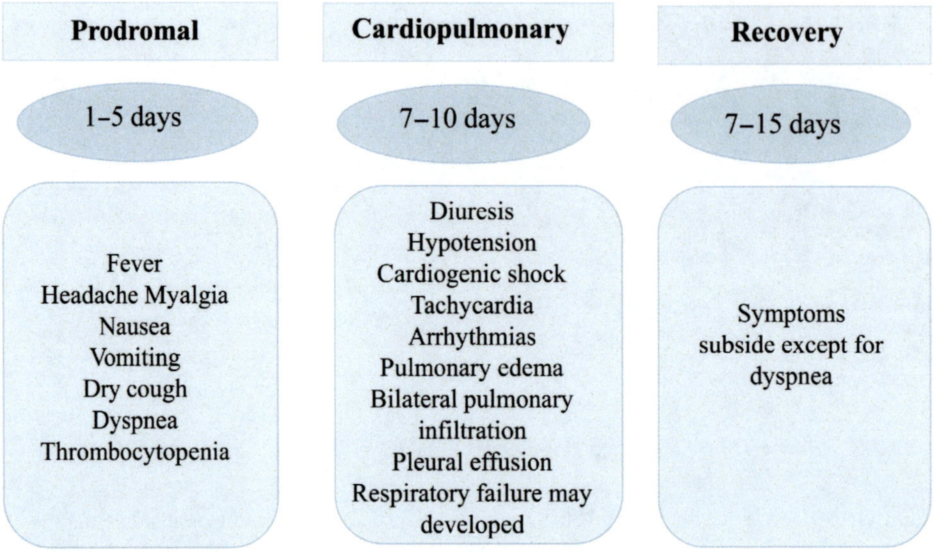

FIGURE 5.6 Clinical course and symptoms of the three phases of HPS.

The phases of HPS/HCPS are as described below:

1. *Prodromal phase*: during this phase, patient develops symptoms such as myalgia, headache, fever, nausea, vomiting, and dry cough, dyspnea, and fast progressing bilateral interstitial edema, along with thrombocytopenia.
2. *Cardiopulmonary phase*: this phase encompasses a few days of patient experiencing cardiopulmonary malfunction, including tachycardia, arrhythmias, and cardiogenic shock – death can also occur during this phase. Pulmonary capillary leakage can occur during this phase, and can cause respiratory failure, hypotension, pulmonary edema, bilateral pulmonary infiltration, in addition to pleural effusion. Some patients also develop acute kidney injury and proteinuria. Mechanical ventilation is often necessitated. However, mortality rates can reaching to 60%, as opposed to 15% for HFRS, and 1% for NE according to the outbreaks, even with appropriate care.
3. *Recovery phase*: during this phase, all symptoms diminish. However, dyspnea can continue for up to 1–2 years.

The distinguishing features of HFRS, NE and HPS/HCPS are provided in Table 5.3. As can be observed they shared a common feature, but each attacks different organ.

TABLE 5.3
Comparison of HFRS, NE and HPS/HCPS

	HFRS	NE	HCPS/HPS
Clinical phases	Febrile Hypotensive Oliguria Polyuria Recovery	Febrile Hypotensive Oliguria Polyuria Recovery	Prodromal, Cardiopulmonary Recovery
Main target organ	Kidneys	Kidneys	Lungs
Highest mortality rate	15%	1.0%	60%

5.8 TREATMENT

Thus far, there is no specific safe and effective pharmacological antiviral therapeutic drug or vaccine approved by FDA to counter HFRS and HPS/HCPS. In fact, the only present alternative for Hantavirus syndromes management is symptomatic and depends majorly on high quality supportive care to increase the survival rates. Still, there are a number of antivirals candidate for Hantavirus syndromes management. Fapilavir, ribavirin, lactoferrin, and vandetanib are among them. However, the efficacy of these antivirals were examined in *in vitro* and *in vivo* studies, and so, human clinical trials need to be conducted to achieve sufficient knowledge on their clinical efficacy and safety in human. Still, considering the lethalness of Hantavirus, mortality rates are high even with best treatment. Also, as the virus affects various organs, interprofessional team comprising experts in infectious disease, neurology, hematology, pulmonology, and intensivists, respiratory care therapists, and pharmacists is required to achieve the best management of the disease [17, 26, 66, 67].

The main treatment measures [1, 6, 8, 19, 26] are:

- Patients must directly admitted to an emergency department or intensive care unit based on his health status to have careful evaluation and caring, if the preliminary signs and symptoms show high potential of Hantavirus infection.
- Antipyretics and analgesics must use as needed.
- The volume status and cardiac function must carefully monitored.
- Fluid and electrolyte must balance so that patient would have stable blood pressure. As Hantavirus can cause capillary leakage, fluids are carefully administrated.
- Inotropic and vasopressor supports must be provided as necessary.
- Hypoxemic patients must give oxygen, because severe HPS can quickly lead to respiratory failure; equipment and supplies for intubation and mechanical ventilation must be available at all times.
- For advanced HPS cases, extracorporeal membrane oxygenation may be used as a rescue therapy – the earlier intervention the higher success rate up to more than two third of the patients.
- In some patients, bleeding and DIC status can be controlled by blood transfusion.
- Other patients may require intermittent extracorporeal blood purification to manage uremia.
- Also the constant renal replacement management is a must to control the multi-organ pulmonary edema.
- In patients with acute thrombocytopenia; platelet transfusions can be performed.
- After these measures if patients show improvement in the first few days, they may be extubated within the first week, without suffering from major long-term complications. However, dyspnea, fatigue, and myalgias may persist. Still, despite the appropriate treatment, about one-third of HSP patients will perish in the first two days after admission.

As mentioned earlier, there are currently no FDA-approved vaccines or specific treatments for Hantavirus infections. As also mentioned earlier, Fapilavir, ibavirin, lactoferrin, and vandetanib are among the antivirals that are showing potential in treating HFRS, NE and HPS/HCPS. Studies have presented with similar potent activity of ribavirin and favipiravir in Hantavirus management *in vitro* and *in vivo*. Both drugs have been administered as prophylactic or postexposure treatment, or in the early infection. Notably, ribavirin when used for HFRS treatment with high doses may be lead to hemolytic anemia. On the other hand, favipiravir is less likely to cause hemolytic anemia, and so, it is better than ribavirin in this regard. However, the demonstrated efficacy of ribavirin is enhanced when combined with low dose of favipiravir, providing a superior treatment option than

TABLE 5.4
Potential Antiviral and Immunotherapies for Hantavirus Infections

Drug candidate	Function	Target	Disease
Ribavirin/Purine nucleoside analog	Blocking viral replication	RdRp	HFRS/HPS
Lactoferrin	Block viral entry	Viral GP	HFRS
Favipiravir/Pyrazine derivatives	Block viral entry	RdRp	HPS
ETAR/Nucleoside analog	Inhibit viral entry	RdRp	HFRS/HPS
Icatibant	Improve vascular function	Bradykinin B2 receptor	HFRS
Corticosteroid	Rebuild immune homeostasis	Immunotherapy	HFRS/HPS
Vandetanib/Tyrosine kinase inhibitor	Improve vascular function	VEGF/vascular function	HPS
Human immune sera/Human polyclonal antibody	Block viral entry	Viral GP	HPS
MIB22 and JL16/Human monoclonal antibodies	Block viral entry	Viral GP	HPS
ADI-42898/Human monoclonal antibodies	Block viral entry	Viral GP	HPS
Domain III and stem peptides	Block viral entry	Gc glycoprotein	HPS
LVRNLAWC/CQATTARNC/Cyclic nonapeptides	Block viral entry	Host receptors	HPS
TNF-α/Pro-inflammatory cytokines	Increase systemic toxicity	Vascular function	HFRS/HPS
RANTES, IP-10, IL-6, and IL-8/Pro-inflammatory chemokines	Immunomodulators	Microvascular endothelium	HFRS

ribavirin or favipiravir alone. However, modern day antiviral methods employ small interfering RNA (siRNA) and other immune-based treatments are also candidate for Hantavirus treatment. Efficacy of siRNA has been reported *in vitro* and *in vivo* on a vast range of viruses. Yet, siRNAs encounter some challenges as the low biological stability. Other studies were demonstrating that immunotherapy that employs neutralizing antibodies created from Hantavirus might increase the Hantavirus' survival. Additionally, in animal models, monoclonal and polyclonal antibodies seem to be effective against Hantavirus infections. Trials have been performed involving potential antivirals and immunotherapies against HFRS and HPS in conjunction with supportive care. Several examples of potential antiviral and immunotherapies for Hantavirus infections are listed in Table 5.4 [1, 16, 23, 25, 68–71].

5.9 DIAGNOSIS OF HANTAVIRUS INFECTIONS

Manifestation of Hantavirus infection is comparable to acute respiratory distress syndrome and the clinical signs and symptoms of many other diseases and Hantavirus are common, thus it is normally difficult to differentiate. For this reason, a high-index knowledge is necessary, so that patient would have the greatest survival. However, diagnosis is not so simple and challenging. It is crucial that Hantavirus infections are affirmed through clinical signs and symptoms, epidemiological data, as well as the laboratory findings. Among the main diagnostics [6, 27, 53, 72–74]:

1. Chest x-ray to show if there is bilateral pulmonary edema. About 33% patients would have this in the initial radiograph, but almost all admitted patients will develop interstitial edema two days after admission, while nearly 67% of admitted patients will experience pleural effusions, bibasilar opacities or perihilar opacities.
2. Patient may manifest thrombocytopenia on their blood count.
3. Circulating immunoblasts and elevated hematocrit are among other vital laboratory findings.

4. Peripheral blood smear on admitted patient may show myelocytes, metamyelocytes, and promyelocytes with severe thrombocytopenia and hypocarbia.
5. Patient may develop hyponatremia in addition to somewhat prolonged activated partial thromboplastin time, a diminished protein level, slight increase in the level of low-density lipoprotein, as well as microscopic hematuria.
6. Immunofluorescent or immunoblot assay may be used in performing the diagnosis, such as the enzyme-linked immunosorbent assay.
7. For Hantavirus antibodies identification (IgM and IgG); serological examinations are the most commonly used method.
8. Other appropriate method is quantitative polymerase chain reaction.

5.10 PREVENTION AND CONTROL OF HANTAVIRUSES

Preventive measures of this disease is crucial, and public health officials are the ones with the best abilities to impose the measures required. Education and actions are necessary in preventing and controlling this disease. Having deep understanding of the virus and taking the appropriate actions would help block the transmission of the virus and protect the humans. Areas at risk and factors contributing to the transmission need to be identified through methods such as epidemiological surveillance and analysis. Among the measures to mitigate Hantavirus risk are as follows [1, 6, 8, 23, 75, 76]:

1. Minimizing human-rodent contact – this is the best preventative approach – because rodents are the main carrier of Hantavirus.
2. Disinfecting the living surroundings of human, this can prevent risks like contamination to food and other items from the rodent source.
3. Aerating homes and letting the sun light to enter into the surrounding - ultraviolet light particularly, can neutralize the virus and curb the spread of the infection.
4. Taking protective measures in situations where human-rodent contact cannot be avoided. Those involved must wear protective gear.

Thus far, there are no globally approved vaccines for Hantavirus, with the exception of mainland China and the Republic of Korea. Long-term results are yet to be available as well. There are indeed some vaccine candidates against Hantavirus, and some have been trialed on humans. Hantavax is one of them; it is an inactivated vaccine was commercial produced in 1990 to inhibit Hantavirus infection in South Korea. Clinical trial results showed good endurance of the vaccine in human volunteers. Another vaccine, a bivalent inactivated vaccine, was introduced to the market in 1994 in mainland China against Hantaan virus and Seoul virus. In 2005, the vaccine was approved for use in the country. Yearly, about 2 million doses of HFRS inactivated bivalent vaccine are used in mainland China, and the vaccine showed vaccine-induced immunity for up to 33 months. Still, safer vaccine with more effectiveness against Hantavirus is needed [1, 6, 8, 23, 75, 76].

As mentioned earlier various Hantavirus candidate are under investigation, however eradicating Hantavirus associated infections is fraught with difficulties. Such as the variation of Hantavirus genetics, adaptability and the severity of the associated syndromes, which complicates the advancement of general vaccines and treatments [77].

REFERENCES

1. Afzal, S., Ali, L., Batool, A., Afzal, M., Kanwal, N., Hassan, M., Safdar, M., Ahmad, A., & Yang, J. (2023). Hantavirus: An overview and advancements in therapeutic approaches for infection. *Frontiers in Microbiology*, *14*, 1233433. https://doi.org/10.3389/fmicb.2023.1233433

2. Vial, P. A., Ferrés, M., Vial, C., Klingström, J., Ahlm, C., López, R., Le Corre, N., & Mertz, G. J. (2023). Hantavirus in humans: A review of clinical aspects and management. *The Lancet. Infectious Diseases*, *23*(9), e371–e382. https://doi.org/10.1016/S1473-3099(23)00128-7
3. Vaheri, A., Henttonen, H., & Mustonen, J. (2021). Hantavirus research in Finland: Highlights and perspectives. *Viruses*, *13*(8), 1452. https://doi.org/10.3390/v13081452
4. Hou, Y., Li, Q., Huang, X., Wang, J., Hou, J., Sun, Y., Wu, X., Dian, Z., Wang, B., & Xia, X. (2024). Distribution and genetic characterization of Hantaviruses in bats and rodents from Yunnan. *PLOS Neglected Tropical Diseases*, *18*(8), e0012437. https://doi.org/10.1371/journal.pntd.0012437
5. Koehler, F. C., Di Cristanziano, V., Späth, M. R., Hoyer-Allo, K. J. R., Wanken, M., Müller, R.-U., & Burst, V. (2022). The kidney in Hantavirus infection-epidemiology, virology, pathophysiology, clinical presentation, diagnosis and management. *Clinical Kidney Journal*, *15*(7), 1231–1252. https://doi.org/10.1093/ckj/sfac008
6. Moore, R. A., & Griffen, D.. (2025). Hantavirus Pulmonary Syndrome. In *StatPearls*. StatPearls Publishing. http://www.ncbi.nlm.nih.gov/books/NBK513243/
7. Akram, S. M., Mangat, R., & Huang, B.. (2025). Hantavirus Cardiopulmonary Syndrome (Archived). In *StatPearls*. StatPearls Publishing. http://www.ncbi.nlm.nih.gov/books/NBK459378/
8. Chen, R.-X., Gong, H.-Y., Wang, X., Sun, M.-H., Ji, Y.-F., Tan, S.-M., Chen, J.-M., Shao, J.-W., & Liao, M. (2023). Zoonotic *Hantaviridae* with global public health significance. *Viruses*, *15*(8), 1705. https://doi.org/10.3390/v15081705
9. Kuhn, J. H., & Schmaljohn, C. S. (2023). A brief history of *Bunyaviral* family *Hantaviridae*. *Diseases (Basel, Switzerland)*, *11*(1), 38. https://doi.org/10.3390/diseases11010038
10. Fox, J. G., Newcomer, C. E., & Rozmiarek, H. (2002). Selected Zoonoses. In *Laboratory Animal Medicine* (pp. 1059–1105). Elsevier. https://doi.org/10.1016/B978-012263951-7/50028-4
11. Jeeva, S., Mir, S., Velasquez, A., Weathers, B. A., Leka, A., Wu, S., Sevarany, A. T., & Mir, M. (2019). Hantavirus RdRp requires a host cell factor for cap snatching. *Journal of Virology*, *93*(5), e02088–18. https://doi.org/10.1128/JVI.02088-18
12. Deng, X., Tian, S., Yu, Z., Wang, L., Liang, R., Li, Y., Xiang, R., Jiang, S., Ying, T., & Yu, F. (2020). Development of small-molecule inhibitors against Hantaviruses. *Microbes and Infection*, *22*(6–7), 272–277. https://doi.org/10.1016/j.micinf.2020.05.011
13. Serris, A., Stass, R., Bignon, E. A., Muena, N. A., Manuguerra, J.-C., Jangra, R. K., Li, S., Chandran, K., Tischler, N. D., Huiskonen, J. T., Rey, F. A., & Guardado-Calvo, P. (2020). The Hantavirus surface glycoprotein lattice and its fusion control mechanism. *Cell*, *183*(2), 442–456.e16. https://doi.org/10.1016/j.cell.2020.08.023
14. Abudurexiti, A., Adkins, S., Alioto, D., Alkhovsky, S. V., Avšič-Županc, T., Ballinger, M. J., Bente, D. A., Beer, M., Bergeron, É, Blair, C. D., Briese, T., Buchmeier, M. J., Burt, F. J., Calisher, C. H., Cháng, C., Charrel, R. N., Choi, I. R., Clegg, J. C. S., de la Torre, J. C., … Kuhn, J. H. (2019). Taxonomy of the order *Bunyavirales*: Update 2019. *Archives of Virology*, *164*(7), 1949–1965. https://doi.org/10.1007/s00705-019-04253-6
15. Hofmann, J., Loyen, M., Faber, M., & Krüger, D. H.. (2022). [Hantavirus disease: An update]. *Deutsche Medizinische Wochenschrift (1946)*, *147*(6), 312–318. https://doi.org/10.1055/a-1664-7259
16. Engdahl, T. B., & Crowe, J. E. (2020). Humoral immunity to Hantavirus infection. *mSphere*, *5*(4), e00482–20. https://doi.org/10.1128/mSphere.00482-20
17. Meier, K., Thorkelsson, S. R., Quemin, E. R. J., & Rosenthal, M. (2021). Hantavirus replication cycle—An updated structural virology perspective. *Viruses*, *13*(8), 1561. https://doi.org/10.3390/v13081561
18. Plyusnina, A., Ferenczi, E., Rácz, G. R., Nemirov, K., Lundkvist, A., Vaheri, A., Vapalahti, O., & Plyusnin, A. (2009). Co-circulation of three pathogenic Hantaviruses: Puumala, Dobrava, and Saaremaa in Hungary. *Journal of Medical Virology*, *81*(12), 2045–2052. https://doi.org/10.1002/jmv.21635
19. Avšič-Županc, T., Saksida, A., & Korva, M. (2019). Hantavirus infections. *Clinical Microbiology and Infection: The Official Publication of the European Society of Clinical Microbiology and Infectious Diseases*, *21S*, e6–e16. https://doi.org/10.1111/1469-0691.12291
20. Outinen, T. K., Mäkelä, S., Pörsti, I., Vaheri, A., & Mustonen, J. (2021). Severity biomarkers in Puumala Hantavirus infection. *Viruses*, *14*(1), 45. https://doi.org/10.3390/v14010045
21. Reynes, J.-M., Carli, D., Thomas, D., & Castel, G. (2019). Puumala Hantavirus genotypes in humans, France, 2012-2016. *Emerging Infectious Diseases*, *25*(1), 140–143. https://doi.org/10.3201/eid2501.180270

22. Clement, J., LeDuc, J. W., McElhinney, L. M., Reynes, J.-M., Van Ranst, M., & Calisher, C. H. (2019). Clinical characteristics of ratborne Seoul Hantavirus disease. *Emerging Infectious Diseases*, *25*(2), 387–388. https://doi.org/10.3201/eid2502.181643
23. Mir, S. (2021). Hantavirus induced kidney disease. *Frontiers in Medicine*, *8*, 795340. https://doi.org/10.3389/fmed.2021.795340
24. Figueiredo, L. T., Martins Machado, A., Mazzoni Campos, G., Garcia De Figueiredo, G., Sabino Dos Santo Junior, G., De Pádua, M., Jabur Badra, S., & Marciel De Souza, W. (2012). Araraquara, the most virulent among all Hantavirus. *International Journal of Infectious Diseases*, *16*, e82. https://doi.org/10.1016/j.ijid.2012.05.196
25. Engdahl, T. B., Kuzmina, N. A., Ronk, A. J., Mire, C. E., Hyde, M. A., Kose, N., Josleyn, M. D., Sutton, R. E., Mehta, A., Wolters, R. M., Lloyd, N. M., Valdivieso, F. R., Ksiazek, T. G., Hooper, J. W., Bukreyev, A., & Crowe, J. E. (2021). Broad and potently neutralizing monoclonal antibodies isolated from human survivors of new world Hantavirus infection. *Cell Reports*, *35*(5), 109086. https://doi.org/10.1016/j.celrep.2021.109086
26. Essex, K., Mullen, J., Lauria, M. J., & Braude, D. A. (2023). Management of Hantavirus cardiopulmonary syndrome in critical care transport: A review. *Air Medical Journal*, *42*(6), 483–487. https://doi.org/10.1016/j.amj.2023.07.011
27. Ashique, S., Sandhu, N. K., Das, S., Haque, S. N., & Koley, K. (2022). Global comprehensive outlook of Hantavirus contagion on humans: A review. *Infectious Disorders Drug Targets*, *22*(3), e050122199975. https://doi.org/10.2174/1871526522666220105110819
28. Mittler, E., Dieterle, M. E., Kleinfelter, L. M., Slough, M. M., Chandran, K., & Jangra, R. K. (2019). Hantavirus entry: Perspectives and recent advances. *Advances in Virus Research*, *104*, 185–224. https://doi.org/10.1016/bs.aivir.2019.07.002
29. Shang, C., Zhang, Q. F., Yin, Q. L., Li, D. X., & Li, J. D. (2020). [Influence factors related epidemics on Hantavirus disease]. *Zhonghua Liu Xing Bing Xue Za Zhi = Zhonghua Liuxingbingxue Zazhi*, *41*(6), 968–974. https://doi.org/10.3760/cma.j.cn112338-20190916-00678
30. Saavedra, F., Díaz, F. E., Retamal-Díaz, A., Covián, C., González, P. A., & Kalergis, A. M. (2021). Immune response during Hantavirus diseases: Implications for immunotherapies and vaccine design. *Immunology*, *163*(3), 262–277. https://doi.org/10.1111/imm.13322
31. Strella, T., Echazarreta, S. E., Couto, E. M., Ferrer, M. F., Quipildor, M., Lavarra, E., Herrera, M. P., Cunto, E., Chediack, V., Saul, P. A., Blanco, M. E., Balasini, C., Lugo, A. V., Flores M, M. A., González, A. L., Lares, M. B., García, M. C., Amico, J., Colque, ÁM., … Staneloni, M. I. (2025). [Controversies on Hantavirus]. *Medicina*, *85*(2), 363–375.
32. Riccò, M., Peruzzi, S., Ranzieri, S., & Magnavita, N. (2021). Occupational Hantavirus infections in agricultural and forestry workers: A systematic review and metanalysis. *Viruses*, *13*(11), 2150. https://doi.org/10.3390/v13112150
33. Tortosa, F., Perre, F., Tognetti, C., Lossetti, L., Carrasco, G., Guaresti, G., Iglesias, A., Espasandin, Y., & Izcovich, A. (2024). Seroprevalence of Hantavirus infection in non-epidemic settings over four decades: A systematic review and meta-analysis. *BMC Public Health*, *24*(1), 2553. https://doi.org/10.1186/s12889-024-20014-w
34. Ferro, I., Bellomo, C. M., López, W., Coelho, R., Alonso, D., Bruno, A., Córdoba, F. E., & Martinez, V. P. (2020). Hantavirus pulmonary syndrome outbreaks associated with climate variability in northwestern Argentina, 1997-2017. *PLOS Neglected Tropical Diseases*, *14*(11), e0008786. https://doi.org/10.1371/journal.pntd.0008786
35. Madai, M., Horváth, G., Herczeg, R., Somogyi, B., Zana, B., Földes, F., Kemenesi, G., Kurucz, K., Papp, H., Zeghbib, S., & Jakab, F. (2021). Effectiveness regarding Hantavirus detection in rodent tissue samples and urine. *Viruses*, *13*(4), 570. https://doi.org/10.3390/v13040570
36. Wells, R. M., Sosa Estani, S., Yadon, Z. E., Enria, D., Padula, P., Pini, N., Mills, J. N., Peters, C. J., & Segura, E. L. (1997). An unusual Hantavirus outbreak in southern Argentina: Person-to-person transmission? Hantavirus pulmonary syndrome study group for Patagonia. *Emerging Infectious Diseases*, *3*(2), 171–174. https://doi.org/10.3201/eid0302.970210
37. Martínez, V. P., Di Paola, N., Alonso, D. O., Pérez-Sautu, U., Bellomo, C. M., Iglesias, A. A., Coelho, R. M., López, B., Periolo, N., Larson, P. A., Nagle, E. R., Chitty, J. A., Pratt, C. B., Díaz, J., Cisterna, D., Campos, J., Sharma, H., Dighero-Kemp, B., Biondo, E., … Palacios, G.. (2020). "Super-spreaders" and person-to-person transmission of Andes virus in Argentina. *New England Journal of Medicine*, *383*(23), 2230–2241. https://doi.org/10.1056/NEJMoa2009040
38. Riesle-Sbarbaro, S. A., Kirchoff, N., Hansen-Kant, K., Stern, A., Kurth, A., & Prescott, J. B. (2023). Human-to-human transmission of Andes virus modeled in Syrian hamsters. *Emerging Infectious Diseases*, *29*(10), 2159–2163. https://doi.org/10.3201/eid2910.230544

39. Toledo, J., Haby, M. M., Reveiz, L., Sosa Leon, L., Angerami, R., & Aldighieri, S. (2022). Evidence for human-to-human transmission of Hantavirus: A systematic review. *The Journal of Infectious Diseases, 226*(8), 1362–1371. https://doi.org/10.1093/infdis/jiab461
40. Alonso, D. O., Pérez-Sautu, U., Bellomo, C. M., Prieto, K., Iglesias, A., Coelho, R., Periolo, N., Domenech, I., Talmon, G., Hansen, R., Palacios, G., & Martinez, V. P. (2020). Person-to-person transmission of Andes virus in Hantavirus pulmonary syndrome, Argentina, 2014. *Emerging Infectious Diseases, 26*(4), 756–759. https://doi.org/10.3201/eid2604.190799
41. Jacob, A. T., Ziegler, B. M., Farha, S. M., Vivian, L. R., Zilinski, C. A., Armstrong, A. R., Burdette, A. J., Beachboard, D. C., & Stobart, C. C. (2023). Sin Nombre virus and the emergence of other Hantaviruses: A review of the biology, ecology, and disease of a zoonotic pathogen. *Biology, 12*(11), 1413. https://doi.org/10.3390/biology12111413
42. Alburkat, H., Smura, T., Bouilloud, M., Pradel, J., Anfray, G., Berthier, K., Dutra, L., Loiseau, A., Niamsap, T., Olander, V., Sepulveda, D., Venkat, V., Charbonnel, N., Castel, G., & Sironen, T. (2024). Evolution and genetic characterization of Seoul virus in wild rats *Rattus norvegicus* from an urban park in Lyon, France 2020-2022. *PLOS Neglected Tropical Diseases, 18*(5), e0012142. https://doi.org/10.1371/journal.pntd.0012142
43. Tkachenko, E., Kurashova, S., Balkina, A., Ivanov, A., Egorova, M., Leonovich, O., Popova, Y., Teodorovich, R., Belyakova, A., Tkachenko, P., Trankvilevsky, D., Blinova, E., Ishmukhametov, A., & Dzagurova, T. (2023). Cases of hemorrhagic fever with renal syndrome in Russia during 2000-2022. *Viruses, 15*(7), 1537. https://doi.org/10.3390/v15071537
44. Machtinger, E. T., & Williams, S. C. (2020). Practical guide to trapping *Peromyscus leucopus* (*Rodentia: Cricetidae*) and *Peromyscus maniculatus* for vector and vector-borne pathogen surveillance and ecology. *Journal of Insect Science (Online), 20*(6), 5. https://doi.org/10.1093/jisesa/ieaa028
45. Pérez-Umphrey, A. A., Settlecowski, A. E., Elbers, J. P., Williams, S. T., Jonsson, C. B., Bonisoli-Alquati, A., Snider, A. M., & Taylor, S. S. (2023). Genetic variants associated with Hantavirus infection in a reservoir host are related to regulation of inflammation and immune surveillance. *Infection, Genetics and Evolution: Journal of Molecular Epidemiology and Evolutionary Genetics in Infectious Diseases, 116*, 105525. https://doi.org/10.1016/j.meegid.2023.105525
46. Mull, N., Erdin, M., Smura, T., Sironen, T., & Forbes, K. M. (2023). Novel Ozark *Orthohantavirus* in hispid cotton rats (*Sigmodon hispidus*), Arkansas, USA. *Emerging Infectious Diseases, 29*(12), 2576–2578. https://doi.org/10.3201/eid2912.230549
47. Hardcastle, K., Scott, D., Safronetz, D., Brining, D. L., Ebihara, H., Feldmann, H., & LaCasse, R. A. (2016). Laguna Negra virus infection causes Hantavirus pulmonary syndrome in Turkish hamsters (*Mesocricetus brandti*). *Veterinary Pathology, 53*(1), 182–189. https://doi.org/10.1177/0300985815570071
48. Salgado, R., Barja, I., Hernández, M. D. C., Lucero, B., Castro-Arellano, I., Bonacic, C., & Rubio, A. V. (2022). Activity patterns and interactions of rodents in an assemblage composed by native species and the introduced black rat: Implications for pathogen transmission. *BMC Zoology, 7*(1), 48. https://doi.org/10.1186/s40850-022-00152-7
49. Gomes, A. P. N., Costa, N. A., Gentile, R., Vilela, R. V., & Maldonado, A. (2020). Morphological and genetic description of *Moniliformis necromysi* sp. n. (*Archiacanthocephala*) from the wild rodent *Necromys lasiurus* (*Cricetidae: Sigmondontinae*) in Brazil. *Journal of Helminthology, 94*, e138. https://doi.org/10.1017/S0022149X20000188
50. Schoch, C. L., Ciufo, S., Domrachev, M., Hotton, C. L., Kannan, S., Khovanskaya, R., Leipe, D., Mcveigh, R., O'Neill, K., Robbertse, B., Sharma, S., Soussov, V., Sullivan, J. P., Sun, L., Turner, S., & Karsch-Mizrachi, I. (2020). NCBI taxonomy: A comprehensive update on curation, resources and tools. *Database: The Journal of Biological Databases and Curation, 2020*, baaa062. https://doi.org/10.1093/database/baaa062
51. Kabwe, E., Davidyuk, Y., Shamsutdinov, A., Garanina, E., Martynova, E., Kitaeva, K., Malisheni, M., Isaeva, G., Savitskaya, T., Urbanowicz, R. A., Morzunov, S., Katongo, C., Rizvanov, A., & Khaiboullina, S. (2020). *Orthohantaviruses*, emerging zoonotic pathogens. *Pathogens, 9*(9), 775. https://doi.org/10.3390/pathogens9090775
52. Lebecque, O., & Dupont, M. (2020). Puumala Hantavirus: An imaging review. *Acta Radiologica (Stockholm, Sweden: 1987), 61*(8), 1072–1079. https://doi.org/10.1177/0284185119889564
53. Brügger, B., & Chuard, C.. (2022). [Hantavirus infections]. *Revue Medicale Suisse, 18*(799), 1900–1903. https://doi.org/10.53738/REVMED.2022.18.799.1900

54. Mustonen, J., Vaheri, A., Pörsti, I., & Mäkelä, S. (2022). Long-term consequences of Puumala Hantavirus infection. *Viruses*, *14*(3), 598. https://doi.org/10.3390/v14030598
55. Riccò, M., Peruzzi, S., Ranzieri, S., Balzarini, F., Valente, M., Marchesi, F., & Bragazzi, N. L. (2021). Hantavirus infections in Italy: Not reported doesn't mean inexistent. *Acta Bio-Medica: Atenei Parmensis*, *92*(4), e2021324. https://doi.org/10.23750/abm.v92i4.10661
56. Welke, R.-W., Sperber, H. S., Bergmann, R., Koikkarah, A., Menke, L., Sieben, C., Krüger, D. H., Chiantia, S., Herrmann, A., & Schwarzer, R. (2022). Characterization of Hantavirus N protein intracellular dynamics and localization. *Viruses*, *14*(3), 457. https://doi.org/10.3390/v14030457
57. Madrières, S., Castel, G., Murri, S., Vulin, J., Marianneau, P., & Charbonnel, N. (2019). The needs for developing experiments on reservoirs in Hantavirus research: Accomplishments, challenges and promises for the future. *Viruses*, *11*(7), 664. https://doi.org/10.3390/v11070664
58. Dospital, C., Arancibia-Avila, P., & Araneda-Flores, J. (2023). Epidemiological profile of Hantavirus in the Ñuble region period 2002-2018, Chile. *Brazilian Journal of Biology = Revista Brasleira De Biologia*, *84*, e269097. https://doi.org/10.1590/1519-6984.269097
59. Jiang, H., Huang, C., Bai, X., Zhang, F., Lin, B., Wang, S., Jia, Z., Wang, J., Liu, J., Dang, S., Zhao, Y., Dou, X., Cui, F., Zhang, W., Lian, J., Wang, G., & Gao, Z., & Infectious Disease Branch of the Chinese Preventive Medicine Association, Infectious Diseases Branch of the Chinese Medical Association (2022). Expert consensus on the prevention and treatment of hemorrhagic fever with renal syndrome. *Infectious Diseases & Immunity*, *2*(4), 224–232. https://doi.org/10.1097/ID9.0000000000000054
60. Jiang, H., Zheng, X., Wang, L., Du, H., Wang, P., & Bai, X. (2017). Hantavirus infection: A global zoonotic challenge. *Virologica Sinica*, *32*(1), 32–43. https://doi.org/10.1007/s12250-016-3899-x
61. Sehgal, A., Mehta, S., Sahay, K., Martynova, E., Rizvanov, A., Baranwal, M., Chandy, S., Khaiboullina, S., Kabwe, E., & Davidyuk, Y. (2023). Hemorrhagic fever with renal syndrome in Asia: History, pathogenesis, diagnosis, treatment, and prevention. *Viruses*, *15*(2), 561. https://doi.org/10.3390/v15020561
62. Lupuşoru, G., Andronesi, A. G., Lupuşoru, M., Ailincăi, I., Sfeatcu, R., Văcăroiu, I., Banu, M., Achim, C., & Ismail, G. (2023). Hantavirus infections in the South-Eastern European countries: A study of two cases and literature review. *Experimental and Therapeutic Medicine*, *26*(3), 430. https://doi.org/10.3892/etm.2023.12129
63. Züst, R., Ackermann-Gäumann, R., Liechti, N., Siegrist, D., Ryter, S., Portmann, J., Lenz, N., Beuret, C., Koller, R., Staehelin, C., Kuenzli, A. B., Marschall, J., Rothenberger, S., & Engler, O. (2023). Presence and persistence of Andes virus RNA in human semen. *Viruses*, *15*(11), 2266. https://doi.org/10.3390/v15112266
64. Hjertqvist, M., Klein, S. L., Ahlm, C., & Klingstrom, J. (2010). Mortality rate patterns for hemorrhagic fever with renal syndrome caused by Puumala virus. *Emerging Infectious Diseases*, *16*(10), 1584–1586. https://doi.org/10.3201/eid1610.100242
65. Bisen, P. S., & Raghuvanshi, R. (2013). *Emerging Epidemics: Management and Control* (1st ed.). Wiley. https://doi.org/10.1002/9781118393277
66. Gowen, B. B., & Hickerson, B. T. (2017). Hemorrhagic fever of bunyavirus etiology: Disease models and progress towards new therapies. *Journal of Microbiology (Seoul, Korea)*, *55*(3), 183–195. https://doi.org/10.1007/s12275-017-7029-8
67. Saúl, P. A., Sánchez-Cunto, M., Gregori-Sabelli, R. A., Chacón, N. R., Chediack, V., & Cunto, E. R.. (2021). [Controversies on corticosteroid therapy in Hantavirus cardiopulmonary syndrome]. *Medicina*, *81*(4), 617–623.
68. Szabó, R. (2017). Antiviral therapy and prevention against Hantavirus infections. *Acta Virologica*, *61*(1), 3–12. https://doi.org/10.4149/av_2017_01_3
69. Dheerasekara, K., Sumathipala, S., & Muthugala, R. (2020). Hantavirus infections—Treatment and prevention. *Current Treatment Options in Infectious Diseases*, *12*(4), 410–421. https://doi.org/10.1007/s40506-020-00236-3
70. Mittler, E., Serris, A., Esterman, E. S., Florez, C., Polanco, L. C., O'Brien, C. M., Slough, M. M., Tynell, J., Gröning, R., Sun, Y., Abelson, D. M., Wec, A. Z., Haslwanter, D., Keller, M., Ye, C., Bakken, R. R., Jangra, R. K., Dye, J. M., Ahlm, C., … Guardado-Calvo, P. (2023). Structural and mechanistic basis of neutralization by a pan-Hantavirus protective antibody. *Science Translational Medicine*, *15*(700), eadg1855. https://doi.org/10.1126/scitranslmed.adg1855
71. Mayor, J., Engler, O., & Rothenberger, S. (2021). Antiviral efficacy of ribavirin and favipiravir against Hantaan virus. *Microorganisms*, *9*(6), 1306. https://doi.org/10.3390/microorganisms9061306

72. Lukman, N., Kosasih, H., Ibrahim, I. N., Pradana, A. A., Neal, A., & Karyana, M. (2019). A review of Hantavirus research in Indonesia: Prevalence in humans and rodents, and the discovery of Serang virus. *Viruses*, *11*(8), 698. https://doi.org/10.3390/v11080698
73. Kennedy, M., Mctabi, O., & Rickman, C. (2024). Hantavirus cardiopulmonary syndrome: Another reason to avoid mice. *Journal of Investigative Medicine High Impact Case Reports*, *12*, 23247096241274572. https://doi.org/10.1177/23247096241274572
74. Şahin, A. M., Çetin, S., Şenel, ., Erdem-Çakır, T., Aydın, E., & Yetkin, M. A. (2024). The role of clinical and laboratory finding in the differential diagnosis of Hantavirus and leptospirosis infections. *Journal of Vector Borne Diseases*, *61*(4), 601–606. https://doi.org/10.4103/JVBD.jvbd_54_24
75. Mir, M. A. (2010). Hantaviruses. *Clinics in Laboratory Medicine*, *30*(1), 67–91. https://doi.org/10.1016/j.cll.2010.01.004
76. Chai, S., Wang, L., Du, H., & Jiang, H. (2025). Achievement and challenges in *Orthohantavirus* vaccines. *Vaccines*, *13*(2), 198. https://doi.org/10.3390/vaccines13020198
77. Qamar, M. T. U., Ahmad, S., Khan, A., & Wei, D. (2024). Editorial: Immunotherapeutics development against Hantaviruses. *Frontiers in Immunology*, *15*, 1377137. https://doi.org/10.3389/fimmu.2024.1377137

6 Dengue Virus
A Global Health Challenge

Lorca Alzoubi, Yassmen Hamzat, Alaa Alqudah, and Alaa A. A. Aljabali

6.1 INTRODUCTION TO DENGUE VIRUS

Dengue virus (DENV) is a mosquito-borne flavivirus, classified in the family *Flaviviridae*, genus Flavivirus. DENV is endemic in tropical and subtropical regions around the world and, although asymptomatic or mild in most cases, the disease can cause a wide spectrum of symptoms that, in rarer cases, lead to hospitalization and death. Therefore, DENV represents a major global health challenge [1].

The permanent threat resulting from the global expansion of the DENV transmission and the emergence of viral outbreaks require harmonized and innovative approaches at global level. A multidisciplinary effort would improve the knowledge on the mechanisms of DENV transmission, the pathogenesis of dengue disease and identify other human, virological, and epidemiological factors related to the disease, improving surveillance, control, and prevention strategies for DENV disease.

6.2 EPIDEMIOLOGY OF DENGUE

Dengue has become the fastest-spreading vector-borne viral disease globally and is a major international public health concern [2]. More than half of the world's population now lives in dengue endemic countries, leading to an increase in reported cases each year. Dengue outbreaks affect most countries in tropical and subtropical regions, with nearly 50 million people infected by the virus annually.

Dengue is caused by a single positive-stranded RNA arbovirus in the Flavivirus genus of the *Flaviviridae* family [3]. Clinically, dengue infection ranges from an undifferentiated fever or febrile illness to the classic presentation of dengue fever (DF), and a few patients develop severe manifestations such as dengue haemorrhagic fever (DHF) and dengue shock syndrome (DSS). Typical DF is a self-limited febrile illness with symptoms lasting 5–7 days, and only a small portion of patients develop the more severe forms. Several hypotheses explaining dengue infection pathogenesis have been proposed.

6.2.1 GLOBAL PREVALENCE AND INCIDENCE

DENV infection poses a global health threat with repeated outbreaks in recent decades. DENVs belong to the genus Flavivirus in the family *Flaviviridae*. There are four antigenically distinct serotypes, namely DENV1, DENV2, DENV3, and DENV4. DENVs are transmitted to humans by mosquitoes [2], with an estimated 100 million infections occurring each year [4]. DHF and DSS have become life-threatening public health problems in many countries. DENVs infect and replicate in target cells, including dendritic cells, macrophages, lymphocytes, endothelial cells, and hepatocytes. The incubation period varies from 3 to 14 days. Typical manifestations include sudden fever for 2–7 days, headache, rash, nausea, oedema, and muscle pain. A proportion of infected individuals develop more severe clinical progress due to immune enhancement effects, with characteristics such

DOI: 10.1201/9781003607816-6

as haemorrhagic fever and shock syndrome. The transmission of DENVs involves mosquitoes as primary vectors and humans as amplifying hosts. Four DENV serotypes exist, and because infection with one serotype provides serotype-specific immunity, subsequent infections with heterologous serotypes pose a greater risk of developing severe dengue. Historically, DENVs have existed as sylvatic cycles that involved monkeys and humans across large geographic areas in Africa and Asia for centuries, with the first reported dengue outbreak in 1779. Worldwide incidence has increased 30-fold during the last 50 years, and epidemiological data from the Global Burden of Disease Study 2017 indicate that about 100 million dengue infections occurred in 2017, resulting in approximately 40,000 deaths. Over 75% of individuals live in urban centres conducive to vector transmission, and dengue outbreaks caused by all four serotypes have increased since the 1980s, affecting Asia, South America, and the Caribbean. Dengue epidemiology parallels vector distribution, with dengue transmission as near to humans as possible; in most cases, transmission is considered human-mosquito-human [5, 6].

6.2.2 Geographic Distribution and Risk Factors

Dengue is mainly distributed in tropical and subtropical areas, with global warming inducing an expanded infectious zone [2]. DENV infection is a major problem both for imported dengue cases and local outbreaks in Japan. No other candidate virus has shown a spread from sylvatic cycles and led to large-scale endemic cycles [7]. Factors such as poor housing, high human population density, and a deteriorated water, sewer, and waste management system also facilitate transmission. Specific endemic zones of arthropod-borne virus disease transmission are mostly limited to tropical and subtropical areas, but the risk of importation exists wherever viraemic travellers and competent mosquito vectors are present.

6.2.3 Trends over Time

According to the Global Burden of Disease Study 2019, DF, the world's most widespread mosquito-borne viral disease, imposes a heavy global burden. The disease's spread over the past decade poses a serious public health challenge. The predicted population at risk expands steadily as shown in Table 6.1.

Large dengue outbreaks have occurred in many low- and middle-income countries in the Americas, Asia, and the Western Pacific.

TABLE 6.1
Global Epidemiology Snapshot of Dengue by WHO Region—Recent Burden, Vectors, and Seasonality

WHO region	2024/Recent situation highlights	Dominant vectors	Typical seasonality
Americas	>7 million cases by end-April 2024 Cases already surpass 2023 annual high	*Aedes aegypti*, *Aedes albopictus*	Peak during rainy/warm months Sub-regional variation [8]
Southeast Asia	Long-standing hyperendemic transmission with periodic large waves	*Ae. aegypti*, *Ae. albopictus*	Monsoon-linked peaks (May–Oct in many countries) [8]
Western Pacific	Recurrent outbreaks Urban/peri-urban amplification	*Ae. aegypti*, *Ae. albopictus*	Warm/rainy seasons Island-specific patterns [8]
Africa	Expanding detection and reporting Multiple countries affected	*Ae. Aegypti*	Rainy seasons Under-ascertainment likely [8]
Europe	Imported + Limited local transmission clusters reported in 2024	*Ae. albopictus* (competent)	Summer/early autumn in receptive areas [8]

DENV exists in four serotypes (DEN-1 to -4), transmitted mainly by *Aedes aegypti* mosquitoes. Non-vector transmission may take place during blood transfusion, solid organ transplantation, and needle-stick injuries. In hyperendemic regions, infections mainly affect adolescents and adults; however, older age groups are at greater risk for severe dengue.

Accelerating factors include climate, transportation, population density, poverty, and WASH (water, sanitation, and hygiene).

The first dengue vaccine, Dengvaxia®, licensed since 2015 for persons aged 9–45 years in 20 countries, has an outstanding safety and efficacy record. Vector-control programmes remain the most important approach to preventing and controlling dengue worldwide. Current vector-control interventions fail to curb increasing dengue incidence, although combined approaches could significantly reduce cases and spread. The World Health Organization (WHO) developed a global strategy for dengue prevention and control spanning the years 2012–2020 [9, 10].

Despite substantial burden estimates, detailed trends and age-distribution data at regional and national levels during recent decades remain limited. The Global Burden of Disease Study of 2017 provided estimates of dengue burden by country from 1990 to 2017, while the Global Burden of Disease Study of 2019 focused on global incidences and public-health challenges. An analysis of dengue incidence from 1990 to 2019 complements these results by offering country-level data that help to design targeted prevention strategies worldwide [11].

6.3 PATHOGENESIS OF DENGUE INFECTION

DENV is a single-stranded RNA virus of genus Flavivirus in family *Flaviviridae* and the etiologic agent of dengue infection [12]. The virus comprises four serotypes, DENV-1 to DENV-4, with DENV-2 and DENV-3 typically linked to the more severe form of the disease. Following the bite of an infectious mosquito, the virus incubates in humans for around 4–7 days before undertaking a brief viraemic phase lasting 4–5 days. Infection can trigger a wide spectrum of clinical presentations, ranging from unapparent infection, non-specific febrile illness and classic DF to the more severe and potentially fatal DHF or shock syndrome [13]. The association with Aedes is long established, but the global spread of dengue appears, in part, to account for changes in human behaviours. The expansion of large urban centres combined with water storage, substandard housing and inadequate waste management are conducive to mosquito proliferation. Although a better understanding of interactions between mosquitoes and different dengue serotypes promises a greater ability to predict and control epidemics, a range of other tools have attracted attention. Genetically modified mosquitoes that are less able to transmit dengue have been developed and shown to reduce larval populations; cyclopoid copepods are voracious invertebrate predators of mosquito larvae. Communicable disease specialists and donors have long called for increased attention to dengue, yet emulation of HIV or SARS funding has failed, and operational research remains limited. Successfully tackling dengue will almost certainly require a coordinated multidisciplinary approach, sufficient funding for research and attention to the considerable public-health impact of infection [14]. Major advances appear possible against this pathogen that was once described as "the most important human arboviral infection in the world."

6.3.1 Virology of Dengue Virus

DENV represents a significant and emerging arboviral threat to global health. It belongs to the genus Flavivirus of the family *Flaviviridae* as does Zika virus and is classified into four antigenically related serotypes. The single-stranded positive-sense RNA genome is approximately 11 kilobases [15]. Virion particles are small (~50 nm) and spherical, comprising an icosahedral nucleocapsid surrounded by a lipid envelope [16]. Competent mosquito vectors, primarily *Ae. aegypti* and Aedes *albopictus*, are highly adapted to urban human environments. After an infected mosquito bites a susceptible human, the virus replicates and spreads into the bloodstream, producing viraemia that subsequently facilitates transmission back to susceptible mosquitoes. Following attachment and entry into susceptible cells, translation of the genome produces a single polypeptide that is cleaved into three structural and seven non-structural proteins. The structural proteins constitute the virus particle, while the non-structural proteins perform essential functions that regulate replication and assembly of the virus [17].

6.3.2 Immune Response to Infection

The DENVs are arthropod-borne agents of remarkable global public health importance. An estimated 400 million individuals are infected annually [18]. Approximately 100 million infections manifest as symptomatic disease, ranging from mild DF to severe haemorrhagic fever and shock syndrome. Four antigenically related DENV serotypes co-circulate throughout the tropical and subtropical world. Aedes *aegypti* and *Ae. albopictus* mosquitoes transmit the viruses.

DENVs are enveloped single-stranded positive-sense RNA viruses in the family *Flaviviridae.* Virions circulate in the bloodstream as a heterogeneous population of mature and partially mature particles defined by the degree of envelope (E) protein organization on the surface. The RNA genome is ~11 kb and encodes 10 proteins: three structural (capsid, premembrane/membrane, envelope) and seven non-structural (NS1, NS2a, NS2b, NS3, NS4a, NS4b, NS5) [19].

The innate immune response to DENV infection is crucial for controlling virus replication. The DENV envelope (E) protein mediates attachment to the host-cell membrane. The humoral immune response to DENV is complex and sometimes exacerbates pathogenicity. Nevertheless, it remains an essential component of host protection. Immunoglobulin-G responses to three evolutionarily stable epitopes of the E protein were measured in human serum samples collected from dengue-infected individuals at different time points after infection. Seroconversion was demonstrated by an increase in E-protein-specific antibodies on day 10 relative to day 3 after symptoms onset. Responses varied widely among individuals, with a large, highly variable, and cross-reactive antibody population directed towards two highly conserved epitopes and a smaller domain III-specific immunoglobulin-G population that correlated significantly with DENV-2 neutralization [16].

6.3.3 Clinical Manifestations

The clinical features of dengue infection have a wide spectrum of presentation ranging from inapparent or sub-clinical infection to undifferentiated fever or classical DF and the most severe forms of the disease that include DHF and DSS [20]. Most of the infections in dengue endemic regions occur sub-clinically but only about 10–25% infected individuals suffer from the clinical form of the disease. Undifferentiated forms of DF are represented by mild febrile illnesses associated with fever, headache, rashes, myalgia and joint pain. Sometime febrile condition of DF also exhibits non-specific symptoms including rashes, dehydration, weakness and influenza-like symptoms. Even though uncomplicated mosquito-borne DF is rarely life threatening, the socioeconomic burden caused by the illness is still huge; millions are sufferers of the fever worldwide. DHF represents the complicated disease associated with widespread haemorrhages. Organ haemorrhages together with plasma leakage collectively led to the condition known as DSS, the most severe form of dengue infection [21].

6.4 DENGUE TRANSMISSION DYNAMICS

Transmission of dengue depends on three parameters: virus, vector, and human host. Mosquito vectors become infected during a viraemic blood meal by either a symptomatic or asymptomatic human, which then supports virus replication and dissemination, and viral delivery to a new host during a successive bite [22]. Both primary and secondary infections of the human host and immune landscape within a population (early-life dengue exposure and cross-reactive immunity) drive transmission [23]. Various factors, including human population growth, urbanization, inadequate vector control, and increased travel and trade, have contributed to the spread of the virus.

DENV is transmitted primarily by the bite of infected female *Ae. aegypti* or, to a lesser extent, female *Ae. albopictus* mosquitoes. Following an infected blood meal, the virus replicates in the midgut epithelial cells before disseminating to the rest of the body, including the salivary glands. Transmission can occur during subsequent bites after roughly nine days or more, continuing for the duration of the vector's life. Transmission intensity exhibits marked seasonal variation, with cases in Thailand and Nicaragua peaking during the rainy season (May to October) and reaching a minimum during the dry season (November to April). Dengue transmission is also sensitive to temperature, and Aedes mosquito vectors preferentially feed during the early morning and late afternoon [24].

6.4.1 Vector Biology and Control

The spread of DENV is a result of vector dynamics, human activities, and environmental factors. The disease exhibits a complex urban transmission cycle involving humans and the *Ae. aegypti* mosquito. This vector species shows nearly complete dependence on humans for blood meals and oviposition sites; adult females rarely feed on other animals or lay eggs in non-human-made containers [25]. Consequently, domestic and peridomestic human activities provide the essential conditions for the virus to establish and persist in urban areas. Although *Ae. aegypti* plays a central role in transmission dynamics, other species such as *Ae. albopictus* and certain sylvatic Aedes and Haemagogus species contribute to DENV transmission in developing countries and specific localities with distinctive enzootic/spillover cycles [26].

Vector-control programmes typically rely on larval surveys to estimate mosquito abundance and identify potential transmitters. However, using larval indices entails limitations: there is no clear relationship between these indices and adult mosquito density; weather conditions and the larval stage of mosquitoes are not considered in transmission assessments; and variations in mosquito longevity affecting vectorial capacity are ignored [27]. These gaps contribute to the persistence of dengue outbreaks despite control efforts.

Strategies focusing on environmental management such as eliminating water-holding containers, implementing solid waste management, and limiting insecticide application have demonstrated success in reducing *Ae. aegypti* populations. Complemented by community engagement and legislative support, these measures form the backbone of effective control programmes exemplified by Singapore's experience. Biological control methods under development include paratransgenesis and the introduction of Wolbachia bacteria into *Ae. aegypti* populations, which reduce vector competence and suppress dengue transmission [28].

6.4.2 Human-Virus-Vector Interactions

Despite nearly a century of investigation, human dengue infections remain a puzzle and many relevant aspects of virus–vector interaction and infection mechanisms are not fully understood. DENVs are predominantly transmitted by *Ae. aegypti* in urban areas and by *Ae. albopictus* to a lesser extent. *Aedes aegypti* originated in Africa and colonized Australasia and the Americas along the routes of international trade in the mid-18th century. *Aedes albopictus* originated in Asia and spread globally only in the past three decades; it is now the predominant species in many tropical and temperate urban and peri-urban areas [29]. The emergence of this vector in turn significantly altered the transmission dynamics of the virus, although the mechanisms involved remain unclear [30]. Following infection of a susceptible individual by the bite of an infected mosquito, the DENV must replicate in many susceptible cells and tissues. The incubation period ranges from 3 to 14 days, with an average of 4–7 days. Viraemia lasts for 4–5 days only, starting approximately 1 day before to 2 or 3 days after the onset of fever. During the incubation period, the virus replicates in lymph node macrophages and in monocytes circulating in the blood. Subsequently, the virus spreads to the liver and other temporary sites of replication. Viraemia typically reaches a low titre (104–105 viral particles per millilitre). Because of the low titre, the inferred transmission period of a patient to mosquitoes has a duration of 2 days at most, and can end before or during the onset of fever.

Together dengue pathogenesis, emergence and transmission dynamics contribute to perpetuation of the global challenge that dengue poses [31].

6.4.3 Environmental Influences on Transmission

The determinants that affect dengue transmission are multifactorial and encompass climatic, ecological, social, and cultural factors. Human actions and behaviours such as urbanization, travel and migration, and consumption are important drivers of the ongoing resurgence of dengue; these must be considered in any effective public health response. Dengue's global spread and the thus increased risk of pandemics can be attributed to rapid population growth, increasing urban expansion, increased worldwide air travel, increasing globalization, and climate change. Coupled with the threat of other arboviruses, such as Zika, dengue represents an ongoing unresolved public health challenge. Recurrent outbreaks, sporadic epidemics, and continual viral circulation are a global problem, and the majority of the world's population is now at risk of infection [32].

6.5 CLINICAL MANAGEMENT OF DENGUE

DENV: clinical management of dengue The clinical management of dengue is challenging, since its signs and symptoms mimic other diseases, particularly in the early stages. Diagnosis must be confirmed by laboratory tests such as virus isolation, polymerase chain reaction, or detection of specific antibodies. There is no specific antiviral treatment for dengue infection; supportive care is the current standard of care. Mild dengue cases can be treated on an outpatient basis with oral rehydration therapy and analgesics such as paracetamol. Patients should be monitored daily for signs of progression to severe disease, which typically develops around defervescence. Dengue vasculopathy manifesting as DHF and DSS is characterized by increased vascular permeability, plasma leakage, haemoconcentration, and bleeding. Severe plasma leakage may lead to hypovolemic shock, multiorgan failure, and death if not promptly managed. Early recognition of these clinical signs is crucial for initiating appropriate interventions, including fluid replacement and close monitoring of vital parameters. The WHO classifies dengue into two main categories: dengue and severe dengue. Updated guidelines emphasize early diagnosis, appropriate treatment, and preventive strategies to reduce morbidity and mortality as shown in Table 6.2 [33, 34].

TABLE 6.2
WHO-aligned Clinical spectrum and Early Warning Signs for Progression to Severe Dengue—With Management Priorities

Category	Core features	Early warning signs (monitor closely)	Immediate management priorities
Dengue (without warning signs)	Acute fever, headache, myalgia/arthralgia, rash, nausea	None prominent Maintain vigilance days 3–7	Oral fluids, paracetamol (avoid NSAIDs), review daily [9]
Dengue with warning signs	Abdominal pain/tenderness, persistent vomiting, mucosal bleed, lethargy/restlessness, hepatomegaly, rising Hct with falling platelets	Same Often around defervescence	Early IV fluids, close vitals/urine output monitoring, labs [9]
Severe dengue	Severe plasma leakage (shock/respiratory distress), severe bleeding, severe organ involvement	Any decompensation sign	Rapid isotonic fluid resuscitation, critical care, blood products as indicated [9]

6.5.1 Diagnosis and Differential Diagnosis

The clinical symptoms of DF develop between 3 and 15 days after the infective mosquito bite, with most cases developing symptoms within 4–7 days. DENV infection should be suspected when a high fever begins in a person residing or recently returning from an endemic area. The usual clinical presentation includes a sudden onset of fever, headache (retro-orbital and intense), myalgia, arthralgia, rash, leukopenia, and haemorrhagic manifestations, although the full set of symptoms is not always present in each clinical case, making differential diagnosis difficult. In dengue-endemic areas, the period of highest transmission often coincides with heavy rainfall, leading to an increase in the number of febrile illnesses, notably with Influenza and Arboviruses, raising the need for differential diagnostic tools [34]. A fever of more than 38.5°C (usually elevated around 40°C) accompanied by irradiation malaise and generalized aches and pains lasts for about 2–7 days. The recovery phase or defervescence is often accompanied by itching, a slow heart rate (in spite of raised body temperature), conjunctival injection, and abdominal tenderness. Profound weakness, which is also common, may have a prolonged course lasting several weeks after the febrile period [35].

6.5.2 Treatment Protocols

Currently no clinically approved antiviral therapy exists to treat DENV infection, but several vaccine candidates are in clinical trials, and experimental antiviral drugs have been tested in the laboratory and in animal models [36]. In the clinical context, several adjunctive therapies have been tested, including chloroquine, corticosteroids, intravenous immunoglobulin, and platelet transfusions, without benefiting patients or modifying the course of disease. Field trials of antivirals such as balapiravir and celgosivir conducted in Vietnam and Singapore also failed to reduce viraemia or influence clinical outcomes despite satisfying safety and pharmacokinetic criteria. In vitro and animal model experiments suggested a therapeutic potential for lovastatin, but clinical trials failed to show efficacy in patients. Acute haemorrhage in the critical phase of DSS may require blood-product replacement, but critical treatment remains vigilant fluid management and careful clinical monitoring in a high dependency setting [34].

In practice, all dengue cases require optimal supportive care, with patient triage to identify who can be treated as outpatients and who requires hospitalization. In many endemic regions, primary health services are weak and unable to cope with the burden of patients demanding hospital treatment early in illness, which rapidly exhausts hospital facilities. After admission, care needs to be taken to identify signs of progression to severe dengue and initiation of the critical phase. During the critical phase of plasma leakage, DSS may be prevented, if intravenous fluid therapy is initiated carefully and adjusted to clinical disease severity. Once the critical period is over, the patient generally recovers quickly and can be discharged shortly after treatment ends. Hospitalization with safe maintenance of the circulation allows many patients who might otherwise die from severe plasma leakage to survive [37].

6.5.3 Management of Severe Cases

Dengue remains a global health threat and is no longer confined solely to tropical and subtropical regions. With a prevalence of 3.6 billion people at risk across 124 countries, it is the most rapidly spreading mosquito-borne viral disease. The increasing burden is documented by a 9-fold increase in reported dengue cases over the past two decades rising from 505,430 in 2000 to 4.2 million in 2019, with the greatest number of cases reported from Asia. Despite ongoing efforts, there is still no specific treatment or effective vaccine, and pathogenesis is not completely understood. Effective prevention depends on managing mosquito populations and human–virus interactions at the molecular and genetic levels [38].

DENV is a mosquito-borne member of the family *Flaviviridae* that causes DF. Factors such as the widespread distribution of Aedes mosquitoes, the inactivity duration of DENV in unfavourable conditions, and the rapid circulation of DENVs in humans make control activities challenging, and the disease continues to generate high morbidity during outbreaks. The latest dengue vaccines have limited efficacy among infants, dengue-naive recipients, and those who receive only a single dose. Development of a safe and cost-effective tetravalent dengue vaccine suitable for all participants remains of global concern. Emphasis should be placed on educating people and communities, creating awareness, and involving the community in monitoring and reporting mosquito breeding. Proper clinical management of dengue-infected patients can reduce mortality rates.

6.6 PREVENTIVE STRATEGIES

No fully effective and broadly applicable vaccine has yet been introduced, due largely to difficulties in addressing the four DENV serotypes and the risk of antibody-dependent enhancement. Dengvaxia, a recombinant live-attenuated tetravalent vaccine, is recommended only for individuals aged 9–45 years with previous dengue infection, as it can increase the severity of primary infections in non-immune recipients. Novel vaccine candidates are in phase III trials but remain unavailable. Applied prevention primarily targets vector control and breeding site reduction [39]. Continued vigilance and investment are essential to minimize the dengue impact across broad ecological contexts [39].

6.6.1 Vaccination Efforts

Only a single dengue vaccine, the live-attenuated chimeric tetravalent vaccine developed by Sanofi Pasteur (CYD-TDV), has been approved in some countries [40]. Owing to the challenges posed by the four antigenically distinct DENV serotypes, further progress in vaccine development remains necessary to ensure comprehensive protection across diverse populations. Recent licensing of the live-attenuated DENVax vaccines (Takeda) in Indonesia, and of the attenuated tetravalent vaccine TV003/TV005 (NIH) in Brazil, highlights continuing efforts to provide safe and effective immunization options [41]. Nonetheless, the potential for more severe disease episodes post-vaccination warrants careful consideration, particularly in regions where previous exposure to flaviviruses is uncommon [42]. This unfavourable risk–benefit balance poses a substantial hurdle in extending vaccine access to a broad at-risk demographic as shown in Table 6.3.

TABLE 6.3
Dengue Vaccines—Product Characteristics, WHO Recommendations (May 2024), and Regulatory Status

Vaccine	Platform/valency	WHO recommendation (2024)	Regulatory/PQ status	Key notes
TAK-003 (Qdenga, Takeda)	Live-attenuated Tetravalent	Recommended for children 6–16 years in high-transmission settings Programmatic use guided by local burden and cost-effectiveness	WHO prequalified (May 2024)	Lower efficacy in <6 years Use per national policy & WHO SAGE [43]
CYD-TDV (Dengvaxia, Sanofi Pasteur)	Live-attenuated chimeric Tetravalent	Use only in seropositive individuals (to avoid increased risk in seronegatives) Age ranges per label/country	Previously licensed in several countries	Pre-vaccination screening recommended where used [43]
TV003/TV005 (NIH-Butantan)	Live-attenuated Tetravalent	Under evaluation No WHO policy recommendation yet	Clinical development Not WHO-PQ	Country-specific trials ongoing [43]

6.6.2 Public Health Interventions

In the absence of specific antiviral therapies, DENV control relies heavily on efficient vector management and, when available, vaccination [27]. Despite numerous control and prevention initiatives, dengue outbreaks persist globally [31].

Integrated vector management under the "One Health" approach can address human, animal, and environmental aspects of DENV transmission, guiding early mitigation strategies and strengthening sustainable control policies. Improvement of community awareness and engagement, establishment of early systems, and enhancement of social mobilization and technical support are also crucial for pandemic preparedness and effective decision-making during outbreaks. Vector adulticide and larvicide are effective only during specific phases of the epidemic cycle. Inefficient vector control, combined with unpredictable climatic and socioenvironmental changes, frequent density of human population, rising travelling activities, neglect of preventive measures, limited health practices, and absence of an efficacious dengue vaccine, has made dengue a continuing major public health problem see Table 6.4.

6.6.3 Community Engagement and Education

Two recent community engaged projects, conducted in Yogyakarta, Indonesia and in Queensland, Australia, demonstrate the role of community engagement and education as a key element in preventing dengue transmission and outbreaks. An innovative community-centred dengue ecosystem management intervention in Yogyakarta provided dengue prevention materials to community leaders, facilitators, schoolchildren, and householders [44]. A short informative video on breeding places and larvae and pupae elimination developed with community members was used during education sessions. A poster, showing how humans cause environmental problems and interact with mosquitoes, was developed with community forums and health authorities and displayed publicly. A pocketbook on dengue prevention for community cadres and householders was created and disseminated. Children in the school programme created their own media and wall magazines related to dengue prevention. On the other side of the world, the ÎIEliminate DengueÎI Program in Queensland used community engagement as a key component of its open release trials involving Wolbachia-infected mosquitoes [45]. In-depth qualitative case studies identified four foundational features: enabling conditions; leadership; core commitments and guiding values; and formative social science research. These informed five operational practices: building the community engagement

TABLE 6.4
Evidence for Major Dengue Vector-Control Interventions—Trial Outcomes and Implementation Notes

Intervention	Best evidence/exemplar	Outcome summary	Implementation notes
***Wolbachia* (wMel)-infected *Aedes aegypti* releases**	Cluster-randomized trial, Yogyakarta, Indonesia (*NEJM* 2021)	~77% reduction in virologically confirmed dengue incidence in treated clusters	Requires staged releases, community engagement, entomological monitoring [7]
Integrated vector management (source reduction, environmental management, enforcement)	Singapore national programme synthesis	Sustained suppression of *Ae. aegypti* indices with multi-sectoral approach	Continuous source reduction, legislation, public education, data-driven ops [10]
Chemical control (adulticides/ larvicides)	WHO guidance and country programmes	Phase-specific benefits Diminishing returns with resistance	Needs resistance surveillance Rotate actives Combine with non-chemical tools [10]

team; integrating engagement into management; discerning stakeholder communities; maintaining a community presence; and socializing the technology and research strategy. Together, these practices contributed to stakeholders' support for the trials.

6.7 ECONOMIC IMPACT OF DENGUE

Since 1990, the recorded number of cases requiring medical care due to dengue has increased more than threefold. The direct and indirect economic costs to affected countries are estimated to exceed US$0.5 billion annually [46]. The macroeconomic impact throughout the economy is less well understood, yet it has the potential to cause a cascading negative effect.

6.7.1 Healthcare Costs

Dengue infection represents a considerable disease burden with major economic consequences worldwide. In 2006, the average cost-per-case was estimated to be US$514 for ambulatory care and US$1,394 for hospitalized episodes [46]. Thirty-eight countries accounting for 70% of the total DALY (disability-adjusted life year) loss were identified based on the geographical distribution of dengue. Concomitantly, dengue causes many missed work and school days. Societal cost of illness from dengue depends on local disease burden in terms of number of cases, and markedly on spending per episode. For example, in American countries the costs-per-case were three times higher than in Asian countries. In both locations, spending was greater for hospitalized than outpatient episodes. Other confounding factors influencing societal cost of illness were incomplete or delayed information in official reports, difficulties of capturing presentations to a variety of health systems and the inclusion or absence of a burden for persistent fatigue [47].

For international travellers, an outpatient dengue episode was estimated to cost US$376–1,135, and a hospitalized episode US$1,469–12,491 depending on the region of travel and the type of insurance [48]. For European travellers the costs per episode were estimated to be around US$547–736. The average length of stay in hospitals was 6.6–9.2 days, and the average workdays lost was 14.5 days. Travellers are predominantly high-income earners returning to high-income countries, where even a small number of cases may have a significant economic impact. Since vaccination against DENV is now available in a few countries, to whom and when to vaccinate are important questions that must consider the specific costs of illness in these countries.

6.7.2 Socioeconomic Burden

Direct medical costs from dengue illness exert enormous pressure on individual households and national health systems within endemic countries. The insurable population in dengue-endemic locations is limited: even in Thailand less than one-third of the 65 million inhabitants can claim access to public medical infrastructure, while in Malaysia two-thirds do. Far more of the population faces out-of-pocket obligations for treatment, a resurgent challenge in anywhere housing informal worker economies. Past studies on dengue's societal impact have largely relied on cost-of-illness estimates that focus on direct medical costs. Much of the financial burden for informal workers remains hidden, and the exposure of this population to dengue infection costs is largely unknown. During a 2013–15 outbreak in Singapore 8% (52/660) of a local sample of dengue-convalescent individuals lost their jobs because the illness cut into absence quotas. Nearly a quarter were temporarily unemployed and approximately half had to take unpaid leave. Despite a subsidy scheme providing a 75% offset against dengue-related medical expenses, nearly two-thirds of affected respondents incurred costs exceeding SGD 500 (US$691.72) and therefore faced significant exposure to lost income. Informal workers would generally not have access to paid leave, amplifying the vulnerability of this group during a dengue infection [46].

The vulnerability of the informal population to dengue-related financial hardship has become increasingly concerning. Absence from work undermined current consumption habits and limited the ability to buffer shocks into the future from reduced income and prolonged health recovery. Failure to treat or to opt for self-medication because of these shocks' risks illness progression and increased opportunity costs. Although the economic burden of dengue extends well beyond healthcare costs, all quantitative appraisals currently in use ignore these broader societal and macroeconomic effects. The ongoing SARS-CoV-2 pandemic has exposed the profound societal economic impact of infectious, Wolbachia-based vector control [49].

6.7.3 Cost-Effectiveness of Interventions

Determining the relative cost-effectiveness of vaccines and vector control can be complex because dengue vaccines and vector control reduce disease in very different ways [50]. Existing and alternative approaches can be modelled and projected for costs and effectiveness in a range of settings [46]. Empirical data from the "Aedes-free" intervention in Girardot, Colombia, allow another real-world check on earlier model results.

Vertical programmes that address only a single pathogen or that employ a single intervention often fail to reach vulnerable populations and do not ensure access to care. Intervention cost-effectiveness and budget impact assessments can help estimate how best to allocate scarce resources, as well as assess the potential health and economic returns to various stakeholders across settings to support budgeting, planning, financing and policy assurance decisions [51].

6.8 CHALLENGES IN DENGUE CONTROL

Dengue is a vector-borne disease transmitted primarily by female *Ae. aegypti* mosquitoes. Its increasing incidence worldwide can be attributed to ineffective vector control, unplanned urbanization, increased travel, and the failure of mosquito-control programmes. Vector resistance to insecticides impedes control efforts, necessitating the development of new agents and strategies. Formulating policies for integrated vector management requires political will, intersectoral coordination, and sustained funding. Dengue prevention and control demand a multidisciplinary approach sensitive to local contexts. Significant information gaps persist regarding vector biology and behaviour, transmission dynamics and pathogenesis, socioeconomic impact, and vaccine development [8, 9, 39].

6.8.1 Resistance to Vector Control Measures

Resistance to vector control is a critical challenge to effective dengue control. Large-scale efforts to reduce dengue through vector control have relied primarily on insecticides and habitat management, yet resistance to insecticides is spreading rapidly and threatening efforts to contain the vector populations that transmit dengue to humans [52]. Resistance compromises mosquito control efforts and reduces the effectiveness of those compounds that remain in use; in India, for example, extensive, long-term use of DDT in indoor spraying exerted indirect selection pressure on mosquito species, resulting in near-complete resistance of *Stegomyia albopicta* and *St. aegypti* populations to DDT [53]. Resistance to other insecticides varied, and most populations remained susceptible to deltamethrin. Effective vector control thus requires systematic monitoring and early detection of resistance in the targeted populations, allowing the implementation of evidence-based decision making and alternative control measures whenever new resistant populations are detected.

6.8.2 Global Health Policy and Funding

6.8.2.1 Global Health Policy and Funding

An estimated 4 billion people still live in DENV endemic settings that fuel transmission of this rapidly expanding disease across the tropics [8–10]. A century of research has laid a foundation of understanding regarding the disease's complex tenets; however, dengue remains a hyper-endemic vector-borne disease, of global significance, infecting several million people and causing tens of thousands of deaths every year. The absence of effective countermeasures, the absence of effective vector control and challenges in global health policy and funding continue to contribute to the rising global burden of this viral scourge [5, 49].

A complex data and policy landscape towards vaccine development and implementation, vector control and travel health have challenged the ultimate understanding of efficacy, with scientific research and global funding lagging behind the rapid expansion. National and subnational health policy measures also remain difficult to ascertain, while many governments continue to struggle with development of meaningful and effective frameworks to support ongoing research and vector control efforts. Competing public health priorities, particularly in low- and middle-income countries, are forcing new decisions where dengue-specific action accounts for only a very modest fraction of total health expenditures, a trend that is also observed among many international donors and global health stakeholders. Extracting meaningful data, assessing the true trajectory of risk, while supporting vaccine development and ensuring equitable access remain particularly challenging when surveillance efforts are continually exposed to factors that corrupt reporting and stifle efforts to quantify burden and measure incidence.

From a global health policy perspective, the example of DENV provides an outstanding opportunity to carefully evaluate the landscape surrounding an urgent and emerging threat that continues to expand, despite considerable research effort. Dengue thus remains a catastrophic threat to global health security, with vectors and disease continuing to expand across the tropics and a global population that remains exposed with little protection.

6.8.3 Data Gaps and Research Needs

Despite the availability of comprehensive surveillance data from WHO on dengue infection, critical information concerning exposure to the virus or immunologically active dengue infections is lacking, yet is vital for guiding and assessing dengue control efforts. The WHO DengueNet website (http://apps.who.int/globalatlas) [54], provides a relatively stable indicator of dengue incidence since 1955. Nevertheless, incomplete reporting, aggregated data presentation (weekly or monthly time resolution and provincial spatial resolution in many countries), and missing source code for government data collection and aggregation methods limit the utility of these data for disentangling the influence of environmental, epidemiological, virological, socioeconomic, and anthropological drivers of transmission. In most endemic countries, rates of dengue infection in febrile patients attending health clinics remain unknown, as does the rate of exposure among the asymptomatic or subclinical fraction of the MGVP population. To address these data gaps and to accelerate the development of dengue diagnostics, new vaccine candidates, and anti-viral drugs, the Development Group for Dengue Vaccines and Antivirals (DGDVA) was established in 2013 at the International Centre for Genetic Engineering and Biotechnology, New Delhi. Gaps also persist in the availability of epidemiological data linked to vector-control programmes and in detailed and timely virological and entomological information essential for outbreak prediction [55]. To overcome these barriers, a concerted effort is necessary to fill these gaps promptly and to develop mechanisms that incentivize the sharing of epidemiological, virological, entomological, and vector control data equipped with other drivers at appropriate temporal and spatial scales. Continuing support for international collaboration and coordination among health organizations and governments is critical. Shifts towards community-based and public-sector programmes in underserved populations and countries, together with more efforts to understand drivers of immunity and population susceptibility, would enable better evaluation of new vaccine formulations and immunization strategies, as well as sustain and extend efforts to tackle this global health challenge. The establishment and engagement of the DGDVA will foster coordination between academia, industry, public health organizations, and funding agencies to promote open data sharing and collaborative efforts, ultimately accelerating the development and deployment of diagnostics, vaccines, and anti-viral drugs.

6.9 FUTURE DIRECTIONS IN DENGUE RESEARCH

The dengue vaccine candidate, Dengvaxia®, developed by Sanofi Pasteur (France), attained licensure in several endemic countries including Mexico, the Philippines, Brazil, Thailand, and Singapore. Dengvaxia® is a recombinant live-attenuated tetravalent dengue vaccine (CYD-TDV) administered as a three-dose regimen at months 0, 6, and 12. Clinical trials across multiple Asian and Latin American locations demonstrated safety and moderate efficacy in children and adolescents. Overall vaccine efficacy was estimated at 60% against virologically confirmed dengue infection, with enhanced protection against hospitalization (approximately 80%) in individuals aged 2–16 years. Pooled analyses of long-term follow-up data corroborated efficacy against hospitalized dengue in children 9 years or older, supporting licensure for use in the 9–45 years age group [56].

The availability of an efficacious vaccine is expected to substantially mitigate the public health burden of dengue in endemic regions. However, vaccine deployment may shift the epidemiological profile of the disease, underscoring the need to study a broad age range due to age-related differences in clinical manifestations. Investigations into additional vaccine candidates offering comparable or improved protection at lower overall costs remain imperative.

6.9.1 Innovative Vaccine Development

Over the past five decades, the number of reported dengue cases globally has increased 30-fold, making dengue the most important mosquito-borne viral disease affecting humanity. DENV is a single-stranded positive-sense RNA virus belonging to the family *Flaviviridae* and genus Flavivirus. It has four distinct serotypes that are transmitted by female mosquitos, mainly *Ae. aegypti* and *Ae. albopictus*. DENVs circulate in approximately 100 countries in tropical and sub-tropical regions. Despite the availability of one licensed vaccine and numerous vaccine candidates at different stages of clinical development, it is difficult to make a dengue vaccine that confers safe and long-term protection for all who are at risk [57]. Since dengue infection phenomena and dengue epidemiology are complex, a combination of strategies should be adopted to tackle the problem rather than relying on a single approach.

A key player in current vaccine research is the secreted non-structural protein 1 (NS1) [58]. NS1 is secreted from DENV-infected cells and binds to the surface of uninfected cells. Because secreted NS1 plays several pathogenic roles, it is a contributor to severe dengue symptoms. Immunization with NS1 prevents DENV infection and protects against DENV-induced pathologies in vivo. In addition, NS1 is highly conserved among the four DENV serotypes. Therefore, NS1 is a promising vaccine candidate that could be used alone or combined with a conventional dengue vaccine. Along with classical dengue vaccines, NS1-based vaccine strategies have a high potential as next-generation dengue vaccines and.

6.9.2 Genomic and Molecular Approaches

DENV is the fastest spreading arboviral disease worldwide and remains a major global health challenge. It belongs to the genus Flavivirus, in the family *Flaviviridae*. There are four closely related, but antigenically distinct, serotypes (DENV-1 to DENV-4) circulating in a wide range of geographic areas. They infect circa 400 million individuals per year, with several millions of people being symptomatic, requiring medical attention, and causing more than 20,000 deaths. Toxic shock and haemorrhagic fever mainly caused by antibody-dependent enhancement of infection are the causes of death in young children or the elderly.

The DENV genome is a single-stranded, positive-sense RNA of approximately 10.7 kb. It includes a single open reading frame (ORF) encoding for a polyprotein that is co- and post-translationally processed. The ORF is flanked with 5′ and 3′ untranslated regions (UTRs). In the 5′ UTR, the conserved type I cap structure m^{7}GpppAm_mG is placed at the 5′ terminus but a canonical poly (A) tail does not exist at the 3′ UTR. The shared flanking sequences between the ORF and UTR provide crucial RNA structural elements to regulate viral replication and translation. A unified global genotyping framework of DENV serotype-1 has been proposed to facilitate contextualization of spatiotemporal epidemiological data [59]. In Southern Vietnam, DENV evolution over the last decade and the utility of DENV-NS1 as a diagnostic marker have been studied [60]. Asian I DENV-2 viruses, for example, may exhibit enhanced fitness in *Ae. aegypti* mosquitoes, potentially reflected in a shorter extrinsic incubation period or a lower infectious dose. Alternatively, the apparent fitness advantage could stem from a higher number of susceptible hosts. Investigations into neutralizing antibody titres against Asian I versus American strains may shed light on host population susceptibility. Linking genotype to virological or clinical phenotype remains a critical research focus; ongoing analyses of 590 DENV-1 genomes from patients with varying disease severity aim to establish such connections. The predictive value of neutralizing antibody assays for clinical outcomes is still uncertain, emphasizing the need for quantitative evaluation in large prospective cohort studies.

6.9.3 Global Collaborations and Initiatives

The DENV remains one of the most widespread arboviruses of public health importance in tropical and subtropical countries. In 2011, the European Commission, within its Seventh Framework Programme, launched funding for an international and interdisciplinary consortium involving 13 partners from Europe, Asia and America to address the issue of dengue control and prevention [61].

Dengue is endemic in most countries in the tropics and sub-tropics, and a main cause of morbidity and mortality in the Asia-Pacific region [62]. Despite the implementation of dengue surveillance over the last century by many countries, the general increasing trend of regional and global incidence (1980–2013) shows that worldwide, current dengue surveillance and control strategies cannot prevent either large dengue epidemics or the dissemination of the disease to new areas.

6.10 CONCLUSION

DF remains a significant global public health threat, acting as an impediment to social and economic development. The disease continues to push countries towards overstretched healthcare systems and high public health costs. It triggers fatal outbreaks in previously non-affected countries and increasingly affects neighbouring countries. The epidemiology of dengue infection is complex, with four antigenically distinct serotypes. Those who have recovered from one DENV serotype are not fully protected against others. Changes in vectorial capacity, climate, and viral evolution have led to an urbanization and globalization of dengue, culminating in a near-pandemic situation blitzing multiple continents beyond historical boundaries. Transmitted primarily by the daytime-basking mosquitoes *Ae. aegypti* and *Ae. albopictus*, DENV infects an estimated 390 million individuals every year. A sizable fraction of the infected individuals exhibits clinical symptoms, which range from the self-resolving DF to the haemorrhagic and shock syndrome. Both clinical and mosquito transmission data offer insight into the underlying spatial disease dynamics that are crucial for both the development of infection control policies and disease forecasting. Influenza-like symptoms make identifying dengue difficult, and the absence of a reliable diagnostic test calls for an increased effort in development. While specific treatment is not yet available, the management of the disease relies on the adequate maintenance of the patient in a tropical setting of increasing resistance to antibiotics. Improvements in dengue control include the use of vaccines, active community participation, appropriate clinical management, and strong political commitment. Urban to peri-urban transmission is dominant, but sylvatic dengue is widespread. Resistance to insecticides, lack of harmonization between stakeholders, insufficient funds at a community level, presence of unnoticed dengue breeding sites, misinformation, and little community participation limit the effectiveness of control programmes. Lack of consolidated databases (environment, epidemiology, socioeconomic factors, and demography) makes developing risk maps difficult. Vaccine development, "Anti-Dengue Drug" discovery, and persistent mosquito control efforts remain research imperatives. Enhancing harmonization between stakeholders and improving research capacities will reduce the dengue threat in a sustainable and holistic manner.

REFERENCES

1. Jahan, F. Dengue fever (DF) in Pakistan. *Asia Pacific Family Medicine* 2011, *10*, 1.
2. Guo, C.; Zhou, Z.; Wen, Z.; Liu, Y.; Zeng, C.; Xiao, D.; Ou, M.; Han, Y.; Huang, S.; Liu, D. Global epidemiology of dengue outbreaks in 1990–2015: A systematic review and meta-analysis. *Frontiers in Cellular and Infection Microbiology* 2017, *7*, 317.
3. Horstick, O.; Tozan, Y.; Wilder-Smith, A. Reviewing dengue: Still a neglected tropical disease? *PLOS Negl Trop Dis* 2015, *9*, e0003632. https://doi.org/10.1371/journal.pntd.0003632.
4. Zeng, Z.; Zhan, J.; Chen, L.; Chen, H.; Cheng, S. Global, regional, and national dengue burden from 1990 to 2017: A systematic analysis based on the global burden of disease study 2017. *EClinicalMedicine* 2021, *32*.
5. Hanley, K.A.; Cecilia, H.; Azar, S.R.; Moehn, B.A.; Gass, J.T.; Oliveira da Silva, N.I.; Yu, W.; Yun, R.; Althouse, B.M.; Vasilakis, N. et al. Trade-offs shaping transmission of sylvatic dengue and Zika viruses in monkey hosts. *Nature Communications* 2024, *15*, 2682. https://doi.org/10.1038/s41467-024-46810-x.
6. Yu, X.; Cheng, G. Contribution of phylogenetics to understanding the evolution and epidemiology of dengue virus. *Animal Model Exp Med* 2022, *5*, 410–417. https://doi.org/10.1002/ame2.12283.
7. Utarini, A.; Indriani, C.; Ahmad, R.A.; Tantowijoyo, W.; Arguni, E.; Ansari, M.R.; Supriyati, E.; Wardana, D.S.; Meitika, Y.; Ernesia, I. Efficacy of Wolbachia-infected mosquito deployments for the control of dengue. *New England Journal of Medicine* 2021, *384*, 2177–2186.
8. Organization, W.H. Dengue – Global situation. https://www.who.int/emergencies/disease-outbreak-news/item/2024-DON518?utm_source=chatgpt.com (accessed on 16/09/2025).
9. Organization, W.H. Dengue and severe dengue. https://www.who.int/health-topics/dengue-and-severe-dengue?utm_source=chatgpt.com#tab=tab_1 (accessed on 19/05/2025).
10. Organization, W.H. Vector-borne diseases. https://www.who.int/news-room/fact-sheets/detail/vector-borne-diseases?utm_source=chatgpt.com (accessed on 19/09/2025).
11. Du, M.; Jing, W.; Liu, M.; Liu, J. The global trends and regional differences in incidence of dengue infection from 1990 to 2019: An analysis from the global burden of disease study 2019. *Infectious Diseases and Therapy* 2021, *10*, 1625–1643.
12. Murugesan, A.; Manoharan, M. *Dengue Virus*; Emerging and Reemerging Viral Pathogens. 2020:281–359. https://doi.org/10.1016/B978-0-12-819400-3.00016-8. Epub 2019 Sep 20.
13. Whitehorn, J.; Simmons, C.P. The pathogenesis of dengue. *Vaccine* 2011, *29*, 7221–7228.
14. Miller, J.D.; Hutchins, M. The impacts of urbanisation and climate change on urban flooding and urban water quality: A review of the evidence concerning the United Kingdom. *Journal of Hydrology: Regional Studies* 2017, *12*, 345–362, https://doi.org/https://doi.org/10.1016/j.ejrh.2017.06.006.
15. Castro, L.R. Can the mosquito bite? The multispecies transmutation of Wolbachia mosquitoes as biotechnologies of epidemic control in Rio de Janeiro. *Engaging Science, Technology, and Society* 2025, *11*, 75–100.
16. Burrell, C.J.; Howard, C.R.; Murphy, F.A. *Virion Structure and Composition*; Fenner and White's Medical Virology. 2017:27–37. https://doi.org/10.1016/B978-0-12-375156-0.00003-5. Epub 2016 Nov 11.
17. Souza-Neto, J.A.; Powell, J.R.; Bonizzoni, M. *Aedes aegypti* vector competence studies: A review. *Infect Genet Evol* 2019, *67*, 191–209. https://doi.org/10.1016/j.meegid.2018.11.009.
18. King, C.A.; Wegman, A.D.; Endy, T.P. Mobilization and activation of the innate immune response to dengue virus. *Frontiers in Cellular and Infection Microbiology* 2020, *10*, 574417.
19. Nanaware, N.; Banerjee, A.; Mullick Bagchi, S.; Bagchi, P.; Mukherjee, A. Dengue virus infection: A tale of viral exploitations and host responses. *Viruses* 2021, *13*. https://doi.org/10.3390/v13101967.
20. Hasan, S.; Jamdar, S.F.; Alalowi, M.; Al Beaiji, S.M.A.A. Dengue virus: A global human threat: Review of literature. *Journal of International Society of Preventive and Community Dentistry* 2016, *6*, 1–6.
21. Wang, W.-H.; Urbina, A.N.; Chang, M.R.; Assavalapsakul, W.; Lu, P.-L.; Chen, Y.-H.; Wang, S.-F. Dengue hemorrhagic fever – A systemic literature review of current perspectives on pathogenesis, prevention and control. *Journal of Microbiology, Immunology and Infection* 2020, *53*, 963–978, https://doi.org/https://doi.org/10.1016/j.jmii.2020.03.007.

22. Espín Jaramillo, E.A. Detection of leptospirosis and dengue in patients with acute undifferentiated febrile illness in the northern coast of Ecuador. Quito (Ecuador): Universidad San Francisco de Quito; 2015. Available from: http://repositorio.usfq.edu.ec/handle/23000/4488
23. Golding, M.A.J.; Noble, S.A.A.; Khouri, N.K.; Layne-Yarde, R.N.A.; Ali, I.; Sandiford, S.L. Natural vertical transmission of dengue virus in Latin America and the Caribbean: Highlighting its detection limitations and potential significance. *Parasites & Vectors* 2023, *16*, 442.
24. Taracena-Agarwal, M.L.; Hixson, B.; Nandakumar, S.; Girard-Mejia, A.P.; Chen, R.Y.; Huot, L.; Padilla, N.; Buchon, N. The midgut epithelium of mosquitoes adjusts cell proliferation and endoreplication to respond to physiological challenges. *BMC Biol* 2024, *22*, 22. https://doi.org/10.1186/s12915-023-01769-x.
25. Rather, I.A.; Parray, H.A.; Lone, J.B.; Paek, W.K.; Lim, J.; Bajpai, V.K.; Park, Y.-H. Prevention and control strategies to counter dengue virus infection. *Frontiers in Cellular and Infection Microbiology* 2017, *7*, 336.
26. Sim, S.; Ng, L.C.; Lindsay, S.W.; Wilson, A.L. A greener vision for vector control: The example of the Singapore dengue control programme. *PLOS Neglected Tropical Diseases* 2020, *14*, e0008428.
27. Brady, O.J.; Godfray, H.C.J.; Tatem, A.J.; Gething, P.W.; Cohen, J.M.; McKenzie, F.E.; Perkins, T.A.; Reiner, R.C.; Tusting, L.S.; Sinka, M.E. et al. Vectorial capacity and vector control: Reconsidering sensitivity to parameters for malaria elimination. *Transactions of The Royal Society of Tropical Medicine and Hygiene* 2016, *110*, 107–117. https://doi.org/10.1093/trstmh/trv113.
28. Codeço, C.T.; Lima, A.W.S.; Araújo, S.C.; Lima, J.B.P.; Maciel-de-Freitas, R.; Honório, N.A.; Galardo, A.K.R.; Braga, I.A.; Coelho, G.E.; Valle, D. Surveillance of *Aedes aegypti*: Comparison of house index with four alternative traps. *PLOS Neglected Tropical Diseases* 2015, *9*, e0003475. https://doi.org/10.1371/journal.pntd.0003475.
29. The, L Dengue: The threat to health now and in the future. *The Lancet* 2024, *404*, 311. https://doi.org/10.1016/S0140-6736(24)01542-3.
30. Tan, C.C.S.; van Dorp, L.; Balloux, F. The evolutionary drivers and correlates of viral host jumps. *Nature Ecology & Evolution* 2024, *8*, 960–971. https://doi.org/10.1038/s41559-024-02353-4.
31. Tambo, E.; Chen, J.-H.; Zhou, X.-N.; Khater, E.I.M. Outwitting dengue threat and epidemics resurgence in Asia-Pacific countries: Strengthening integrated dengue surveillance, monitoring and response systems. *Infectious Diseases of Poverty* 2016, *5*, 56.
32. Matysiak, A.; Roess, A. Interrelationship between climatic, ecologic, social, and cultural determinants affecting dengue emergence and transmission in Puerto Rico and their implications for Zika response. *Journal of Tropical Medicine* 2017, *2017*, 8947067.
33. Zeyaullah, M.; Muzammil, K.; AlShahrani, A.M.; Khan, N.; Ahmad, I.; Alam, M.S.; Ahmad, R.; Khan, W.H. Preparedness for the dengue epidemic: Vaccine as a viable approach. *Vaccines* 2022, *10*, 1940.
34. Halstead, S.B.; Lum, L.C.S. Assessing the prognosis of dengue-infected patients. *F1000 Medicine Reports* 2009, *1*, 73.
35. Kalayanarooj, S. Clinical manifestations and management of dengue/DHF/DSS. *Trop Med Health* 2011, *39*, 83–87. https://doi.org/10.2149/tmh.2011-S10.
36. Yacoub, S.; Mongkolsapaya, J.; Screaton, G. Recent advances in understanding dengue. *F1000Research* 2016, *5*, F1000-Faculty.
37. Adhikari, S.; Bhusal, S.; Hossain, M.S.; Basnyat, B. Assessment and management of dengue. *BMJ* 2025, *388*.
38. Zhang, W.X.; Zhao, T.Y.; Wang, C.C.; He, Y.; Lu, H.Z.; Zhang, H.T.; Wang, L.M.; Zhang, M.; Li, C.X.; Deng, S.Q. Assessing the global dengue burden: Incidence, mortality, and disability trends over three decades. *PLOS Neglected Tropical Diseases* 2025, *19*, e0012932. https://doi.org/10.1371/journal.pntd.0012932.
39. Gomez, D.H.; Willoquet, J. Dengue en las Américas: desafíos para su prevención y control. *Cad Saúde Pública* 2009, *25*, 19–31.
40. Khetarpal, N.; Khanna I. Dengue fever: Causes, complications, and vaccine strategies. *Journal of Immunology Research* 2016, *2016*, 6803098. https://doi.org/10.1155/2016/6803098.
41. Ahmed, I.; Ahamed, R.; Nahar, S.; Bari, L.F.; Dewan, S.M.R. Immunization against dengue virus infection is coercive: A timely call. *Health Science Reports* 2024, *7*, e2170.
42. Hladish, T.J.; Pearson, C.A.B.; Chao, D.L.; Rojas, D.P.; Recchia, G.L.; Gómez-Dantés, H.; Halloran, M.E.; Pulliam, J.R.C.; Longini, I.M. Projected impact of dengue vaccination in Yucatán, Mexico. *PLOS Neglected Tropical Diseases* 2016, *10*, e0004661.
43. World Health Organization. WHO position paper on dengue vaccines – May 2024. http://who.int/publications/i/item/who-wer-9918-203-224?utm_source=chatgpt.com (accessed on 19/09/2025).
44. Tana, S.; Umniyati, S.; Petzold, M.; Kroeger, A.; Sommerfeld, J. Building and analyzing an innovative community-centered dengue-ecosystem management intervention in Yogyakarta, Indonesia. *Pathogens and Global Health* 2012, *106*, 469–478.

45. Kolopack, P.A.; Parsons, J.A.; Lavery, J.V. What makes community engagement effective?: Lessons from the eliminate dengue program in Queensland Australia. *PLOS Neglected Tropical Diseases* 2015, *9*, e0003713.
46. Shepard, D.S.; Undurraga, E.A.; Betancourt-Cravioto, M.; Guzman, M.G.; Halstead, S.B.; Harris, E.; Mudin, R.N.; Murray, K.O.; Tapia-Conyer, R.; Gubler, D.J. Approaches to refining estimates of global burden and economics of dengue. *PLOS Neglected Tropical Diseases* 2014, *8*, e3306.
47. Castro, M.C.; Wilson, M.E.; Bloom, D.E. Disease and economic burdens of dengue. *The Lancet Infectious Diseases* 2017, *17*, e70–e78. https://doi.org/10.1016/S1473-3099(16)30545-X.
48. Tozan, Y.; Headley, T.Y.; Sewe, M.O.; Schwartz, E.; Shemesh, T.; Cramer, J.P.; Eberhardt, K.A.; Ramharter, M.; Harrison, N.; Leder, K. A prospective study on the impact and out-of-pocket costs of dengue illness in international travelers. *The American Journal of Tropical Medicine and Hygiene* 2019, *100*, 1525.
49. Marczell, K.; García, E.; Roiz, J.; Sachdev, R.; Towle, P.; Shen, J.; Sruamsiri, R.; da Silva, B.M.; Hanley, R. The macroeconomic impact of a dengue outbreak: Case studies from Thailand and Brazil. *PLOS Neglected Tropical Diseases* 2024, *18*, e0012201.
50. O'Mahony, J.F.; Paulden, M. Appraising the cost-effectiveness of vaccines in the UK: Insights from the Department of Health Consultation on the revision of methods guidelines. *Vaccine* 2019, *37*, 2831–2837, https://doi.org/https://doi.org/10.1016/j.vaccine.2019.03.072.
51. Taborda, A.; Chamorro, C.; Quintero, J.; Carrasquilla, G.; Londoño, D. Cost-effectiveness of a dengue vector control intervention in Colombia. *The American Journal of Tropical Medicine and Hygiene* 2022, *107*, 180.
52. Rather, I.A.; Parray, H.A.; Lone, J.B.; Paek, W.K.; Lim, J.; Bajpai, V.K.; Park, Y.H. Prevention and control strategies to counter dengue virus infection. *Front Cell Infect Microbiol* 2017, *7*, 336. https://doi.org/10.3389/fcimb.2017.00336.
53. Yadav, K.; Rabha, B.; Dhiman, S.; Veer, V. Multi-insecticide susceptibility evaluation of dengue vectors *Stegomyia albopicta* and *St. aegypti* in Assam, India. *Parasites & Vectors* 2015, *8*, 143.
54. Ruberto, I.; Marques, E.; Burke, D.S.; Van Panhuis, W.G. The availability and consistency of dengue surveillance data provided online by the World Health Organization. *PLOS Neglected Tropical Diseases* 2015, *9*, e0003511.
55. Olliaro, P.; Fouque, F.; Kroeger, A.; Bowman, L.; Velayudhan, R.; Santelli, A.C.; Garcia, D.; Skewes Ramm, R.; Sulaiman, L.H.; Tejeda, G.S. Improved tools and strategies for the prevention and control of arboviral diseases: A research-to-policy forum. *PLOS Neglected Tropical Diseases* 2018, *12*, e0005967.
56. Coronel-MartÍnez, D.L.; Park, J.; López-Medina, E.; Capeding, M.R.; Cadena Bonfanti, A.A.; Montalbán, M.C.; Ramírez, I.; Gonzales, M.L.A.; DiazGranados, C.A.; Zambrano, B. et al. Immunogenicity and safety of simplified vaccination schedules for the CYD-TDV dengue vaccine in healthy individuals aged 9–50 years (CYD65): A randomised, controlled, phase 2, non-inferiority study. *The Lancet Infectious Diseases* 2021, *21*, 517–528. https://doi.org/10.1016/S1473-3099(20)30767-2.
57. Moi, M.L.; Takasaki, T.; Kurane, I. Human antibody response to dengue virus: Implications for dengue vaccine design. *Tropical Medicine and Health* 2016, *44*, 1.
58. Liu, Y.; Liu, J.; Cheng, G. Vaccines and immunization strategies for dengue prevention. *Emerging Microbes & Infections* 2016, *5*, 1–6.
59. Li, L.; Guo, X.; Zhang, X.; Zhao, L.; Li, L.; Wang, Y.; Xie, T.; Yin, Q.; Jing, Q.; Hu, T. A unified global genotyping framework of dengue virus serotype-1 for a stratified coordinated surveillance strategy of dengue epidemics. *Infectious Diseases of Poverty* 2022, *11*, 107.
60. Hang, V.T.T. *Dengue Virus Evolution in Southern Vietnam Over the Last Ten Years and Utility of DENV-NSI as a Diagnostic Marker*; Open University (United Kingdom): 2010.
61. Wilder-Smith, A.; Renhorn, K.-E.; Tissera, H.; Abu Bakar, S.; Alphey, L.; Kittayapong, P.; Lindsay, S.; Logan, J.; Hatz, C.; Reiter, P. DengueTools: Innovative tools and strategies for the surveillance and control of dengue. *Global Health Action* 2012, *5*, 17273.
62. eClinicalMedicine Dengue as a growing global health concern. *EClinicalMedicine* 2024, *77*, 102975. https://doi.org/10.1016/j.eclinm.2024.102975.

7 Viral Evolution and Adaptation

Jasmin Šutković

7.1 INTRODUCTION

The most well-known characteristic of viruses as infection agents is clearly their quick development and adaptation to shifting environmental conditions and human hosts worldwide. Mutations rates are elevated in viruses, making them remarkably capable for adaptation and evolutionary advancement. In addition, occurrences of recombination and intricate relationships with host organisms helped in this process (1). All of the mentioned characteristics enabled the viruses to survive in oscillating environments, break the barriers of species and evade immune defenses, thereby presenting considerable obstacles to public health and disease management (2).

Therefore, the development of efficient antiviral drugs, vaccines, and therapeutic interventions requires a thorough understanding of viral evolution and adaptation. The real factors supporting viral development and adaptation are genetic diversity, which results from mutations, genetic recombination, and reassortment, is the real factor propelling viral development and adaptation (3). These changes in nucleotide sequences can enhance the fitness of viruses, improving their abilities to target and attach to hosts and achieve more efficient transmission (4). One of the most commonly found viruses is influenza virus, representing an excellent example of viral evolution, where new strains emerge continuously. Other examples are the adaptation of coronaviruses to human hosts, and the development of drug-resistant variants of the human immunodeficiency virus (HIV) and hepatitis C virus (HCV) (5–8). Furthermore, the interaction mechanism between the viruses and hoist plays a crucial role in shaping viral adaptation. Viruses must compete with host immune host reaction, cellular mechanisms, and ecological obstacles to successfully establish infections. This continuous competition between viral evolution and host immune responses often results in ongoing cycles of adaptation, commonly referred to as an evolutionary arms race (9).

Understanding the mechanisms of viral evolution and their interactions with host organisms is essential for forecasting potential viral outbreaks and for the design of effective therapeutic strategies. It's known that rapid mutation rates of viruses complicate the development of vaccines and the design of antiviral medications, hindering the management of epidemic situations. This chapter explores the adaptive processes of viruses, emphasizing genetic modifications, host interactions, and the implications for public health. Emphasizing the molecular and environmental influences that propel viral evolution, we aim to explain viral persistence, survival and transmission, thereby offering strategies to combat viral diseases.

7.2 GENETIC VARIABILITY OF VIRUSES

Viruses demonstrate significant genetic diversity, which plays a critical role in their adaptability, the expansion of their host range, and their capacity to evade immune defenses (10). Factors such as elevated mutation rates, recombination, reassortment, genetic drift, and selection pressures are instrumental in shaping viral evolution and enabling their survival across various environments (3, 11).

7.3 GENETIC VARIABILITY OF VIRUSES

Mutation is a primary driver of viral genetic diversity, and its rate varies significantly between RNA and DNA viruses. In RNA viruses, mutation rates are much higher than in DNA organisms, explaining their enhanced virulence and evolvability. However, RNA viruses cause the most devastating

 DOI: 10.1201/9781003607816-7

pandemics (12). RNA viruses typically exhibit higher mutation rates compared to DNA viruses due to the lack of proofreading mechanisms in their RNA-dependent RNA polymerases (RdRp) (13).

The mutation rate of RNA viruses ranges from 10^{-3} to 10^{-5} mutations per nucleotide per replication cycle (13, 14). This high mutation rate leads to the formation of viral quasispecies, a population of genetically related variants that enhance adaptability and resistance to antiviral drugs (12). Quasispecies was first explained by Eigen and Schuster, in 1970, explaining is as a mathematical framework called "precellular RNA world" (15).

Examples of highly mutable RNA viruses include:

- *Influenza virus*: The elevated mutation rate of the influenza virus promotes seasonal antigenic drift, requiring frequent modifications for vaccine production (2)
- *HIV*: The rapid mutation rate in HIV results in immune evasion, swift evolutionary changes, and the development of drug resistance, creating difficulties in antiviral treatment (16).
- *SARS-CoV-2*: The virus responsible for COVID-19 has exhibited considerable genetic diversity, resulting in the emergence of various variants that possess modified transmissibility and capabilities for immune evasion (17, 18). For example, the antigenic shift in Omicron led to the emergence of distinct SARS-CoV-2 variants, resulting in reinfections even among fully vaccinated or boosted individuals (19).

Viruses that typically are less diverse and evolve more slowly are DNA viruses, although they also include important pathogens such as herpesviruses and poxviruses (20). DNA viruses generally have lower mutation rates, around 10^{-6} to 10^{-8} mutations per nucleotide per replication cycle, due to the proofreading activity of DNA polymerases. However, herpesviruses and poxviruses DNA viruses exhibit genetic variability through recombination and other evolutionary mechanisms (10). Random mutations and fluctuations in the genome are not solely responsible for quasispecies formation but also, they occur by recombination and reassortment (21).

Other examples of DNA viruses with notable genetic variability include herpesviruses and hepatitis B virus (HBV). Herpesviruses undergo recombination, allowing them to evade host immunity and establish lifelong infections (22). However, these viruses are known to induce lifelong immunity after infection (23). HBV uses reverse transcription (despite being a DNA virus), and therefore introduces mutations, resulting in drug resistance (24).

The differences in mutation rates play a crucial role in shaping the evolutionary paths of viruses, affecting their capacity for antigenic drift and the development of drug-resistant variants.

7.3.1 Recombination and Reassortment Processes

Recombination refers to a genetic exchange mechanism whereby two viral genomes merge to create a new genome. This process takes place when two closely related viruses simultaneously infect a single host cell, resulting in the generation of recombinant genomes through template switching during the replication phase (25).

There are three major types of recombination; these are:

- *Homologous recombination*: Occurs when similar sequences from two viral genomes are exchanged, as seen in coronaviruses (26, 27).
- *Non-homologous recombination*: Occurs when unrelated sequences are combined, potentially leading to novel functionalities (26, 28).
- *Copy-choice recombination*: Occurs when RdRp switches templates during replication (29).

RNA viruses usually engage in recombination through either homologous or non-homologous mechanisms, both of which are crucial for their biological functions and evolutionary processes (30).

Homologous recombination in DNA viruses' functions via the same molecular processes as those involved in cellular DNA repair. This includes mechanisms such as strand invasion, the formation of Holliday junctions, and break-induced replication (31). Recombination plays a crucial role in creating genetic diversity within viral populations (32). RNA viruses experience genome recombination through a more distinct molecular mechanism, through a process known as "copy choice". Copy choice mechanism involves template switching during the replication of the genome. In this process, the RNA replicase complex detaches from one genomic template and reattaches to another. It is likely that complementary regions within the genomes facilitate the positioning of the replicase complex on the new template. The rate of copy choice recombination varies according to the specific mechanisms employed by different virus families for genome replication and transcription (31). For example, RNA viruses such as coronaviruses, frequently undergo recombination, which contributes to their emergence as novel viruses (27), as presented in Figure 7.1.

However, interactions between distinct and genetically diverse genomic regions, or between unrelated RNA molecules, can result in what is termed "non-homologous" (or illegitimate) recombination. This type of RNA recombination occurs between regions that share minimal sequence similarity, often leading to harmful genotypes. This may explain its significantly lower frequency compared to homologous recombination (33). For example, defective interfering (DI) RNAs (DI RNAs) of Sindbis virus has evolved from larger and some smaller viral species. It includes some cytopathic forms of bovine viral diarrhea virus, cellular RNA sequences at their 5′ ends, RNAs from potato leaf roll virus containing tobacco chloroplast RNA, or an influenza virus with an insertion of ribosomal RNA into the hemagglutinin (HA) (34, 35).

On the other hand, majority of RNA viruses undergo recombination via a copy-choice mechanism. This process involves the production of recombinant RNA molecules as the RdRp transitions from one template to another during the replication of the genome (30). For example, the RNA copy-choice recombination of CHIKV is responsible for genome diversification. Furthermore, RNA recombination has the potential to produce novel viral variants, as seen within mosquito vectors. This process serves as an effective mechanism for CHIKV to navigate stringent genetic bottlenecks and may even confer benefits for viral replication (36).

During the replication of the genome the copy-choice recombination plays a crucial role in generating recombinant RNA molecules by enabling the RdRp to switch between different templates. In the context of a specific picornavirus, certain unique sequences are particularly susceptible to copy-choice recombination. These sequences, referred to as recombination hotspots, are linked to both the secondary structures of RNA and the composition of nucleotide bases (37).

Reassortment represents a distinctive process found in segmented viruses, including influenza viruses, whereby complete genome segments are swapped between viruses that co-infect a host (38). This phenomenon can lead to the creation of highly pathogenic and antigenically innovative strains, exemplified by the rise of pandemic influenza viruses, such as the H1N1 strain identified in 2009 (39). Single-stranded and segmented genomes are associated with a broader range of hosts. However, they are prevalent among plant viruses, implying an evolutionary advantage. Research indicates that segmented genomes facilitate increased mutation rates, enhance particle stability, and promote genetic exchange via reassortment (40). For example, in influenza viruses the segmented genome enables antigenic shift, leading to pandemics (41).

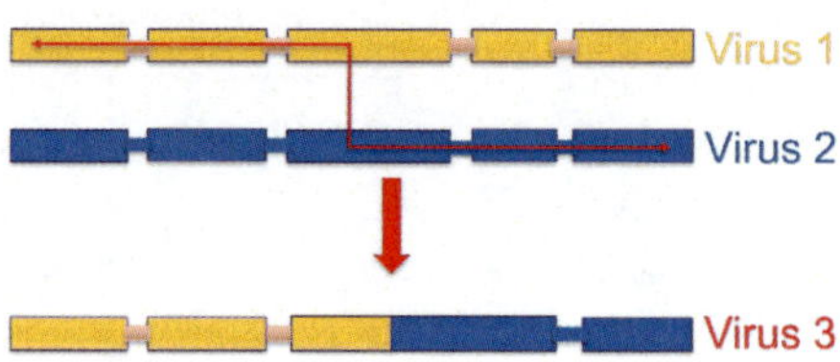

FIGURE 7.1 A simple presentation of viral RNA homologous recombination.

Another example is seen in a rotavirus, where segmentation leads to diverse strains affecting humans and animals (42).

To conclude, recombination and reassortment play a crucial role in the evolution of viruses, enabling them to switch hosts, undergo antigenic changes, and develop resistance against immune responses or antiviral treatments.

7.3.2 Role of Genetic Drift and Selection Pressure

Genetic drift is defined as a random change in allele frequencies within a viral population, especially common in smaller populations (10). Genetic drift significantly influences viral evolution by stabilizing neutral or harmful mutations within the population, thereby affecting antigenic variation and general viral fitness (43). As the RNA viruses are characterized by high mutation rates, as with the HIV and influenza viruses, the genetic drift impacts the effectiveness of vaccines and causes immune evasion (44).

The occurrence of genetic drift is higher in populations with higher incidence of certain viruses. For example, in populations affected by HIV, genetic drift has been noted to occur at rates roughly ten times greater than what would be expected in an ideal population. This particular drift is explained by the asynchronous pattern of infection, which provides a reproductive advantage to cells infected early on, resulting in random variations in the frequencies of viral genes (45).

Another example where genetic drift is seen is the influenza virus. Herem the genetic drift results in continuous antigenic changes that require regular updates to vaccines. For instance, the mixing of avian influenza viruses, for instance, has led to the emergence of new variants, such as the H5N1 clade 2.3.4.4b, which has shown mutations enhancing replication in human cells. In this study, the researchers found that these viruses have significantly evolved and mutated into new subtypes like H5N1, H5N6, and H5N8 (46).

Among the environmental factors that influence viral survival, selection pressure represents an crucial factor leading to viral mutations and adaptations, promoting genetic mutations that improve the virus's ability to survive, replicate, and spread (47). Several factors cause and force selection pressure, including host immune responses, antiviral drugs, and environmental factors. Immune responses from the host impose selective pressures on viruses, prompting them to mutate and evolve to avoid detection by the immune system. The adaptive immune response, by neutralizing antibodies and cytotoxic T lymphocytes (CTLs), focuses on viral proteins, forcing the virus to mutate in order to avoid recognition by the host immune response, as shown in influenza A virus (48). Further, HIV has showed to have elevated mutation rate of its reverse transcriptase enzyme, facilitates swift antigenic variation, enabling the virus to evade destruction by CTLs of the host (49). Similarly, variants of concern of SARS-CoV-2 virus, including Omicron, have developed mutations in the spike protein that improve their ability to evade the immune response and increase the likelihood of reinfection (50).

Antiviral drug overuse can drive viral evolution through the selection of drug-resistant mutations. For instance, the prolonged use of antivirals, particularly among immunocompromised patients, can result in the development of drug-resistant viral strains. This occurs because continuous viral replication while the drug is present favors the selection of resistant variants (51). Certain antiviral drugs, including molnupiravir, operate by promoting mutations within the viral genome as it replicates. Although many of these mutations are harmful to the virus, an overabundance of mutagenesis can unexpectedly result in the development of new viral variants with uncertain characteristics. Recent research has associated molnupiravir treatment with a heightened frequency of mutations in SARS-CoV-2, prompting worries about the possibility of creating new variants that could escape immune responses or exhibit changed transmissibility (52).

Environmental factors play an important role in shaping viral mutation rates and evolutionary paths by modifying the selective pressures that act upon viral populations. For example, a study investigated the H1N1 influenza A virus, revealing that the minimum temperature has a nonlinear association with viral mutations, reaching its highest point at approximately 15°C. Furthermore, factors like precipitation and population density were identified as correlating with mutation rates, indicating that both climatic and demographic factors may influence the evolution of the virus (53).

Another example, with the vesicular stomatitis virus suggests that environmental variability, encompassing factors such as the type of host cell and the duration of infection, significantly affects phenotypic diversity (54).

It is certain that climate change has a multivariable impact on the transmission of viruses. With climate change comes the potential of increasing viral interactions promoting mutations. For instance, rising global temperatures and increased climate-related changes have been linked to the increased prevalence of vector-borne diseases, including viruses (55). In addition, disinfection methods, while vital for lowering viral transmission, can impose selective pressures on the evolution of RNA viruses. For instance, exposure to disinfectants such as chlorine dioxide has been found to favor viral variants that exhibit reduced sensitivity, underscoring the importance of managing disinfection strategies carefully to mitigate the risk of developing resistant strains (56).

The interaction between genetic drift and selection pressure is shown to affect the diversity of viruses and plays a crucial role in the success of various viral strains across different hosts and environments. Gaining a more profound insight into these evolutionary mechanisms is vital for creating effective control strategies, such as vaccines, antiviral medications, and surveillance systems aimed at reducing viral threats.

7.4 HOST-VIRAL INTERACTIONS

The interactions between hosts and viruses represent a complicated and evolving relationship that influences the results of viral infections. Viruses belong to intracellular pathogens that rely on their hosts' cellular processes and entry mechanisms for both survival and reproduction (58).

7.4.1 Viral Entry Mechanisms

One of the most vital stages in the infection cycle is the virus's entry into the host cells. Viral surface proteins and host cell receptors interact precisely during this process, which are followed by mechanisms that enable the transfer of the viral genome into the cytoplasm of the host cell (58), and these mechanisms are:

- *Receptor binding stage*: The entry of a virus into a host cell starts with its attachment to specific receptors on the cell's surface. This interaction is marked by a high degree of specificity, which is essential for determining the virus's tropism. For example, the influenza virus utilizes HA to attach to sialic acid residues located on the epithelial cells of the respiratory tract (59). In a similar mechanism, the HIV specifically targets CD4 receptors along with co-receptors such as CCR5 or CXCR4 found on T cells, as demonstrated in Figure 7.2 (60).

  The most common identified Viral receptors include sialylated glycans, cell adhesion molecules such as those from the immunoglobulin superfamily and integrins, along with phosphatidylserine receptors. The redundancy observed in receptor engagement implies that viruses may target certain receptors or "common locks" to take the advantage of their cellular functions, suggesting an evolutionary conservation of these mechanisms (61).
- *Membrane fusion*: Enveloped viruses frequently use membrane fusion to introduce their genetic material into host cells. This mechanism is driven by fusion proteins that undergo structural changes upon interacting with specific receptors (62). For instance, the fusion protein in paramyxoviruses plays a crucial role in the fusion of the viral membrane with that of the host (63).
- *Endocytosis*: Some viruses utilize pore formation or endocytosis as alternative strategies for cell entry (64).

  For example, in reoviruses a special class of protein, called fusion-associated small transmembrane (FAST) proteins promote cell-cell fusion, thereby aiding the spread of

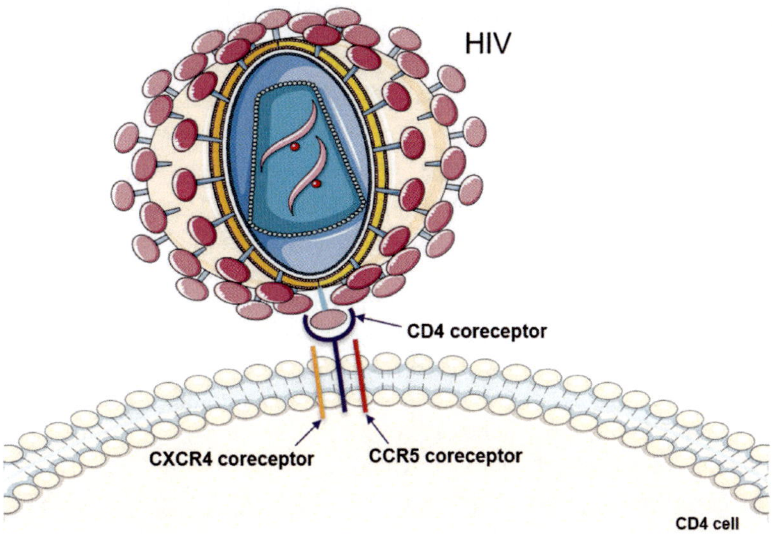

FIGURE 7.2 HIV-specific receptor binding.

the virus to adjacent cells. Although these proteins do not play a role in viral entry and reoviruses are classified as non-enveloped viruses, they are significant because they FAST proteins are small and non-structural proteins. Furthermore, when purified FAST proteins are incorporated into liposome membranes, they demonstrate the ability to induce fusion, confirming their role as genuine fusogen proteins (65).

Receptor-mediated endocytosis is a common strategy, as seen in dengue virus, which binds to specific receptors and is internalized into endosomes. Once inside, the acidic environment triggers conformational changes in viral proteins, enabling the release of the viral genome into the cytoplasm (66).

- *Direct injection*: Certain bacteriophages, as seen in myovirus bacteriophages, avoid the need for membrane fusion and endocytosis by directly injecting their genetic material into the host cell, as presented in Figure 7.3. This process is facilitated by unique structures, including tail fibers and a syringe-like device, this is known as reversible binding (67).

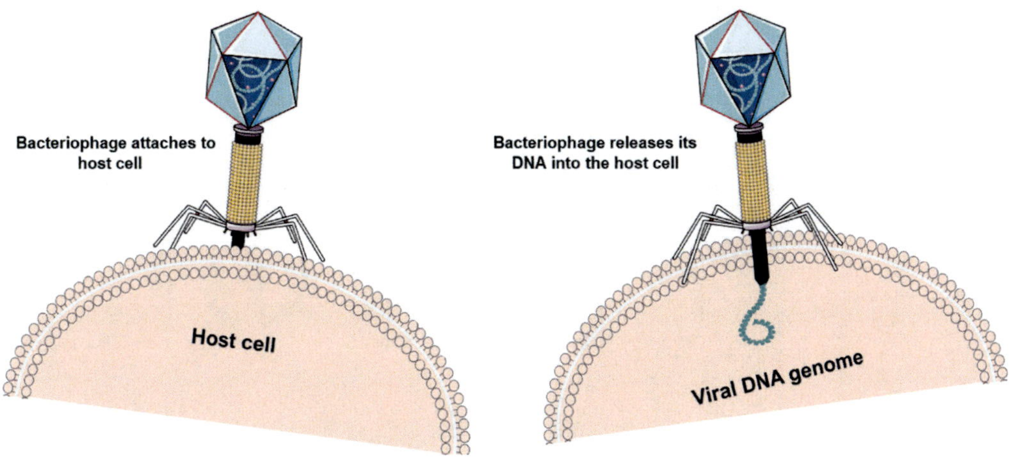

FIGURE 7.3 Example of myovirus bacteriophage direct entry into the human host cell.

7.4.2 Evasion of Host Immune Responses

Viruses have developed advanced strategies to initiate infection and maintain their presence within the host by evading or inhibiting the host's immune response. These mechanisms encompass antigenic variation, immune suppression, and the establishment of immune privilege (68).

- Antigenic variation enables viruses to avoid detection by the immune system through modifications to their surface proteins. For instance, influenza viruses experience both antigenic drift and shift, resulting in alterations to the HA and neuraminidase proteins. This adaptability allows the virus to avoid neutralizing antibodies, contributing to the occurrence of repeated infections (69).
- Viruses can directly inhibit the immune response of their hosts by interfering with essential signaling pathways. For example, the Epstein-Barr virus (EBV) generates proteins that resemble host cytokines, which in turn alters the activity of immune cells (70). Similarly, HIV disrupts the functionality of cluster of differentiation 4 cell surface glycoprotein (CD4+) T cells, resulting in progressive decline of the immune system and leads to the acquired immunodeficiency syndrome (AIDS) (71).
- Certain viruses can enter a state of latency in areas of the body that are considered immune-privileged, such as neurons, where the immune system's monitoring is reduced. A prime example of this is the herpes simplex virus (HSV), which resides in sensory ganglia and can reactivate when conditions become favorable (72). Recent studies show that while sensory neurons have impaired intrinsic immunity to HSV infection, paracrine IFN signaling can potentiate a potent antiviral response (73).

The innate immune system acts as the primary barrier against viral infections. Nevertheless, viruses have developed strategies to evade these immune responses. Viruses can evade, suppress, or manipulate the host's innate immune system to their own advantage, subverse it (74). For example, the HCV acts on the interferons by inhibition, targeting key components of the signaling pathway (75).

In the example of SARS-CoV-2, the virus enters the cell via two distinct mechanisms:

- TMPRSS2-Dependent Pathway (Cell Surface Entry):

  The virus attaches to the ACE2 receptor located on the cell membrane. Subsequently, the host enzyme TMPRSS2 cleaves the viral spike (S) protein, thereby activating it for fusion with the cell membrane. This process facilitates the direct entry of the viral genome into the cell at the surface, presented in Figure 7.4 (76, 77).

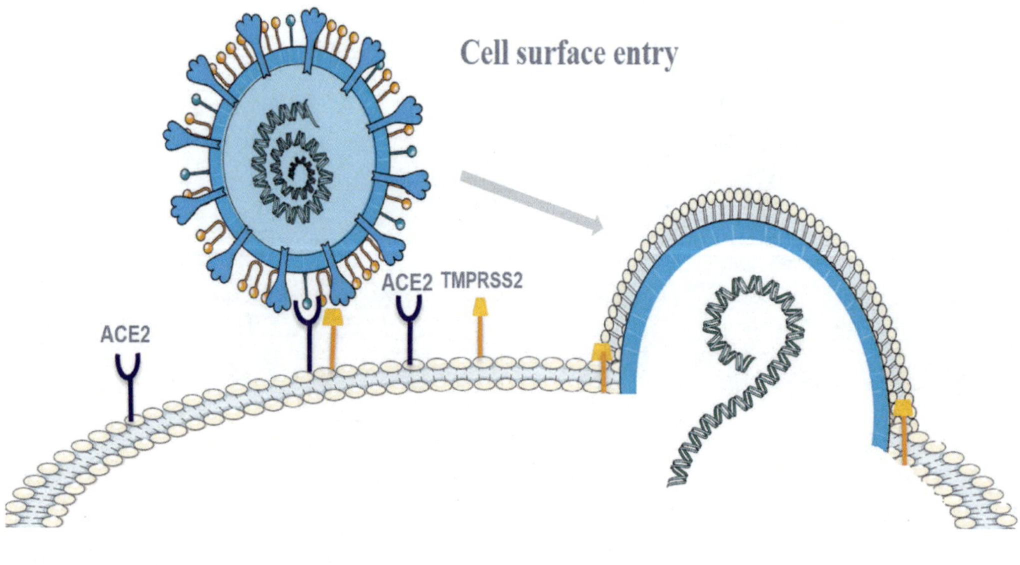

FIGURE 7.4 Cell surface entry SARS-CoV-2 entry.

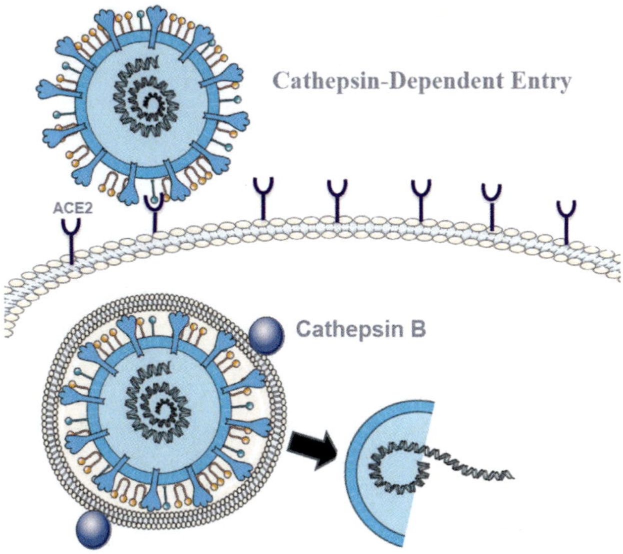

FIGURE 7.5 Cathepsin-dependent SARS-CoV-2 entry.

- Endosomal Pathway (Cathepsin-Dependent Entry):

  In the absence of TMPRSS2, the virus is taken up into endosomes. Within these endosomes, host proteases such as cathepsin B activate the spike protein, which results in fusion with the endosomal membrane and the subsequent release of the viral genome into the cytoplasm, as presented in Figure 7.5 (78).

7.4.3 Co-evolution of Viruses and Hosts

The molecular processes that facilitate viral invasion reveal the important strategies employed for successful infection and replication. Understanding these processes is crucial for the development of antiviral therapies and vaccines (79).

The complex relationship between viruses and their hosts has significantly influenced the field of evolutionary biology, marked by an ongoing "arms race" in which both sides evolve tactics to gain the upper hand over one another. The evolutionary connections among host species and hosts may account for differences in transmission rates, virulence, and the composition of viral communities among these hosts (80, 81). However, it is still unclear if there are any correlations in the infection characteristics of related viruses when they infect novel hosts.

7.4.3.1 The Evolutionary Arms Race: Host Defenses vs. Viral Strategies

Viruses are obligatory intracellular parasites, purely depending on the cellular machinery of their hosts for their replication processes. Over time, hosts have developed various defense mechanisms designed to identify and eliminate viral threats and viruses in return, have evolved strategies to circumvent or inhibit these defenses, creating an ongoing cycle of adaptation and counter-adaptation. A notable example of this dynamic is the interaction between viral pathogens and host cell surface receptors (82). For example, viral glycoproteins are located on the surface of enveloped viruses, where they engage with host cell receptors to facilitate viral entry and interact with antibodies. These proteins are crucial for the virus's infectivity and are subject to various functional limitations (83). These proteins are required to have a correct protein folding, incorporation into virions, oligomerization, binding to one or more cellular receptors, and undergo conformational changes necessary for membrane fusion, which enables the virus to enter the host cell (84). Despite these complications and limitations, certain viral glycoproteins exhibit a degree of mutational tolerance, as demonstrated by deep mutual scanning (DMS) screening, that identified variants capable of supporting viral replication in cell culture. For example, in the HIV-1 envelope protein (env) virus, known to be highly tolerant to mutations, a slight mutation in the env fusion peptide didn't affect its core function, mediating viral entry into the cell, so maintaining its functional integrity (82). Similarly, on the receptor binding site of influenza HA and the SARS-CoV-2 receptor-binding domains, similar mutations are found by DMS, where both viruses have tolerated the mutations and even enhanced protein binding domain functions (85, 86). Many viruses exploit specific receptors to enter host cells. For example, HIV uses its CD4 receptor along with co-receptors like CCR5 or CXCR4 to infect target cells (87). To cope with these viral processes, hosts have evolved restriction factors—proteins that inhibit various stages of the viral life cycle. These protein factors play a crucial role at each stage of the HIV-1 replication cycle, either facilitating effective infection and replication or significantly diminishing the likelihood of successful outcomes (88). Recent studies have demonstrated that genes encoding these restriction factors often exhibit signs of positive selection, fighting viral infection (89). However, viruses can overcome these restriction factors. A good example is seen in the Vif protein of HIV, which specifically targets the host restriction factor APOBEC3G for degradation. This action inhibits the incorporation of APOBEC3G into virions, thus facilitating the generation of infectious particles (90). This example illustrates the ongoing evolutionary struggle between hosts and viruses, where each modification made by the host is countered by a corresponding response from the virus.

7.4.3.2 Host Genetic Factors Influencing Exposure: Case of CCR5-Δ32 Mutations

Genetic variations among individuals play a crucial role in determining their vulnerability impact to viral infections and the advancement of related diseases. One of the confirmed and well-studied mutations is the mutation in the CCR5 gene, during HIV infections. This specific mutation, known as CCR5-Δ32 mutation, consists of a deletion of 32 base pairs within the CCR5 gene, resulting in the formation of a shortened, non-functional receptor that does not successfully localize to the cell membrane (91, 92). Homozygous individual with the CCR5-Δ32 allele, demonstrate a significant resistance to HIV-1 infection. This resistance is attributed to the lack of functional CCR5 co-receptors, which are essential for the virus to enter cells (93). In contrast, heterozygous individuals, who have one copy of the mutation, experience slower progression to AIDS when compared to individuals who do not carry the mutation (94).

The frequency of the CCR5-Δ32 allele varies geographically, with higher percentages observed in European populations (~10%) and rarity in African and Asian populations. In Brazil, the CCR5-Δ32 allele frequency is intermediate (4–6%) (95). This distribution indicates that historical selective pressures, potentially arising from previous epidemics, may have promoted the preservation of this mutation within specific populations (96). Further, CCR5-Δ32 mutation increase susceptibility to other pathogens as well. Research suggests that individuals possessing the CCR5-Δ32 genotype face an elevated risk of experiencing severe complications from West Nile Virus (WNV) infection (97). Studies have shown that mice lacking CCR5 show higher mortality rates when infected with WNV, indicating that CCR5 plays a vital role in the immune response to this virus, but not for viral transmission (98, 99). Similarly, studies involving humans showed increased CCR5-Δ32 mutations in patients who succumbed to WNV encephalitis, pointing out the essential function of CCR5 in the body's defense mechanisms (100). In addition to CCR5, various genetic mutations also play a role in determining susceptibility to HIV and the progression of the disease. Variations in genes that code for chemokine receptors and their corresponding ligands have been associated with the modulation of HIV pathogenesis (101).

TABLE 7.1
Common Mutations in Human Viruses and Their Impacts

Viruses	Gene/ Allele	Mutation/ Variant	Effects	References
HIV	CCR2	64I	Associated with delayed progression to AIDS	[102]
	CX3CR1	I249M280 haplotype	Linked to accelerated progression to AIDS	[103]
	HLA	Various alleles (B27, B57)	Affects strength and effectiveness of cytotoxic T-cell responses	[104]
Influenza	Hemagglutinin	K166Q	Leads to reduced antibody recognition	[105]
		S145N	Affects antigenicity and immune evasion.	[105]
		N225G/D	Modifies receptor-binding specificity	[106]
	Neuraminidase	H275Y	Confers resistance to oseltamivir (antiviral medication)	[107]
		E119V	Confers resistance to multiple neuraminidase inhibitors (oseltamivir, zanamivir, peramivir)	[108]
SARS-CoV-2	Spike (S)	D614G	Increases the number of spikes that can facilitate infection	[109]
		N501Y	Increases ACE2 receptor affinity	[110]
		E484K	Reduces neutralization by antibodies	[111]
		L452R	Enhances ACE2 binding and viral replication	[112]

7.4.3.3 Polymorphisms in Other Viruses

In general, most of the discovered polymorphisms influence susceptibility to viral infections. In Table 7.1 we present a few genetic polymorphisms examples found in HIV, influenza, and SARS-CoV-2.

7.5 PATHOGENESIS AND DISEASE OUTCOMES

In 1977, the world witnessed the elimination of smallpox alongside starting the modern era of genomics. In the last half century, seven epidemic viruses of significant international impact were mobilized, which led to increasingly extensive virus genome sequencing. These sequencing initiatives have illuminated the rapid adaptation of viruses to new hosts, particularly humans, offering valuable insights into the molecular processes that drive viral evolution (113).

7.5.1 How Viral Adaptation Impacts Disease Severity

Viral adaptation plays a crucial role in determining the severity of diseases via multiple interrelated mechanisms. In the case of influenza virus, several scientific studies have confirmed that mutations in the HA, non-structural protein 1 (NS1) and polymerase basic protein 2 (PB2) significantly increase the virus virulence and leading increase of clinical symptoms (114, 115). In the example of SARS-CoV-2's spike protein, the mutations in D614G and N501Y, enhanced the ACE2 binding, increased binding affinity to host and increasing the total viral load (116). By a different mechanism, viral mutations can help avoid recognition by antibodies or immune cells, reducing vaccine effectiveness or prolonging infection. An example is the Omicron variant of SARS-CoV-2, which showed significant immune escape properties (117).

Viruses can exhibit intense and accelerated replication as a result of adaptive mutations, which may result in more pronounced symptoms in humans. For example, the D614G mutation in SARS-CoV-2 has been associated with enhanced replication and increased infectivity (109, 116, 118).

The process of adaptation may enable viruses to invade various or more susceptible tissues, thereby increasing the severity of diseases. A good example is the Zika virus, which has evolved to specifically target neural progenitor cells, resulting in microcephaly in developing fetuses (119).

Further, due to adaptive mutations in viruses, the efficiency of viral transmission can be significantly increased (e.g., aerosolization), which indirectly affects severity due to healthcare system strain. In the case of SARS-CoV-2, the high mutation rate increased the total aerosol transmission potential (120).

Viruses show sometime host specific adaptations. For example, when viruses transfer from animals to humans, subsequent adaptations can either enhance or decrease the severity of the infection based on the interactions between the host and the pathogen (121). When viruses transition across species boundaries, this phenomenon is known as zoonotic spillover. These adaptations can affect the severity of the disease in various ways, either exacerbating or diminishing symptoms, contingent upon multiple host-pathogen interactions (122). In the case of H5N1 avian influenza, the virus has shown to have a high mortality in humans but low transmissibility, due to lack of adaptation (123, 124, 125).

7.6 CLIMATE CHANGE AND GLOBALIZATION IN THE ERA OF VIRAL INFECTIONS

Over 10,000 viruses can infect humans; however, the majority are currently circulating unnoticed among wild mammals (126). Changes in climate and land provide sufficient opportunities for viral transmission among wildlife species that were previously geographically isolated (127). A recent study has predicted that some species may cluster in new combinations at high altitudes, within biodiversity hotspots, and in regions of high human population density in Asia and Africa, resulting in an estimated 4,000 instances of cross-species transmission of their associated viruses (128). Bats, due to their unique dispersal capabilities and their abilities to protect themselves from viruses, are responsible for most novel viral transmissions and are likely to propagate viruses mutations and evolution, which will eventually facilitate future emergence in humans (129). Climate change is in progress, supporting ecological transition, and some studies confirm that global warming in the twenty-first century will not reduce future viral sharing (128).

The twenty-first century has experienced several significant infectious disease outbreaks, with COVID-19 being the most prominent, strongly impacting lives and economies on a global scale (130). Earlier outbreaks such as SARS (2003), swine flu (2009), MERS (2012), Ebola (2013–2016), and Zika (2015) resulted in considerable morbidity and mortality, frequently transcending national boundaries. Rapid global transformations, including an significant increase in air travel since 2000, urbanization overtaking rural living since 2007, persistent population growth, and escalating climate change—have amplified the risks associated with pandemics (128). Human actions such as overcrowding, inadequate sanitation, and environmental alterations including deforestation play a crucial role in the spread of viral diseases. Further, viral diseases can adversely affect the environment by contaminating water supplies, exacerbating deforestation, and diminishing biodiversity (131).

7.7 CONCLUSIONS

Viral genetic diversity and rapid adaptability are a crucial characteristic that allow them to survive, evolve, and flourish in various hosts and environments. The ongoing occurrence of new viral strains, sometimes exhibiting greater virulence or transmissibility, highlights the complex and evolving nature of viral evolution. Evolution is primarily driven by genetic processes such as point mutations, recombination, and reassortment, which collectively produce a diverse range of viral variants capable of evading immune detection, resisting antiviral therapies, and adapting to different hosts. Mutations, particularly in RNA viruses such as HIV, SARS-CoV-2 (COVID-19), hepatitis B and C, and influenza, are especially significant due to high mutation rates. These mutations usually modify viral proteins (exp: HA, NS1, PB2), which play a role in host recognition or immune evasion, thereby influencing disease severity and transmission potential. As an example, mutations in the HIV envelope protein will obstruct vaccine development and treatment strategies, while mutations in the SARS-CoV-2 spike protein have led to variants with stronger transmissibility and partial resistance to known antibodies. Besides mutations, genetic recombination and reassortment speed up the exchange of genetic material among viruses, resulting in new combinations and varieties. This is particularly observable in influenza viruses, where reassortment among avian, swine, and human strains can lead to the emergence of pandemic strains, as demonstrated during the 2009 H1N1 pandemic. Given examples demonstrate that viruses can acquire genetic characteristics that improve infectivity, expand host range, or provide drug resistance traits that can significantly alter disease outcomes.

While selection pressure, which is primarily driven by host immunological responses, antiviral medications, or immunizations, encourages the survival of viral variants with advantageous qualities, genetic drift refers to random changes in genetic material over time, particularly within viral populations. More "fit" virus strains become more common and persistent over time as a result of this dynamic shift, typically by decreasing the efficacy of current therapeutic interventions.

Furthermore, the host-virus interactions are important as they affect how viruses behave when they enter a host. Important factors like host genetics, existing medical conditions and immune health, influence how viruses replicate, how the immune system responds, and how diseases progress. These interactions can lead to outcomes that range from no symptoms to severe illness or death. Additionally, changes in climate and the environment complicate how viruses adapt. Climate change changes the habitats and behaviors of disease-carrying animals, which expands the areas where diseases can spread.

Last, the urban growth, deforestation, and global travel bring humans closer to wildlife, increasing the chances of diseases jumping from animals to humans. As a result, new and returning viral diseases are becoming more common and harder to manage. Understanding how viruses evolve and adapt is vital for predicting and tackling current and future viral threats. A combined approach that includes virology, epidemiology, ecology, and global health policy is required to monitor viral mutations and adaptations, develop responses, and protect populations in a world that is increasingly interconnected and environmentally unstable.

REFERENCES

1. Villarreal, L. P. (2008). Evolution of viruses. *Encyclopedia of Virology* (pp. 174–184). Academic Press. https://doi.org/10.1016/B978-012374410-4.00706-8
2. Elena, S. F., Solé, R., & Sardanyés, J. (2021). Phase transitions in virology. *Reports on Progress in Physics*, *84*(11), 115901. https://doi.org/10.1088/1361-6633/ac2ab0
3. Garcia-Ruiz, H., & LaTourrette, K. (2022). Determinants of virus variation, evolution, and host adaptation. *Pathogens*, *11*(9), 1039. https://doi.org/10.3390/pathogens11091039
4. Yu, X., Zhu, Y., Yin, G., Wang, Y., Shi, X., & Cheng, G. (2024). Exploiting hosts and vectors: Viral strategies for facilitating transmission. *EMBO Reports*, *25*(8), 3187–3201. https://doi.org/10.1038/s44319-024-00214-6
5. Pavia, G., Scarpa, F., Ciccozzi, A., Romano, C., Branda, F., Quirino, A., … & Ciccozzi, M. (2024). Changing and evolution of influenza virus: Is it a trivial flu? *Chemotherapy*, *69*(3), 185–193. https://doi.org/10.1159/000538382
6. Forni, D., Cagliani, R., Arrigoni, F., Benvenuti, M., Mozzi, A., Pozzoli, U., … & Sironi, M. (2021). Adaptation of the endemic coronaviruses HCoV-OC43 and HCoV-229E to the human host. *Virus Evolution*, *7*(2), veab061. https://doi.org/10.1093/ve/veab061
7. Nastri, B. M., Pagliano, P., Zannella, C., Folliero, V., Masullo, A., Rinaldi, L., … & Franci, G. (2023). HIV and drug-resistant subtypes. *Microorganisms*, *11*(1), 221. https://doi.org/10.3390/microorganisms11010221
8. Galli, A., & Bukh, J. (2023). Mechanisms and consequences of genetic variation in hepatitis C virus (HCV). In *Viral Fitness and Evolution: Population Dynamics and Adaptive Mechanisms* (pp. 237–264). Springer International Publishing. https://doi.org/10.1007/978-3-031-15640-3_7
9. Correa, A. M., Howard-Varona, C., Coy, S. R., Buchan, A., Sullivan, M. B., & Weitz, J. S. (2021). Revisiting the rules of life for viruses of microorganisms. *Nature Reviews Microbiology*, *19*(8), 501–513. https://doi.org/10.1038/s41579-021-00530-x

10. Sanjuan, R., & Domingo-Calap, P. (2021). Genetic diversity and evolution of viral populations. *Encyclopedia of Virology, 53*. https://doi.org/10.1016/B978-0-12-809633-8.20958-8
11. Greenbaum, B. D., & Ghedin, E. (2015). Viral evolution: beyond drift and shift. *Current Opinion in Microbiology, 26*, 109–115. https://doi.org/10.1016/j.mib.2015.06.015
12. Singh, K., Mehta, D., Dumka, S., Chauhan, A. S., & Kumar, S. (2023). Quasispecies nature of RNA viruses: Lessons from the past. *Vaccines, 11*(2), 308. https://doi.org/10.3390/vaccines11020308
13. Domingo, E., García-Crespo, C., Lobo-Vega, R., & Perales, C. (2021). Mutation rates, mutation frequencies, and proofreading-repair activities in RNA virus genetics. *Viruses, 13*(9), 1882. https://doi.org/10.3390/v13091882
14. Duffy, S. (2018). Why are RNA virus mutation rates so damn high? *PLOS Biology, 16*(8), e3000003. https://doi.org/10.1371/journal.pbio.3000003
15. Lauring, A. S., & Andino, R. (2010). Quasispecies theory and the behavior of RNA viruses. *PLOS Pathogens, 6*(7), e1001005.
16. Khaja, U. M., Bhat, A. H., Ahmed, M., Ali, A., & Ganie, S. A. (2023). Pharmacogenomics in viral diseases. In *Pharmacogenomics* (pp. 247–269). Academic Press. https://doi.org/10.1016/B978-0-443-15336-5.00006-3
17. Thakur, V., Bhola, S., Thakur, P., Patel, S. K. S., Kulshrestha, S., Ratho, R. K., & Kumar, P. (2022). Waves and variants of SARS-CoV-2: Understanding the causes and effect of the COVID-19 catastrophe. *Infection, 50*(2), 309–325. https://doi.org/10.1007/s15010-021-01752-0
18. Amicone, M., Borges, V., Alves, M. J., Isidro, J., Zé-Zé, L., Duarte, S., … & Gordo, I. (2022). Mutation rate of SARS-CoV-2 and emergence of mutators during experimental evolution. *Evolution, Medicine, and Public Health, 10*(1), 142–155. https://doi.org/10.1093/emph/eoac010
19. Carreño, J. M., Alshammary, H., Tcheou, J., Singh, G., Raskin, A. J., Kawabata, H., … & Krammer, F. (2022). Activity of convalescent and vaccine serum against SARS-CoV-2 Omicron. *Nature, 602*(7898), 682–688. https://doi.org/10.1038/s41586-022-04399-5
20. Holmes, E. C., & Drummond, A. J. (2007). The evolutionary genetics of viral emergence. *Wildlife and Emerging Zoonotic Diseases: The biology, Circumstances and Consequences of Cross-Species Transmission, 315*, 51–66. https://doi.org/10.1007/978-3-540-70962-6_3
21. Domingo, E., García-Crespo, C., & Perales, C. (2021). Historical perspective on the discovery of the quasispecies concept. *Annual Review of Virology, 8*(1), 51–72. https://doi.org/10.1146/annurev-virology-091919-105900
22. Soltani, S., Shenagari, M., & Emadi, M. S. (2023). Exploring the replication mechanisms of DNA and RNA viruses. In *Viral Replication Cycle-From Pathogenesis and Immune Response to Diagnosis and Therapy*. IntechOpen. https://doi.org/10.5772/intechopen.1003767
23. Kamel, M. S., Munds, R. A., & Verma, M. S. (2023). The quest for immunity: Exploring human herpesviruses as vaccine vectors. *International Journal of Molecular Sciences, 24*(22), 16112. https://doi.org/10.3390/ijms242216112
24. Qu, B., & Brown, R. J. (2021). Strategies to inhibit hepatitis B virus at the transcript level. *Viruses, 13*(7), 1327. https://doi.org/10.3390/v13071327
25. Sztuba-Solińska, J., Urbanowicz, A., Figlerowicz, M., & Bujarski, J. J. (2011). RNA-RNA recombination in plant virus replication and evolution. *Annual Review of Phytopathology, 49*(1), 415–443. https://doi.org/10.1146/annurev-phyto-072910-095351
26. Pérez-Losada, M., Arenas, M., Galán, J. C., Palero, F., & González-Candelas, F. (2015). Recombination in viruses: Mechanisms, methods of study, and evolutionary consequences. *Infection, Genetics and Evolution, 30*, 296–307. https://doi.org/10.1016/j.meegid.2014.12.022
27. Wells, H. L., Bonavita, C. M., Navarrete-Macias, I., Vilchez, B., Rasmussen, A. L., & Anthony, S. J. (2023). The coronavirus recombination pathway. *Cell Host & Microbe, 31*(6), 874–889. https://doi.org/10.1016/j.chom.2023.05.003
28. Vallée, G., Norris, P., Paszkowski, P., Noyce, R. S., & Evans, D. H. (2021). Vaccinia virus gene acquisition through nonhomologous recombination. *Journal of Virology, 95*(14), 10–1128. https://doi.org/10.1128/jvi.00318-21
29. Arnold, J. J., Martinez, A., Jain, A., Liu, X., Moustafa, I. M., & Cameron, C. E. (2025). Mechanism of forced-copy-choice RNA recombination by enteroviral RNA-dependent RNA polymerases. *bioRxiv*, 2025-02. https://doi.org/10.1101/2025.02.07.637143
30. Wang, H., Cui, X., Cai, X., & An, T. (2022). Recombination in positive-strand RNA viruses. *Frontiers in Microbiology, 13*, 870759. https://doi.org/10.3389/fmicb.2022.870759
31. Payne, S. (2017). Virus evolution and genetics. In *Viruses: From Understanding to Investigation* (pp. 81–86). Elsevier.

32. Domingo, E. (2010). Mechanisms of viral emergence. *Veterinary Research*, *41*(6), 38. https://doi.org/10.1051/vetres/2010010
33. Simon-Loriere, E., & Holmes, E. C. (2011). Why do RNA viruses recombine? *Nature Reviews Microbiology*, *9*(8), 617–626. https://doi.org/10.1038/nrmicro2614
34. Domingo, E. (2019). Molecular basis of genetic variation of viruses: Error-prone replication. *Virus as Populations*, *35*. https://doi.org/10.1016/B978-0-12-816331-3.00002-7
35. Domingo, E., Escarmís, C., Menéndez-Arias, L., Perales, C., Herrera, M., Novella, I. S., & Holland, J. J. (2008). Viral quasispecies: Dynamics, interactions, and pathogenesis. In E. Domingo, C. Parrish, & J. J. Holland (Eds.), *Origin and Evolution of Viruses* (pp. 87–118). Elsevier. https://doi.org/10.1016/B978-0-12-374153-0.00004-7
36. Filomatori, C. V., Bardossy, E. S., Merwaiss, F., Suzuki, Y., Henrion, A., Saleh, M. C., & Alvarez, D. E. (2019). RNA recombination at Chikungunya virus 3′UTR as an evolutionary mechanism that provides adaptability. *PLOS Pathogens*, *15*(4), e1007706. https://doi.org/10.1371/journal.ppat.1007706
37. Runckel, C., Westesson, O., Andino, R., & DeRisi, J. L. (2013). Identification and manipulation of the molecular determinants influencing poliovirus recombination. *PLOS Pathogens*, *9*(2), e1003164. https://doi.org/10.1371/journal.ppat.1003164
38. Webster, R. G., Bean, W. J., Gorman, O. T., Chambers, T. M., & Kawaoka, Y. (1992). Evolution and ecology of influenza A viruses. *Microbiological Reviews*, *56*(1), 152–179. https://doi.org/10.1128/mr.56.1.152-179.1992
39. Neumann, G., Noda, T., & Kawaoka, Y. (2009). Emergence and pandemic potential of swine-origin H1N1 influenza virus. *Nature*, *459*(7249), 931–939. https://doi.org/10.1038/nature08157
40. Moury, B., Fabre, F., Hébrard, E., & Froissart, R. (2017). Determinants of host species range in plant viruses. *Journal of General Virology*, *98*(4), 862–873. https://doi.org/10.1099/jgv.0.000742
41. Lowen, A. C. (2017). Constraints, drivers, and implications of influenza A virus reassortment. *Annual Review of Virology*, *4*(1), 105–121. https://doi.org/10.1146/annurev-virology-101416-041726
42. McDonald, S. M., Nelson, M. I., Turner, P. E., & Patton, J. T. (2016). Reassortment in segmented RNA viruses: Mechanisms and outcomes. *Nature Reviews Microbiology*, *14*(7), 448–460. https://doi.org/10.1038/nrmicro.2016.46
43. Duggal, N. K., & Emerman, M. (2012). Evolutionary conflicts between viruses and restriction factors shape immunity. *Nature Reviews Immunology*, *12*(10), 687–695. https://doi.org/10.1038/nri3295
44. Parrish, C. R., Holmes, E. C., Morens, D. M., Park, E. C., Burke, D. S., Calisher, C. H., … & Daszak, P. (2008). Cross-species virus transmission and the emergence of new epidemic diseases. *Microbiology and Molecular Biology Reviews*, *72*(3), 457–470. https://doi.org/10.1128/mmbr.00004-08
45. Voronin, Y., Holte, S., Overbaugh, J., & Emerman, M. (2009). Genetic drift of HIV populations in culture. *PLOS Genetics*, *5*(3), e1000431. https://doi.org/10.1371/journal.pgen.1000431
46. Lin, S., Chen, J., Li, K., Liu, Y., Fu, S., Xie, S., … & Jia, W. (2024). Evolutionary dynamics and comparative pathogenicity of clade 2.3. 4.4 b H5 subtype avian influenza viruses, China, 2021–2022. *Virologica Sinica*, *39*(3), 358–368. https://doi.org/10.1016/j.virs.2024.04.004
47. Spielman, S. J., Weaver, S., Shank, S. D., Magalis, B. R., Li, M., & Kosakovsky Pond, S. L. (2019). Evolution of viral genomes: Interplay between selection, recombination, and other forces. *Evolutionary Genomics: Statistical and Computational Methods*, *1910*, 427–468. https://doi.org/10.1007/978-1-4939-9074-0_14
48. McMichael, A. J., Borrow, P., Tomaras, G. D., Goonetilleke, N., & Haynes, B. F. (2010). The immune response during acute HIV-1 infection: Clues for vaccine development. *Nature Reviews Immunology*, *10*(1), 11–23. https://doi.org/10.1038/nri2674
49. Hill, A. L., Rosenbloom, D. I., & Nowak, M. A. (2012). Evolutionary dynamics of HIV at multiple spatial and temporal scales. *Journal of Molecular Medicine*, *90*, 543–561. https://doi.org/10.1007/s00109-012-0892-1
50. Harvey, W. T., Carabelli, A. M., Jackson, B., Gupta, R. K., Thomson, E. C., Harrison, E. M., … & Robertson, D. L. (2021). SARS-CoV-2 variants, spike mutations and immune escape. *Nature Reviews Microbiology*, *19*(7), 409–424. https://doi.org/10.1038/s41579-021-00573-0
51. Strasfeld, L., & Chou, S. (2010). Antiviral drug resistance: Mechanisms and clinical implications. *Infectious Disease Clinics of North America*, *24*(2), 413. https://doi.org/10.1016/j.idc.2010.01.001
52. Sanderson, T., Hisner, R., Donovan-Banfield, I. A., Hartman, H., Løchen, A., Peacock, T. P., & Ruis, C. (2023). A molnupiravir-associated mutational signature in global SARS-CoV-2 genomes. *Nature*, *623*(7987), 594–600. https://doi.org/10.1038/s41586-023-06649-6

53. Jiang, D., Wang, Q., Bai, Z., Qi, H., Ma, J., Liu, W., ... & Li, J. (2021). Could the environment affect the mutation of H1N1 influenza virus?. In *Advances in Clinical Immunology, Medical Microbiology, COVID-19, and Big Data* (pp. 255–267). Jenny Stanford Publishing. https://doi.org/10.3390/ijerph17093092
54. Smith-Tsurkan, S. D., Herr, R. A., Khuder, S., Wilke, C. O., & Novella, I. S. (2013). The role of environmental factors on the evolution of phenotypic diversity in vesicular stomatitis virus populations. *Journal of General Virology*, *94*(4), 860–868. https://doi.org/10.1099/vir.0.048082-0
55. Ali, A., Shaikh, A., Sethi, I., & Surani, S. (2024). Climate change and the emergence and exacerbation of infectious diseases: A review. *World Journal of Virology*, *13*(4), 96476. https://dx.doi.org/10.5501/wjv.v13.i4.96476
56. Rachmadi, A. T., Kitajima, M., Watanabe, K., Okabe, S., & Sano, D. (2018). Disinfection as a selection pressure on RNA virus evolution. https://doi.org/10.1021/acs.est.8b00590
57. Kaminskyy, V., & Zhivotovsky, B. (2010). To kill or be killed: How viruses interact with the cell death machinery. *Journal of Internal Medicine*, *267*(5), 473–482. https://doi.org/10.1111/j.1365-2796.2010.02222.x
58. Thorley, J. A., McKeating, J. A., & Rappoport, J. Z. (2010). Mechanisms of viral entry: Sneaking in the front door. *Protoplasma*, *244*, 15–24. https://doi.org/10.1007/s00709-010-0152-6
59. Skehel, J. J., & Wiley, D. C. (2000). Receptor binding and membrane fusion in virus entry: The influenza hemagglutinin. *Annual Review of Biochemistry*, *69*(1), 531–569. https://doi.org/10.1146/annurev.biochem.69.1.531
60. Vicenzi, E., Liò, P., & Poli, G. (2013). The puzzling role of CXCR4 in human immunodeficiency virus infection. *Theranostics*, *3*(1), 18. https://doi.org/10.7150/thno.5392
61. Maginnis, M. S. (2018). Virus–receptor interactions: The key to cellular invasion. *Journal of Molecular Biology*, *430*(17), 2590–2611. https://doi.org/10.1016/j.jmb.2018.06.024
62. Podbilewicz, B. (2014). Virus and cell fusion mechanisms. *Annual Review of Cell and Developmental Biology*, *30*(1), 111–139. https://doi.org/10.1146/annurev-cellbio-101512-122422).
63. Jardetzky, T. S., & Lamb, R. A. (2014). Activation of paramyxovirus membrane fusion and virus entry. *Current Opinion in Virology*, *5*, 24–33. https://doi.org/10.1016/j.coviro.2014.01.005
64. Mas, V., & Melero, J. A. (2024). Entry of enveloped viruses into host cells: Membrane fusion. *Structure and Physics of Viruses: An Integrated Guide*, *105*, 567–592. https://doi.org/10.1007/978-94-007-6552-8_16
65. Top, D., de Antueno, R., Salsman, J., Corcoran, J., Mader, J., Hoskin, D., ... & Duncan, R. (2005). Liposome reconstitution of a minimal protein-mediated membrane fusion machine. *The EMBO Journal*, *24*(17), 2980–2988. https://doi.org/10.1038/sj.emboj.7600767
66. Cruz-Oliveira, C., Freire, J. M., Conceição, T. M., Higa, L. M., Castanho, M. A., & Da Poian, A. T. (2015). Receptors and routes of dengue virus entry into the host cells. *FEMS Microbiology Reviews*, *39*(2), 155–170. https://doi.org/10.1093/femsre/fuu004
67. Maghsoodi, A., Chatterjee, A., Andricioaei, I., & Perkins, N. C. (2019). How the phage T4 injection machinery works including energetics, forces, and dynamic pathway. *Proceedings of the National Academy of Sciences*, *116*(50), 25097–25105. https://doi.org/10.1073/pnas.1909298116
68. Strumillo, S. T., Kartavykh, D., de Carvalho, F. F. Jr, Cruz, N. C., de Souza Teodoro, A. C., Sobhie Diaz, R., & Curcio, M. F. (2021). Host–virus interaction and viral evasion. *Cell Biology International*, *45*(6), 1124–1147. https://doi.org/10.1002/cbin.11565
69. Kim, H., Webster, R. G., & Webby, R. J. (2018). Influenza virus: Dealing with a drifting and shifting pathogen. *Viral Immunology*, *31*(2), 174–183. https://doi.org/10.1089/vim.2017.0141
70. Ressing, M. E., van Gent, M., Gram, A. M., Hooykaas, M. J., Piersma, S. J., & Wiertz, E. J. (2015). Immune evasion by Epstein-Barr virus. *Epstein Barr Virus Volume 2: One Herpes Virus: Many Diseases*, *391*, 355–381. https://doi.org/10.1007/978-3-319-22834-1_12
71. Okoye, A. A., & Picker, L. J. (2013). CD 4+ T-cell depletion in HIV infection: Mechanisms of immunological failure. *Immunological Reviews*, *254*(1), 54–64. https://doi.org/10.1111/imr.12066
72. Gasilina, A. (2016). *Herpes Simplex Virus-1: Crosstalk Between the Host Immunity and the Virus during Infection, Latency and Reanimation* (Master's thesis, University of Cincinnati). https://rave.ohiolink.edu/etdc/view?acc_num=ucin1460444615
73. Rosato, P. C., & Leib, D. A. (2015). Neurons versus herpes simplex virus: The innate immune interactions that contribute to a host–pathogen standoff. *Future Virology*, *10*(6), 699–714. https://doi.org/10.2217/fvl.15.45
74. Tortorella, D., Gewurz, B. E., Furman, M. H., Schust, D. J., & Ploegh, H. L. (2000). Viral subversion of the immune system. *Annual Review of Immunology*, *18*(1), 861–926. https://doi.org/10.1146/annurev.immunol.18.1.861
75. Bode, J. G., Brenndörfer, E. D., & Häussinger, D. (2007). Subversion of innate host antiviral strategies by the hepatitis C virus. *Archives of Biochemistry and Biophysics*, *462*(2), 254–265. https://doi.org/10.1016/j.abb.2007.03.033

76. Jackson, C. B., Farzan, M., Chen, B., & Choe, H. (2022). Mechanisms of SARS-CoV-2 entry into cells. *Nature Reviews Molecular Cell Biology, 23*(1), 3–20. https://doi.org/10.1038/s41580-021-00418-x
77. Hoffmann, M., Kleine-Weber, H., Schroeder, S., Krüger, N., Herrler, T., Erichsen, S., … & Pöhlmann, S. (2020). SARS-CoV-2 cell entry depends on ACE2 and TMPRSS2 and is blocked by a clinically proven protease inhibitor. *Cell, 181*(2), 271–280. https://doi.org/10.1016/j.cell.2020.02.052
78. Zhao, M. M., Yang, W. L., Yang, F. Y., Zhang, L., Huang, W. J., Hou, W., … & Yang, J. K. (2021). Cathepsin L plays a key role in SARS-CoV-2 infection in humans and humanized mice and is a promising target for new drug development. *Signal Transduction and Targeted Therapy, 6*(1), 134. https://doi.org/10.1038/s41392-021-00558-8
79. Kausar, S., Said Khan, F., Ishaq Mujeeb Ur Rehman, M., Akram, M., Riaz, M., Rasool, G., … & Malik, A. (2021). A review: Mechanism of action of antiviral drugs. *International Journal of Immunopathology and Pharmacology, 35*, 20587384211002621. https://doi.org/10.1177/20587384211002621
80. Imrie, R. M., Roberts, K. E., & Longdon, B. (2021). Between virus correlations in the outcome of infection across host species: Evidence of virus by host species interactions. *Evolution Letters, 5*(5), 472–483. https://doi.org/10.1002/evl3.247
81. Han, A. X., de Jong, S. P., & Russell, C. A. (2023). Co-evolution of immunity and seasonal influenza viruses. *Nature Reviews Microbiology, 21*(12), 805–817. https://doi.org/10.1038/s41579-023-00945-8
82. Tenthorey, J. L., Emerman, M., & Malik, H. S. (2022). Evolutionary landscapes of host-virus arms races. *Annual Review of Immunology, 40*(1), 271–294. https://doi.org/10.1146/annurev-immunol-072621-084422
83. Li, Y., Liu, D., Wang, Y., Su, W., Liu, G., & Dong, W. (2021). The importance of glycans of viral and host proteins in enveloped virus infection. *Frontiers in Immunology, 12*, 638573. https://doi.org/10.3389/fimmu.2021.638573
84. Checkley, M. A., Luttge, B. G., & Freed, E. O. (2011). HIV-1 envelope glycoprotein biosynthesis, trafficking, and incorporation. *Journal of Molecular Biology, 410*(4), 582–608. https://doi.org/10.1016/j.jmb.2011.04.042).
85. Starr, T. N., Greaney, A. J., Hilton, S. K., Ellis, D., Crawford, K. H., Dingens, A. S., … & Bloom, J. D. (2020). Deep mutational scanning of SARS-CoV-2 receptor binding domain reveals constraints on folding and ACE2 binding. *Cell, 182*(5), 1295–1310. https://doi.org/10.1016/j.cell.2020.08.012
86. Doud, M. B., & Bloom, J. D. (2016). Accurate measurement of the effects of all amino-acid mutations on influenza hemagglutinin. *Viruses, 8*(6), 155.
87. Svanberg, C. (2023). *HIV-1 Exploitation of Dendritic Cell Functionality and Initial Responses in Mucosal Tissues Elucidation of Influence of HIV-1 Complement Opsonization, and HIV-1-HSV-2 Co-Infection*. Linkopings Universitet (Sweden). https://doi.org/10.3384/9789180751230
88. Moezpoor, M. R., & Stevenson, M. (2024). Help or hinder: Protein host factors that impact HIV-1 replication. *Viruses, 16*(8), 1281. https://doi.org/10.3390/v16081281
89. Coffin, J. M. (2013). Virions at the gates: Receptors and the host–virus arms race. *PLOS Biology, 11*(5), e1001574. https://doi.org/10.1371/journal.pbio.1001574
90. Desimmie, B. A., Delviks-Frankenberrry, K. A., Burdick, R. C., Qi, D., Izumi, T., & Pathak, V. K. (2014). Multiple APOBEC3 restriction factors for HIV-1 and one Vif to rule them all. *Journal of Molecular Biology, 426*(6), 1220–1245. https://doi.org/10.1016/j.jmb.2013.10.033
91. Galvani, A. P., & Novembre, J. (2005). The evolutionary history of the CCR5-Δ32 HIV-resistance mutation. *Microbes and Infection, 7*(2), 302–309. https://doi.org/10.1016/j.micinf.2004.12.006
92. Petronić, T., Alfilali, M., & Ašić, A. (2024). The CCR5Δ32 allele as an HIV infection resistance marker: Possible evolutionary theories of origin. *Gene Reports, 34*, 101856. https://doi.org/10.1016/j.genrep.2023.101856
93. Saifullah, M., Laghzaoui, O., Ozyahyalar, H., & Irfan, A. (2024). The CRISPR-Cas9 induced CCR5 Δ32 mutation as a potent gene therapy methodology for resistance to HIV-1 variant: A review. *European Review for Medical & Pharmacological Sciences, 28*(6). https://doi.org/10.26355/eurrev_202403_35751
94. Mohamed, H., Gurrola, T., Berman, R., Collins, M., Sariyer, I. K., Nonnemacher, M. R., & Wigdahl, B. (2022). Targeting CCR5 as a component of an HIV-1 therapeutic strategy. *Frontiers in Immunology, 12*, 816515. https://doi.org/10.3389/fimmu.2021.816515
95. Kulmann-Leal, B., Ellwanger, J. H., & Chies, J. A. B. (2021). CCR5Δ32 in Brazil: Impacts of a European genetic variant on a highly admixed population. *Frontiers in Immunology, 12*, 758358. https://doi.org/10.3389/fimmu.2021.758358
96. OBrien, S. J. (2024). Legacy of a magic gene—CCR5-Δ 32: From discovery to clinical benefit in a generation. *Proceedings of the National Academy of Sciences, 121*(12), e2321907121. https://doi.org/10.1073/pnas.2321907121

97. Glass, W. G., Lim, J. K., Cholera, R., Pletnev, A. G., Gao, J. L., & Murphy, P. M. (2005). Chemokine receptor CCR5 promotes leukocyte trafficking to the brain and survival in West Nile virus infection. *The Journal of Experimental Medicine*, *202*(8), 1087–1098. https://doi.org/10.1084/jem.20042530
98. Lim, J. K., McDermott, D. H., Lisco, A., Foster, G. A., Krysztof, D., Follmann, D., ... & Murphy, P. M. (2010). CCR5 deficiency is a risk factor for early clinical manifestations of West Nile virus infection but not for viral transmission. *The Journal of Infectious Diseases*, *201*(2), 178–185. https://doi.org/10.1086/649426
99. Glass, W. G., McDermott, D. H., Lim, J. K., Lekhong, S., Yu, S. F., Frank, W. A., ... & Murphy, P. M. (2006). CCR5 deficiency increases risk of symptomatic West Nile virus infection. *The Journal of Experimental Medicine*, *203*(1), 35–40. https://doi.org/10.1084/jem.20051970
100. Jasinska, A. J., Pandrea, I., & Apetrei, C. (2022). CCR5 as a coreceptor for human immunodeficiency virus and simian immunodeficiency viruses: A prototypic love-hate affair. *Frontiers in Immunology*, *13*, 835994. https://doi.org/10.3389/fimmu.2022.835994
101. Kou, J., & Kuang, Y. Q. (2019). Mutations in chemokine receptors and AIDS. *Progress in Molecular Biology and Translational Science*, *161*, 113–124. https://doi.org/10.1016/bs.pmbts.2018.10.001
102. Smith, M. W., Dean, M., Carrington, M., Winkler, C., Huttley, G. A., Lomb, D. A., ... & O'Brien, S. J. (1997). Contrasting genetic influence of CCR2 and CCR5 variants on HIV-1 infection and disease progression. *Science*, *277*(5328), 959–965. https://doi.org/10.1126/science.277.5328.959
103. Faure, S., Meyer, L., Costagliola, D., Vaneensberghe, C., Genin, E., Autran, B., ... & Combadière, C. (2000). Rapid progression to AIDS in HIV+ individuals with a structural variant of the chemokine receptor CX3CR1. *Science*, *287*(5461), 2274–2277. https://doi.org/10.1126/science.287.5461.2274
104. Carrington, M., & O'Brien, S. J. (2003). The influence of HLA genotype on AIDS. *Annual Review of Medicine*, *54*(1), 535–551. https://doi.org/10.1146/annurev.med.54.101601.152346
105. McCarthy, D. P., Donald, P. F., Scharlemann, J. P., Buchanan, G. M., Balmford, A., Green, J. M., ... & Butchart, S. H. (2012). Financial costs of meeting global biodiversity conservation targets: Current spending and unmet needs. *Science*, *338*(6109), 946–949. https://doi.org/10.1126/science.1229803
106. Wu, N. C., Thompson, A. J., Xie, J., Lin, C. W., Nycholat, C. M., Zhu, X., ... & Wilson, I. A. (2018). A complex epistatic network limits the mutational reversibility in the influenza hemagglutinin receptor-binding site. *Nature Communications*, *9*(1), 1264. https://doi.org/10.1038/s41467-018-03663-5
107. DeVries, A., Wotton, J., Lees, C., Boxrud, D., Uyeki, T., & Lynfield, R. (2012). Neuraminidase H275Y and hemagglutinin D222G mutations in a fatal case of 2009 pandemic influenza A (H1N1) virus infection. *Influenza and other Respiratory Viruses*, *6*(6), e85–e88. https://doi.org/10.1111/j.1750-2659.2011.00329.x
108. Abed, Y., Baz, M., & Boivin, G. (2006). Impact of neuraminidase mutations conferring influenza resistance to neuraminidase inhibitors in the N1 and N2 genetic backgrounds. *Antiviral Therapy*, *11*(8), 971–976. https://doi.org/10.1177/135965350601100804
109. Zhang, J., Cai, Y., Xiao, T., Lu, J., Peng, H., Sterling, S. M., ... & Chen, B. (2021). Structural impact on SARS-CoV-2 spike protein by D614G substitution. *Science*, *372*(6541), 525–530. https://doi.org/10.1126/science.abf2303
110. Dang, S., Ren, L., & Wang, J. (2022). Functional mutations of SARS-CoV-2: Implications to viral transmission, pathogenicity and immune escape. *Chinese Medical Journal*, *135*(10), 1213–1222. https://doi.org/10.1097/CM9.0000000000002158
111. Yang, W. T., Huang, W. H., Liao, T. L., Hsiao, T. H., Chuang, H. N., & Liu, P. Y. (2022). SARS-CoV-2 E484K mutation narrative review: Epidemiology, immune escape, clinical implications, and future considerations. *Infection and Drug Resistance*, *15*, 373–385. https://doi.org/10.2147/IDR.S344099
112. Pondé, R. A. A. (2022). Physicochemical effect of the N501Y, E484K/Q, K417N/T, L452R and T478K mutations on the SARS-CoV-2 spike protein RBD and its influence on agent fitness and on attributes developed by emerging variants of concern. *Virology*, *572*, 44–54. https://doi.org/10.1016/j.virol.2022.05.003
113. Rochman, N. D., Wolf, Y. I., & Koonin, E. V. (2022). Molecular adaptations during viral epidemics. *EMBO Reports*, *23*(8), e55393. https://doi.org/10.15252/embr.202255393
114. Goka, E. A., Vallely, P. J., Mutton, K. J., & Klapper, P. E. (2014). Mutations associated with severity of the pandemic influenza A (H1N1) pdm09 in humans: A systematic review and meta-analysis of epidemiological evidence. *Archives of Virology*, *159*, 3167–3183. https://doi.org/10.1007/s00705-014-2179-z
115. Pereira, L. A., Lapinscki, B. A., Santos, J. S., Debur, M. C., Petterle, R. R., Nogueira, M. B., ... & Raboni, S. M. (2024). Influenza A infections: Predictors of disease severity. *Brazilian Journal of Microbiology*, *55*(1), 75–86. https://doi.org/10.1007/s42770-023-01186-w

116. Liu, Y., Liu, J., Plante, K. S., Plante, J. A., Xie, X., Zhang, X., … & Weaver, S. C. (2022). The N501Y spike substitution enhances SARS-CoV-2 infection and transmission. *Nature*, *602*(7896), 294–299. https://doi.org/10.1038/s41586-021-04245-0
117. Dejnirattisai, W., Huo, J., Zhou, D., Zahradník, J., Supasa, P., Liu, C., … & Murphy, D. (2022). SARS-CoV-2 Omicron-B. 1.1. 529 leads to widespread escape from neutralizing antibody responses. *Cell*, *185*(3), 467–484. https://doi.org/10.1016/j.cell.2021.12.046
118. Korber, B., Fischer, W. M., Gnanakaran, S., Yoon, H., Theiler, J., Abfalterer, W., … & Montefiori, D. C. (2020). Tracking changes in SARS-CoV-2 spike: Evidence that D614G increases infectivity of the COVID-19 virus. *Cell*, *182*(4), 812–827. https://doi.org/10.1016/j.cell.2020.06.043
119. Tang, H., Hammack, C., Ogden, S. C., Wen, Z., Qian, X., Li, Y., … & Ming, G. L. (2016). Zika virus infects human cortical neural progenitors and attenuates their growth. *Cell Stem Cell*, *18*(5), 587–590. https://dx.doi.org/10.1016/j.stem.2016.02.016
120. Zhou, B., Thao, T. T. N., Hoffmann, D., Taddeo, A., Ebert, N., Labroussaa, F., … & Beer, M. (2021). SARS-CoV-2 spike D614G change enhances replication and transmission. *Nature*, *592*(7852), 122–127. https://doi.org/10.1038/s41586-021-03361-1
121. Ellwanger, J. H., & Chies, J. A. B. (2021). Zoonotic spillover: Understanding basic aspects for better prevention. *Genetics and Molecular Biology*, *44*(1 Suppl 1), e20200355. https://doi.org/10.1590/1678-4685-GMB-2020-0355
122. Fauziah, I., Nugroho, H. A., Yanthi, N. D., Tiffarent, R., & Saputra, S. (2024). Potential zoonotic spillover at the human–animal interface: A mini-review. *Veterinary World*, *17*(2), 289. https://doi.org/10.14202/vetworld.2024.289-302
123. Tosheva, I. I., Filaire, F., Rijnink, W. F., de Meulder, D., van Kekem, B., Bestebroer, T. M., … & Herfst, S. (2025). Influenza A (H5N1) shedding in air corresponds to transmissibility in mammals. *Nature Microbiology*, *10*(1), 14–19. https://doi.org/10.1038/s41564-024-01885-6
124. Garg, S., Reinhart, K., Couture, A., Kniss, K., Davis, C. T., Kirby, M. K., … & Olsen, S. J. (2025). Highly pathogenic avian influenza A (H5N1) virus infections in humans. *New England Journal of Medicine*, *392*(9), 843–854. https://doi.org/10.1056/NEJMoa2414610
125. Krammer, F., Hermann, E., & Rasmussen, A. L. (2025). Highly pathogenic avian influenza H5N1: History, current situation, and outlook. *Journal of Virology*, *99*(4), e02209–24. https://doi.org/10.1128/jvi.02209-24
126. Guo, Y., Su, C., Liang, H., Jiang, X., Yang, R., Ye, J., … & Xu, L. (2025). Virome diversity and potential sharing of wild mammals in a biodiversity hotspot, Yunnan, China. *Virology Journal*, *22*(1), 79. https://doi.org/10.1186/s12985-025-02702-0
127. Carlson, C. J., Albery, G. F., Merow, C., Trisos, C. H., Zipfel, C. M., Eskew, E. A., … & Bansal, S. (2022). Climate change increases cross-species viral transmission risk. *Nature*, *607*(7919), 555–562. https://doi.org/10.1038/s41586-022-04788-w
128. Baker, R. E., Mahmud, A. S., Miller, I. F., Rajeev, M., Rasambainarivo, F., Rice, B. L., … & Metcalf, C. J. E. (2022). Infectious disease in an era of global change. *Nature Reviews Microbiology*, *20*(4), 193–205. https://doi.org/10.1038/s41579-021-00639-z
129. Apoorva, & Singh, S. K. (2024). A tale of endurance: Bats, viruses and immune dynamics. *Future Microbiology*, *19*(9), 841–856. https://doi.org/10.2217/fmb-2023-0233
130. Naseer, S., Khalid, S., Parveen, S., Abbass, K., Song, H., & Achim, M. V. (2023). COVID-19 outbreak: Impact on global economy. *Frontiers in Public Health*, *10*, 1009393. https://doi.org/10.3389/fpubh.2022.1009393
131. do Amaral, C. G., André, E. P., Cilli, E. M., da Costa, V. G., & Sanches, P. R. S. (2024). Viral diseases and the environment relationship. *Environmental Pollution*, *362*, 124845. https://doi.org/10.1016/j.envpol.2024.124845

8 Diagnostic Challenges and Innovations

Nawal Abd El-Baky, Amro A. Amara, and Elrashdy M. Redwan

8.1 INTRODUCTION

Since the global recognition of the term "neglected tropical diseases" in 2005, they have attracted increasing research interest, in addition to enhanced political will and more funding resources to achieve control and ultimate elimination of these infections [1–3]. The list of these diseases has remarkably expanded from 13 neglected tropical diseases, which comprise three bacterial, and seven helminths, besides three protozoal infections to reach over 40 diseases in the last decade that highlights infections caused by neglected viruses, in addition to neglected ectoparasitic and fungal infections [4].

The list of neglected tropical diseases recognized by WHO includes the following viral infections: Rift Valley fever, Enterovirus 71 infection, Arboviral infections (e.g., Japanese encephalitis, chikungunya, dengue, Zika, jungle yellow fever), rabies, non-HIV retrovirus infections (e.g., HTLV-1 and HTLV-2), monkeypox, and Viral hemorrhagic fevers [5].

However, the Ebola virus has moved from this list of neglected viral threats to a global emergency state in the last decade [6]. This extremely virulent virus started as endemic in some African regions causing severe viral hemorrhagic fever in patients. What makes the situation worse is its easy transmission via various fluids of the human body. Moreover, its diagnosis was challenging due to the similarity between Ebola infection and septic shock in provoking a strong inflammatory response and causing immunosuppression. The virus remained neglected for nearly four decades despite the occurrence of definite sporadic outbreaks in Central Africa due to the local government authorities' disinterest in acknowledging this infectious agent as an emergency state in their countries. Nevertheless, March 2014 has witnessed the worst outbreak of Ebola in some West African Coast countries [7]. On August of the same year, WHO declared this outbreak as a global emergency of public health [8].

A lesson from Ebola epidemic and SARS-CoV-2 pandemic, we learned the hardest way that early viral diagnosis and understanding the infection pattern of a virus threat is an obligation. On general regard, medical viruses are diagnosed via three main approaches: Immunological-based, biosensor-based, and molecular-based methodologies [9]. Immunological-based methodologies rely on identifying certain viral antigens or antiviral antibodies in patients' specimens by enzyme-linked antibodies. While, biosensor-based approaches employ an analytical device that comprises biological elements besides physicochemical components, which detect viral antigens by giving a signal. Viral genome amplification is the principle upon which nucleic acid-based procedures are based.

Diagnosis of Ebola as an example involves ELISA immunological assay or viral RNA detection by inverse polymerase chain reaction (RT-PCR) either in patient blood or other body fluids. Ebola virus can be detected in the patient's blood 3 days after symptoms are onset. IgM evaluation exhibited low sensitivity for early Ebola diagnosis, nonetheless, can be used to monitor the immunity of the patients [10, 11]. In 2015, WHO approved the first antigen rapid Ebola test (ReEBOV™ Kit produced by Corgenix Medical Corporation, Broomfield, USA). This test qualitatively detects ZEBOV VP40 antigen via immunochromatography, providing results within 15 minutes with 92% sensitivity in patients infected with *Zaire ebolavirus* [12].

 DOI: 10.1201/9781003607816-8

Research is a necessity to develop new diagnostics for rapid detection of neglected viruses so as to intervene rapidly with neglected viral infections to avert, treat and manage their complications. In 2018, WHO Global Observatory on Health R&D reported presence of 31 diagnostics in the pipeline for all neglected tropical infections [13].

In this chapter, we emphasize neglected viral infections, reviewing current and emerging diagnostic methods and highlights 2030 agenda of roadmap for neglected viral infections diagnostics.

8.2 CURRENT DIAGNOSTIC METHODS FOR NEGLECTED VIRUSES

Table 8.1 summarizes certain important neglected viral infections, including causative viruses, their classification, endemic areas, and current diagnostic methods. In the following subsections, diagnostic approaches for Lyssaviruses (e.g., rabies virus), Arboviruses (e.g., chikungunya, dengue, yellow fever, Zika, Oropouche, and Mayaro viruses), monkeypox, and hantavirus will be reviewed.

TABLE 8.1
Current Diagnostic Methods of Some Neglected Viruses

Neglected viral disease	Causative virus	Classification	Endemic areas	Current diagnostic methods	Reference(s)
Japanese encephalitis	Japanese encephalitis virus	Genus: *Flavivirus*	Indian subcontinent, South-East Asia, Northern Australia	Serological diagnosis	[14, 15]
Dengue	Dengue fever virus	Genus: *Flavivirus*	Indian subcontinent, Central/South America, South-East Asia, Northern Australia, and infrequently in some African regions	Virological (direct) and Serological (indirect) diagnosis	[16]
Yellow fever	Yellow fever virus	Genus: *Flavivirus*	South America, Central/ West Africa, infrequently in East of Africa	Virological (direct) and Serological (indirect) diagnosis	[16, 17]
Rift Valley fever	Rift Valley fever virus	Genus: *Phlebovirus*	Whole Africa and Arabian Peninsula	Serological and molecular methods	[18]
Rabies	Rabies virus	Genus: *Lyssavirus*	Globally distributed but occurs predominately in South/ Central America, Asia, Africa, and Europe (in the East)	Microscopic, cultural, molecular, and serological approaches	[19, 20]
Viral hemorrhagic fevers	Hanta viruses Lassa virus Crimean-Congo hemorrhagic fever virus (HFV) Omsk HFV Ebola and Marburg viruses	*Bunyaviridae* *Arenaviridae* *Bunyaviridae* *Flaviviridae* *Filoviridae*	Globally distributed West Africa Arabian Peninsula, South-Eastern Europe, Africa, Central Asia West Siberia Some African regions	Serological and molecular methods	[21–25]
Monkeypox	Monkeypox virus	Genus: *Orthopoxvirus*	Central and western Africa areas	Skin biopsy, PCR, transmission electron microscopy, and immunological-based methods	[26]

8.2.1 Diagnostic Approaches for Rabies Virus

Rabies virus leads to fatal encephalitis, resulting in loss of consciousness, respiratory paralysis and lastly death [27]. The virus is typically transmitted by infected dogs' saliva. It causes complex neuropathogenetic mechanisms in infected humans [28]. Thus, as soon as rabies symptoms develop, it is virtually 100% fatal. Rabies cannot be diagnosed before symptoms manifest [29]. Therefore, antemortem rabies diagnosis in human is a need in primary facilities of health care in addition to postmortem rabies diagnosis in animals to advance post-bite therapy.

8.2.1.1 Sampling for Suspected Rabies

Expert consultation of WHO on rabies summarized how to sample for suspected rabies [30]. Secretions such as serum, cerebrospinal fluid, saliva, or tears and skin biopsy specimens (nuchal skin with hair follicles) from vital patients are the typical diagnostic materials for rabies [31]. Three specimens of saliva are needed in 3–6 h intervals because of the intermitting shedding of rabies virus. Samples are transported into sterile containers and stored on dry ice but not for the long term. If transport at cooling conditions is not available, samples are kept in 50% glycerin-saline for stabilization. Brain biopsies are used for postmortem rabies diagnosis.

8.2.1.2 Microscopic Diagnosis

For microscopic diagnosis, no less than 20 skin biopsies from vital patients are tested by direct fluorescence antibody (DFA) [32]. Postmortem diagnosis depends on brain tissue examination [32]. Exposure of epitopes of rabies virus is performed by pre-treatment of formalin-fixed tissue of the brain with proteinase K [33]. Direct rapid immunohistochemical testing (DRIT) is an effective substitute for DFA [34]. In this test, biotinylated monoclonal antibodies against viral nucleocapsid proteins are employed. Viral replication and transcription occur in Negri bodies. Thus, they contain nucleocapsids of rabies virus. Negri bodies can be detected via Seller's staining [35, 36].

DFA assay takes from 3 to 4 h. It has about 100% specificity and sensitivity. DFA gives positive results in almost 50% of specimens in the course of the first 7 days with symptoms if several specimens are stained. Later disease stages exhibit increased positivity [32]. DRIT displayed equivalent specificity and sensitivity to DFA [34]. Detection of Negri bodies was possible in a small fraction of infected cells [37, 38]. This test exhibited merely 50%–80% sensitivity albeit properly done [39].

DFA or DRIT are current diagnostic methods for rabies virus [32, 34]. Postmortem DFA diagnosis performed on tissue of the brain remains the standard, by which other approaches of diagnosis are estimated [40].

8.2.1.3 Cultural Diagnosis

Lyssaviruses are cultured in neuroblastoma cells of mice under conditions of Biosafety Level 3 [40]. Following sample inoculation and 2–4 days incubation, the growth of Lyssaviruses is identified using antibodies against lyssavirus labeled with immunofluorescence [41]. Rabies virus can also grow in mice for its detection in a laboratory [42]. Airborne transmission of rabies in laboratory animals has been reported [43].

Cultural diagnosis explicitly validates the existence of active lyssavirus and permits augmentation for succeeding assessments [44]. Cultural diagnosis is rarely performed. Moreover, cell culture is preferred to mice inoculation, therefore mice inoculation was abandoned [42].

8.2.1.4 Molecular Diagnosis

RT-PCR and loop-mediated isothermal amplification (LAMP) are highly sensitive approaches for antemortem lyssavirus detection from cerebrospinal fluids, saliva, and tissue and permit discrimination of the virus on the molecular level [45]. The main obstacle in molecular diagnosis of rabies is the high diversity among variants of lyssavirus, which leads to the nonexistence of universal (non-degenerate) primers. Thus, RT-PCR is done using various degenerate primers groups. As an example, rabies N-gene detection among different virus variants found in the USA involves six forward degenerated primers and one reverse primer [40]. Typing of rabies is dependent on sequences of N-gene [46], accompanied by P-, G-, and L-genes sequencing [44].

The nucleic acid extraction approach, specimen material, as well as involved protocol of PCR greatly affect the reliability of PCR diagnosis of rabies. Brain tissue resulted in the unsurpassed sensitivity [45]. Furthermore, RT-PCR can recognize geographic viral origin [47]. Rabies in decomposed brain samples can be successfully diagnosed by RT-PCR [48]. PCR is the best diagnostic approach for rabies if primers specific to regions of the virus are obtainable and thus have broadly substituted other approaches [40, 45].

8.2.1.5 Serological Diagnosis

Anti-rabies G protein neutralizing antibodies are evaluated through rapid fluorescent focus inhibition testing/fluorescence antibody virus neutralization (RFFIT/FAVN) [49]. Serosurveillance of rabies was performed using robust neutralization testing [50]. Immunofluorescence testing was carried out to evaluate IgM and IgG from cerebrospinal fluid or blood [40]. ELISA assays available for purchase or in-house reference ELISA to detect antibodies against human rabies virus have been developed [51].

RFFIT/FAVN can be used to assess state of vaccination [52]. All rabies patients have neutralizing antibodies by nearly two weeks of infection, half of them by about a week, and fewer of them by day 6. Any titers of neutralizing antibodies detected in cerebrospinal fluid are relevant for diagnosis. Yet, definite oligoclonal bands should be found in cerebrospinal fluid to validate central nervous system infection [53]. Serological diagnosis has very restricted value for acute infection diagnosis as detectable antibodies occur latter [53].

8.2.2 Diagnostic Approaches for Arbovirus Infections

A laboratory recommendation for the diagnosis of chikungunya, dengue, yellow fever, Zika, Oropouche, and Mayaro viruses in the Americas was published by Pan American Health Organization [54]. These recommendations included direct virological diagnostic methods as well as indirect serological diagnostics for these viruses as follows.

8.2.2.1 Diagnostic Approaches for Chikungunya, Dengue, Yellow Fever, and Zika Viruses

8.2.2.1.1 Direct Virological Diagnostic Methods: Molecular Diagnosis

Viral ribonucleic acid is amplified then specific sequences are detected. Mostly, RT- PCR is used to amplify ribonucleic acid of chikungunya, dengue, yellow fever, and Zika viruses. Since all of these viruses have an RNA genome, RT-PCR is employed to synthesize cDNA from RNA of these viruses and then amplify their genome. However, other assays like transcription-mediated amplification and reverse transcription LAMP (RT-LAMP) can amplify the viral genome as well for molecular diagnosis of Zika virus [55, 56].

The typical diagnostic materials for these viruses include fresh or formalin-fixed tissue specimens and fluids for instance cerebrospinal fluid, plasma, serum, saliva, and urine. Tissue samples are preserved in solutions that stabilize RNA then disrupted and homogenized, while fluids are directly lysed and extracted. RNA is extracted from these diagnostic materials and purified prior to amplification. Various commercial kits and in-house procedures that vary depending on diagnostic materials used for viral detection were developed for disruption, deparaffinization, homogenization, and RNA purification. For extraction of RNA, the use of RNase-free reagents/supplies is a must. Moreover, thoroughly cleaned equipment or surfaces, and powder-free gloves should be used in RNA extraction.

Some of RT-PCR assays are multiplex assays that permit the detection of numerous arboviruses simultaneously [54]. As an example, the real-time multiplex RT-PCR for the detection of four types of dengue virus simultaneously, which targets NS5 sequence of dengue virus type-1, envelope sequence of dengue virus type-2, and prM sequences of both dengue virus type-3 and dengue virus type-4. Another example is the real-time triplex RT-PCR detecting chikungunya, dengue, and Zika viruses simultaneously, which targets the nsP1 sequence of the chikungunya virus, 5′ untranslated region (5′ UTR) sequence of the dengue virus, and the envelope sequence of Zika virus [54].

Molecular diagnosis of chikungunya, dengue, yellow fever, and Zika viruses is highly sensitive, and also vastly specific in terms of the amplified viral genome sequence that discriminates the target virus. Nevertheless, their actual sensitivity is affected by the specific viral RNA levels to be detected in a definite diagnostic specimen, and the decrease in viremia level of the four viruses with time from onset of symptoms. The most appropriate periods for molecular diagnosis of chikungunya, dengue, yellow fever, and Zika viruses in biological samples are demonstrated in Table 8.2. Postmortem molecular detection of chikungunya, dengue, yellow fever, and Zika viruses could be attempted on placenta, liver, brain, and kidney specimens [54].

8.2.2.1.2 Direct Virological Diagnostic Methods: Viral Antigens Detection

Detection of viral glycoproteins is based principally on their existence and abundance in biological specimens. The NS1 protein of the dengue virus represents the best antigen of arboviruses for detection in serum. Viral antigen detection by immunohistochemistry in tissue sections is also convenient for postmortem diagnosis, particularly for fatal cases of yellow fever. The most appropriate tissues for antigen detection by immunohistochemistry for fatal cases of yellow fever are the kidney and liver [59].

The NS1 protein of the dengue virus is detected by ELISA. Steadily, detection sensitivity becomes lower in secondary infections. Commonly, NS1 protein abundance in blood declines during infection, and accordingly the sensitivity of its detection declines. Despite this difference in temporal kinetics of NS1 between primary and secondary infections of dengue virus, this test cannot determine whether the patient is exhibiting an acute primary dengue virus infection or a secondary dengue virus infection. The sensitivity of NS1 detection varies among different serotypes of dengue virus [62].

Rapid immunochromatographic NS1 detection assays are available, but their sensitivity is less than that of ELISA [63]. So far, antigens for the detection of chikungunya, yellow fever, and Zika viruses in serum or other biological specimens have not been confirmed. Since viral antigens are

TABLE 8.2
The Periods for Detection of Chikungunya, Dengue, Yellow Fever, and Zika Viruses by Molecular Assays in Diagnostic Specimens

Arbovirus	Diagnostic material	Days after onset of symptoms*	References
Chikungunya	Serum	1–8	[57]
Dengue	Serum	1–7	[58]
Yellow fever	Serum	1–10	[59]
Zika	Cerebrospinal fluid#	Any day	[60]
	Serum	1–4 or 1–5€	[60]
	Whole blood	1–28	[61]
	Urine	1–15	[60]
	Saliva	1–4 to 5	[60]

* On general regard, the viremia of the four viruses decreases with time from onset of symptoms, influencing their molecular diagnosis sensitivity.

\# It should be collected for clinical reasons not just to identify the etiological agent.

€ Pregnant women have longer periods for detection.

detected by antibodies, and, taking into account the similarity in antigens among viruses belonging to the same genus, there is a possibility for cross-reactivity in detection assays for viral antigens. This cross-reactivity will affect the specificity of the assay. Especially, the performance of ELISA and rapid detection assays for NS1 of dengue virus should be reassessed in the context of co-circulation of various flaviviruses such as yellow fever and Zika viruses. Up to now, two reports have confirmed the high specificity and differential diagnosis of ELISA test for NS1 of dengue virus in Zika cases that have been confirmed in the laboratory [64, 65]. Moreover, monoclonal antibodies employed for immunohistochemistry testing should be cautiously chosen and estimated to certify their specificity.

8.2.2.1.3 Direct Virological Diagnostic Methods: Virus Isolation

Isolation of chikungunya, dengue, yellow fever, and Zika viruses is mainly carried out in cell cultures (i.e., Vero mammalian cell line and C6/36 mosquito cell line) [66, 67]. Otherwise, these viruses can be inoculated in suckling mice and other rodents. Isolation of arbovirus can be performed from total blood, serum, cerebrospinal fluid, mosquitoes, and animal or human tissues. Tissue samples for postmortem diagnosis must be collected within a day of death.

Mostly, virus isolation is not a chosen approach for the diagnosis of arboviral infections, nor is it required to confirm the diagnosis. The limitations of using this diagnostic approach manifest themselves in costs, technical complexity, and biosafety requirements. Furthermore, isolated viruses should be further identified by RT-PCR or immunofluorescence. Yet, genetical and phenotypical characterization of these viruses requires their isolation. Additionally, isolated strains can act as reference controls for molecular or serological assays. Virus isolation can also help to identify new viruses, thus becoming a vital tool for emerging/reemerging viruses' surveillance.

8.2.2.1.4 Indirect Serological Diagnostic Methods: IgM Detection

Serological diagnosis is employed to detect immune reactions of patients to viral infection, predominantly antibody response in serum or case of neurological diseases antibody response in cerebrospinal fluid. ELISA assays or indirect immunofluorescence assays are used to detect IgM antibodies against diverse arboviruses. Many laboratories in several countries have developed their in-house ELISA procedures that can capture IgM including antigens created from recombinant proteins of viruses or inactivated viruses.

Pan American Health Organization in the USA in collaboration with WHO developed ELISA procedures that can capture IgM for chikungunya, dengue, and Zika viruses, while Centers for Disease Control and Prevention (CDC) in USA developed that for yellow fever virus [57, 68, 69]. Instituto de medicina Tropical Pedro Kourí in Cuba developed ELISA procedure that can capture IgM for the dengue virus [70, 71]. Besides, optimized versions of ELISA that can capture IgM for yellow fever virus developed by CDC in the USA and ELISA that can capture IgM for dengue virus developed by Instituto de medicina Tropical Pedro Kourí in Cuba for easier implementation have been reported [69, 71].

Though cross-reactivity of ELISA kits for dengue virus with Zika virus was not explored after introduction of Zika virus in the Americas [72, 73]. Consequently, further assessments are required to evaluate specificity of these kits for Zika virus infections. As in the case of molecular diagnostics, other serological IgM detection assays besides reference assays can always be developed and applied but need to be evaluated before applied in routine surveillance.

In cases of neurological disease, IgM is evaluated in cerebrospinal fluid as well as serum whatever available, but it is endorsed to process cerebrospinal fluid and serum specimens in parallel. The cerebrospinal fluid diagnostic specimen should be either pure or at low dilution. IgM detection in cerebrospinal fluid proves the recent occurrence of central nervous system infection, continually bearing in mind the possible IgM persistence in addition to the likely cross-reactivity among viruses belonging to the same genus.

The limitations for IgM detection appear in the fact that a positive IgM detection in a single specimen is just presumptive but not necessarily an indication for an acute infection, as detected IgM can be evidence for a recent infection. Additionally, detected IgM may be an outcome of recent vaccination.

8.2.2.1.5 Indirect Serological Diagnostic Methods: IgG Detection

Various techniques are used to detect IgG antibodies including ELISA, fixation of complement, immunofluorescence, microsphere immunoassays, or inhibition of hemagglutination. Since IgG against viruses lives for long, this limits its diagnostic value in a sole sample. To confirm infection, a fourfold or superior IgG titers rise between acute and convalescent specimens or IgG seroconversion must be detected [69]. The detection of IgG titers rise necessitates quantitative assay, whereas in most cases marketable assays for IgG are qualitative. Likewise, within-genus IgG cross-reactivity remains a problem. Nevertheless, IgG to NS1 protein was reported to have a greater specificity than that to envelope protein [74].

8.2.2.1.6 Indirect Serological Diagnostic Methods: Neutralizing Antibodies Detection

The standard serological diagnostic method for arboviral infections is plaque reduction neutralization assay that detects neutralizing antibodies [75]. Antibodies from serum of infected persons can neutralize infection of virus in cell cultures, which forms the basis for this assay. The assay can measure the antibody titer required to decrease viral plaques number by 90%. The limitations for this assay are time and resources consumption as well as need for cell culture facilities. Moreover, the availability of diverse viral strains, various cell lines, and other conditions that affect the assay makes standardization of the test conditions a necessity [75, 76].

To confirm infection, a fourfold or more upsurge in neutralizing antibodies titers between acute and convalescent specimens or a seroconversion must be detected. Nonetheless, cross-reactivity of neutralizing antibody remains a possibility as occurs with other serological methods, predominantly in patients who suffer a secondary infection with flavivirus. In those patients, concurrent neutralizing antibody titers rise against many flaviviruses can take place [68, 77]. This makes it necessary to detect neutralizing antibodies against a panel of related flaviviruses in the context of epidemiological state simultaneously. Yet, it should be taken into account that titers increase in secondary infection would be predominately against the virus causing the primary infection rather than the virus causing a secondary infection, another reason for difficulty in results interpretation [78].

8.2.2.1.7 When to Use Indirect Serological Diagnostic Methods

Indirect serological diagnostic methods are the best option for arboviral infections diagnosis in some cases as: 1) diagnosis after the end of viremic phase; 2) fine biological specimens that have been collected in the right time window to serve virological diagnosis are unavailable; 3) if laboratory equipment and areas specific for virological diagnosis are not available; 4) serological diagnosis is simpler to perform, thus implemented easily in a subnational laboratories network; 5) in-house ELISA kits are more cost-effective than amplification/extraction reagents used in virological diagnostics; 6) the two methods if combined together will improve specificity as well as sensitivity of diagnosis [59, 79]; and 7) newborns who have inherited Zika syndrome rarely exhibit the virus or any of its components in their biological samples, making serological diagnosis crucial.

8.2.2.2 Diagnostic Approaches for Oropouche and Mayaro Viruses

8.2.2.2.1 Diagnostic Approaches for Oropouche Virus

Oropouche fever clinical diagnosis is problematic. Elevated transaminases in addition to leukopenia detected in some Oropouche fever patients are non-specific for Oropouche virus diagnosis [80].

Molecular tests are preferred for its diagnosis. During laboratory diagnosis of cases of Oropouche fever, it should be taken into consideration that this virus has a genome comprising three segments (S, M, and L) [80]. Viral isolation or molecular diagnosis are used to identify Oropouche virus during the acute fever phase (2–7 days) [80, 81]. Molecular diagnosis can identify the S segment in genome of Oropouche virus in cerebrospinal fluid and serum samples [81, 82]. To single out Oropouche virus isolates from natural reassortment Oropouche viruses, molecular diagnosis should target the M segment [83]. Isolation of Oropouche virus is done into Vero mammalian cell line and C6/36 mosquito cell line or can be intracerebral inoculated in suckling mice [80].

To detect antibodies in serum, convalescence phase should be reached (day 5 of the symptoms onset) [81]. Various in-house procedures were developed for serologic diagnosis of Oropouche virus, for instance complement fixation assay, plaque reduction neutralization test, immunofluorescence, ELISA that capture IgM and IgG, and inhibition of hemagglutination [80, 81]. Cerebrospinal fluid can also be diagnostic material for detection of IgM and IgG [80].

8.2.2.2.2 Diagnostic Approaches for Mayaro Virus

Mayaro virus RNA can be detected in serum using RT-PCR [84]. Isolation of Mayaro virus is done by inoculation of patient specimen (total blood, serum, or tissues) into Vero mammalian cell line or C6/36 mosquito cell line or can be intracerebral inoculated in suckling mice [85]. Unfortunately, the infection of Mayaro virus has brief viremia period (generally not more than 6 days after onset of symptoms), which limits virological diagnosis. In contrast, IgM antibodies to Mayaro virus persist for up to 3 months, and IgG persist for years [85]. Various in-house procedures were developed for serologic diagnosis of Mayaro virus [85]. However, serologic diagnosis procedures, counting ELISA assays that capture IgM exhibited cross-reactivity with other alphaviruses [86]. Plaque reduction neutralization test in BHK-21 or Vero cell lines is the gold standard of Mayaro virus serological diagnosis, yet cross-reactivity remains a possibility to worry about. Antibody titer rise between acute- and convalescent-phase specimens or seroconversion is required to confirm Mayaro virus serological diagnosis [87]. This makes molecular diagnosis is the preferred option, but with the condition that the required samples are available.

8.2.3 Diagnostic Approaches for Monkeypox

Samples for monkeypox diagnosis are collected from lesions in the skin (for instance vesicles, pustules, and dry crusts), stored in a dry sterilized container and kept cold [26]. Skin biopsy is a good substitute to collect monkeypox samples. The most obvious challenge for confirmation of monkeypox diagnosis appears in the common systemic rash symptoms among bacterial skin infections, measles, syphilis, scabies, and allergies associated with medication [88]. Chicken pox and smallpox had better be included in the differential diagnosis for monkeypox. Monkeypox virus can be detected in pustular swab via PCR and transmission electron microscopy and diagnosis can be confirmed by immunofluorescence assay, tissue culture, and ELISA [89]. The issue of poor diagnostic facilities in prevalent regions of monkeypox infection requires to be addressed to boost early detection against monkeypox outbreaks.

8.2.4 Diagnostic Approaches for Hantavirus

Hantavirus causes hemorrhagic fever associated with renal disorder in Asia and Europe and hantavirus pulmonary disorder in the Americas [24]. Hantavirus infection diagnosis needs clinical and epidemiological evidence along with laboratory assessments. Hantavirus RNA or antigens should be detected for early diagnosis [90]. Viral RNA is detected by RT-PCR for clinical or postmortem diagnosis. Hantavirus N-protein represents the best choice for early detection as it is highly abundant and has conserved structure in infected cells/virions, therefore IgM or IgG antibodies against this protein are commonly detected for serologic diagnosis [90]. Isolation of this virus is challenging and necessitates biosafety 3 and 4 laboratories. The differential diagnosis for hantavirus in tropical nations comprise influenza, numerous uncharacteristic pneumonia conditions, heart failure, dengue, malaria, arenavirus, and rickettsia [90].

8.3 INNOVATIVE DIAGNOSTICS FOR NEGLECTED VIRUSES

To achieve innovation in diagnostics for neglected viruses especially in prevalence and priority areas, and consequently better control and elimination of neglected viruses, some points should be addressed as follows: 1) perceive resemblance features for different neglected viruses regarding sentinel populations, geographic overlap, and therapy approaches will let programs for neglected tropical diseases to establish multiplex and cost-effective platforms for surveillance; 2) Novel amplification technologies for nucleic acid accomplishing sample in-answer out at the point-of-care propose enhanced performance over current methodologies in addition to the probability to do multiple pathogens testing with a single specimen; 3) linking information come from diagnostic laboratories and point-of-care assessment readers/devices makes automated surveillance possible, health systems more competent; and 4) partnerships between public- and private-product development are important in influencing diagnostic modernization in other priority area [91].

As mentioned above, serological diagnosis of all neglected viruses needs a detection of seroconversion or IgG titers rise between acute and convalescent serum samples. Such paired serum specimens are difficult to obtain. It was reported that more than 90% of suspected cases of dengue are diagnosed from a single specimen of serum collected in the first 10 days of dengue infection (acute phase) by combining either ELISA that detect NS1 antigen and ELISA that capture IgM or RT-PCR and ELISA that capture IgM [79]. Thus, test combinations can be employed for diagnosis of neglected viruses using a single sample of blood collected throughout the acute phase of infection. This proposes a marvelous advance over current struggles with emergencies of global health, most of which are a result of viral outbreaks.

If cultural growth should be performed in biosafety level 3 and 4 laboratories, nevertheless such high-standards infrastructure are not existing, the alternatives will be molecular and serological diagnosis from inactivated specimens. Multiple successive assessments via several approaches for diagnosis in a stepwise approach can advance specificity as well as sensitivity. Development of "rapid diagnostic tests" that provide easy use and cost-effectiveness affects control measures for neglected viral diseases as with the case of antigen rapid Ebola test (ReEBOV™ Kit) for Ebola diagnosis, nonetheless, should not impede developing countries from "state of the art" diagnostics [20].

High throughput sequencing is emerging as the future gold standard in molecular viral infection diagnosis [92]. Undoubtedly, it is more precise and universal than conventional clinical assessments at profiling pathogens, but not until now have been broadly accepted as a diagnostic tool. What limits its acceptance are the difficulty of specimen preparation and data interpretation besides its high cost. To overcome those obstacles, purpose-specific pipelines should be assembled, universal procedures to prepare the library can be designed and assessed, and computational infrastructure to fit the financial capacities and needs of modern healthcare centers should be built. High throughput sequencing can generate data that not only improve diagnostics assisting in choosing the accurate therapy, but also enable genetics, epidemiology, and virology research [92].

8.4 A 2021–2030 ROAD MAP FOR NEGLECTED VIRAL DISEASES

WHO has reported a 2021–2030 road map for all neglected tropical diseases to end the neglect and achieve the goals for sustainable development [93]. The map comprises two neglected viral infections, rabies, dengue, and chikungunya. One of the top intended targets of this road map is to accelerate programmatic action to attain effective diagnostics. According to the map, effective diagnostics are essential for infection confirmation, screening, mapping, surveillance, and evaluation and monitoring, helping programs for neglected diseases to reach the 2030 disease targets [93]. Effective diagnostics support treatment selection at the level of individuals and inform decisions on discontinuing mass treatment or altering treatment incidence at the level of the population.

Better diagnostics ensure elimination or certify eradication by early detection and management of disease progress besides identifying the causing pathogen, consequently reducing morbidity. Countries can monitor trends of the disease by the help of effective diagnostics, enhancing control program effectiveness. The problem with neglected diseases lies in the fact that they are endemic among marginalized populations, because of the unobtainability or insufficiency of diagnostic tools that result in pragmatic misdiagnosis cases. The pressing priority is early diagnosis at the point of care of patients with neglected diseases.

Substantial advancement has been accomplished in novel point-of-care diagnostics. Innovative diagnostic tools and approaches for neglected diseases are becoming accessible, with nonstop engagement of vital partners. While immunological-based point-of-care diagnostics exist for the diagnosis of rabies, chikungunya, and dengue, they are considered to be suboptimal [94]. For chikungunya and dengue, available point-of-care diagnostics are incompetent as they are cross-reactive; dengue point-of-care test cross-reacts with Zika virus and chikungunya point-of-care test cross-reacts with dengue virus [94]. Chikungunya and dengue are frequently misdiagnosed as they happen as a seasonal outbreak within Zika-endemic areas, and, consequently, new multiplexing point-of-care tests that can discriminate between these viral infections should be developed to accelerate diagnosis and diminish delays in treatment.

For many neglected infections, it was found that lack of disease awareness and test unavailability result in key diagnostic insufficiencies. For instance, rabies diagnosis is delayed attributable to the fondness of traditional healing practices since there is test unavailability and lack of disease awareness. Moreover, lessons learned from effective diagnostic approaches should be adapted and applied to neglected diseases. The best example is control programs for malaria that can be adapted to chikungunya and dengue [94].

Strengthening diagnostics priority for some neglected viral infections according to the WHO (2021–2030) road map for neglected diseases is presented in Table 8.3. For these diseases, diagnostics are either inexistent or insufficient for 2030 targets. Overall, around 39% of investment of research and development for neglected diseases is dedicated to vaccines and medicines, while only 5% for diagnostics [93]. Even if effective and accurate diagnostic approaches exist, they may not be accessible or affordable. Laboratory equipment, infrastructure, and trained personnel can be inadequate.

Microscopy is the most broadly accepted diagnostic method for neglected diseases, but it involves trained technicians and a laboratory, besides its sensitivity is frequently relatively low. On the other hand, nucleic acid assays or culturing pathogens are vastly specific but are costly, technically demanding, and time consuming. The best thinking in that case is not to abandon effective procedures until improved alternatives become confirmed, reachable, and affordable. The priorities for 2030 targets comprise field kits, multiplex diagnostic platforms, more sensitive non-invasive diagnostics, and strengthening of the capacity of the laboratory network [93].

While the diagnostic requirements for programs of neglected diseases are wide-ranging, there is cause to be hopeful about the predictions of novel tools development to meet the requirements of programs. The last two decades have witnessed massive development of novel molecular biology procedures, involving innovative analytic approaches to secretome and proteomes and rapid and cost-effective sequencing of genome. These procedures may lead to potential new biomarkers identification. The COVID-19 pandemic has exemplified the comparative rapidity at which innovative diagnostics can be developed when partnerships are intensified, and resources are abundant. The community of neglected diseases will have to make enticements for developers of tests in the lack of new resources by stating strong procedures for regulatory agreement and implementation of tests by programs. Furthermore, early payment purchase obligations and new funding plans must be well-thought-out to diminish the market risk for manufacturers of tests. The dramatic expansion in

TABLE 8.3
Strengthening Diagnostics is Priority for Some Neglected Viral Infections, Existing Point-of-Care Tests, and Future Implementation Requirements for Them

Disease	Screening, confirm diagnosis, surveillance	Existing point-of-care tests	Strengthening diagnostics priority	Implementation requirements
Rabies	*Screening and surveillance*: not applicable. *Confirm diagnosis*: diagnostics are available but need key modifications.	Numerous marketable immunological tests (e.g., Antigen Rapid rabies Ag test and Vet-o-test rabies Ag) [95].	1. Antemortem diagnostics should be developed to be applied in primary facilities for health care. 2. Postmortem diagnosis in animals (dogs) should be validated (e.g., rapid diagnostic test combined with the collection of non-invasive specimens).	1. Quality evaluation for all existing point-of-care assays. 2. Field-movable point-of-care assay in humans.
Dengue	*Screening, confirm diagnosis, and surveillance*: Suitable diagnostics are available, but alterations are mandatory to achieve 2030 targets.	Several marketable antibody and antigen-based assays (e.g., ASSURE Dengue IgA Rapid antigen-based test and SD BIOLINE Dengue Duo Rapid Test) [96].	1. Quality guarantee for rapid point-of-care diagnostics should be improved. 2. PCR assays should be developed to confirm diagnosis.	1. Innovative identification approach for circulating biomarkers. 2. Field-movable point-of-care assay that does not cross-react with Zika virus. 3. Multiplexed assay with Zika virus.
Chikungunya	The same as dengue	Chikungunya IgM Combo Rapid antigen-based Test CE and SD BIOLINE antigen-based chikungunya [97].	The same as dengue.	1. Innovative identification approach for circulating biomarker. 2. Multiplexed assay with dengue virus. 3. Field-movable point-of-care assay that does not cross-react with dengue.
Chikungunya/ Dengue/Zika viral infections	-	-	Combination rapid diagnostic test should be developed to allow differential diagnosis.	-

country-level testing capacity caused by the pandemic should be employed in introducing molecular diagnostics in locations where this capacity has been restricted [98].

8.5 CONCLUSION

While viruses are continually evolving at a disturbing rate and our struggles with earlier viral epidemics and the late COVID-19 pandemic have pushed the global healthcare systems and programs to move faster to develop approaches to diagnose, control, eliminate, hinder spreading, and provide cutting-edge therapies for these viral diseases, yet the outcome is still disappointing. The situation is quite worse with neglected viruses as they are not adequately noticed by local health systems of prevalence areas or the global health systems till the alarm of a new epidemic begins to ring. This was obvious in the case of the Ebola virus, which started as a neglected virus and then moved to the list of global emergency states. Consequently, the exploration for informative, rapid, and reliable diagnostic methods is becoming as vital as ever. PCR and serological diagnosis are subjected to continuous optimization, but so far provided very restricted data. High-throughput sequencing programs may offer a promising future solution for the molecular diagnosis of neglected viral infections. WHO has also prepared a 2021–2030 road map to end the neglect, one of its top goals is accelerating programmatic action for improving diagnostics.

REFERENCES

1. Molyneux, D. H. (2004). "Neglected" diseases but unrecognised successes – Challenge of infectious disease control. The Lancet, 364(9431), 380–383.
2. Molyneux, D. H., Hotez, P. J., & Fenwick, A. (2005). "Rapid-impact interventions": How a policy of integrated control for Africa's neglected tropical diseases could benefit the poor. PLOS Medicine, 2(11), e336.
3. Molyneux, D. H., Asamoa-Bah, A., Fenwick, A., Savioli, L., & Hotez, P. (2021). The history of the neglected tropical disease movement. Transactions of the Royal Society of Tropical Medicine and Hygiene, 115(2), 169–175. https://doi.org/10.1093/trstmh/trab015
4. Utzinger, J., Becker, S. L., Knopp, S., Blum, J., Neumayr, A. L., Keiser, J., & Hatz, C. F. (2012). Neglected tropical diseases: Diagnosis, clinical management, treatment and control. Swiss Medical Weekly [Internet], 142(4748), w13727. https://smw.ch/index.php/smw/article/view/1615
5. WHO. (2025). *Neglected tropical diseases (NTDs)*. Questions & Answers. WHO; 2025. Available from (https://www.who.int/news-room/questions-and-answers/item/neglected-tropical-diseases). World Health Organization. Archived from the original (https://www.who.int/news-room/questions-and-answers/item/neglected-tropical-diseases). "The 17 neglected tropical diseases" (https://www.who.int/health-topics/neglected-tropical-diseases?utm_source=chatgpt.com#tab=tab_1). World Health Organization. Archived from the original (https://www.who.int/neglected_diseases/diseases/en/).
6. Pacheco, D. A., Rodrigues, A. A., & Silva, C. M. (2016). Ebola virus – From neglected threat to global emergency state. Revista da Associacao Medica Brasileira (1992), 62(5), 458–467. https://doi.org/10.1590/1806-9282.62.05.458
7. WHO. (Updated January 2016). Ebola virus disease. https://www.who.int/mediacentre/factsheets/fs103/en/
8. BBC. (Updated August 2014). WHO: Ebola 'an international emergency'. https://www.bbc.com/news/world-africa-28702356
9. Harsh, & Tripathi, P. (2023). Medical viruses: Diagnostic techniques. Virology Journal, 20, 143. https://doi.org/10.1186/s12985-023-02108-w
10. Towner, J. S., Rollin, P. E., Bausch, D. G., Sanchez, A., Crary, S. M., Vincent, M., Lee, W. F., Spiropoulou, C. F., Ksiazek, T. G., Lukwiya, M., Kaducu, F., Downing, R., & Nichol, S. T. (2004). Rapid diagnosis of Ebola hemorrhagic fever by reverse transcription-PCR in an outbreak setting and assessment of patient viral load as a predictor of outcome. Journal of Virology, 78(8), 4330–4341. https://doi.org/10.1128/jvi.78.8.4330-4341.2004
11. CDC. (Updated April 2015). Ebola virus disease: Diagnosis. https://www.cdc.gov/vhf/ebola/diagnosis/index.html
12. WHO. (Cited February 2015). First antigen rapid test for Ebola through emergency assessment and eligible for procurement. https://www.who.int/medicines/ebola-treatment/1st_antigen_RT_Ebola/en/
13. World Health Organization. (2018). Health products in the pipeline for infectious diseases.
14. Erlanger, T. E., Weiss, S., Keiser, J., Utzinger, J., & Wiedenmayer, K. (2009). Past, present, and future of Japanese encephalitis. Emerging Infectious Diseases, 15(1), 1–7. https://doi.org/10.3201/eid1501.080311
15. Campbell, G. L., Hills, S. L., Fischer, M., Jacobson, J. A., Hoke, C. H., Hombach, J. M., Marfin, A. A., Solomon, T., Tsai, T. F., Tsu, V. D., & Ginsburg, A. S. (2011). Estimated global incidence of Japanese encephalitis: A systematic review. Bulletin of the World Health Organization, 89(10), 766–774, 774A-774E. https://doi.org/10.2471/BLT.10.085233

16. Rogers, D. J., Wilson, A. J., Hay, S. I., & Graham, A. J. (2006). The global distribution of yellow fever and dengue. Advances in Parasitology, 62, 181–220. https://doi.org/10.1016/S0065-308X(05)62006-4
17. Barnett, E. D. (2007). Yellow fever: Epidemiology and prevention. Clinical Infectious Diseases, 44(6), 850–856. https://doi.org/10.1086/511869
18. Chevalier, V., Pépin, M., Plée, L., & Lancelot, R. (2010). Rift valley fever – A threat for Europe? Euro Surveillance, 15(10), 19506. Erratum in: Euro Surveillance, 2010, 15(11), pii: 19516.
19. Wilde, H., Briggs, D. J., Meslin, F. X., Hemachudha, T., & Sitprija, V. (2003). Rabies update for travel medicine advisors. Clinical Infectious Diseases, 37(1), 96–100. https://doi.org/10.1086/375605
20. Schwarz, N. G., Loderstaedt, U., Hahn, A., Hinz, R., Zautner, A. E., Eibach, D., Fischer, M., Hagen, R. M., & Frickmann, H. (2017). Microbiological laboratory diagnostics of neglected zoonotic diseases (NZDs). Acta Tropica, 165, 40–65. https://doi.org/10.1016/j.actatropica.2015.09.003
21. Ergönül, Ö (2006). Crimean-Congo haemorrhagic fever. The Lancet Infectious Diseases, 6(4), 203–214. https://doi.org/10.1016/S1473-3099(06)70435-2
22. Ogbu, O., Ajuluchukwu, E., & Uneke, C. J. (2007). Lassa fever in West African subregion: An overview. Journal of Vector Borne Diseases, 44(1), 1–11.
23. Gould, E. A., & Solomon, T. (2008). Pathogenic flaviviruses. The Lancet, 371(9611), 500–509. https://doi.org/10.1016/S0140-6736(08)60238-X
24. Jonsson, C. B., Figueiredo, L. T., & Vapalahti, O. (2010). A global perspective on hantavirus ecology, epidemiology, and disease. Clinical Microbiology Reviews, 23(2), 412–41. https://doi.org/10.1128/CMR.00062-09
25. Leroy, E. M., Gonzalez, J. P., & Baize, S. (2011). Ebola and Marburg haemorrhagic fever viruses: Major scientific advances, but a relatively minor public health threat for Africa. Clinical Microbiology and Infection, 17(7), 964–976. https://doi.org/10.1111/j.1469-0691.2011.03535.x
26. Oladoye, M. J. (2021). Monkeypox: A neglected viral zoonotic disease. European Journal of Medical and Educational Technologies, 14(2), em2108. https://doi.org/10.30935/ejmets/10911
27. Nigg, A. J., & Walker, P. L. (2009). Overview, prevention, and treatment of rabies. Pharmacotherapy, 29(10), 1182–1195. https://doi.org/10.1592/phco.29.10.1182
28. Hemachudha, T., Ugolini, G., Wacharapluesadee, S., Sungkarat, W., Shuangshoti, S., & Laothamatas, J. (2013). Human rabies: Neuropathogenesis, diagnosis, and management. The Lancet Neurology, 12(5), 498–513. https://doi.org/10.1016/S1474-4422(13)70038-3
29. WHO. (2014, March 16). About rabies. https://www.who.int/rabies/about/en/; https://web.archive.org/web/20140130053158/http://www.who.int/rabies/about/en/
30. WHO Team (Control of Neglected Tropical Diseases (NTD)). (2013). WHO Expert Consultation on Rabies: WHO TRS N°982. Second report. The WHO Technical Report Series, 982, 1–139.
31. Dacheux, L., Reynes, J. M., Buchy, P., Sivuth, O., Diop, B. M., Rousset, D., Rathat, C., Jolly, N., Dufourcq, J. B., Nareth, C., Diop, S., Iehlé, C., Rajerison, R., Sadorge, C., & Bourhy, H. (2008). A reliable diagnosis of human rabies based on analysis of skin biopsy specimens. Clinical Infectious Diseases, 47(11), 1410–1417. https://doi.org/10.1086/592969
32. Rudd, R. J., Smith, J. S., Yager, P. A., Orciari, L. A., & Trimarchi, C. V. (2005). A need for standardized rabies-virus diagnostic procedures: Effect of cover-glass mountant on the reliability of antigen detection by the fluorescent antibody test. Virus Research, 111(1), 83–88. https://doi.org/10.1016/j.virusres.2005.03.014
33. Whitfield, S. G., Fekadu, M., Shaddock, J. H., Niezgoda, M., Warner, C. K., & Messenger, S. L., & Rabies Working Group (2001). A comparative study of the fluorescent antibody test for rabies diagnosis in fresh and formalin-fixed brain tissue specimens. Journal of Virological Methods, 95(1-2), 145–151. https://doi.org/10.1016/s0166-0934(01)00304-4
34. Lembo, T., Niezgoda, M., Velasco-Villa, A., Cleaveland, S., Ernest, E., & Rupprecht, C. E. (2006). Evaluation of a direct, rapid immunohistochemical test for rabies diagnosis. Emerging Infectious Diseases, 12(2), 310–313. https://doi.org/10.3201/eid1202.050812
35. De Brito, T., Araujo, & Mde, F. (1973). Ultrastructure of the Negri body in human rabies. Journal of the Neurological Sciences, 20(4), 363–372. https://doi.org/10.1016/0022-510x(73)90170-6
36. Lahaye, X., Vidy, A., Pomier, C., Obiang, L., Harper, F., Gaudin, Y., & Blondel, D. (2009). Functional characterization of Negri bodies (NBs) in rabies virus-infected cells: Evidence that NBs are sites of viral transcription and replication. Journal of Virology, 83(16), 7948–7958. https://doi.org/10.1128/JVI.00554-09
37. Jackson, A. C., & Fenton, M. B. (2001). Human rabies and bat bites. The Lancet, 357(9269), 1714. https://doi.org/10.1016/S0140-6736(00)04852-2

38. Jackson, A. C., Ye, H., Ridaura-Sanz, C., & Lopez-Corella, E. (2001). Quantitative study of the infection in brain neurons in human rabies. Journal of Medical Virology, 65(3), 614–618.
39. Perl, D. P., & Good, P. F. (1991). The Pathology of Rabies in the central nervous system. In: G. M. Baer (Ed.), The Natural History of Rabies, 2nd ed. (pp. 163–190, 163–190). CRC Press, Boca Raton, FL.
40. Orciari, L. A., & Rupprecht, C. E. (2007). Rabies Virus. In: P. R. Murray, E. J. Baron, J. H. Jorgensen, M. L. Landry, M. A. Pfaller (Eds.), Manual of Clinical Microbiology, 9th ed. (pp. 1470–1477). ASM Press, Washington, DC.
41. Webster, W. A., & Casey, G. A. (1996). Virus Isolation in Neuroblastoma Cell Culture. In: F. X. Meslin, M. M. Kaplan, H. Koprowski (Eds.), Laboratory Techniques in Rabies, 4th ed. (pp. 96–104). World Health Organization, Geneva, Switzerland.
42. Valleca, V. M., & Forester, F. T. (1981). The Mouse Inoculation Test. In: Laboratory Methods for the Detecting Rabies (pp. 91–100). U.S. Government Printing Office, Washington, DC.
43. Winkler, W. G., Fashinell, T. R., Leffingwell, L., Howard, P., & Conomy, P. (1973). Airborne rabies transmission in a laboratory worker. JAMA, 226(10), 1219–1221.
44. Nadin-Davis, S. A., Muldoon, F., & Wandeler, A. I. (2006). A molecular epidemiological analysis of the incursion of the raccoon strain of rabies virus into Canada. Epidemiology and Infection, 134(3), 534–547. https://doi.org/10.1017/S0950268805005108
45. Mani, R. S., & Madhusudana, S. N. (2013). Laboratory diagnosis of human rabies: Recent advances. Scientific World Journal, 2013, 569712. https://doi.org/10.1155/2013/569712
46. Kuzmin, I. V., Botvinkin, A. D., McElhinney, L. M., Smith, J. S., Orciari, L. A., Hughes, G. J., Fooks, A. R., & Rupprecht, C. E. (2004). Molecular epidemiology of terrestrial rabies in the former Soviet Union. Journal of Wildlife Diseases, 40(4), 617–631. https://doi.org/10.7589/0090-3558-40.4.617
47. Nadin-Davis, S. A. (1998). Polymerase chain reaction protocols for rabies virus discrimination. Journal of Virological Methods, 75(1), 1–8. https://doi.org/10.1016/s0166-0934(98)00106-2
48. Whitby, J. E., Johnstone, P., & Sillero-Zubiri, C. (1997). Rabies virus in the decomposed brain of an Ethiopian wolf detected by nested reverse transcription-polymerase chain reaction. Journal of Wildlife Diseases, 33(4), 912–915. https://doi.org/10.7589/0090-3558-33.4.912
49. Briggs, D. J., Smith, J. S., Mueller, F. L., Schwenke, J., Davis, R. D., Gordon, C. R., Schweitzer, K., Orciari, L. A., Yager, P. A., & Rupprecht, C. E. (1998). A comparison of two serological methods for detecting the immune response after rabies vaccination in dogs and cats being exported to rabies-free areas. Biologicals, 26(4), 347–355. https://doi.org/10.1006/biol.1998.0162
50. Wright, E., McNabb, S., Goddard, T., Horton, D. L., Lembo, T., Nel, L. H., Weiss, R. A., Cleaveland, S., & Fooks, A. R. (2009). A robust lentiviral pseudotype neutralization assay for in-field serosurveillance of rabies and lyssaviruses in Africa. Vaccine, 27(51), 7178–7186. https://doi.org/10.1016/j.vaccine.2009.09.024
51. Welch, R. J., Anderson, B. L., & Litwin, C. M. (2009). An evaluation of two commercially available ELISAs and one in-house reference laboratory ELISA for the determination of human anti-rabies virus antibodies. Journal of Medical Microbiology, 58(Pt 6), 806–810. https://doi.org/10.1099/jmm.0.006064-0
52. Moore, S. M., & Hanlon, C. A. (2010). Rabies-specific antibodies: Measuring surrogates of protection against a fatal disease. PLOS Neglected Tropical Diseases, 4(3), e595. https://doi.org/10.1371/journal.pntd.0000595
53. Alvarez, L., Fajardo, R., Lopez, E., Pedroza, R., Hemachudha, T., Kamolvarin, N., Cortes, G., & Baer, G. M. (1994). Partial recovery from rabies in a nine-year-old boy. Pediatric Infectious Disease Journal, 13(12), 1154–1155. https://doi.org/10.1097/00006454-199412000-00020
54. Pan American Health Organization. (2023). Recommendations for Laboratory Detection and Diagnosis of Arbovirus Infections in the Region of the Americas. PAHO, Washington, DC. https://doi.org/10.37774/9789275125878
55. Ren, P., Ortiz, D. A., Terzian, A. C. B., Colombo, T. E., Nogueira, M. L., Vasilakis, N., & Loeffelholz, M. J. (2017). Evaluation of Aptima Zika virus assay. Journal of Clinical Microbiology, 55(7), 2198–2203. https://doi.org/10.1128/JCM.00603-17
56. Kurosaki, Y., Martins, D. B. G., Kimura, M., Catena, A. D. S., Borba, M. A. C. S. M., Mattos, S., Abe, D. S., Yoshikawa, H., de Lima Filho, R., & Yasuda, J. L. (2017). Development and evaluation of a rapid molecular diagnostic test for Zika virus infection by reverse transcription loop-mediated isothermal amplification. Scientific Reports, 7(1), 13503. https://doi.org/10.1038/s41598-017-13836-9
57. Pan American Health Organization, Centers for Disease Control and Prevention of the United States of America. (2011). Preparedness and Response for Chikungunya Virus Introduction in the Americas. PAHO, Washington, DC. https://www.paho.org/hq/dmdocuments/2012/CHIKV-English.pdf

58. United States Centers for Disease Control and Prevention. (2017). Dengue: Laboratory Guidance and Diagnostic Testing. CDC, Atlanta. https://www.cdc.gov/dengue/clinicallab/laboratory.html
59. Pan American Health Organization. (2018). Laboratory Diagnosis of Yellow Fever Virus Infection. PAHO, Washington, DC. https://www.paho.org/en/documents/laboratory-diagnosis-yellow-fever-virus-infection
60. Pan American Health Organization. (2018). Guidelines for Surveillance of Zika Virus Disease and its Complications. 2018 edition. PAHO, Washington, DC. https://iris.paho.org/xmlui/handle/123456789/49518
61. Mansuy, J. M., Mengelle, C., Pasquier, C., Chapuy-Regaud, S., Delobel, P., Martin-Blondel, G., & Izopet, J. (2017). Zika virus infection and prolonged viremia in whole-blood specimens. Emerging Infectious Diseases, 23(5), 863–865. https://doi.org/10.3201/eid2305.161631
62. Hermann, L. L., Thaisomboonsuk, B., Poolpanichupatam, Y., Jarman, R. G., Kalayanarooj, S., Nisalak, A., Yoon, I. K., & Fernandez, S. (2014). Evaluation of a dengue NS1 antigen detection assay sensitivity and specificity for the diagnosis of acute dengue virus infection. PLOS Neglected Tropical Diseases, 8(10), e3193. https://doi.org/10.1371/journal.pntd.0003193
63. Guzman, M. G., Jaenisch, T., Gaczkowski, R., Ty Hang, V. T., Sekaran, S. D., Kroeger, A., Vazquez, S., Ruiz, D., Martinez, E., Mercado, J. C., Balmaseda, A., Harris, E., Dimano, E., Leano, P. S., Yoksan, S., Villegas, E., Benduzu, H., Villalobos, I., Farrar, J., & Simmons, C. P. (2010). Multi-country evaluation of the sensitivity and specificity of two commercially-available NS1 ELISA assays for dengue diagnosis. PLOS Neglected Tropical Diseases, 4(8), e811. https://doi.org/10.1371/journal.pntd.0000811
64. Matheus, S., Boukhari, R., Labeau, B., Ernault, V., Bremand, L., Kazanji, M., & Rousset, D. (2016). Specificity of dengue NS1 antigen in differential diagnosis of dengue and Zika virus infection. Emerging Infectious Diseases, 22(9), 1691–1693. https://doi.org/10.3201/eid2209.160725
65. Lima, M. D. R. Q., Chouin-Carneiro, T., Azeredo, E., Barbosa, L. S., Souza, T. M. A., Silva, J. B. C. D., Nunes, P. C. G., Dal Fabbro, M., Facco, I. H. R., Venâncio-da-Cunha, R., & Dos Santos, F. B. (2019). The inability of a dengue NS1 ELISA to detect Zika infections. Scientific Reports, 9(1), 18596. https://doi.org/10.1038/s41598-019-55160-4
66. Medina, F., Medina, J. F., Colon, C., Vergne, E., Santiago, G. A., & Muñoz-Jordan, J. L. (2012). Dengue virus: Isolation, propagation, quantification, and storage. Current Protocols in Microbiology, Chapter 15, Unit 15D 2. https://doi.org/10.1002/9780471729259.mc15d02s27
67. Medina, F. A., Torres, G., Acevedo, J., Fonseca, S., Casiano, L., De Leon-Rodriguez, C. M., Santiago, G. A., Doyle, K., Sharp, T. M., Alvarado, L. I., Paz-Bailey, G., & Muñoz-Jordán, J. L. (2019). Duration of the presence of infectious Zika virus in semen and serum. The Journal of Infectious Diseases, 219(1), 31–40. https://doi.org/10.1093/infdis/jiy462
68. Lanciotti, R. S., Kosoy, O. L., Laven, J. J., Velez, J. O., Lambert, A. J., Johnson, A. J., Stanfield, S. M., & Duffy, M. R. (2008). Genetic and serologic properties of Zika virus associated with an epidemic, Yap State, Micronesia, 2007. Emerging Infectious Diseases, 14(8), 1232–1239. https://doi.org/10.3201/eid1408.080287
69. Basile, A. J., Goodman, C., Horiuchi, K., Laven, J., Panella, A. J., Kosoy, O., Lanciotti, R. S., & Johnson, B. W. (2015). Development and validation of an ELISA kit (YF MAC-HD) to detect IgM to yellow fever virus. Journal of Virological Methods, 225, 41–48. https://doi.org/10.1016/j.jviromet.2015.08.025
70. Vazquez, S., Saenz, E., Huelva, G., Gonzalez, A., Kouri, G., & Guzman, M. (1998). Detección de IgM contra el virus del dengue en sangre entera absorbida en papel de filtro [Detection of IgM against the dengue++ virus in whole blood absorbed on filter paper]. Revista Panamericana de Salud Publica, 3(3), 174–178. Spanish. https://doi.org/10.1590/s1020-49891998000300005
71. Pedro Kourí Institute of Tropical medicine. (2013). Laboratory Techniques for the Diagnosis and Characterization of Dengue Viruses. Ministry of Public Health, Havana.
72. Hunsperger, E. A., Yoksan, S., Buchy, P., Nguyen, V. C., Sekaran, S. D., Enria, D. A., Pelegrino, J. L., Vázquez, S., Artsob, H., Drebot, M., Gubler, D. J., Halstead, S. B., Guzmán, M. G., Margolis, H. S., Nathanson, C. M., Rizzo Lic, N. R., Bessoff, K. E., Kliks, S., & Peeling, R. W. (2009). Evaluation of commercially available antidengue virus immunoglobulin M tests. Emerging Infectious Diseases, 15(3), 436–440. https://doi.org/10.3201/eid1503.080923
73. Hunsperger, E. A., Yoksan, S., Buchy, P., Nguyen, V. C., Sekaran, S. D., Enria, D. A., Vazquez, S., Cartozian, E., Pelegrino, J. L., Artsob, H., Guzman, M. G., Olliaro, P., Zwang, J., Guillerm, M., Kliks, S., Halstead, S., Peeling, R. W., & Margolis, H. S. (2014). Evaluation of commercially available diagnostic tests for the detection of dengue virus NS1 antigen and anti-dengue virus IgM antibody. PLOS Neglected Tropical Diseases, 8(10), e3171. https://doi.org/10.1371/journal.pntd.0003171

74. Balmaseda, A., Stettler, K., Medialdea-Carrera, R., Collado, D., Jin, X., Zambrana, J. V., Jaconi, S., Cameroni, E., Saborio, S., Rovida, F., Percivalle, E., Ijaz, S., Dicks, S., Ushiro-Lumb, I., Barzon, L., Siqueira, P., Brown, D. W. G., Baldanti, F., Tedder, R., Zambon, M., de Filippis, A. M. B., Harris, E., & Corti, D. (2017). Antibody-based assay discriminates Zika virus infection from other flaviviruses. Proceedings of the National Academy of Sciences of the United States of America, 114(31), 8384–8389. https://doi.org/10.1073/pnas.1704984114
75. Roehrig, J. T., Hombach, J., & Barrett, A. D. (2008). Guidelines for plaque-reduction neutralization testing of human antibodies to dengue viruses. Viral Immunology, 21(2), 123–132. https://doi.org/10.1089/vim.2008.0007
76. Salje, H., Rodriguez-Barraquer, I., Rainwater-Lovett, K., Nisalak, A., Thaisomboonsuk, B., Thomas, S. J., Fernandez, S., Jarman, R. G., Yoon, I. K., & Cummings, D. A. (2014). Variability in dengue titer estimates from plaque reduction neutralization tests poses a challenge to epidemiological studies and vaccine development. PLOS Neglected Tropical Diseases, 8(6), e2952. https://doi.org/10.1371/journal.pntd.0002952
77. Mansfield, K. L., Horton, D. L., Johnson, N., Li, L., Barrett, A. D. T., Smith, D. J., Galbraith, S. E., Solomon, T., & Fooks, A. R. (2011). Flavivirus-induced antibody cross-reactivity. Journal of General Virology, 92(Pt 12), 2821–2829. https://doi.org/10.1099/vir.0.031641-0
78. Zompi, S., & Harris, E. (2013). Original antigenic sin in dengue revisited. Proceedings of the National Academy of Sciences of the United States of America, 110(22), 8761–8762. https://doi.org/10.1073/pnas.1306333110
79. Hunsperger, E. A., Munoz-Jordan, J., Beltran, M., Colon, C., Carrion, J., Vazquez, J., Acosta, L. N., Medina-Izquierdo, J. F., Horiuchi, K., Biggerstaff, B. J., & Margolis, H. S. (2016). Performance of dengue diagnostic tests in a single-specimen diagnostic algorithm. The Journal of Infectious Diseases, 214(6), 836–844. https://doi.org/10.1093/infdis/jiw103. Erratum in: The Journal of Infectious Diseases, 2017, 215(11), 1774. https://doi.org/10.1093/infdis/jix219
80. Sakkas, H., Bozidis, P., Franks, A., & Papadopoulou, C. (2018). Oropouche fever: A review. Viruses, 10(4). https://doi.org/10.3390/v10040175
81. Romero-Alvarez, D., & Escobar, L. E. (2018). Oropouche fever, an emergent disease from the Americas. Microbes and Infection, 20(3), 135–146. https://doi.org/10.1016/j.micinf.2017.11.013
82. Naveca, F. G., Nascimento, V. A. D., Souza, V. C., Nunes, B. T. D., Rodrigues, D. S. G., & Vasconcelos, P. (2017). Multiplexed reverse transcription real-time polymerase chain reaction for simultaneous detection of Mayaro, Oropouche, and Oropouche-like viruses. Memorias do Instituto Oswaldo Cruz, 112(7), 510–513. https://doi.org/10.1590/0074-02760160062
83. Nunes, M. R. T., de Souza, W. M., Savji, N., Figueiredo, M. L., Cardoso, J. F., da Silva, S. P., da Silva de Lima, C. P., Vasconcelos, H. B., Rodrigues, S. G., Ian Lipkin, W., Vasconcelos, P. F. C., & Palacios, G. (2018). Oropouche Orthobunyavirus: Genetic characterization of full-length genomes and development of molecular methods to discriminate natural reassortments. Infection, Genetics and Evolution, 68, 16–22. https://doi.org/10.1016/j.meegid.2018.11.020
84. Waggoner, J. J., Rojas, A., Mohamed-Hadley, A., de Guillen, Y. A., & Pinsky, B. A. (2018). Real-time RT-PCR for Mayaro virus detection in plasma and urine. Journal of Clinical Virology, 98, 1–4. https://doi.org/10.1016/j.jcv.2017.11.006
85. Mota, M. T. O., ., Ribeiro, M. R., Vedovello, D., & Nogueira, M. L. (2015). Mayaro virus: A neglected arbovirus of the Americas. Future Virology, 10(9), 1109–1122. https://doi.org/10.2217/fvl.15.76
86. Calisher, C. H., el-Kafrawi, A. O., Al-Deen Mahmud, M. I., Travassos da Rosa, A. P., Bartz, C. R., Brummer-Korvenkontio, M., Haksohusodo, S., & Suharyono, W. (1986). Complex-specific immunoglobulin M antibody patterns in humans infected with alphaviruses. Journal of Clinical Microbiology, 23(1), 155–159. https://doi.org/10.1128/jcm.23.1.155-159.1986
87. Acosta-Ampudia, Y., Monsalve, D. M., Rodriguez, Y., Pacheco, Y., Anaya, J. M., & Ramirez-Santana, C. (2018). Mayaro: An emerging viral threat? Emerging Microbes and Infections, 7(1), 163. https://doi.org/10.1038/s41426-018-0163-5
88. Damon, I. K. (2011). Status of human monkeypox: Clinical disease, epidemiology and research. Vaccine, 29, D54–D59. https://doi.org/10.1016/j.vaccine.2011.04.014
89. Reynolds, M. G., Carroll, D. S., & Karem, K. L. (2012). Factors affecting the likelihood of monkeypox's emergence and spread in the post-smallpox era. Current Opinion in Virology, 2(3), 335–343. https://doi.org/10.1016/j.coviro.2012.02.004
90. Mattar, S., Guzmán, C., & Figueiredo, L. T. (2015). Diagnosis of hantavirus infection in humans. Expert Review of Anti-infective Therapy, 13(8), 939–946. https://doi.org/10.1586/14787210.2015.1047825

91. Peeling, R. W., Boeras, D. I., & Nkengasong, J. (2017). Re-imagining the future of diagnosis of neglected tropical diseases. Computational and Structural Biotechnology Journal, 15, 271–274. https://doi.org/10.1016/j.csbj.2017.02.003
92. Kiselev, D., Matsvay, A., Abramov, I., Dedkov, V., Shipulin, G., & Khafizov, K. (2020). Current trends in diagnostics of viral infections of unknown etiology. Viruses, 12(2), 211. https://doi.org/10.3390/v12020211
93. World Health Organization. (2020). Ending the neglect to attain the Sustainable Development Goals: A road map for neglected tropical diseases 2021–2030. https://www.who.int/publications/i/item/9789240052932
94. Bharadwaj, M., Bengtson, M., Golverdingen, M., Waling, L., & Dekker, C. (2021). Diagnosing point-of-care diagnostics for neglected tropical diseases. PLOS Neglected Tropical Diseases, 15(6), e0009405. https://doi.org/10.1371/journal.pntd.0009405
95. Le´chenne, M., Naïssengar, K., Lepelletier, A., Alfaroukh, I. O., Bourhy, H., Zinsstag, J., & Dacheux, L. (2016). Validation of a rapid rabies diagnostic tool for field surveillance in developing countries. PLOS Neglected Tropical Diseases, e0005010, 10. https://doi.org/10.1371/journal.pntd.0005010
96. Blacksell, S. D., Jarman, R. G., Bailey, M. S., Tanganuchitcharnchai, A., Jenjaroen, K., Gibbons, R. V., Paris, D. H., Premaratna, R., de Silva, H. J., Lalloo, D. G., & Day, N. P. (2011). Evaluation of six commercial point-of-care tests for diagnosis of acute dengue infections: The need for combining NS1 antigen and IgM/IgG antibody detection to achieve acceptable levels of accuracy. Clinical and Vaccine Immunology, 18, 2095–2101. https://doi.org/10.1128/CVI.05285-11
97. Vongsouvath, M., Bharucha, T., Seephonelee, M., de Lamballerie, X., Newton, P. N., & Dubot-Pe´rès, A. (2020). Harnessing dengue rapid diagnostic tests for the combined surveillance of dengue, Zika, and chikungunya viruses in Laos. American Journal of Tropical Medicine and Hygiene, 102, 1244–1248. https://doi.org/10.4269/ajtmh.19-0881
98. Souza, A. A., Ducker, C., Argaw, D., King, J. D., Solomon, A. W., Biamonte, M. A., Coler, R. N., Cruz, I., Lejon, V., Levecke, B., Marchini, F. K., Marks, M., Millet, P., Njenga, S. M., Noordin, R., Paulussen, R., Sreekumar, E., & Lammie, P. J. (2021). Diagnostics and the neglected tropical diseases roadmap: Setting the agenda for 2030. Transactions of the Royal Society of Tropical Medicine and Hygiene, 115(2), 129–135. https://doi.org/10.1093/trstmh/traa118

9 Immunotherapies and Vaccines

Altijana Hromić-Jahjefendić and Maryam Golzardi

9.1 INTRODUCTION

Vaccines play a critical role in global health by preventing infection and transmission of multiple diseases worldwide [1]. The World Health Organization estimates that immunizations prevent the death of 2–3 million people every year [2]. In addition, vaccines have a significant economic impact by reducing costs of illnesses and hospitalizations of over $500 billion [3].

Most vaccines in clinical use focus on training a patient's own immune system to recognize and combat specific deadly diseases. Vaccination programs have led not only to a significant decrease in morbidity and mortality, particularly during childhood, but also to complete eradication of some diseases (e.g., smallpox). More recently, the swift development of vaccines against severe acute respiratory syndrome coronavirus 2 (SARS-CoV-2), especially messenger RNA (mRNA) vaccines, can be credited to revolutionary vaccine technologies that can rapidly counter threats from emerging pathogens [4].

Discovery of live attenuated/killed whole organism vaccines is one of the most important findings in the medical field, which has eradicated many life-threatening diseases such as smallpox, rubella, polio, mumps, tetanus, and diphtheria. However, rapid emergence of deadly diseases such as SARS-CoV-2, H1N1, Ebola generate a challenge for conventional vaccines, development of which takes around decade and even more time is needed to scale up the manufacturing [5].

Nanotechnology represents a cutting-edge approach, able to address some of the most relevant vaccinology concerns. Nanomaterials (NMs) with sizes between 1 and 100 nm have emerged as innovative tools in many biomedical applications, from diagnostics to drug delivery. Many NMs are characterized by high biocompatibility, morphological customization, reliable functionalization, and prolonged circulation in blood [6]. Among many different nanoparticles which include lipids, self-assembled proteins, biopolymers and inorganic materials [7], properly functionalized gold nanoparticles (AuNPs), represent one of the most promising NMs in the development of a new generation of vaccines.

On the other hand, owing to the high prevalence of viral diseases, and emerging drug resistance in microorganisms, particularly in parasites and bacteria have sparked the development of a novel vaccine candidate using artificial intelligence (AI – a term given by McCarthy in 1956) [8] and this provides an alternative to the traditional approaches because it takes years to generate an effective vaccine since conventional approaches are considerably costly and time-consuming. In an effort to target an important endemic infection, international health organizations are also speeding up their vaccine research program. Substantial information, i.e., proteomic and genomic data have been developed in the research and medical fields by recent advances in high throughput experimental processes and such data allow AI to provide incredibly accurate forecasts and prediction [8].

Production of therapeutic antibodies represents a challenge and possibility in recent times. The application of mAb as immunotherapeutic agents with high-specificity profiles has revolutionized the treatment for several life-threatening diseases such as diverse types of cancer, autoimmune diseases, metabolic disorders, and infectious diseases, including those caused by SARS-CoV-2 and Influenza virus [9]. With the purpose of enhancing effectiveness and therapeutic efficacy, further mAb engineering remains the primary approach of different drug developing companies. However, antibodies must face several cellular and non-cellular barriers after entering the human body to reach their target antigens and that specifies their application purpose [9].

DOI: 10.1201/9781003607816-9

Current approaches towards vaccination are constantly under change. Introducing modern techniques like chimeric antigen receptor (CAR)-T therapy for breast cancer or integration of human papillomavirus (HPV) immunization among young female population or introducing various immunotherapy approaches in cancer treatment represent only one step in the upgrading current vaccination and disease treatment mode. However, more upgrades are needed in order to enhance vaccination status among children as well as adults [10].

9.2 VACCINE DEVELOPMENT CHALLENGES

Vaccine development is a complex and multifaceted process that has been instrumental in reducing morbidity and mortality from infectious diseases worldwide. However, despite these successes, numerous challenges persist, hindering the creation of effective vaccines against various pathogens [11]. This section will explore the key challenges facing vaccine development, including an incomplete understanding of immunity, host and pathogen variability, societal skepticism, and logistical hurdles. The COVID-19 pandemic has highlighted both the potential for rapid vaccine development and the ongoing challenges in this field.

9.2.1 Incomplete Understanding of Immunity and Pathogen Variability

One of the primary challenges in vaccine development is the incomplete understanding of how immunity develops and is maintained. The immune system is highly complex, involving intricate interactions between various cell types and molecular pathways [12]. Understanding these processes is crucial for designing vaccines that can elicit protective immune responses. For instance, the development of vaccines against rapidly evolving pathogens like influenza and HIV is complicated by the need to induce broad and durable immunity, which is difficult to achieve with current technologies [13].

Host and pathogen genetic variability also pose significant challenges. Pathogens can evolve rapidly, leading to changes in their surface antigens, which are the targets of the immune response. This means that vaccines may need to be updated frequently to remain effective, as seen with influenza vaccines [14]. Additionally, genetic differences among human populations can affect how individuals respond to vaccines, necessitating the development of vaccines that are effective across diverse populations [15].

The COVID-19 pandemic has underscored these challenges. The emergence of SARS-CoV-2 variants has raised concerns about the long-term efficacy of vaccines, as some variants can evade immune responses elicited by current vaccines [16]. The high mutation rate of RNA viruses like SARS-CoV-2 complicates the development of universal vaccines that can protect against all variants. Understanding the biological basis of immune evasion is crucial for combating these obstacles [17].

9.2.2 Societal and Logistical Challenges

Societal skepticism regarding vaccine safety is another major hurdle. Misinformation and misconceptions about vaccines have led to increased vaccine hesitancy, which can undermine public health efforts to control infectious diseases [18]. This skepticism often stems from a lack of understanding about vaccine development and safety testing processes. Addressing these concerns requires effective communication and education about the benefits and risks of vaccines [18].

Logistical challenges also play a significant role. The development of vaccines is a costly and time-consuming process, typically requiring around $700 million to $1 billion and taking about 10 years from concept to licensure [19]. The need for maintaining a cold chain for vaccine distribution is particularly problematic in resource-limited settings, where infrastructure may be inadequate. Furthermore, there is a limited number of vaccine manufacturers, which can lead to supply chain vulnerabilities during global health emergencies [20].

The COVID-19 pandemic highlighted the potential for rapid vaccine development, with vaccines being developed in less than a year after the identification of the virus [21]. However, this accelerated process also underscored logistical challenges, such as the need for ultra-cold storage for mRNA vaccines, which can be difficult to maintain in some regions. The pandemic has shown that substantial investment and collaboration are necessary to overcome these logistical hurdles [22].

9.2.3 Innovative Solutions and Future Directions

Despite these challenges, innovative approaches are being explored to improve vaccine development. Advances in genomics, transcriptomics, and bioinformatics are enabling the rapid identification of vaccine antigens and the design of more effective vaccines [23]. For example, the use of mRNA technology in COVID-19 vaccines demonstrated how new platforms can accelerate vaccine development and improve efficacy [24, 25].

Innovative vaccine delivery systems, such as exosomes, are also being investigated for their potential to enhance vaccine efficacy and stability [26]. Exosomes are naturally occurring vesicles that can be engineered to carry antigens, potentially offering a novel method for vaccine delivery that could improve immune responses [27].

Moreover, collaborative efforts like the Coalition for Epidemic Preparedness Innovations (CEPI) are working to address specific challenges in vaccine development by promoting global coordination and investment in vaccine research [28]. These initiatives highlight the importance of international collaboration and coordinated research strategies in overcoming the complex challenges facing vaccine development [29].

The COVID-19 pandemic has accelerated the development of new vaccine technologies, including mRNA and viral vector platforms, which have shown promising results [30]. However, challenges remain, such as the need for booster shots to maintain immunity and the potential for escape mutations that could reduce vaccine effectiveness [30]. Addressing these challenges will require continued investment in vaccine research and development.

9.2.4 Reflection on COVID-19

The COVID-19 pandemic has been a pivotal moment in vaccine development, demonstrating both the potential for rapid innovation and the ongoing challenges in this field. The development of COVID-19 vaccines in record time was facilitated by decades of scientific research into vaccine platforms and technologies, as well as unprecedented financial investments and global collaboration [31, 32].

However, the pandemic has also highlighted the need for continued research into vaccine technologies that can address emerging variants and provide long-term immunity [33]. The development of universal vaccines that can protect against multiple variants of a virus remains a significant challenge, requiring a deeper understanding of immune evasion mechanisms and the development of novel vaccine strategies [34].

In conclusion, vaccine development is fraught with challenges ranging from scientific and logistical hurdles to societal skepticism. However, ongoing research and innovative approaches offer promising solutions to these challenges. The COVID-19 pandemic has underscored the importance of continued investment in vaccine research and development, as well as global collaboration to address the evolving landscape of infectious diseases [35].

9.2.5 Immunotherapies and Vaccines

Immunotherapies have emerged as a transformative force in modern medicine, revolutionizing the treatment landscape for a variety of diseases, particularly cancer [36]. By harnessing the power of the immune system, immunotherapies offer targeted and durable responses that were once unattainable with traditional therapies. While vaccines traditionally focused on prophylaxis against infectious diseases, the convergence of immunotherapy and vaccinology is creating new paradigms for disease management. This synergy is particularly evident in cancer, where immunotherapeutic strategies are being explored to both treat existing tumors and prevent recurrence [37].

A central focus of cancer immunotherapy is the modulation of immune checkpoints, which act as brakes on the immune system, preventing excessive inflammation and autoimmunity [38]. Monoclonal antibodies targeting these checkpoints, such as cytotoxic T-lymphocyte–associated protein 4 (CTLA-4), programmed cell death protein 1 (PD-1), and programmed death-ligand (PD-L1), have demonstrated remarkable success in unleashing T cell-mediated anti-tumor immunity. By blocking these inhibitory signals, checkpoint inhibitors allow T cells to recognize and destroy cancer cells that would otherwise evade immune surveillance [39]. These therapies have shown durable responses in a subset of patients with various cancers, transforming the standard of care for melanoma, lung cancer, and other malignancies [40].

Beyond checkpoint inhibition, other immunotherapeutic strategies aim to directly stimulate anti-tumor immune responses. Adoptive cell therapies, such as CAR-T cell therapy, involve engineering a patient's own T cells to express a CAR that specifically targets a tumor-associated antigen (TAA) [41, 42]. These CAR-T cells are then infused back into the patient, where they can recognize and kill cancer cells with high precision. CAR-T cell therapy has shown remarkable success in treating hematologic malignancies, such as leukemia and lymphoma, and is being explored for solid tumors as well [43].

Cancer vaccines represent another promising avenue for immunotherapy, aiming to elicit de novo T cell responses against tumor-specific antigens [44]. While early cancer vaccine trials yielded limited success, recent advances in antigen discovery and vaccine delivery technologies have reignited interest in this approach. Personalized cancer vaccines, tailored to an individual's unique tumor mutations, hold particular promise for stimulating potent and durable anti-tumor immunity. These vaccines can be administered alone or in combination with other immunotherapies, such as checkpoint inhibitors, to enhance their efficacy [45].

Immunotherapies are also being explored for their potential to treat autoimmune diseases, which result from aberrant immune responses against self-antigens [46]. By selectively suppressing or modulating specific immune pathways, immunotherapies can restore immune tolerance and alleviate the symptoms of autoimmune disorders. Monoclonal antibodies targeting cytokines, such as TNF-α and IL-17, have shown significant efficacy in treating rheumatoid arthritis, psoriasis, and other autoimmune conditions [47]. Furthermore, cell-based therapies, such as regulatory T cell therapy, are being investigated for their potential to restore immune homeostasis and prevent autoimmune disease progression [48].

9.3 VACCINE DEVELOPMENT STRATEGIES

Vaccine development is a dynamic field that has seen significant advancements in recent years, particularly during the COVID-19 pandemic. The rapid development of vaccines against SARS-CoV-2 highlighted the potential for innovation and collaboration in addressing global health challenges [49]. This section will explore key strategies in vaccine development, including collaborative partnerships, innovative technologies, and approaches to address emerging challenges.

9.3.1 Collaborative Partnerships and Regulatory Innovations

Collaborative partnerships have become a cornerstone of vaccine development. The COVID-19 pandemic demonstrated how alliances among companies, academia, governments, and not-for-profit organizations can accelerate research and development and manufacturing [50]. For example, partnerships like the National Institutes of Health/Moderna and the University of Oxford/AstraZeneca collaborations were instrumental in the rapid development of COVID-19 vaccines [51].

These partnerships not only facilitate the sharing of resources and expertise but also help in navigating regulatory hurdles. Global regulatory cooperation can streamline administrative processes, as seen during the pandemic with the formation of COVID-19 working groups by the International Coalition of Medicines Regulatory Authorities. Such collaborations can establish governing protocols, agree on approaches to adapt vaccines to address variants, and improve regulatory agility [52].

9.3.2 Innovative Technologies

Innovative technologies are revolutionizing vaccine development. Nucleic acid vaccines, such as mRNA and DNA vaccines, have gained prominence due to their ease of design, production, and potential for rapid response to emerging diseases [53]. The COVID-19 pandemic showcased the effectiveness of mRNA vaccines, which were developed and approved in record time [54].

Other innovative approaches include microarray patches (MAPs) and heat-stable vaccines. MAPs offer a novel delivery method that could simplify vaccination processes, especially in resource-constrained settings. Heat-stable vaccines eliminate the need for cold chain storage, making them ideal for areas with limited infrastructure [55, 56].

9.3.3 Addressing Emerging Challenges

Emerging challenges in vaccine development include addressing vaccine hesitancy, improving vaccine accessibility, and developing vaccines against complex pathogens. The Immunization Agenda 2030 emphasizes the need for innovative strategies to mitigate vaccine misinformation and reach underserved populations. This includes implementing research and context-specific approaches to improve vaccine delivery in challenging settings [57].

Innovative delivery systems and technologies, such as barcodes on primary packaging, can enhance vaccine management and tracking, ensuring that vaccines reach those who need them most. Additionally, advances in diagnostics and surveillance systems are crucial for identifying and addressing emerging health threats [58].

The future of vaccine development will likely involve continued innovation in technologies and strategies. Recoding strategies, which involve altering the genetic structure of microorganisms, offer a promising approach for developing safer and more effective vaccines [31]. Furthermore, collaborative efforts like the Vaccine Innovation Prioritisation Strategy (VIPS) are working to accelerate the development of priority vaccine innovations, focusing on equitable access and improved effectiveness in low- and middle-income countries [59].

9.4 EMERGING TECHNOLOGIES IN VACCINE DEVELOPMENT

The field of vaccine development is rapidly evolving with the emergence of new technologies that offer faster, more effective, and safer vaccine production. These technologies are crucial for addressing the challenges posed by emerging infectious diseases and improving global health security. This section will explore some of the key emerging technologies in vaccine development.

9.4.1 Nucleic Acid Vaccines: mRNA and DNA

Nucleic acid vaccines, including mRNA and DNA vaccines, have gained significant attention due to their rapid development capabilities and strong immunogenicity [60]. mRNA vaccines, in particular, have shown remarkable success during the COVID-19 pandemic, with vaccines like those from Moderna and Pfizer-BioNTech demonstrating high efficacy in preventing severe disease [61]. These vaccines use mRNA to instruct cells to produce specific antigens, which then trigger an immune response.

The mRNA platform allows for rapid design and production, making it ideal for responding to emerging pathogens. Recent advancements in mRNA vaccines have focused on optimizing their delivery systems and modifying nucleotides to enhance translation efficiency and reduce immunogenicity [62]. For instance, the use of modified nucleosides like N1-methylpseudouridine (N1mΨ-UTP) has significantly improved the performance of mRNA vaccines by increasing protein expression while minimizing immune stimulation [63]. This modification has been crucial in the development of effective COVID-19 vaccines, such as Pfizer-BioNTech's BNT162b2, which has shown high efficacy in preventing COVID-19 across various populations [64].

DNA vaccines involve delivering a plasmid that encodes for specific antigens, which are then produced by the host cells. This approach can induce both humoral and cell-mediated immune responses and is noted for its stability and safety profile [65]. DNA vaccines have been explored for various infectious diseases and cancers, with ongoing research aimed at enhancing their immunogenicity through optimization of plasmid design and delivery methods [66].

Both mRNA and DNA vaccines offer the advantage of rapid development and flexibility in design, making them valuable tools in responding to emerging health threats [62]. The COVID-19 pandemic has highlighted the potential of mRNA vaccines, with their ability to be quickly adapted to new variants and produced at scale. However, challenges remain, such as ensuring global access to these vaccines and addressing concerns related to their long-term safety and efficacy [67].

Innovations in nucleic acid vaccines are ongoing, with research focusing on improving delivery systems and enhancing immune responses [68]. For example, lipid nanoparticles (LNPs) have been used to deliver mRNA vaccines, protecting the mRNA from degradation and facilitating its uptake by cells [69]. The integration of AI and machine learning into vaccine design is also being explored to optimize antigen selection and predict immunogenicity, potentially expanding the range of diseases treatable with nucleic acid vaccines [70].

9.4.2 Viral Vector and Subunit Vaccines

Viral vector vaccines have emerged as a significant strategy in vaccine development, particularly highlighted by their use in COVID-19 vaccines [71]. These vaccines utilize a weakened virus to deliver genetic material encoding for antigens, which are then expressed by host cells to stimulate an immune response. Viral vectors, such as adenoviruses, have been used in vaccines like AstraZeneca's and Johnson & Johnson's, offering a balance between safety and efficacy [72]. The adenovirus vector platform has been instrumental in rapidly developing vaccines against emerging pathogens, leveraging its ability to induce robust humoral and cellular immune responses [73].

One of the most successful viral vector platforms is the vesicular stomatitis virus (VSV) vector [74]. The rVSV-ZEBOV vaccine, which uses a VSV vector to express the Ebola glycoprotein, has been approved for use in humans and has demonstrated high efficacy in preventing Ebola infection [75]. This platform is being explored for other diseases, including filoviruses like Sudan virus and Bundibugyo virus, as well as coronaviruses and Zika virus, with promising results in preclinical studies.

Subunit vaccines, on the other hand, contain only specific components of a pathogen, such as proteins or sugars, rather than the entire microorganism. They are often produced using recombinant DNA technology and require adjuvants to enhance their immunogenicity. Subunit vaccines offer a safer alternative to live attenuated vaccines, as they cannot cause disease themselves. However, they may require booster shots to maintain immunity and can be less effective without adjuvants [76].

Recent advancements in viral vector vaccines include the development of multivalent vaccines, which can protect against multiple pathogens simultaneously [77]. For example, a multivalent VSV-vectored vaccine expressing antigens from several filoviruses has shown complete protection in animal models. This approach is particularly valuable in regions where multiple severe pathogens co-circulate, offering a potential solution for complex epidemiological scenarios [78].

Moreover, heterologous viral vector vaccines, which involve using different vectors in prime/boost regimens, have been explored to enhance immune responses [79]. This strategy can induce higher magnitude and more durable immunity compared to homologous vector vaccination. The use of viral vectors in combination with other vaccine platforms, such as mRNA or DNA vaccines, is also being investigated to optimize vaccine efficacy and address emerging variants [80].

9.4.3 Nanoparticle and Viral-Like Particle Vaccines

Nanoparticle vaccines represent a novel approach in vaccine development, leveraging nanoparticles to deliver antigens and enhance immune responses [81]. These vaccines can be engineered to target specific cells or tissues, potentially increasing vaccine efficacy by ensuring that antigens are delivered directly to immune cells like dendritic cells, which are crucial for initiating immune responses [81]. Nanoparticles, due to their unique properties such as high surface area and tunable size, can be functionalized to improve antigen presentation and stimulate robust immune responses [82].

Recent advancements in nanoparticle-based vaccines have shown significant promise in addressing various infectious diseases. For instance, nanoparticles have been explored for delivering antigens against Japanese encephalitis virus (JEV), demonstrating enhanced humoral and cellular immune responses in animal models [83]. Additionally, nanoparticle-based vaccines for COVID-19 are being developed, with over 26 candidates in clinical trials, offering diverse formats such as LNPs, virus-like particles (VLPs), and protein nanoparticles [84].

VLP vaccines are another innovative approach, where non-infectious particles mimic the structure of viruses but lack genetic material. VLPs can stimulate strong immune responses and have been successfully used in vaccines against diseases like HPV [85]. The HPV vaccine, which uses VLPs to mimic the outer shell of the virus, has been highly effective in preventing HPV-related cancers by inducing robust antibody responses against the virus [86].

The integration of nanoparticles with other vaccine platforms, such as mRNA vaccines, has further enhanced their potential. LNPs, for example, have revolutionized the delivery of mRNA vaccines by protecting the fragile mRNA molecules and facilitating their efficient uptake by cells [82]. This combination has been instrumental in the development of highly effective COVID-19 vaccines, demonstrating the power of nanotechnology in vaccine development [87].

Moreover, nanoparticles can function as both delivery systems and adjuvants, enhancing the stability and immunogenicity of vaccines. They can be engineered to carry multiple antigens, enabling the development of multivalent vaccines that can protect against several pathogens simultaneously [88]. This feature is particularly valuable in regions where multiple diseases co-circulate, offering a potential solution for complex epidemiological scenarios.

9.4.4 Innovative Manufacturing Technologies

The development of innovative manufacturing technologies is crucial for enhancing vaccine production capacity and reducing response times to emerging outbreaks. One such technology is the RNAbox, supported by the CEPI, which aims to scale up mRNA vaccine production locally [89]. This approach enables rapid response to pandemics by making vaccines more accessible worldwide, particularly in low- and middle-income countries where traditional manufacturing infrastructure may be limited.

The RNAbox technology uses a bespoke manufacturing process designed to overcome the need for centralized vaccine production, instead allowing for local manufacturing at small production sites globally. This continuous production process can create 7–10 times more mRNA at once, enabling more efficient use of raw materials and aligning with the 100 Days Mission to accelerate vaccine development timelines [89].

Recent advancements in mRNA vaccine manufacturing have also focused on automating the production of synthetic DNA templates, which are essential for mRNA vaccine production. CEPI's partnership with DNA Script aims to reduce the time needed to produce these templates from weeks to days, further accelerating mRNA vaccine development timelines and enhancing readiness in regions with less established vaccine manufacturing infrastructure [90].

Moreover, innovations in biopharmaceutical manufacturing, such as the use of advanced robotics and digital twinning, have significantly enhanced efficiency and precision in vaccine production. These technologies enable real-time data analysis and optimization of production cycles, ensuring that vaccines can be produced at scale while maintaining quality standards.

The summary of innovations in vaccine development can be seen in Table 9.1.

TABLE 9.1
Different Vaccine Development Strategies

Vaccine Technology	Description	Key Features	Examples/ Applications	References
		Nucleic Acid Vaccines		
mRNA Vaccines	Use messenger RNA to instruct cells to produce specific antigens, triggering an immune response.	Rapid design and production, optimized delivery systems, modified nucleotides for enhanced translation and reduced immunogenicity.	COVID-19 vaccines (Moderna, Pfizer-BioNTech), potential for rapid adaptation to new variants.	[62, 91]
DNA Vaccines	Deliver a plasmid encoding specific antigens, which are then produced by host cells.	Induce humoral and cell-mediated immune responses, noted for stability and safety profile, potential for long-term immunity.	Exploration for various infectious diseases and cancers.	[68]
		Viral Vector Vaccines		
Adenovirus Vector Vaccines	Utilize a weakened adenovirus to deliver genetic material encoding for antigens.	Balance between safety and efficacy, can induce robust humoral and cellular immune responses, potential for rapid vaccine development.	COVID-19 vaccines (AstraZeneca, Johnson & Johnson).	[92]
VSV Vector Vaccines	Use Vesicular Stomatitis Virus to express a target protein.	Induce robust humoral and cellular immune responses, potential for rapid vaccine development.	Ebola vaccine.	[93]
Subunit Vaccines	Contain only specific components of a pathogen (e.g., proteins or sugars).	Safer alternative to live attenuated vaccines, require adjuvants to enhance immunogenicity.	N/A	[94]
		Nanoparticle Vaccines		
General Nanoparticle Vaccines	Utilize nanoparticles to deliver antigens, enhancing immune responses and improving vaccine stability. Can be engineered to target specific cells or tissues.	Targeted delivery to immune cells, improved antigen presentation, stimulation of robust immune responses, potential for multivalent vaccines.	Exploration for Japanese encephalitis virus (JEV), COVID-19 vaccine candidates (LNPs, VLPs, protein nanoparticles).	[81, 95,96]
Viral-Like Particle (VLP) Vaccines	Non-infectious particles mimicking the structure of viruses but lacking genetic material.	Stimulate strong immune responses, safe and effective, can induce robust antibody responses against the virus.	Human papillomavirus (HPV) vaccine.	[97]
LNP Vaccines	Combine lipid nanoparticles with mRNA	protects the mRNA from degradation and facilitating its uptake by cells	COVID-19 vaccine (Pfizer, Moderna)	[98]

9.5 THERAPEUTIC ANTIBODIES

Therapeutic antibodies have emerged as a cornerstone in the treatment of a wide range of diseases, from cancer and autoimmune conditions to infections and cardiovascular diseases. These biological drugs have revolutionized the field of medicine by offering targeted therapies with high specificity and efficacy [99]. The development of therapeutic antibodies has been facilitated by advances in antibody engineering and manufacturing technologies, enabling the production of large quantities of specific monoclonal antibodies with demonstrated clinical efficacy [100].

9.5.1 Mechanisms of Action and Applications

Therapeutic antibodies exert their effects through several mechanisms, including neutralization, antibody-dependent cellular cytotoxicity (ADCC), and complement-dependent cytotoxicity (CDC) [101]. Neutralization involves binding to specific ligands or receptors on cell surfaces, thereby blocking pathophysiological signaling pathways. This mechanism is crucial in cancer treatment, where antibodies like trastuzumab target the human epidermal growth factor receptor 2 to inhibit tumor growth by disrupting receptor dimerization and internalization [102].

ADCC and CDC are immune-mediated mechanisms where antibodies bind to target cells, marking them for destruction by immune cells or complement proteins, respectively [103]. Rituximab, for example, uses ADCC to target CD20-positive B cells in non-Hodgkin's lymphoma, inducing cell death through the recruitment of immune effector cells like macrophages and natural killer cells4. Additionally, rituximab can induce CDC, further enhancing its anti-tumor effects [104].

The versatility of therapeutic antibodies allows them to be engineered into various formats, such as bispecific antibodies, which can bind to two different antigens simultaneously, enhancing their therapeutic effects [105]. Bispecific antibodies like blinatumomab have been approved for the treatment of certain leukemias by redirecting T cells to target cancer cells [106]. This dual specificity enables bispecific antibodies to engage multiple immune cells or pathways, potentially improving treatment outcomes in complex diseases [107].

Therapeutic antibodies have been approved for numerous indications, with the "big five" being infliximab, adalimumab, trastuzumab, bevacizumab, and rituximab. These antibodies are primarily used in treating autoimmune diseases like rheumatoid arthritis and Crohn's disease, as well as various cancers [108]. The use of antibodies in oncology has been particularly successful, with drugs like bevacizumab targeting vascular endothelial growth factor A to inhibit tumor angiogenesis, thereby reducing tumor growth and metastasis [109].

In oncology, therapeutic antibodies can enhance cytotoxic T lymphocyte-mediated anti-tumor immune responses by blocking co-inhibitory receptors or activating co-stimulatory receptors [108]. For instance, antibodies targeting PD-1/PD-L1 have shown significant efficacy in treating various cancers by releasing the brakes on the immune system, allowing T cells to recognize and attack tumor cells more effectively [110].

Moreover, therapeutic antibodies can modulate the immune response by altering the cytokine milieu, which is crucial for creating an environment conducive to anti-tumor immunity [111]. This ability to manipulate immune responses makes therapeutic antibodies valuable tools in cancer immunotherapy, where they can be used alone or in combination with other treatments like chemotherapy or checkpoint inhibitors [112]. The summary of antibody-dependent cellular phagocytosis (ADCP) among CDC and ADCC can be seen in Figure 9.1.

9.5.2 Therapeutic Antibodies in Immunotherapies and Vaccines

Therapeutic antibodies have become a cornerstone in immunotherapies, particularly in cancer treatment, by enhancing immune responses against tumors. These antibodies can target immune

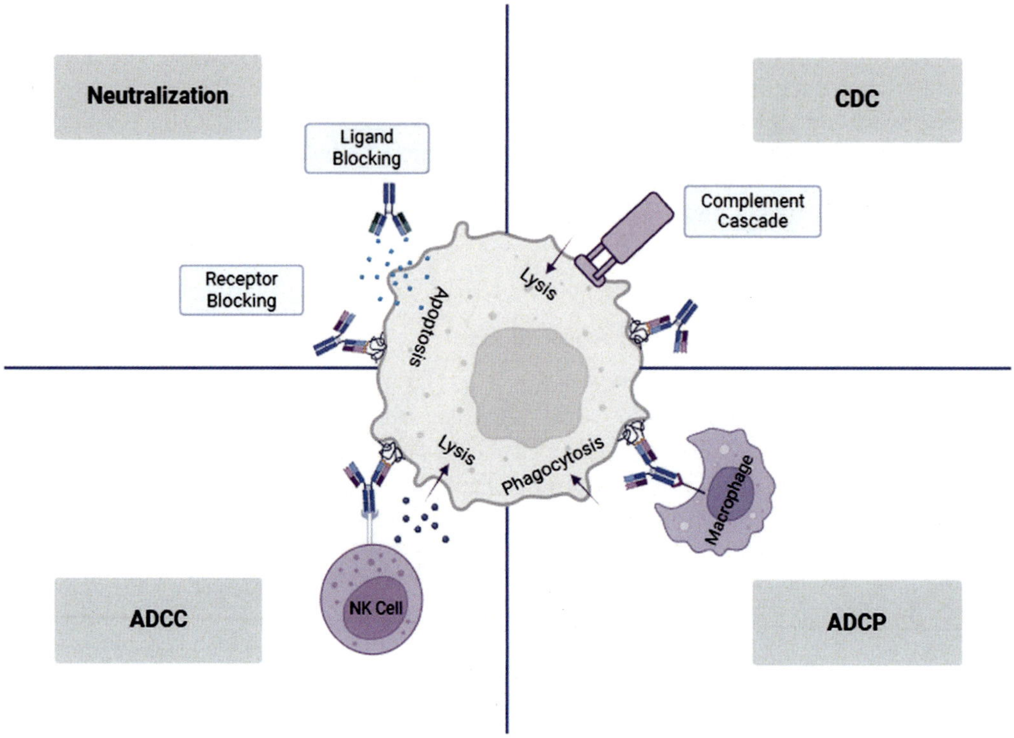

FIGURE 9.1 Antibody-dependent cellular phagocytosis (ADCP) among CDC and ADCC. (Figure was generated using BioRender.com.)

checkpoints or engage effector cells like T cells and natural killer cells to combat cancer [113]. Bispecific antibodies, such as blinatumomab, have been approved for the treatment of certain leukemias by redirecting T cells to target cancer cells [114].

In the context of vaccines, therapeutic antibodies can provide immediate protection against infections, especially in individuals who may not respond well to traditional vaccines. For example, monoclonal antibodies against SARS-CoV-2 have been used to treat COVID-19 by neutralizing the virus and preventing severe disease [115]. These antibodies can serve as a bridge until vaccines become effective or in scenarios where vaccines are not available [13].

Moreover, therapeutic antibodies can be engineered to enhance vaccine efficacy by targeting specific immune cells or modulating the immune response to improve vaccine-induced immunity. This integration of therapeutic antibodies with vaccines represents a promising strategy for enhancing immunotherapeutic outcomes [116].

9.5.3 Therapeutic Cancer Vaccines

Therapeutic cancer vaccines have shown significant promise in eliciting de novo T cell responses targeting tumor antigens, including TAAs and tumor-specific antigens [117]. These vaccines aim to amplify and diversify the intrinsic repertoire of tumor-specific T cells, enhancing the body's natural ability to recognize and attack cancer cells. However, challenges remain, such as tumor immune microenvironment suppression and optimal candidate identification for vaccine development [118].

Recent advancements in therapeutic cancer vaccines include the use of mRNA vaccines that target TAAs and neoantigens. For instance, an mRNA vaccine comprising four TAAs has demonstrated the capacity to elicit robust and durable immune responses directed against these antigens, with or without immune checkpoint inhibitors, in patients with unresectable melanoma [119]. Neoantigen-based vaccines are particularly promising for stimulating cytotoxic T cells to mount effective anti-tumor responses, as they target mutations unique to cancer cells, thereby minimizing off-target effects on healthy tissues [120].

The concept of therapeutic cancer vaccines predates innovations in checkpoint blockade immunotherapies and has seen a resurgence in recent years due to advances in understanding tumor mutational burden and neoantigen-reactive T cells [121]. mRNA-based vaccines, following their success in preventing COVID-19, have re-ignited enthusiasm for cancer vaccination strategies by offering rapid design and production capabilities [122]. These vaccines can be engineered to target specific antigens, enhancing their potential to induce anti-tumor immunity.

Moreover, therapeutic cancer vaccines have been explored in various clinical trials, with some showing promising results. Sipuleucel-T, an autologous dendritic cell vaccine, was the first therapeutic cancer vaccine approved by the US FDA, demonstrating improved survival in patients with metastatic prostate cancer [123]. Other strategies involve employing TAAs like prostate acid phosphatase in prostate cancer, highlighting the potential for personalized immunotherapy approaches [124].

Despite these advancements, challenges persist, including tumor-induced immunosuppression and the need for better understanding of the underlying immunological mechanisms [125]. Combining therapeutic cancer vaccines with other treatments, such as chemoradiotherapy or immunotherapy, offers a promising strategy to enhance their efficacy [126]. For example, in non-small cell lung cancer, therapeutic vaccines are being explored in combination with checkpoint inhibitors to improve survival outcomes [127].

9.5.4 Integration of Therapeutic Antibodies with Vaccines

The integration of therapeutic antibodies with vaccines offers a dual approach to enhancing immunotherapeutic outcomes. Therapeutic antibodies can be used to modulate the immune response, potentially improving vaccine efficacy by enhancing antigen presentation or activating immune cells [44]. For example, antibodies targeting immune checkpoints like PD-1/PD-L1 can be combined with vaccines to enhance T cell responses against tumor antigens [128].

In the context of infectious diseases, therapeutic antibodies can provide immediate protection while vaccines are being developed or in cases where vaccines are not effective. This strategy has been particularly relevant during the COVID-19 pandemic, where monoclonal antibodies have been used to treat severe cases and prevent hospitalization [129].

Future research will likely focus on optimizing the integration of therapeutic antibodies with vaccines to address complex diseases. Advances in AI and machine learning are being explored to predict clinical efficacy and optimize antibody design, potentially expanding the range of diseases treatable with therapeutic antibodies [23].

Moreover, the development of novel antibody formats, such as bispecific and trispecific antibodies, holds promise for treating complex diseases by targeting multiple pathways simultaneously. These advancements underscore the potential for therapeutic antibodies to enhance vaccine efficacy and improve patient outcomes in immunotherapies [130].

Despite the successes of therapeutic antibodies in immunotherapies and vaccines, challenges remain. One of the primary concerns is the development of resistance or loss of efficacy over time, which can occur due to the formation of antibodies against the therapeutic antibodies themselves [131]. Additionally, the high cost of these treatments can limit access, particularly in resource-constrained settings.

However, the integration of therapeutic antibodies with vaccines offers opportunities to overcome these challenges. By enhancing immune responses and providing immediate protection, therapeutic antibodies can improve patient outcomes and expand the reach of immunotherapies.

9.5.5 Engineering and Manufacturing Technologies

The development of therapeutic antibodies has been significantly advanced by antibody engineering technologies. These technologies allow for the modification of antibodies to improve their specificity, affinity, and stability. Humanization and chimerization techniques have reduced the immunogenicity of antibodies, making them more suitable for human use [132].

Recombinant manufacturing technologies have enabled the large-scale production of monoclonal antibodies, facilitating their widespread use in medicine. The development of new formats, such as single-domain antibodies and antibody-drug conjugates (ADCs), has expanded the therapeutic options for diseases like cancer. ADCs, for instance, combine the specificity of antibodies with the potency of cytotoxic drugs, allowing for targeted delivery to cancer cells while minimizing damage to healthy tissues [133].

Despite the successes of therapeutic antibodies, challenges remain. One of the primary concerns is the development of resistance or loss of efficacy over time, which can occur due to the formation of antibodies against the therapeutic antibodies themselves. Additionally, the high cost of these treatments can limit access, particularly in resource-constrained settings.

Future research is expected to focus on overcoming these challenges by developing more effective and affordable antibody therapies. Advances in AI and machine learning are being explored to predict clinical efficacy and optimize antibody design. Furthermore, the development of novel antibody formats, such as bispecific and trispecific antibodies, holds promise for treating complex diseases by targeting multiple pathways simultaneously [134].

9.6 INNOVATIVE APPROACHES TO IMMUNIZATION

Immunization is one of the most effective strategies for preventing infectious diseases and has saved countless lives worldwide. However, despite these successes, challenges persist, particularly in achieving equitable vaccine coverage and addressing emerging pathogens. Innovative approaches to immunization are crucial for overcoming these barriers and ensuring that vaccines reach those who need them most. This section will explore some of the key innovative strategies being developed to enhance immunization efforts.

9.6.1 Vaccine-Product Innovations

Vaccine-product innovations are essential for addressing barriers to immunization, particularly in resource-constrained settings. The VIPS has identified several priority innovations, including MAPs, heat-stable and controlled-temperature chain enabled liquid vaccine formulations, and barcodes on primary packaging [135]. These innovations aim to improve vaccine delivery, reduce waste, and enhance equity in immunization coverage.

MAPs offer a novel delivery method that could simplify vaccination processes, especially in areas with limited infrastructure. MAPs consist of thousands of tiny projections that deliver vaccines just below the skin surface, eliminating the need for needles and syringes [136]. This needle-free approach not only reduces the risk of needlestick injuries but also makes vaccination more acceptable to individuals who fear injections. Recent studies have demonstrated the safety and efficacy of MAPs in delivering measles and rubella vaccines, with results showing strong immune responses comparable to traditional injection methods [137].

Heat-stable vaccines are another critical innovation, as they eliminate the need for cold chain storage, making them ideal for regions where maintaining a cold chain is challenging. This feature is particularly beneficial in low-resource settings, where infrastructure limitations often hinder vaccine distribution. Heat-stable vaccines can be stored at room temperature, reducing logistical costs and increasing the reach of immunization programs [56].

Barcodes on primary packaging can enhance vaccine management and tracking, ensuring that vaccines reach those who need them most. This technology helps in monitoring vaccine distribution, reducing waste, and improving supply chain efficiency. By leveraging these innovations, healthcare systems can better manage vaccine stockpiles and ensure timely delivery to remote or underserved areas [138].

The Immunization Agenda 2030 and Gavi 5.0 strategy emphasize the need for such innovations to achieve equitable vaccine coverage [57]. The COVID-19 pandemic has highlighted the urgency of developing vaccine products that can address context-specific barriers to immunization, providing an opportunity to leverage unprecedented investment and momentum in vaccine innovation. The pandemic has shown that substantial investment and collaboration are necessary to overcome logistical hurdles and ensure global access to vaccines [57].

Innovations like MAPs have the potential to significantly increase immunization coverage by overcoming traditional barriers such as cold chain requirements and needle phobia. UNICEF and other global health organizations are exploring the use of MAPs to improve vaccine equity, particularly in low- and middle-income countries [139]. These efforts underscore the importance of continued research and development in vaccine-product innovations to address emerging challenges and achieve global health goals.

9.6.2 Innovative Vaccine Platforms and Technologies

Innovative vaccine platforms and technologies are revolutionizing the field of immunization by offering faster development timelines and improved efficacy compared to traditional vaccine platforms. Advances in mRNA and viral vector technologies have been particularly significant, as demonstrated by their role in the rapid development of COVID-19 vaccines. mRNA vaccines, for instance, have shown remarkable success in eliciting strong immune responses against SARS-CoV-2, with their ability to be rapidly designed and produced making them ideal for responding to emerging pathogens [140].

The use of mRNA and viral vector technologies has not only accelerated vaccine development but also enabled the production of billions of vaccine doses per year, transforming the scale of vaccine manufacturing [141]. These platforms have been supported by advancements in bioprocess technology and analytical tools, which have improved the characterization and quality control of vaccines [141].

Moreover, novel vaccine approaches, such as DNA vaccines and nanoparticle-based vaccines, are being explored for their potential to enhance immune responses and improve vaccine stability. DNA vaccines utilize stable plasmids to deliver genetic material encoding for antigens, which are then expressed by host cells to stimulate an immune response [65]. Nanoparticle-based vaccines, including liposomes and LNPs, serve as antigen delivery systems and adjuvants, enhancing immunogenicity by protecting antigens from degradation and targeting specific immune cells [88].

These innovative platforms can be engineered to target specific immune cells or modulate the immune response, potentially improving vaccine efficacy. For example, nanoparticles can be designed to deliver antigens to dendritic cells, which are crucial for initiating immune responses4. The integration of novel technologies, such as AI, into vaccine development is also being explored to optimize vaccine design and predict immunogenicity, further enhancing the potential of these innovative platforms [142].

Despite these innovations, challenges remain. The development of new vaccine technologies requires significant investment and collaboration among stakeholders. Ensuring equitable access to these innovations, particularly in low- and middle-income countries, is crucial for achieving global health goals. Future research will likely focus on optimizing these innovative approaches to address emerging challenges.

9.7 CONCLUSIONS AND FUTURE DIRECTIONS

Vaccine development stands as a cornerstone of global health, playing a critical role in preventing infectious diseases and reducing morbidity and mortality worldwide. From the eradication of smallpox to the rapid development of mRNA vaccines against SARS-CoV-2, vaccines have consistently demonstrated their transformative potential in safeguarding public health and bolstering economic stability by averting costs associated with illnesses and hospitalizations. However, despite these successes, the field of vaccinology faces a complex array of challenges, ranging from scientific and technological hurdles to societal skepticism and logistical obstacles. Addressing these challenges requires a multifaceted approach that leverages innovative technologies, fosters collaborative partnerships, and prioritizes equitable access to vaccines for all populations.

One of the primary challenges in vaccine development lies in the incomplete understanding of the intricacies of the human immune system and the mechanisms of pathogen variability. The immune system is a complex network of cells and molecular pathways, and a thorough understanding of these processes is essential for designing vaccines that can elicit robust and durable protective immune responses. Moreover, the rapid evolution of pathogens, exemplified by influenza and HIV, necessitates the development of vaccines that can induce broad immunity against a range of viral strains. The emergence of SARS-CoV-2 variants during the COVID-19 pandemic underscored the urgency of developing universal vaccines that can provide protection against multiple variants of a virus. Overcoming these scientific and technological hurdles requires continued investment in basic and translational research to unravel the complexities of immunity and pathogen evolution.

Societal skepticism regarding vaccine safety represents another significant challenge to vaccine development efforts. Misinformation and misconceptions about vaccines have fueled vaccine hesitancy, undermining public health initiatives to control infectious diseases. Addressing these concerns requires effective communication strategies to educate the public about the rigorous safety testing processes that vaccines undergo and the overwhelming evidence supporting their benefits. Building trust in vaccines requires transparent and evidence-based communication from healthcare professionals, public health organizations, and policymakers.

Logistical challenges, such as the high costs of vaccine development and the complexities of maintaining a cold chain for vaccine distribution, also pose significant barriers to vaccine accessibility, particularly in resource-limited settings. The development of a new vaccine can cost hundreds of millions of dollars and take years to complete, limiting the capacity of many organizations to invest in vaccine research and development. Furthermore, the need for cold chain storage presents significant challenges in regions with limited infrastructure, hindering the delivery of vaccines to vulnerable populations. Overcoming these logistical challenges requires increased investment in vaccine research and development, as well as the development of innovative vaccine delivery systems, such as exosomes, MAPs, and heat-stable vaccines that do not require cold chain storage.

Despite these challenges, the field of vaccine development is undergoing a period of rapid innovation, driven by technological advances and collaborative partnerships. Nucleic acid vaccines, such as mRNA and DNA vaccines, have emerged as promising platforms due to their ease of design, production, and potential for rapid response to emerging diseases. The success of mRNA vaccines against COVID-19 demonstrated the transformative potential of these platforms in accelerating vaccine development and improving efficacy. Moreover, collaborative partnerships among companies, academia, governments, and not-for-profit organizations have become increasingly important in vaccine development efforts. These partnerships facilitate the sharing of resources and expertise, streamline regulatory processes, and promote global coordination in addressing emerging health threats.

Looking to the future, the field of vaccine development will likely involve continued innovation in technologies and strategies. Recoding strategies, which involve altering the genetic structure of microorganisms, offer a promising approach for developing safer and more effective vaccines.

Innovative delivery systems and technologies, such as barcodes on primary packaging, can enhance vaccine management and tracking, ensuring that vaccines reach those who need them most. Furthermore, collaborative efforts like the VIPS are working to accelerate the development of priority vaccine innovations, focusing on equitable access and improved effectiveness in low- and middle-income countries.

In conclusion, vaccine development is a dynamic and evolving field that is essential for safeguarding global health security. While the field faces significant challenges, ranging from scientific and technological hurdles to societal skepticism and logistical obstacles, ongoing research and innovative approaches offer promising solutions. By leveraging innovative technologies, fostering collaborative partnerships, and prioritizing equitable access to vaccines, the global community can continue to advance the field of vaccine development and address the complex health challenges of the future. The COVID-19 pandemic has underscored the importance of continued investment in vaccine research and development, as well as global collaboration to address the evolving landscape of infectious diseases.

REFERENCES

1. Orenstein, W.A.; Ahmed, R. Simply put: Vaccination saves lives. *Proc Natl Acad Sci U S A* 2017, *114*, 4031–4033. https://doi.org/10.1073/pnas.1704507114
2. Vaccines and immunization. https://www.who.int/health-topics/vaccines-and-immunization (accessed on 3 March 2025).
3. Gebre, M.S.; Brito, L.A.; Tostanoski, L.H.; Edwards, D.K.; Carfi, A.; Barouch, D.H. Novel approaches for vaccine development. *Cell* 2021, *184*, 1589–1603. https://doi.org/10.1016/j.cell.2021.02.030
4. Shetty, S.; Alvarado, P.C.; Pettie, D.; Collier, J.H. Next-generation vaccine development with nanomaterials: Recent advances, possibilities, and challenges. *Annual Review of Biomedical Engineering* 2024, *26*, 273–306. https://doi.org/10.1146/annurev-bioeng-110122-124359
5. Bansal, S.; Prakash, A.; Medhi, B. New generation vaccine: Novel approaches of vaccine design and delivery and current challenges of vaccine development. *Indian J Pharmacol* 2022, *54*, 241–243. https://doi.org/10.4103/ijp.ijp_555_22
6. Niikura, K.; Matsunaga, T.; Suzuki, T.; Kobayashi, S.; Yamaguchi, H.; Orba, Y.; Kawaguchi, A.; Hasegawa, H.; Kajino, K.; Ninomiya, T. et al. Gold nanoparticles as a vaccine platform: Influence of size and shape on immunological responses in vitro and in vivo. *ACS Nano* 2013, *7*, 3926–3938. https://doi.org/10.1021/nn3057005
7. Mateu Ferrando, R.; Lay, L.; Polito, L. Gold nanoparticle-based platforms for vaccine development. *Drug Discovery Today: Technologies* 2020, *38*, 57–67. https://doi.org/10.1016/j.ddtec.2021.02.001
8. Gorki, V.; Medhi, B. Use of artificial intelligence in vaccine development against pathogens: Challenges and future directions. *Indian J Pharmacol* 2024, *56*, 77–79. https://doi.org/10.4103/ijp.ijp_259_24
9. Alejandra, W.-P.; Miriam Irene, J.-P.; Fabio Antonio, G.-S.; Patricia, R.-G.R.; Elizabeth, T.-A.; Aleman-Aguilar, J.P.; Rebeca, G.-V. Production of monoclonal antibodies for therapeutic purposes: A review. *International Immunopharmacology* 2023, *120*, 110376. https://doi.org/10.1016/j.intimp.2023.110376
10. Bergot, A.-S.; Kassianos, A.; Frazer, I.H.; Mittal, D. New approaches to immunotherapy for HPV associated cancers. *Cancers (Basel)* 2011, *3*, 3461–3495. https://doi.org/10.3390/cancers3033461
11. Ghattas, M.; Dwivedi, G.; Lavertu, M.; Alameh, M.-G. Vaccine technologies and platforms for infectious diseases: Current progress, challenges, and opportunities. *Vaccines (Basel)* 2021, *9*, 1490. https://doi.org/10.3390/vaccines9121490
12. Kennedy, R.B.; Ovsyannikova, I.G.; Palese, P.; Poland, G.A. Current challenges in vaccinology. *Front Immunol* 2020, *11*, 1181. https://doi.org/10.3389/fimmu.2020.01181
13. Brisse, M.; Vrba, S.M.; Kirk, N.; Liang, Y.; Ly, H. Emerging concepts and technologies in vaccine development. *Front Immunol* 2020, *11*, 583077. https://doi.org/10.3389/fimmu.2020.583077
14. Pollard, A.J.; Bijker, E.M. A guide to vaccinology: From basic principles to new developments. *Nat Rev Immunol* 2021, *21*, 83–100. https://doi.org/10.1038/s41577-020-00479-7
15. Ryu, H.-E.; Yoon, J.; Choi, J.-E.; Heo, S.-J.; Hong, K.-W.; Jung, D.-H. The human genetic differences in the outcomes of mRNA vaccination against COVID-19: A prospective cohort study. *Vaccines* 2024, *12*, 626. https://doi.org/10.3390/vaccines12060626

16. Zabidi, N.Z.; Liew, H.L.; Farouk, I.A.; Puniyamurti, A.; Yip, A.J.W.; Wijesinghe, V.N.; Low, Z.Y.; Tang, J.W.; Chow, V.T.K.; Lal, S.K. Evolution of SARS-CoV-2 variants: Implications on immune escape, vaccination, therapeutic and diagnostic strategies. *Viruses* 2023, *15*, 944. https://doi.org/10.3390/v15040944
17. Zhao, F.; Zai, X.; Zhang, Z.; Xu, J.; Chen, W. Challenges and developments in universal vaccine design against SARS-CoV-2 variants. *NPJ Vaccines* 2022, *7*, 167. https://doi.org/10.1038/s41541-022-00597-4
18. Romer, D.; Winneg, K.M.; Jamieson, P.E.; Brensinger, C.; Jamieson, K.H. Misinformation about vaccine safety and uptake of COVID-19 vaccines among adults and 5–11-year-olds in the United States. *Vaccine* 2022, *40*, 6463–6470. https://doi.org/10.1016/j.vaccine.2022.09.046
19. Plotkin, S.; Robinson, J.M.; Cunningham, G.; Iqbal, R.; Larsen, S. The complexity and cost of vaccine manufacturing – An overview. *Vaccine* 2017, *35*, 4064–4071. https://doi.org/10.1016/j.vaccine.2017.06.003
20. Feddema, J.J.; Fernald, K.D.S.; Schikan, H.G.C.P.; van de Burgwal, L.H.M. Upscaling vaccine manufacturing capacity – Key bottlenecks and lessons learned. *Vaccine* 2023, *41*, 4359–4368. https://doi.org/10.1016/j.vaccine.2023.05.027
21. Burgos, R.M.; Badowski, M.E.; Drwiega, E.; Ghassemi, S.; Griffith, N.; Herald, F.; Johnson, M.; Smith, R.O.; Michienzi, S.M. The race to a COVID-19 vaccine: Opportunities and challenges in development and distribution. *Drugs Context* 2021, *10*, 2020. https://doi.org/10.7573/dic.2020-12-2
22. Uddin, M.N.; Roni, M.A. Challenges of storage and stability of mRNA-based COVID-19 vaccines. *Vaccines (Basel)* 2021, *9*, 1033. https://doi.org/10.3390/vaccines9091033
23. Olawade, D.B.; Teke, J.; Fapohunda, O.; Weerasinghe, K.; Usman, S.O.; Ige, A.O.; Clement David-Olawade, A. Leveraging artificial intelligence in vaccine development: A narrative review. *Journal of Microbiological Methods* 2024, *224*, 106998. https://doi.org/10.1016/j.mimet.2024.106998.
24. Excler, J.-L.; Saville, M.; Privor-Dumm, L.; Gilbert, S.; Hotez, P.J.; Thompson, D.; Abdool-Karim, S.; Kim, J.H. Factors, enablers and challenges for COVID-19 vaccine development. *BMJ Glob Health* 2023, *8*. https://doi.org/10.1136/bmjgh-2023-011879
25. Mohammadi, D.; Ghasemi, M.; Manouchehrian, N.; Zafarmand, M.; Akbari, M.; Boroumand, A.B. COVID-19 vaccines: Current and future challenges. *Front. Pharmacol.* 2024, *15*. https://doi.org/10.3389/fphar.2024.1434181
26. Tan, D.; Li, G.; Fu, W.; Lei, C. Exosomes: The next frontier in vaccine development and delivery. *Front Immunol* 2024, *15*, 1435426. https://doi.org/10.3389/fimmu.2024.1435426
27. Huda, M.N.; Nurunnabi, M. Potential application of exosomes in vaccine development and delivery. *Pharm Res* 2022, *39*, 2635–2671. https://doi.org/10.1007/s11095-021-03143-4
28. CEPI. https://cepi.net/CEPI (accessed on 3 March 2025).
29. McDonnell, A.; Toxvaerd, F. (2021, August 20). How important is international cooperation in the vaccine rollout? *Economics Observatory*. Retrieved from https://www.economicsobservatory.com/how-important-is-international-cooperation-in-the-vaccine-rollout
30. Zhao, F.; Zai, X.; Zhang, Z.; Xu, J.; Chen, W. Challenges and developments in universal vaccine design against SARS-CoV-2 variants. *NPJ Vaccines* 2022, *7*, 1–12. https://doi.org/10.1038/s41541-022-00597-4
31. Excler, J.-L.; Saville, M.; Privor-Dumm, L.; Gilbert, S.; Hotez, P.J.; Thompson, D.; Abdool-Karim, S.; Kim, J.H. Factors, enablers and challenges for COVID-19 vaccine development. *BMJ Glob Health* 2023, *8*, e011879. https://doi.org/10.1136/bmjgh-2023-011879
32. Carneiro, D.C.; Sousa, J.D.; Monteiro-Cunha, J.P. The COVID-19 vaccine development: A pandemic paradigm. *Virus Res* 2021, *301*, 198454. https://doi.org/10.1016/j.virusres.2021.198454
33. Prieto, A.; Huang, R.; Eusebi, C.A.; Shostak, M.. A brief overview of emerging vaccine technologies for pandemic preparedness. *Rand Health Q* 2023, *11*, 6.
34. Lim, C.M.L.; Komarasamy, T.V.; Adnan, N.A.A.B.; Radhakrishnan, A.K.; Balasubramaniam, V.R.M.T. Recent advances, approaches and challenges in the development of universal influenza vaccines. *Influenza Other Respir Viruses* 2024, *18*, e13276. https://doi.org/10.1111/irv.13276
35. Goje, O.; Kapoor, A. Meeting the challenge of vaccine hesitancy. *CCJM* 2024, *91*, S50–S56. https://doi.org/10.3949/ccjm.91.s1.08
36. Theivendren, P.; Kunjiappan, S.; Pavadai, P.; Ravi, K.; Murugavel, A.; Dayalan, A.; Santhana Krishna Kumar, A. Revolutionizing cancer immunotherapy: Emerging nanotechnology-driven drug delivery systems for enhanced therapeutic efficacy. *ACS Meas Sci Au* 2024, *5*, 31–55. https://doi.org/10.1021/acsmeasuresciau.4c00062
37. Fan, T.; Zhang, M.; Yang, J.; Zhu, Z.; Cao, W.; Dong, C. Therapeutic cancer vaccines: Advancements, challenges and prospects. *Signal Transduct Target Ther* 2023, *8*, 450. https://doi.org/10.1038/s41392-023-01674-3

38. Pardoll, D.M. The blockade of immune checkpoints in cancer immunotherapy. *Nat Rev Cancer* 2012, *12*, 252–264. https://doi.org/10.1038/nrc3239
39. Wojtukiewicz, M.Z.; Rek, M.M.; Karpowicz, K.; Górska, M.; Polityńska, B.; Wojtukiewicz, A.M.; Moniuszko, M.; Radziwon, P.; Tucker, S.C.; Honn, K.V. Inhibitors of immune checkpoints—PD-1, PD-L1, CTLA-4—New opportunities for cancer patients and a New challenge for internists and general practitioners. *Cancer Metastasis Rev* 2021, *40*, 949–982. https://doi.org/10.1007/s10555-021-09976-0
40. Knight, A.; Karapetyan, L.; Kirkwood, J.M. Immunotherapy in melanoma: Recent advances and future directions. *Cancers (Basel)* 2023, *15*, 1106. https://doi.org/10.3390/cancers15041106
41. Alard, E.; Butnariu, A.-B.; Grillo, M.; Kirkham, C.; Zinovkin, D.A.; Newnham, L.; Macciochi, J.; Pranjol, M.Z.I. Advances in anti-cancer immunotherapy: Car-T cell, checkpoint inhibitors, dendritic cell vaccines, and oncolytic viruses, and emerging cellular and molecular targets. *Cancers (Basel)* 2020, *12*, 1826. https://doi.org/10.3390/cancers12071826
42. Eckert, F.; Gaipl, U.S.; Niedermann, G.; Hettich, M.; Schilbach, K.; Huber, S.M.; Zips, D. Beyond checkpoint inhibition – Immunotherapeutical strategies in combination with radiation. *Clin Transl Radiat Oncol* 2017, *2*, 29–35. https://doi.org/10.1016/j.ctro.2016.12.006
43. Uscanga-Palomeque, A.C.; Chávez-Escamilla, A.K.; Alvizo-Báez, C.A.; Saavedra-Alonso, S.; Terrazas-Armendáriz, L.D.; Tamez-Guerra, R.S.; Rodríguez-Padilla, C.; Alcocer-González, J.M. CAR-T cell therapy: From the shop to cancer therapy. *Int J Mol Sci* 2023, *24*, 15688. https://doi.org/10.3390/ijms242115688
44. Kaczmarek, M.; Poznańska, J.; Fechner, F.; Michalska, N.; Paszkowska, S.; Napierała, A.; Mackiewicz, A. Cancer vaccine therapeutics: Limitations and effectiveness—A literature review. *Cells* 2023, *12*, 2159. https://doi.org/10.3390/cells12172159
45. Seclì, L.; Leoni, G.; Ruzza, V.; Siani, L.; Cotugno, G.; Scarselli, E.; D'Alise, A.M. personalized cancer vaccines go viral: viral vectors in the era of personalized immunotherapy of cancer. *Int J Mol Sci* 2023, *24*, 16591. https://doi.org/10.3390/ijms242316591
46. Buckner, J.H. Antigen-specific immunotherapies for autoimmune disease. *Nat Rev Rheumatol* 2025, *21*, 88–97. https://doi.org/10.1038/s41584-024-01201-w
47. Akiyama, S.; Sakuraba, A. Distinct roles of interleukin-17 and T helper 17 cells among autoimmune diseases. *J Transl Autoimmun* 2021, *4*, 100104. https://doi.org/10.1016/j.jtauto.2021.100104
48. Goswami, T.K.; Singh, M.; Dhawan, M.; Mitra, S.; Emran, T.B.; Rabaan, A.A.; Mutair, A.A.; Alawi, Z.A.; Alhumaid, S.; Dhama, K.. Regulatory T cells (Tregs) and their therapeutic potential against autoimmune disorders – Advances and challenges. *Hum Vaccin Immunother 18*, 2035117, https://doi.org/10.1080/21645515.2022.2035117
49. Chakraborty, C.; Bhattacharya, M.; Dhama, K. SARS-CoV-2 vaccines, vaccine development technologies, and significant efforts in vaccine development during the pandemic: The lessons learned might help to fight against the next pandemic. *Vaccines (Basel)* 2023, *11*, 682. https://doi.org/10.3390/vaccines11030682
50. Druedahl, L.C.; Minssen, T.; Price, W.N. Collaboration in times of crisis: A study on COVID-19 vaccine R&D partnerships. *Vaccine* 2021, *39*, 6291–6295. https://doi.org/10.1016/j.vaccine.2021.08.101
51. Ahmed, S.; Khan, S.; Imran, I.; Al Mughairbi, F.; Sheikh, F.S.; Hussain, J.; Khan, A.; Al-Harrasi, A. Vaccine development against COVID-19: Study from pre-clinical phases to clinical trials and global use. *Vaccines (Basel)* 2021, *9*, 836. https://doi.org/10.3390/vaccines9080836
52. Franco, P.; Jain, R.; Rosenkrands-Lange, E.; Hey, C.; Koban, M.U. Regulatory pathways supporting expedited drug development and approval in ICH member countries. *Ther Innov Regul Sci* 2023, *57*, 484–514. https://doi.org/10.1007/s43441-022-00480-3
53. Maruggi, G.; Zhang, C.; Li, J.; Ulmer, J.B.; Yu, D. mRNA as a transformative technology for vaccine development to control infectious diseases. *Mol Ther* 2019, *27*, 757–772. https://doi.org/10.1016/j.ymthe.2019.01.020
54. Chavda, V.P.; Soni, S.; Vora, L.K.; Soni, S.; Khadela, A.; Ajabiya, J. mRNA-based vaccines and therapeutics for COVID-19 and future pandemics. *Vaccines (Basel)* 2022, *10*, 2150. https://doi.org/10.3390/vaccines10122150
55. Goodson, J.L.; Rota, P.A. Innovations in vaccine delivery: Increasing access, coverage, and equity and lessons learnt from measles and rubella elimination. *Drug Deliv Transl Res* 2022, *12*, 959–967. https://doi.org/10.1007/s13346-022-01130-9
56. Ghaemmaghamian, Z.; Zarghami, R.; Walker, G.; O'Reilly, E.; Ziaee, A. Stabilizing vaccines via drying: Quality by design considerations. *Advanced Drug Delivery Reviews* 2022, *187*, 114313. https://doi.org/10.1016/j.addr.2022.114313

57. Lindstrand, A.; Mast, E.; Churchill, S.; Rahimi, N.; Grevendork, J.; Brooks, A.; Magnus, E.; Nandy, R.; O'Brien, K.L. Implementing the immunization agenda 2030: A framework for action through coordinated planning, monitoring & evaluation, ownership & accountability, and communications & advocacy. *Vaccine* 2024, *42*, S15–S27. https://doi.org/10.1016/j.vaccine.2021.09.045
58. Jarrett, S.; Wilmansyah, T.; Bramanti, Y.; Alitamsar, H.; Alamsyah, D.; Krishnamurthy, K.R.; Yang, L.; Pagliusi, S. The role of manufacturers in the implementation of global traceability standards in the supply chain to combat vaccine counterfeiting and enhance safety monitoring. *Vaccine* 2020, *38*, 8318–8325. https://doi.org/10.1016/j.vaccine.2020.11.011
59. Mvundura, M.; Frivold, C.; Janik Osborne, A.; Soni, P.; Robertson, J.; Kumar, S.; Anena, J.; Gueye, A.; Menozzi-Arnaud, M.; Giersing, B. et al. vaccine innovation prioritisation strategy: Findings from three country-stakeholder consultations on vaccine product innovations. *Vaccine* 2021, *39*, 7195–7207. https://doi.org/10.1016/j.vaccine.2021.08.024
60. Li, S.; Li, W.; Jin, Y.; Wu, B.; Wu, Y.. Advancements in the development of nucleic acid vaccines for syphilis prevention and control. *Hum Vaccin Immunother 19*, 2234790, https://doi.org/10.1080/21645515.2023.2234790
61. Szabó, G.T.; Mahiny, A.J.; Vlatkovic, I. COVID-19 mRNA vaccines: Platforms and current developments. *Mol Ther* 2022, *30*, 1850–1868. https://doi.org/10.1016/j.ymthe.2022.02.016
62. Al Fayez, N.; Nassar, M.S.; Alshehri, A.A.; Alnefaie, M.K.; Almughem, F.A.; Alshehri, B.Y.; Alawad, A.O.; Tawfik, E.A. Recent advancement in mRNA vaccine development and applications. *Pharmaceutics* 2023, *15*, 1972. https://doi.org/10.3390/pharmaceutics15071972
63. Kim, K.Q.; Burgute, B.D.; Tzeng, S.-C.; Jing, C.; Jungers, C.; Zhang, J.; Yan, L.L.; Vierstra, R.D.; Djuranovic, S.; Evans, B.S. et al. N1-methylpseudouridine found within COVID-19 mRNA vaccines produces faithful protein products. *Cell Rep* 2022, *40*, 111300. https://doi.org/10.1016/j.celrep.2022.111300
64. Katoto, P.D.M.C.; Tamuzi, J.L.; Brand, A.S.; Marangu, D.M.; Byamungu, L.N.; Wiysonge, C.S.; Gray, G.. Effectiveness of COVID-19 Pfizer-BioNTech (BNT162b2) mRNA vaccination in adolescents aged 12–17 years: A systematic review and meta-analysis. *Hum Vaccin Immunother 19*, 2214495, https://doi.org/10.1080/21645515.2023.2214495
65. Kozak, M.; Hu, J. DNA vaccines: Their formulations, engineering and delivery. *Vaccines (Basel)* 2024, *12*, 71. https://doi.org/10.3390/vaccines12010071
66. Li, L.; Petrovsky, N. Molecular mechanisms for enhanced DNA vaccine immunogenicity. *Expert Rev Vaccines* 2016, *15*, 313–329. https://doi.org/10.1586/14760584.2016.1124762
67. Xu, J.; Wu, Z.; Wass, L.; Larson, H.J.; Lin, L. Mapping global public perspectives on mRNA vaccines and therapeutics. *NPJ Vaccines* 2024, *9*, 218. https://doi.org/10.1038/s41541-024-01019-3
68. Lu, B.; Lim, J.M.; Yu, B.; Song, S.; Neeli, P.; Sobhani, N.; K, P.; Bonam, S.R.; Kurapati, R.; Zheng, J. et al. The next-generation DNA vaccine platforms and delivery systems: Advances, challenges and prospects. *Front Immunol* 2024, *15*, 1332939. https://doi.org/10.3389/fimmu.2024.1332939
69. Reichmuth, A.M.; Oberli, M.A.; Jaklenec, A.; Langer, R.; Blankschtein, D. mRNA vaccine delivery using lipid nanoparticles. *Ther Deliv* 2016, *7*, 319–334. https://doi.org/10.4155/tde-2016-0006
70. Imani, S.; Li, X.; Chen, K.; Maghsoudloo, M.; Jabbarzadeh Kaboli, P.; Hashemi, M.; Khoushab, S.; Li, X. Computational biology and artificial intelligence in mRNA vaccine design for cancer immunotherapy. *Front Cell Infect Microbiol* 2025, *14*, 1501010. https://doi.org/10.3389/fcimb.2024.1501010
71. Deng, S.; Liang, H.; Chen, P.; Li, Y.; Li, Z.; Fan, S.; Wu, K.; Li, X.; Chen, W.; Qin, Y. et al. Viral vector vaccine development and application during the COVID-19 pandemic. *Microorganisms* 2022, *10*, 1450. https://doi.org/10.3390/microorganisms10071450
72. Chavda, V.P.; Bezbaruah, R.; Valu, D.; Patel, B.; Kumar, A.; Prasad, S.; Kakoti, B.B.; Kaushik, A.; Jesawadawala, M. Adenoviral vector-based vaccine platform for COVID-19: Current status. *Vaccines (Basel)* 2023, *11*, 432. https://doi.org/10.3390/vaccines11020432
73. Elkashif, A.; Alhashimi, M.; Sayedahmed, E.E.; Sambhara, S.; Mittal, S.K. Adenoviral vector-based platforms for developing effective vaccines to combat respiratory viral infections. *Clin Transl Immunology* 2021, *10*, e1345. https://doi.org/10.1002/cti2.1345
74. Fathi, A.; Dahlke, C.; Addo, M.M. Recombinant vesicular stomatitis virus vector vaccines for WHO blueprint priority pathogens. *Hum Vaccin Immunother* 2019, *15*, 2269–2285. https://doi.org/10.1080/21645515.2019.1649532
75. Henao-Restrepo, A.M.; Camacho, A.; Longini, I.M.; Watson, C.H.; Edmunds, W.J.; Egger, M.; Carroll, M.W.; Dean, N.E.; Diatta, I.; Doumbia, M. et al. Efficacy and effectiveness of an rVSV-vectored vaccine in preventing Ebola virus disease: Final results from the Guinea ring vaccination,

open-label, cluster-randomised trial (Ebola Ça Suffit!). *Lancet* 2017, *389*, 505–518, https://doi.org/10.1016/S0140-6736(16)32621-6

76. Hou, Y.; Chen, M.; Bian, Y.; Zheng, X.; Tong, R.; Sun, X. Advanced subunit vaccine delivery technologies: From vaccine Cascade obstacles to design strategies. *Acta Pharm Sin B* 2023, *13*, 3321–3338. https://doi.org/10.1016/j.apsb.2023.01.006.
77. Travieso, T.; Li, J.; Mahesh, S.; Mello, J.D.F.R.E.; Blasi, M. The use of viral vectors in vaccine development. *NPJ Vaccines* 2022, *7*, 75. https://doi.org/10.1038/s41541-022-00503-y
78. Callendret, B.; Vellinga, J.; Wunderlich, K.; Rodriguez, A.; Steigerwald, R.; Dirmeier, U.; Cheminay, C.; Volkmann, A.; Brasel, T.; Carrion, R. et al. A prophylactic multivalent vaccine against different filovirus species is immunogenic and provides protection from lethal infections with Ebolavirus and Marburgvirus species in non-human primates. *PLOS One* 2018, *13*, e0192312. https://doi.org/10.1371/journal.pone.0192312
79. Shoushtari, M.; Salehi-Vaziri, M.; Kadkhodazadeh, M.; Teimoori, A.; Arashkia, A.; Roohvand, F.; Teimoori-Toolabi, L.; Pouriayevali, M.H.; Azadmanesh, K. Heterologous prime-boost immunization with adenoviral vector and recombinant subunit vaccines strategies against dengue virus type 2. *International Immunopharmacology* 2025, *148*, 114032. https://doi.org/10.1016/j.intimp.2025.114032
80. Henríquez, R.; Muñoz-Barroso, I. Viral vector- and virus-like particle-based vaccines against infectious diseases: A minireview. *Heliyon* 2024, *10*, e34927. https://doi.org/10.1016/j.heliyon.2024.e34927
81. Márquez, P.G.; Wolman, F.J.; Glisoni, R.J. Nanotechnology platforms for antigen and immunostimulant delivery in vaccine formulations. *Nano Trends* 2024, *8*, 100058. https://doi.org/10.1016/j.nwnano.2024.100058
82. Parvin, N.; Joo, S.W.; Mandal, T.K. Enhancing vaccine efficacy and stability: A review of the utilization of nanoparticles in mRNA vaccines. *Biomolecules* 2024, *14*, 1036. https://doi.org/10.3390/biom14081036
83. Adugna, T.; Niu, Q.; Guan, G.; Du, J.; Yang, J.; Tian, Z.; Yin, H. Advancements in nanoparticle-based vaccine development against Japanese encephalitis virus: A systematic review. *Front Immunol* 2024, *15*, 1505612. https://doi.org/10.3389/fimmu.2024.1505612
84. Vu, M.N.; Kelly, H.G.; Kent, S.J.; Wheatley, A.K. Current and future nanoparticle vaccines for COVID-19. *EBioMedicine* 2021, *74*, 103699. https://doi.org/10.1016/j.ebiom.2021.103699
85. Tariq, H.; Batool, S.; Asif, S.; Ali, M.; Abbasi, B.H. Virus-like particles: Revolutionary platforms for developing vaccines against emerging infectious diseases. *Front Microbiol* 2022, *12*, 790121. https://doi.org/10.3389/fmicb.2021.790121
86. Wang, R.; Pan, W.; Jin, L.; Huang, W.; Li, Y.; Wu, D.; Gao, C.; Ma, D.; Liao, S. Human papillomavirus vaccine against cervical cancer: Opportunity and challenge. *Cancer Letters* 2020, *471*, 88–102. https://doi.org/10.1016/j.canlet.2019.11.039
87. Prabhakar, P.K.; Khurana, N.; Vyas, M.; Sharma, V.; Batiha, G.E.-S.; Kaur, H.; Singh, J.; Kumar, D.; Sharma, N.; Kaushik, A. et al. Aspects of nanotechnology for COVID-19 vaccine development and its delivery applications. *Pharmaceutics* 2023, *15*, 451. https://doi.org/10.3390/pharmaceutics15020451
88. Bezbaruah, R.; Chavda, V.P.; Nongrang, L.; Alom, S.; Deka, K.; Kalita, T.; Ali, F.; Bhattacharjee, B.; Vora, L. Nanoparticle-based delivery systems for vaccines. *Vaccines (Basel)* 2022, *10*, 1946. https://doi.org/10.3390/vaccines10111946
89. New vaccine-making process could transform pandemic response. CEPI. https://cepi.net//new-vaccine-making-process-could-transform-pandemic-response (accessed on 3 March 2025).
90. Pushing mRNA vaccine development timelines to new speeds. CEPI. https://cepi.net//pushing-mrna-vaccine-development-timelines-new-speeds (accessed on 3 March 2025).
91. Gote, V.; Bolla, P.K.; Kommineni, N.; Butreddy, A.; Nukala, P.K.; Palakurthi, S.S.; Khan, W. A comprehensive review of mRNA vaccines. *Int J Mol Sci* 2023, *24*, 2700. https://doi.org/10.3390/ijms24032700
92. Mendonça, S.A.; Lorincz, R.; Boucher, P.; Curiel, D.T. Adenoviral vector vaccine platforms in the SARS-CoV-2 pandemic. *NPJ Vaccines* 2021, *6*, 97. https://doi.org/10.1038/s41541-021-00356-x
93. Ton, C.; Winters, M.A.; Ducoat, R.; Richardson, D.D.; Fuller, K.; Hughes, M.. Development of recombinant vesicular stomatitis virus vaccine platform for rapid response to Ebola and COVID-19 outbreaks. *Vaccine Insights* 2023, *2*(5), 181–190. https://doi.org/10.18609/vac/2023.029
94. Facciolà, A.; Visalli, G.; Laganà, A.; Di Pietro, A. An overview of vaccine adjuvants: Current evidence and future perspectives. *Vaccines (Basel)* 2022, *10*, 819. https://doi.org/10.3390/vaccines10050819

95. Priyanka; Abusalah, M.A.H.; Chopra, H.; Sharma, A.; Mustafa, S.A.; Choudhary, O.P.; Sharma, M.; Dhawan, M.; Khosla, R.; Loshali, A. et al. Nanovaccines: A game changing approach in the fight against infectious diseases. *Biomedicine & Pharmacotherapy* 2023, *167*, 115597. https://doi.org/10.1016/j.biopha.2023.115597
96. Adugna, T.; Niu, Q.; Guan, G.; Du, J.; Yang, J.; Tian, Z.; Yin, H. Advancements in nanoparticle-based vaccine development against Japanese encephalitis virus: A systematic review. *Front. Immunol.* 2024, *15*. https://doi.org/10.3389/fimmu.2024.1505612
97. Gupta, R.; Arora, K.; Roy, S.S.; Joseph, A.; Rastogi, R.; Arora, N.M.; Kundu, P.K. Platforms, advances, and technical challenges in virus-like particles-based vaccines. *Front Immunol* 2023, *14*, 1123805. https://doi.org/10.3389/fimmu.2023.1123805
98. Kapnick, S.M. The nanoparticle-enabled success of COVID-19 mRNA vaccines and the promise of microneedle platforms for pandemic vaccine response. *DNA Cell Biol* 2022, *41*, 25–29. https://doi.org/10.1089/dna.2021.0538
99. Sharma, P.; Joshi, R.V.; Pritchard, R.; Xu, K.; Eicher, M.A. Therapeutic antibodies in medicine. *Molecules* 2023, *28*, 6438. https://doi.org/10.3390/molecules28186438
100. Singh, R.; Chandley, P.; Rohatgi, S. Recent advances in the development of monoclonal antibodies and next-generation antibodies. *Immunohorizons* 2023, *7*, 886–897. https://doi.org/10.4049/immunohorizons.2300102
101. Suzuki, M.; Kato, C.; Kato, A. Therapeutic antibodies: Their mechanisms of action and the pathological findings they induce in toxicity studies. *J Toxicol Pathol* 2015, *28*, 133–139. https://doi.org/10.1293/tox.2015-0031
102. Albagoush, S.A.; Zubair, M.; Limaiem, F.. Tissue evaluation for HER2 tumor marker. In *StatPearls*; StatPearls Publishing: 2025.
103. Macagno, M.; Bandini, S.; Bolli, E.; Bello, A.; Riccardo, F.; Barutello, G.; Merighi, I.F.; Forni, G.; Lamolinara, A.; Del Pizzo, F. et al. Role of ADCC, CDC, and CDCC in vaccine-mediated protection against Her2 mammary carcinogenesis. *Biomedicines* 2022, *10*, 230. https://doi.org/10.3390/biomedicines10020230
104. Weiner, G.J. Rituximab: Mechanism of action. *Semin Hematol* 2010, *47*, 115–123. https://doi.org/10.1053/j.seminhematol.2010.01.011
105. Herrera, M.; Pretelli, G.; Desai, J.; Garralda, E.; Siu, L.L.; Steiner, T.M.; Au, L. Bispecific antibodies: Advancing precision oncology. *Trends in Cancer* 2024, *10*, 893–919. https://doi.org/10.1016/j.trecan.2024.07.002
106. Burt, R.; Warcel, D.; Fielding, A.K. Blinatumomab, a bispecific B-cell and T-cell engaging antibody, in the treatment of B-cell malignancies. *Hum Vaccin Immunother* 2018, *15*, 594–602. https://doi.org/10.1080/21645515.2018.1540828
107. Kim, H.; Kim, Y.K. Bispecific antibodies and CLEM: An analytical approach to advanced cell imaging for therapeutic strategies. *Applied Microscopy* 2025, *55*, 1. https://doi.org/10.1186/s42649-024-00106-y
108. Manso, T.; Kushwaha, A.; Abdollahi, N.; Duroux, P.; Giudicelli, V.; Kossida, S. Mechanisms of action of monoclonal antibodies in oncology integrated in IMGT/mAb-DB. *Front Immunol* 2023, *14*, 1129323. https://doi.org/10.3389/fimmu.2023.1129323
109. Mukherji, S.K. Bevacizumab (Avastin). *AJNR Am J Neuroradiol* 2010, *31*, 235–236. https://doi.org/10.3174/ajnr.A1987
110. Kirthiga Devi, S.S.; Singh, S.; Joga, R.; Patil, S.Y.; Meghana Devi, V.; Chetan Dushantrao, S.; Dwivedi, F.; Kumar, G.; Kumar Jindal, D.; Singh, C. et al. Enhancing cancer immunotherapy: Exploring strategies to target the PD-1/PD-L1 axis and analyzing the associated patent, regulatory, and clinical trial landscape. *European Journal of Pharmaceutics and Biopharmaceutics* 2024, *200*, 114323. https://doi.org/10.1016/j.ejpb.2024.114323
111. Tsao, L.-C.; Force, J.; Hartman, Z.C. Mechanisms of therapeutic anti-tumor monoclonal antibodies. *Cancer Res* 2021, *81*, 4641–4651. https://doi.org/10.1158/0008-5472.CAN-21-1109
112. Kciuk, M.; Yahya, E.B.; Mohamed Ibrahim Mohamed, M.; Rashid, S.; Iqbal, M.O.; Kontek, R.; Abdulsamad, M.A.; Allaq, A.A. Recent advances in molecular mechanisms of cancer immunotherapy. *Cancers (Basel)* 2023, *15*, 2721. https://doi.org/10.3390/cancers15102721
113. Zahavi, D.; Weiner, L. Monoclonal antibodies in cancer therapy. *Antibodies (Basel)* 2020, *9*, 34. https://doi.org/10.3390/antib9030034
114. Shouse, G. Bispecific antibodies for the treatment of hematologic malignancies: The magic is T-cell redirection. *Blood Reviews* 2025, *69*, 101251. https://doi.org/10.1016/j.blre.2024.101251

115. Pantaleo, G.; Correia, B.; Fenwick, C.; Joo, V.S.; Perez, L. Antibodies to combat viral infections: Development strategies and progress. *Nat Rev Drug Discov* 2022, *21*, 676–696. https://doi.org/10.1038/s41573-022-00495-3
116. Dalgleish, A.G. Vaccines versus immunotherapy: Overview of approaches in deciding between options. *Hum Vaccin Immunother* 2015, *10*, 3369–3374. https://doi.org/10.4161/21645515.2014.980707
117. Fan, T.; Zhang, M.; Yang, J.; Zhu, Z.; Cao, W.; Dong, C. Therapeutic cancer vaccines: Advancements, challenges and prospects. *Sig Transduct Target Ther* 2023, *8*, 1–23. https://doi.org/10.1038/s41392-023-01674-3
118. Ruzzi, F.; Riccardo, F.; Conti, L.; Tarone, L.; Semprini, M.S.; Bolli, E.; Barutello, G.; Quaglino, E.; Lollini, P.-L.; Cavallo, F. Cancer vaccines: Target antigens, vaccine platforms and preclinical models. *Molecular Aspects of Medicine* 2025, *101*, 101324. https://doi.org/10.1016/j.mam.2024.101324
119. Zhou, Y.; Wei, Y.; Tian, X.; Wei, X. Cancer vaccines: Current status and future directions. *Journal of Hematology & Oncology* 2025, *18*, 18. https://doi.org/10.1186/s13045-025-01670-w
120. Li, X.; You, J.; Hong, L.; Liu, W.; Guo, P.; Hao, X. Neoantigen cancer vaccines: A new star on the horizon. *Cancer Biol Med* 2024, *21*, 274–311. https://doi.org/10.20892/j.issn.2095-3941.2023.0395
121. Makker, S.; Galley, C.; Bennett, C.L. Cancer vaccines: From an immunology perspective. *Immunother Adv* 2023, *4*, ltad030. https://doi.org/10.1093/immadv/ltad030
122. Sun, H.; Zhang, Y.; Wang, G.; Yang, W.; Xu, Y. mRNA-based therapeutics in cancer treatment. *Pharmaceutics* 2023, *15*, 622. https://doi.org/10.3390/pharmaceutics15020622
123. Cheever, M.A.; Higano, C.S. PROVENGE (Sipuleucel-T) in prostate cancer: The first FDA-approved therapeutic cancer vaccine. *Clin Cancer Res* 2011, *17*, 3520–3526. https://doi.org/10.1158/1078-0432.CCR-10-3126
124. Adamaki, M.; Zoumpourlis, V. Immunotherapy as a precision medicine tool for the treatment of prostate cancer. *Cancers (Basel)* 2021, *13*, 173. https://doi.org/10.3390/cancers13020173
125. Ricci, J.-E. Tumor-induced metabolic immunosuppression: Mechanisms and therapeutic targets. *Cell Reports* 2025, *44*, 115206. https://doi.org/10.1016/j.celrep.2024.115206
126. Liu, J.; Fu, M.; Wang, M.; Wan, D.; Wei, Y.; Wei, X. Cancer vaccines as promising immuno-therapeutics: Platforms and current progress. *Journal of Hematology & Oncology* 2022, *15*, 28. https://doi.org/10.1186/s13045-022-01247-x
127. Chiu, L.-C.; Lin, S.-M.; Lo, Y.-L.; Kuo, S.C.-H.; Yang, C.-T.; Hsu, P.-C. Immunotherapy and vaccination in surgically resectable non-small cell lung cancer (NSCLC). *Vaccines (Basel)* 2021, *9*, 689. https://doi.org/10.3390/vaccines9070689
128. Sharma, P.; Goswami, S.; Raychaudhuri, D.; Siddiqui, B.A.; Singh, P.; Nagarajan, A.; Liu, J.; Subudhi, S.K.; Poon, C.; Gant, K.L. et al. Immune checkpoint therapy—Current perspectives and future directions. *Cell* 2023, *186*, 1652–1669. https://doi.org/10.1016/j.cell.2023.03.006
129. Cowan, J.; Amson, A.; Christofides, A.; Chagla, Z. Monoclonal antibodies as COVID-19 prophylaxis therapy in immunocompromised patient populations. *International Journal of Infectious Diseases* 2023, *134*, 228–238. https://doi.org/10.1016/j.ijid.2023.06.021
130. Lan, H.-R.; Chen, M.; Yao, S.-Y.; Chen, J.-X.; Jin, K.-T. Bispecific antibodies revolutionizing breast cancer treatment: A comprehensive overview. *Front Immunol* 2023, *14*, 1266450. https://doi.org/10.3389/fimmu.2023.1266450
131. Carter, P.J.; Rajpal, A. Designing antibodies as therapeutics. *Cell* 2022, *185*, 2789–2805. https://doi.org/10.1016/j.cell.2022.05.029
132. Almagro, J.C.; Daniels-Wells, T.R.; Perez-Tapia, S.M.; Penichet, M.L. Progress and challenges in the design and clinical development of antibodies for cancer therapy. *Front. Immunol.* 2018, *8*. https://doi.org/10.3389/fimmu.2017.01751
133. Medina Pérez, V.M.; Baselga, M.; Schuhmacher, A.J. Single-domain antibodies as antibody–drug conjugates: From promise to practice—A systematic review. *Cancers (Basel)* 2024, *16*, 2681. https://doi.org/10.3390/cancers16152681
134. Bauer, J.; Rajagopal, N.; Gupta, P.; Gupta, P.; Nixon, A.E.; Kumar, S. How can we discover developable antibody-based biotherapeutics? *Front Mol Biosci* 2023, *10*, 1221626. https://doi.org/10.3389/fmolb.2023.1221626
135. Giersing, B.; Shah, N.; Kristensen, D.; Amorij, J.-P.; Kahn, A.-L.; Gandrup-Marino, K.; Jarrahian, C.; Zehrung, D.; Menozzi-Arnaud, M. Strategies for vaccine-product innovation: Creating an enabling environment for product development to uptake in low- and middle-income countries. *Vaccine* 2021, *39*, 7208–7219. https://doi.org/10.1016/j.vaccine.2021.07.091

136. Peyraud, N.; Zehrung, D.; Jarrahian, C.; Frivold, C.; Orubu, T.; Giersing, B. Potential use of microarray patches for vaccine delivery in low- and middle- income countries. *Vaccine* 2019, *37*, 4427–4434. https://doi.org/10.1016/j.vaccine.2019.03.035 https://doi.org/
137. Mangla, B.; Javed, S.; Sultan, M.H.; Ahsan, W.; Aggarwal, G.; Kohli, K. Nanocarriers-assisted needle-free vaccine delivery through oral and intranasal transmucosal routes: A novel therapeutic conduit. *Front Pharmacol* 2022, *12*, 757761. https://doi.org/10.3389/fphar.2021.757761
138. Pagliusi, S.; Madrid, Y.; Bramanti, Y.; Wilmansyah, T.; Yu, H.; Acebal, A.; Krishnamurthy, K.R.; Raju Pinnamaraju, V.; Jadhav, P.; Park, R. et al. Vaccine traceability: Key learnings from the supply chain initiative by manufacturers from emerging countries. *Vaccine X* 2023, *15*, 100366. https://doi.org/10.1016/j.jvacx.2023.100366
139. Md Khairi, L.N.H.; Fahrni, M.L.; Lazzarino, A.I. The race for global equitable access to COVID-19 vaccines. *Vaccines (Basel)* 2022, *10*, 1306. https://doi.org/10.3390/vaccines10081306
140. Okuyama, R. mRNA and adenoviral vector vaccine platforms utilized in COVID-19 vaccines: Technologies, ecosystem, and future directions. *Vaccines (Basel)* 2023, *11*, 1737. https://doi.org/10.3390/vaccines11121737
141. Buckland, B.; Sanyal, G.; Ranheim, T.; Pollard, D.; Searles, J.A.; Behrens, S.; Pluschkell, S.; Josefsberg, J.; Roberts, C.J. Vaccine process technology—A decade of progress. *Biotechnol Bioeng* 2024, *121*, 2604–2635. https://doi.org/10.1002/bit.28703
142. Asediya, V.S.; Anjaria, P.A.; Mathakiya, R.A.; Koringa, P.G.; Nayak, J.B.; Bisht, D.; Fulmali, D.; Patel, V.A.; Desai, D.N. Vaccine development using artificial intelligence and machine learning: A review. *International Journal of Biological Macromolecules* 2024, *282*, 136643. https://doi.org/10.1016/j.ijbiomac.2024.136643

10 Global Surveillance and Early Warning Systems for Neglected Viruses

Aman Shukla, Vijay Mishra, Alaa A. A. Aljabali, and Yachana Mishra

10.1 INTRODUCTION

Despite their considerable potential to generate outbreaks and long-term disease loads, a class of viral pathogens known as neglected viruses (NVs) receives little attention from academics, public health authorities, and pharmaceutical makers. These viruses are frequently found in low- and middle-income nations, especially in tropical and subtropical areas with limited access to antiviral treatments, vaccinations, and diagnostic tools [1].

Numerous NVs have zoonotic origins and can spread from animals to people directly or through vectors like ticks and mosquitoes. The Rift Valley fever (RVF) virus, the Nipah virus (NiV), the Zika virus, the chikungunya virus, and the Crimean-Congo haemorrhagic fever (CCHF) virus are a few examples [2, 3]. A lack of consistent research and development is a result of these viruses' low geographic and economic influence on high-income nations, even though some of them can occasionally create widespread epidemics [4]. Ignoring these pathogens poses a serious threat to international health readiness. For example, recent Zika and chikungunya epidemics have exposed surveillance system flaws and the pressing need for antiviral medications and vaccinations [2]. Additionally, these viruses are becoming a greater hazard globally as a result of urbanisation, environmental changes, and international travel [3]. A number of NVs have been designated as priority pathogens under the R&D Blueprint for urgent research efforts by organisations like the World Health Organization (WHO) and the Coalition for Epidemic Preparedness Innovations [5]. To lessen the threat of neglected viral illnesses, a multidisciplinary "One Health" strategy that integrates environmental, animal, and human health is crucial. Different NVs adopt various mode of their transmission (Figure 10.1) [6, 7].

10.1.1 Global Impact and Public Health Concerns of Neglected Viruses

A subset of infectious pathogens known as NVs afflict millions of people worldwide, but they get little money for research, surveillance, diagnostics, and vaccine development. Despite posing serious public health risks, especially in low- and middle-income countries (LMICs), many viruses—such as chikungunya, CCHF, Zika, Nipah, and yellow fever—are not given enough attention in international health priorities.

10.1.1.1 High Burden in Low-Income Regions

People in tropical and subtropical regions, where healthcare facilities are already overworked, are the main targets of NVs. For example, the 2006 outbreak of the chikungunya virus on Réunion Island affected both public health and economic activities, infecting nearly 260,000 people, or one-third of the island's population [8]. Similarly, due to inadequate vector management and low vaccination rates, illnesses like CCHF and Japanese encephalitis continue to be common in South Asia and Africa [9, 10].

DOI: 10.1201/9781003607816-10

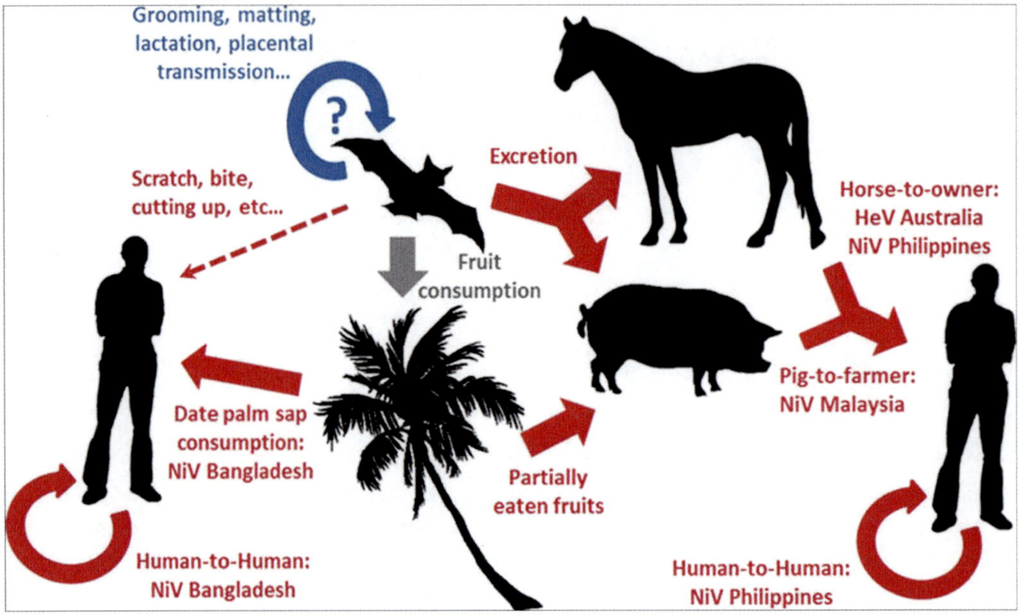

FIGURE 10.1 Schematic representation of the HeV and Nipah virus (NiV) transmission from the natural host, fruit bats, to the susceptible species. Blue arrows represent asymptomatic transmission in the natural reservoir and red arrows indicate the spillover infection causing a disease. The different contributing factors are indicated adjacent to the arrows. (From Ref. [6].)

10.1.1.2 Inadequate Surveillance and Diagnostics

The majority of endemic areas lack the infrastructure necessary for real-time viral tracking, despite the fact that surveillance and early identification are essential for stopping epidemics. NV diseases sometimes resemble other feverish infections such as dengue or malaria, resulting in delayed public health interventions and misdiagnosis [11, 12]. For instance, outbreaks of CCHF are frequently not identified until extensive transmission takes place [13].

10.1.1.3 Epidemic and Pandemic Risk

Pandemic risks come from neglected zoonotic viruses, which are viruses that spread from animals to people. Because of its potential for human-to-human transmission and the lack of an approved treatment or vaccine, the NiV, which has a death rate of up to 75%, is a WHO priority disease [14, 15]. Deforestation, urbanisation, climate change, and human-wildlife interaction all contribute to zoonotic spillovers [16, 17].

10.1.1.4 Economic Impact

NV outbreaks frequently cause significant economic impact. The Zika virus outbreak in 2015–2016 put a strain on public health systems and raised healthcare expenses by causing limitations on foreign travel and a spike in birth abnormalities [18]. Similar to this, the 2017 yellow fever outbreaks in Brazil resulted in widespread fear, a lack of vaccines, and a decline in traveller spending [19].

10.1.1.5 Underfunded Research and Development

Diseases like HIV or TB receive much more funding than NVs. A significant deficit in pipeline development for vaccines, treatments, and diagnostics for neglected viral illnesses was discovered by reviewing R&D portfolios [20, 21]. Thus, readiness is still poor, particularly for viruses that are prone to outbreaks, such as Lassa, RVF, or chikungunya.

10.1.1.6 Challenges in Outbreak Management

Misinformation, logistical obstacles, and a lack of community involvement all hinder effective public health interventions. Moreover, international coordination and data exchange are frequently inadequate during outbreaks [22]. Although it is no longer a NV, the West African Ebola outbreak of 2014 serves as an example of how systematic neglect results in avoidable tragedies [23].

10.2 IMPORTANCE OF SURVEILLANCE AND EARLY WARNING SYSTEMS FOR NEGLECTED VIRUSES

NVs mainly impact people in LMICs, particularly in isolated or underdeveloped regions. Because of their limited geographic dissemination, lack of commercial interest, or irregular epidemic patterns, these viruses are frequently disregarded when discussing global health priorities. Their capacity to significantly increase illness, mortality, and economic disruption, however, emphasises the need for strong surveillance and early warning systems in order to effectively identify and stop their spread. The foundation of disease prevention and control is surveillance. In order to identify and quickly address outbreaks, it entails the methodical gathering, evaluation, interpretation, and distribution of health data. Surveillance is crucial for comprehending the presence, distribution, and emerging trends of NVs, which frequently lack commercial diagnostics or therapies [24]. In many areas, arboviral and haemorrhagic fevers including Kyasanur Forest Disease (KFD), RVF, and CCHF are monitored with the aid of integrated disease surveillance and response (IDSR) systems, such as those advised by the WHO and the Centre for Disease Control (CDC) [25]. Both passive (regular reporting from medical facilities) and active (field-based case detection) monitoring are used in these systems; sentinel surveillance is frequently added, especially for zoonotic viruses that have animal reservoirs. The purpose of early warning systems is to identify anomalous illness trends or environmental factors that might be present before an outbreak occurs. Before there is widespread transmission, these technologies assist health authorities in taking preventive or mitigating measures. When it comes to NVs, early warning systems (EWS) may include syndromic surveillance, particularly in places with limited resources where it is challenging to obtain laboratory confirmation. Rainfall and temperature are examples of ecological and climatic indicators that can be used to predict the occurrence of vector-borne illnesses like Oropouche virus and RVF [26]. Community-based reporting systems, such as local health professionals and mobile health (mHealth) platforms, are crucial in places with inadequate infrastructure [27]. For instance, early warning indicators including a rise in cattle abortions and predictions of heavy rains were present during the 2006–2007 RVF outbreak in East Africa, but action was delayed due to a lack of coordinated surveillance [28]. Numerous NVs are zoonotic, meaning they spread from animals to people either directly or through vectors. For early detection, a One Health strategy that combines environmental, animal, and human health surveillance is essential. Fruit bat populations, animal diseases, and human febrile encephalitis cases are all monitored as part of collaborative surveillance initiatives, as those for the NiV in Bangladesh and India [29]. This integrative method aids in preventing cross-species transmission and expedites risk evaluation.

10.2.1 Role in Disease Control and Prevention for Neglected Viruses

Populations in low-income and resource-constrained areas are disproportionately affected by NVs, a class of pathogens that frequently have limited access to infrastructure for disease management and medical care. NVs are typically endemic or sporadic in distribution and are often linked to environmental and socioeconomic factors, in contrast to internationally prioritised viruses like HIV, hepatitis B, or SARS-CoV-2 [30]. A multimodal strategy including surveillance, community involvement, immunisation, vector management, and intersectoral collaboration is necessary for the effective control and prevention of these diseases.

10.2.1.1 Early Detection and Rapid Response

Since many NV epidemics can spread quickly through vector-borne or zoonotic transmission, early detection is essential. Public health surveillance systems that gather and analyse data in real-time enable the early detection of anomalous trends and enable fast responses [31]. For instance, local authorities have been able to swiftly execute containment measures, such as patient isolation and vector management, thanks to CCHF surveillance in endemic areas [32].

10.2.1.2 Vaccination and Immunisation Campaigns

Some NVs, including Japanese encephalitis and yellow fever, have effective vaccinations, but the majority do not. Routine vaccination campaigns have considerably decreased morbidity and mortality for these in endemic areas [33]. The Eliminate Yellow Fever Epidemics (EYE) strategy's effectiveness in West Africa in containing yellow fever highlights the critical role that ongoing vaccination plays in reducing viral threats [34]. Disease prevention is mostly dependent on other public health interventions because the majority of NVs are only theoretically avoidable by vaccine due to a lack of funding and commercial interest in research and development [35].

10.2.1.3 Vector and Reservoir Control

Dengue, Zika, RVF, and chikungunya are examples of vector-borne viral infections that require strong, multifaceted vector control measures. The extensive use of indoor residual spraying and insecticide-treated nets, which directly protect against adult mosquitoes at rest indoors, is a key component of prevention. Larvicide treatments and source reduction initiatives, such as removing standing water from drains, containers, or waste near dwellings, are complementary to these measures since they interfere with mosquito reproduction cycles. Continuous environmental sanitation practices, which are best carried out as community-wide initiatives, further minimise vector habitats. Examples of these practices include routine gutter cleaning, appropriate trash disposal, and enhanced water storage. Last but not least, public education initiatives are essential for promoting behavioural changes by instructing populations on how to locate and remove breeding grounds and implement personal defence strategies including screened houses, repellents, and protective clothes. Effective management of Aedes-transmitted arboviruses in endemic areas and during outbreaks is supported by these multi-layered treatments [36, 37].

10.2.1.4 Health Education and Community Engagement

Disease prevention relies heavily on behavioural interventions and community awareness. Campaigns for health education can enable vulnerable groups to interact with health systems, adopt protective behaviours, and identify symptoms early. Education on avoiding ticks and using repellents has been crucial in reducing infection rates during outbreaks, such KFD in India [38]. Additionally, community-based interventions improve reporting and adherence to public health policies, particularly in isolated regions with limited official health coverage.

10.2.1.5 Strengthening Health Systems and Infrastructure

Where health systems are weak, NVs flourish. Thus, it is essential to increase capacity for surveillance, diagnostics, epidemic response, and health staff training. Clinical results and epidemic control are enhanced when laboratories are strengthened to support differential diagnosis and point-of-care testing [39]. Additionally, NVs must be given top priority in national health agendas via policy and governance institutions, backed by partnerships and international finance. An increasingly popular sustainable paradigm for managing NVs is the integration of One Health approaches, which integrate the environmental, animal, and human health systems [40].

10.3 KEY COMPONENTS OF SURVEILLANCE SYSTEMS

Public health responses are based on efficient monitoring systems, which allow for the prompt identification, evaluation, and management of infectious disease outbreaks, particularly those brought on by NVs. Because of the intricate ecology and socio-geographic patterns of these viruses, monitoring systems need to be multi-sectoral, thorough, and flexible enough to operate in environments with limited resources.

10.3.1 Case Detection and Reporting

The foundation of any surveillance system is the prompt and precise detection of illness cases. There are two types of case detection: active, which involves proactive outreach to find instances in the community, and passive, which depends on medical facilities to report cases [41]. However, because of their vague symptoms, lack of diagnostic resources, and restricted access to medical treatment in endemic areas, NVs are frequently underdiagnosed [42]. For instance, inadequate community involvement and hospital reporting systems resulted in a significant overestimate of case counts during Lassa fever epidemics in Nigeria [43].

10.3.2 Laboratory Capacity and Diagnostics

Confirming suspected cases, describing infections, and spotting genetic alterations or novel strains all depend on diagnostics. However, advanced molecular testing is frequently restricted to national or reference laboratories, and many NVs lack point-of-care diagnostic techniques [44]. WHO and other international organisations have promoted decentralised laboratory networks, bolstered by mobile testing units and local technician training, to improve detection [45]. For neglected tropical and zoonotic viruses, novel approaches like syndromic panels, CRISPR-based diagnostics, and isothermal amplification are being investigated [46].

10.3.3 Data Collection, Analysis, and Sharing

Data must be precisely and swiftly analysed after collection in order to spot trends, forecast outbreaks, and direct responses. The WHO African Region promotes IDSR frameworks that seek to standardise data gathering for various diseases and exchange knowledge across national boundaries [47]. Although many low-income countries are still under-represented in global data streams, situational awareness has improved thanks to real-time data analytics, geographic information systems (GIS)-based mapping, and cloud-based platforms like HealthMap, ProMED, and GPHIN [48].

10.3.4 Community-Based Surveillance (CBS)

Community health workers and local informants are essential for early diagnosis in disadvantaged or distant areas. As the initial warning for underreported virus outbreaks such as Ebola, Marburg, or Nipah, CBS helps communities recognise anomalous health occurrences [49]. Additionally, community participation promotes culturally sensitive responses, expedites timely care-seeking, and builds trust [50].

10.3.5 Multisectoral and One Health Integration

Effective surveillance necessitates the integration of environmental, animal, and human health data—the foundation of the One Health approach—because many NVs are zoonotic [51]. For viruses such as KFD, CCHF, and RVF, it is particularly important to conduct joint surveillance of human, animal, and wildlife populations [52]. For risk assessment and coordinated action, cross-sectoral cooperation across the environment, wildlife, agricultural, and health ministries is essential.

10.3.6 Risk Communication and Public Engagement

In the absence of prompt risk communication tactics, surveillance systems are insufficient. During epidemics, communities and health professionals are guaranteed to comprehend the hazards and preventive measures through clear and transparent information. The Zika outbreak and COVID-19 pandemic highlighted the importance of squelching false information and interacting with communities via reliable local channels [53].

10.3.7 Surveillance Strategies and Methodologies for Neglected Viruses

The dynamic and frequently evasive characteristics of NVs make it necessary to employ specialised surveillance techniques that take into account certain ecological, epidemiological, and socioeconomic circumstances. For these infections, surveillance systems need to combine old and new methods to guarantee early identification, prompt action, and efficient management.

10.3.7.1 Passive Surveillance

Routinely reporting disease cases to health authorities from healthcare facilities is known as passive surveillance. This approach is popular and economical, particularly in environments with limited resources [54]. It is, however, largely dependent on the standard of clinical diagnosis and the accessibility of diagnostic instruments, both of which are frequently insufficient for viruses that are typically overlooked. Such restrictions have historically resulted in underreporting of diseases such as Lassa fever, chikungunya, and Zika virus [55].

10.3.7.2 Active Surveillance

More proactive, active monitoring entails methodical case detection, such as through field investigations, contact tracking, or recurring community surveys. Although it requires a lot of resources, it is essential for dealing with outbreaks or new dangers. To effectively control epidemics, for instance, Ebola virus disease surveillance in West Africa integrated real-time reporting methods, mobile field laboratories, and active surveillance [56].

10.3.7.3 Syndromic Surveillance

Syndromic surveillance gathers information on clinical symptoms prior to verified diagnoses in order to detect any epidemics early. When laboratory confirmation is unavailable or delayed, this tactic can be helpful. Syndromic surveillance can aid in early warning and alert creation for NVs that manifest as nonspecific febrile diseases, such as CCHF or RVF [57]. Digital tools like as HealthMap and ProMED-mail help with early detection worldwide by keeping an eye on open sources for syndromic reports [58].

10.3.7.4 Event-Based Surveillance

Gathering and analysing unstructured data from a range of sources, including news articles, social media, and community alerts, is known as event-based surveillance. This approach is particularly helpful in areas with no formal health infrastructure and serves as a supplement to traditional surveillance [59]. Even in places lacking organised surveillance systems, event-based monitoring aided in tracking the spread of the Zika virus during the Brazilian outbreak [60].

10.3.7.5 Integrated Disease Surveillance and Response

The goal of IDSR, a WHO-recommended approach in Africa and some regions of Asia, is to combine several disease monitoring systems into a unified framework. It enables a coordinated and effective use of resources and is especially well-suited for monitoring neglected tropical illnesses and viruses [61]. Through cross-border cooperation and real-time reporting, the IDSR system has been crucial in identifying epidemics of Lassa fever, dengue, and yellow fever [62].

10.3.7.6 Genomic and Molecular Surveillance

Viral surveillance has been transformed by cutting-edge molecular methods including metagenomics, next-generation sequencing (NGS), and real-time PCR. Identification and tracking of the Zika virus lineage and emerging NiV strains were made possible in large part by genomic surveillance. Because it makes it possible to identify new viral strains early, monitor mutations and emerging variants, and identify viral evolution and transmission channels, genomic monitoring is essential to contemporary public health. By directly sequencing the genomes of dozens of infected persons and mosquitoes during the Zika virus outbreak, scientists were able to recreate the outbreak's geographic spread and pinpoint its source well before widespread clinical cases appeared. Similarly, whole genome sequencing has been essential to recent studies of the NiV, allowing for the mapping of transmission chains during outbreaks and the real-time monitoring of genetic alterations. These methods help us better understand the behavioural dynamics of viruses and guide prompt outbreak management measures [63, 64].

10.3.7.7 One Health Surveillance

One health-based surveillance is essential because NVs frequently have zoonotic reservoirs. This method combines data on human, veterinary, and environmental health to identify and address epidemics early. One health-based surveillance greatly aids in the surveillance of RVF in Africa or KFD in India [65].

10.3.8 Data Collection and Analysis Techniques

Robust data collecting and sophisticated analytical techniques are necessary for effective surveillance of NVs in order to guarantee prompt discovery, precise outbreak forecasting, and well-informed decision-making. Because of the intricate epidemiology of these viruses, which frequently involves zoonotic cycles, environmental factors, and underreporting, data methods need to be multifaceted and situation-specific.

10.3.8.1 Data Sources

A variety of data sources, such as medical facility records, laboratory diagnostics, community reports, veterinary and animal health records, environmental and entomological data, and digital media monitoring, are used in the surveillance of NVs [66, 67]. To predict epidemics, for example, RVF surveillance combines veterinarian case reports, mosquito density indices, and rainfall data [68].

10.3.8.2 Field-Based Data Collection Techniques

Field monitoring in many endemic areas still relies on semi-digital or manual methods, such as verbal autopsies in cases when death certification is not available, syndromic case definitions for quick field reporting, structured household questionnaires, and mobile platforms like Open Data Kit (ODK) or EpiCollect [69, 70]. Near real-time transmission of outbreak signals from remote areas directly into national databases is now possible due to the increasing usage of GPS-enabled smartphones and tablets that sync with the cloud [71].

10.3.8.3 Laboratory Data and Diagnostic Integration

The accurate identification of viral infections is made possible by molecular diagnostic techniques including genome sequencing, RT-PCR, and ELISA. In order to include collected samples into spatial models, they might be processed and geotagged. Rapid molecular diagnostics, for example, were crucial in linking patient symptoms to viral propagation during the Lassa virus outbreak in Nigeria [72].

10.3.8.4 GIS and Spatial Analysis

GIS-based spatial modelling has been used in Karnataka's Western Ghats to map land cover, tick-pool infectivity, and monkey mortality data—creating hotspot and ecological niche models that reliably predict areas where monkey deaths foreshadow human KFD cases [73].

10.3.8.5 Temporal and Trend Analysis

For identifying patterns and forecasting epidemics, statistical methods including time-series analysis, moving averages, and seasonality decomposition are essential. For viruses such as dengue, which exhibit seasonal patterns associated with temperature and rainfall, these methods are helpful [74].

10.3.8.6 Data Integration and Visualisation Platforms

Modern surveillance relies on integrated software platforms such as DHIS2, HealthMap, ProMED-mail, and ArcGIS Online to bring together diverse data streams—from structured facility and lab data to community reports, media and environmental signals—into dashboards and tools that accelerate outbreak recognition and response. DHIS2 is routinely deployed at national scale (in over 40 countries and used during COVID-19 by 55 ministries) to collect case-based and aggregate data, issue threshold alerts, link to lab systems and support national and sub-national dashboards for rapid action. ProMED-mail and HealthMap scan global informal and official sources (often in multiple languages) to pick up early warning signs and situational awareness for agencies such as WHO, CDC, and ECDC. ArcGIS Online, paired with tools like Survey123, converts GPS-tagged field inputs into real-time maps, graphs and epi-curves—the John Hopkins COVID-19 dashboard during the pandemic being a notable example of how spatial dashboards support global and local public health decision-makers [75].

10.3.8.7 Statistical and Machine Learning Methods

These days, outbreak prediction and case classification are enhanced by advanced analytics, such as regression modelling, Bayesian inference, and machine learning (ML) techniques. To predict danger zones or identify anomalies in incoming data streams, these techniques can be trained on previous epidemic datasets [76, 77].

10.4 TECHNOLOGICAL INNOVATIONS IN SURVEILLANCE

Disease monitoring systems have been transformed by technological advancements, particularly in LMICs where endemic NVs are prevalent. Real-time data sharing, more precise diagnosis, quicker detection, and focused reaction tactics are all made possible by these advancements. In previously under-surveyed areas, the combination of digital health tools, remote sensing, mobile technology, genetic surveillance, and artificial intelligence (AI) has improved public health outcomes and strengthened early warning systems.

10.4.1 mHealth Technologies

Mobile-based illness reporting, contact tracing, and health education have been made easier by the widespread usage of mobile phones, even in distant areas. Health professionals can gather real-time surveillance data in the field using apps like CommCare, mHero, and ODK, which are then promptly sent to national databases [78]. Because they allowed for real-time case tracking and resource allocation, mHealth technologies played a crucial role in managing the 2014 Ebola outbreak in Nigeria [79]. Similarly, dengue and chikungunya outbreaks in India and Southeast Asia have been tracked using mobile-based technologies [80].

10.4.2 Genomic and Molecular Surveillance

Rapid identification and thorough genetic characterisation of neglected viral pathogens are made possible by NGS and metagenomic monitoring. These techniques can be used to rebuild virus emergence and transmission pathways, find co-infections, and identify new strains or mutations [81]. Real-time genome data from field samples was provided by portable platforms such as the Oxford Nanopore MinION, which was employed during the Zika outbreak in Brazil as part of the ZiBRA project. This allowed for the tracking of viral evolution and transmission in real time. These technologies are now essential for real-time molecular epidemiology, particularly in situations with limited resources or remote locations [82].

10.4.3 GIS and Remote Sensing

Viral outbreaks, vector populations, and environmental risk factors can all be spatially mapped thanks to GIS technologies. For arboviruses that are impacted by environmental factors including rainfall, vegetation, and animal movement, such as RVF and KFD, this is particularly helpful [83]. In order to monitor climate-driven vector habitats and provide predictive alarms prior to the occurrence of human infections, satellite data has been integrated with surveillance platforms [84].

10.4.4 Artificial Intelligence and Machine Learning

AI-based models are being utilised more and more to: Identify early outbreak indicators from news and social media, Determine the hotspots for outbreaks, examine big datasets from several sources, including genetic, environmental, and clinical [85]. Compared to traditional reporting methods, projects like BlueDot and HealthMap have proven successful in locating disease clusters and sending out alerts more quickly [86].

10.4.5 Cloud-Based Surveillance Platforms and Dashboards

Data from many sources can be aggregated, stored, and visualised in real time thanks to cloud technologies. Cross-border data exchange and quicker response coordination are made possible by platforms such as DHIS2, SORMAS, and ProMED [87]. SORMAS made it possible for hospitals, labs, and public health units to coordinate easily during Lassa and yellow fever outbreaks in West Africa, which greatly enhanced case management [88].

10.4.6 Internet of Things (IoT) and Smart Sensors

For vector surveillance and early outbreak identification, IoT-based devices including wearable biosensors, environmental monitoring stations, and mosquito traps with real-time sensors are being investigated. Temperature, humidity, and pathogen presence data can be automatically collected and transmitted by these devices [89]. Smart mosquito traps were used in certain Southeast Asian pilot studies to identify the presence of dengue virus in Aedes mosquito populations prior to human outbreaks [90].

10.4.7 Use of AI and ML in Surveillance of Neglected Viruses

10.4.7.1 Early Detection and Outbreak Prediction

Real-time monitoring of data streams from non-traditional sources has been made possible by AI and ML, which have completely changed the early detection of overlooked viral epidemics. AI systems like HealthMap and BlueDot use web scraping, natural language processing, and big data analytics to find outbreak signals across digital platforms like news media, social networks, and health forums, in contrast to traditional surveillance, which frequently has underreporting and time lags. For instance, before official data from health authorities were released, HealthMap was able to identify the Zika virus outbreak in Brazil [91]. By acting more quickly thanks to these technology, public health organisations may be able to stop the spread of illness sooner.

10.4.7.2 Forecasting Disease Spread

Forecasting the temporal and geographic trends of neglected viral infections is a common use of ML algorithms. To forecast future outbreaks and high-risk areas, these models take into account demographic, environmental, and vector-related variables. Algorithms trained on vector density data and climate variables (temperature, rainfall) have shown good prediction accuracy in locating possible hotspots for dengue and chikungunya. Particularly in areas with little health infrastructure, this forecasting capability aids in directing community awareness campaigns, healthcare resource allocation, and vector control actions [92].

10.4.7.3 Genomic Surveillance and Viral Evolution Tracking

AI is also essential for examining virus genetic sequences in order to track variations and track the evolution of the virus. Large amounts of sequencing data may be processed by deep learning algorithms to find new mutations, track down ancestral patterns, and evaluate the possibility of virulence or treatment resistance. This strategy was crucial during the Zika virus outbreak, when real-time genome sequencing assisted in determining the geographic spread and chains of transmission [93]. Our knowledge of the genetic diversity and evolutionary dynamics of viruses such as RVF and CCHF could be greatly improved by applying comparable approaches to these diseases.

10.4.7.4 Decision Support for Public Health Planning

In order to maximise outbreak responses, AI-driven systems are being used more and more as decision support tools. To suggest specific measures, like distributing diagnostic kits or sending out public health alerts, these systems evaluate the results of predictive models and real-time surveillance. These technologies enable the effective use of scarce cash and people, making them especially useful in environments with limited resources. GIS-based monitoring platforms are increasingly being combined with AI-powered decision-support systems, which facilitate the visualisation of outbreak dynamics and the planning of localised responses [94].

10.4.7.5 Challenges and Ethical Considerations

In the context of neglected viral surveillance, AI and ML have a number of drawbacks despite their potential. The infrastructure necessary to facilitate data collecting, storage, and processing is lacking in many low- and middle-income nations, which are the main areas impacted by these infections. The quality of the input data is still a major worry as well. Biassed, delayed, or incomplete data might impair model performance and result in poorly informed decisions. Urgent attention is also needed for ethical issues including algorithmic transparency, data privacy, and fair access to AI tools. Therefore, by combining local epidemiology information and grassroots reporting channels, AI should enhance rather than replace conventional monitoring systems [95].

10.4.8 Integration of Big Data Analytics in the Surveillance of Neglected Viruses

The diagnosis, tracking, and control of NVs have been improved with the use of big data analytics into public health surveillance systems. Big data refers to vast, complex datasets that are generated from a variety of sources, including electronic health records (EHRs), mHealth apps, social media, remote sensing devices, climate data systems, genomic databases, and international disease reporting networks. When analysed with advanced analytics, these data sources offer a comprehensive picture of the dynamics of virus transmission, assisting in overcoming the drawbacks of conventional epidemiological techniques. By making it possible to identify minute trends or irregularities that can point to the start of an outbreak, big data analytics enhances early warning capabilities. For example, vector-borne virus epidemics like RVF and chikungunya in Africa and Asia have been predicted using algorithms that analyse satellite-based environmental data (such as temperature, vegetation indices, and rainfall) [96].

Similar to this, even in environments with limited resources, it is possible to search extensive electronic health data for symptom clusters or anomalous hospital admission rates, offering near real-time alerts on new viral activity [97]. The ability to analyse disease spread and migration patterns in real time is another important benefit of big data analytics. Public health organisations can track population movements and evaluate the risk of virus exportation or introduction into new areas by integrating anonymised data from mobile phones and travel logs. This strategy was successfully used during the Zika virus outbreak, when movement-based modelling guided cross-border public health actions and assisted in tracking the infection's progress throughout the Americas [98]. Additionally, genomic surveillance is improved by big data analytics, especially when it comes to tracking viral evolution, mutations, and the appearance of novel strains. Viral genome alterations over time can be tracked using high-throughput sequencing data and bioinformatics pipelines, enabling risk assessments for virulence, vaccination escape, or the emergence of resistance. Since neglected viral threats like KFD and CCHF may experience quick genetic changes in zoonotic reservoirs, such approaches are becoming more and more important for monitoring these dangers [99]. Additionally, risk categorisation and resource prioritisation are made possible by big data-supported predictive modelling. For instance, dengue and West Nile virus high-transmission zones have been predicted using ML algorithms trained on large datasets, which has helped to strategically allocate vector control measures and diagnostic efforts. In low- and middle-income nations, where public health resources are limited and NVs frequently go unreported or are misdiagnosed, these prognostic abilities are extremely helpful.

The incorporation of big data analytics into neglected viral surveillance presents a number of difficulties, notwithstanding its potential. These include worries regarding data privacy and ethics, lack of standardised reporting formats, data fragmentation, and problems with system interoperability. Moreover, countries where NVs are most prevalent may lack the technical infrastructure, trained personnel, or financial investment required to implement and maintain big data systems effectively. Collaborative international frameworks and public–private partnerships are therefore essential to support capacity building, equitable data access, and sustainable technology transfer. In conclusion, big data analytics represents a transformative tool in global health surveillance, offering unprecedented potential to improve the timely detection, tracking, and response to neglected viral threats. When integrated thoughtfully with existing public health systems, big data can bridge information gaps and empower more proactive and precise health interventions in vulnerable populations [100].

10.5 GLOBAL SURVEILLANCE NETWORKS FOR NEGLECTED VIRUSES

10.5.1 Introduction to Global Surveillance

In order to identify and address neglected viral outbreaks—which frequently go unreported because of their minimal impact in high-income nations—global surveillance systems are crucial. These networks are intended to track the spread of infectious diseases, gather and evaluate information, and send out early warning signals that guide public health measures. Diseases like Lassa fever, chikungunya, and RVF are examples of NVs that frequently strike isolated or underdeveloped areas with inadequate or nonexistent surveillance systems. International surveillance networks are essential in these situations, not only for detecting outbreaks but also for starting cross-border coordinated responses [101].

10.5.2 World Health Organization's Global Outbreak Alert and Response Network (GOARN)

One of the most extensive infectious disease surveillance networks is the GOARN, which was created by the WHO in 2000. More than 250 partners, including national health ministries, labs, universities, and non-governmental organisations, collaborate through GOARN. It is especially made to detect and address illness outbreaks, including those brought on by underappreciated viruses. In impacted areas, GOARN deploys field teams, improves laboratory diagnostic capabilities, and offers technical assistance. Notably, GOARN aided in the response to yellow fever epidemics in the Democratic Republic of the Congo and Angola, two countries with weak internal surveillance systems [102].

10.5.3 The Global Health Security Agenda (GHSA)

In order to strengthen nations' capacities to prevent, identify, and address infectious disease risks, the GHSA was introduced in 2014. GHSA's efforts have greatly improved surveillance infrastructure in LMICs, where neglected viral diseases are common, even though the agency is focused on NVs. GHSA supports the creation of National Action Plans for Health Security and encourages the Joint External Evaluations, which evaluate national health systems. In order to manage infections like the Zika virus and CCHF, these tools assist countries in developing basic capabilities for reporting, laboratory diagnosis, and surveillance [103].

10.5.4 GLEWS—The One Health Model for Zoonotic Surveillance

An innovative programme run in collaboration with the World Organisation for Animal Health, Food and Agriculture Organisation (FAO), and WHO is the Global Early Warning System for Major Animal Diseases, including Zoonoses (GLEWS). GLEWS integrates data on human and animal health to give risk assessments and early warning for zoonotic outbreaks. For neglected zoonotic viruses that frequently start in livestock or animals before spreading to human populations, such as hantaviruses, NiV, and RVF, this method is particularly pertinent. GLEWS's One Health concept guarantees a comprehensive understanding of the disease ecosystem, enabling preventative measures before extensive human transmission takes place [104].

10.5.5 Role of Digital Surveillance Networks

Traditional epidemiological techniques have been enhanced in recent years by computerised surveillance technologies. In order to identify early indicators of infectious disease epidemics, platforms like HealthMap, ProMED-mail, and EpiWatch use crowdsourced reporting, social media monitoring, and internet data mining. These techniques are useful supplements to international surveillance efforts because they frequently detect illness outbreaks before official health organisations issue alerts. HealthMap has been used to follow dengue and chikungunya virus activity in under-reported areas, for instance, and was able to identify early online reports of the 2014 Ebola outbreak [105, 106]. NVs that might not be reported through traditional means can especially benefit from these methods.

10.5.6 Challenges in Global Surveillance for Neglected Viruses

The worldwide surveillance of NVs still faces a number of obstacles in spite of progress. First, political, economic, or social constraints frequently result in underreporting, particularly in areas with poor access to healthcare and laboratory equipment. Second, standardised diagnostic techniques are frequently lacking, which has an impact on cross-border comparability and data quality. Third, prompt response may be hampered by delayed information exchange between nations and agencies. Lastly, diseases that primarily impact the Global South are further marginalised by financial limitations and the unequal allocation of surveillance resources. Improving global health security and real-time illness tracking requires removing these obstacles [107].

10.5.7 WHO and International Collaboration

Global health security is primarily coordinated by the WHO, especially when it comes to surveillance and response to neglected infections. These viruses pose serious hazards to vulnerable people, especially in LMICs, despite frequently lacking the public awareness and funding of more well-known pathogens. The WHO supports a multi-country, multi-sectoral approach to discovering, evaluating, and responding to new and re-emerging viral hazards through its technical know-how, policy leadership, and operational response mechanisms. The International Health Regulations (IHR) are one of the cornerstones of WHO's cooperative efforts. All 196 nations, including WHO Member States, have a legally binding agreement to cooperate for the security of global health under these standards. According to the IHR, nations must develop and preserve their capacity for risk assessment, early identification, notification, and response to public health incidents, particularly those brought on by underutilised viruses like RVF, Lassa fever, and NiV [108]. Information exchange, real-time risk assessments, and coordinated responses to viral threats that could otherwise go undetected because of inadequate surveillance infrastructures are made possible by this worldwide system. The GOARN is a vital operational tool that supports the IHR. More than 250 partner institutions, including academic institutions, public health agencies, laboratories, and non-governmental organisations, are members of this network. By sending interdisciplinary teams to outbreak areas, offering technical assistance, and assisting with laboratory diagnosis, GOARN improves outbreak preparedness worldwide. GOARN has been essential in providing emergency response operations and bolstering local surveillance systems during outbreaks of NVs like chikungunya and Ebola [109].

Crucially, GOARN also provides a platform for surveillance and response capacity building, assisting nations in creating long-lasting mechanisms to control outbreaks in the future. The R&D Blueprint for Action to Prevent Epidemics is another way that WHO supports global health readiness in addition to emergency response. Research and development activities for high-risk pathogens—many of which are NVs—are given priority under this programme. For example, the WHO has listed the NiV and CCHF as priority diseases that need to have their diagnostics, vaccines, and treatments developed more quickly [110]. In order to treat diseases that usually receive little attention or financing, the Blueprint promotes information exchange and resource pooling through worldwide partnership with scientific institutions, regulatory organisations, and corporate stakeholders.

WHO's endorsement of the One Health approach, which acknowledges the interdependence of environmental, animal, and human health, is another example of its collaborative efforts. Numerous NVs, like RVF virus and hantaviruses, are zoonotic, or spread from animal to human. Through the Tripartite Alliance, WHO collaborates closely with the FAO and the World Organisation for Animal Health to promote integrated surveillance systems that monitor hazards of zoonotic diseases across sectors. In areas where human and animal populations coexist closely, this type of intersectoral cooperation is essential for the early detection of outbreaks.

There are still issues in spite of these efforts. Many LMICs lack the infrastructure necessary to fully participate in WHO-led surveillance initiatives or to achieve IHR core competencies. Effective cooperation can also be hampered by political unpredictability, financial limitations, and a lack of technical know-how. To close the gap between global objectives and local execution, it is crucial to maintain investments in health system strengthening, enhance technical assistance from high-income nations, and provide equitable access to data and technologies [111].

In summary, the world's ability to monitor, identify, and combat NVs has greatly increased as a result of WHO's leadership in international cooperation. The organisation has established a strong worldwide surveillance infrastructure through operational networks like GOARN, legally enforceable agreements like the IHR, and intersectoral programmes like One Health. To ensure that even the most neglected infections are not allowed to grow unchecked, however, maintaining these accomplishments calls for political will, sufficient financing, and fair collaborations.

10.5.8 Regional and National Surveillance Initiatives

In order to effectively manage NVs, national and regional monitoring systems are essential parts of the global health architecture. These methods guarantee prompt identification, notification, and action in the event of infectious disease epidemics that might not get enough international attention. Particularly in high-risk and low-resource environments, regional and national operations give the operational foundation required to carry out surveillance, in contrast to global programmes that offer coordination and standardisation.

10.5.8.1 African Surveillance Networks

Since the Africa Centres for Disease Control and Prevention (Africa CDC) were established, the continent has achieved significant strides in disease surveillance. In order to enable member states to identify and address neglected tropical illnesses and viral haemorrhagic fevers like Lassa fever and RVF, Africa CDC was established in 2017 and operates within the IDSR framework [112]. The IDSR approach improves the promptness and efficacy of responses at the local and national levels by combining disease reporting from several sectors.

10.5.8.2 South Asian National Programmes

South Asian nations, such as Bangladesh and India, have created sophisticated national surveillance infrastructures. The Nigeria Centre for Disease Control (NCDC) inaugurated India's Integrated Health Information Platform, which allows primary health centres to report notifiable infections in real time to central authorities [113]. When it comes to viral risks like KFD and Japanese encephalitis, it has proven especially helpful. In addition to spearheading the country's NiV monitoring, Bangladesh's Institute of Epidemiology, Disease Control and Research facilitates intersectoral cooperation and field-level epidemic investigations [114].

10.5.8.3 Latin American and Caribbean Cooperation

The Pan American Health Organisation (PAHO) is crucial in bolstering state surveillance networks throughout Latin America and the Caribbean. PAHO assists member nations with the monitoring of arboviral infections, such as dengue, chikungunya, and the less well-known Mayaro virus, through programmes like the Regional Network for Emerging Infectious infections [115]. This endeavour focuses on training programmes, laboratory standardisation, and cross-border collaboration to fill regional gaps in risk assessment and early warning.

10.5.8.4 Middle Eastern and EMRO Region Efforts

The Eastern Mediterranean Region has established surveillance systems to combat underappreciated viral risks like MERS-CoV and CCHF, particularly in nations like Saudi Arabia, Iran, and Egypt. These nations work together as part of the WHO Eastern Mediterranean Regional Office (EMRO) strategy to improve laboratory and epidemic detection capabilities. Despite regional obstacles such as infrastructural deficiencies and violence, targeted investments have been made in digital illness reporting systems and virology labs [116].

10.6 CHALLENGES AND LIMITATIONS OF GLOBAL SURVEILLANCE AND EARLY WARNING SYSTEMS FOR NEGLECTED VIRUSES

In LMICs, where surveillance systems are underfunded and health infrastructure is unreliable, NVs frequently disproportionately impact populations. The efficient surveillance and early detection of these viral dangers continue to present major obstacles, notwithstanding the many worldwide initiatives to improve global health security. The main obstacles to the effectiveness of worldwide surveillance and early warning systems for NVs are described in the section that follows. These include data problems, technology inequalities, political limitations, and infrastructure deficiencies.

10.6.1 Weak Health Infrastructure in Endemic Regions

The lack of proper health infrastructure in areas where NVs are common is one of the main barriers to efficient surveillance of these viruses. Basic diagnostic facilities, qualified staff, and trustworthy reporting systems are lacking in many sub-Saharan African, South Asian, and Latin American nations. For example, due to inadequate laboratory capacity, diagnostic confirmation is frequently delayed during Lassa fever epidemics in West Africa [117]. Similarly, there are very few high-containment labs for NiV surveillance in Bangladesh [118]. Timely case detection and response are severely hampered by these infrastructure deficiencies.

10.6.2 Inconsistent and Fragmented Reporting Systems

Local, national, and international databases may contain disparate surveillance data for NVs with minimal standardisation or interoperability. Data transmission to central health authorities is delayed in many countries due to the use of semi-digital or paper-based reporting systems [119]. This fragmentation makes it more difficult to perform reliable epidemiological analyses and prevents epidemics from being detected early. Furthermore, the absence of prompt or precise information from impacted areas may cause global surveillance platforms like ProMED-mail or WHO's Disease Outbreak News to underreport outbreaks of NVs [120].

10.6.3 Political and Socioeconomic Constraints

Armed conflict, political unrest, and poor governance can seriously impair response and monitoring activities. Maintaining fundamental health monitoring tasks is a constant issue for nations going through humanitarian crises, such South Sudan, Yemen, or portions of the Democratic Republic of Congo [121]. Furthermore, a lack of political will or a failure to prioritise neglected diseases in national agendas frequently leads to a lack of policy and funding support. In global health investment policies, diseases that do not immediately threaten high-income nations are often ignored.

10.6.4 Limited Laboratory and Genomic Surveillance Capacity

The genomic sequencing and molecular diagnostics resources required to describe newly emerging viral infections are lacking in many LMICs. Surveillance systems cannot track viral changes or variants without such capabilities, which are crucial for predicting and controlling outbreaks. Cross-reactivity with other flaviviruses, such as dengue and yellow fever, made it difficult for countries in the Americas to confirm cases during the 2015–2016 Zika virus outbreak [122]. Underreported instances of Ebola, CCHF, and RVF present similar difficulties in African locations.

10.6.5 Data Sharing and Transparency Issues

Countries frequently hesitate to rapidly share outbreak information because of worries about trade, tourism, or their national image, even though international accords encourage open access to health data. This hesitancy hinders coordinated responses and undermines global readiness [123]. Global efforts to contain the infections were hampered in the early phases of some outbreaks, including SARS-CoV-1 and MERS-CoV, by delays in the sharing of information. For NVs, which are subject to less worldwide monitoring and pressure for transparency, these worries are even more acute.

10.6.6 Underutilisation of Technological Tools

Although advancements in digital epidemiology, ML, and AI have revolutionised global disease monitoring, their usage in surveillance systems for NVs is still quite small. Many impacted nations lack the infrastructure and skills needed for high-tech tools like satellite surveillance, electronic medical records, and mHealth [124]. Furthermore, integrating cutting-edge tools with current workflows and data streams might be challenging for public health organisations operating in low-resource environments.

10.6.7 Community Engagement and Behavioural Barriers

Technology is simply one aspect of effective surveillance; community involvement is another. In many contexts, people are deterred from reporting symptoms or seeking medical attention by cultural beliefs, stigma, and mistrust of authorities. For NVs like Zika, dengue, or chikungunya, this restricts the precision and comprehensiveness of monitoring data [125]. Despite its promise for early outbreak detection, CBS is still underfunded and underdeveloped.

10.6.8 Funding and Resource Constraints

The persistent underfunding of health services and research projects aimed at these illnesses is a significant barrier to the worldwide surveillance of NVs. Because NVs primarily affect poor, rural, or marginalised populations with little political clout and little financial clout, they frequently receive less financial attention from both international donors and national governments than high-profile pathogens like HIV, influenza, or COVID-19 [126, 127]. Despite limited investments in surveillance, laboratory capacity, and vector control programmes, diseases such as RVF, chikungunya, and CCHF persist in endemic areas [128]. The financing gap is made worse by donor-driven priorities, which frequently change quickly in reaction to new concerns around the world. The response capability for additional viral threats was weakened, for example, when resources from routine health services and existing surveillance systems were transferred during the West African Ebola outbreak in 2014–2016 [129]. Similar to this, short funding cycles, unpredictable budgets, and a lack of local ownership of donor programmes can hinder long-term commitments to constructing surveillance infrastructure [130].

Moreover, without outside support, endemic nations usually lack the financial resources to maintain personnel training, laboratory tests, and monitoring operations. Instead of consistent, systematic surveillance, this results in a dependence on sporadic, outside-funded research or outbreak responses. Only 43% of low- and lower-middle-income nations examined by the WHO and its partners have budgeted funding expressly for zoonotic disease surveillance, many of which include NVs [131]. Last but not least, the high expense of implementing cutting-edge technologies—like real-time data platforms, AI-based surveillance tools, and genetic sequencing—further restricts adoption in environments with limited resources. Global health systems run the risk of repeatedly failing to identify and react quickly to neglected viral outbreaks in the absence of consistent funding mechanisms, which could result in needless morbidity, mortality, and financial losses.

10.6.9 Ethical and Legal Considerations

Global surveillance systems must take ethical and legal considerations into account, especially when it comes to neglected diseases, which frequently disproportionately afflict underprivileged communities. The gathering, use, and dissemination of biological and personal data without sufficient informed consent is a common ethical concern, particularly in areas with limited resources and low health literacy. The acquisition of sensitive data, such as genetic or health information, may be necessary for surveillance programmes, which raises concerns of autonomy, privacy, and secrecy. These issues are exacerbated in emergency scenarios when quick data collection might take place without strict ethical supervision [132]. The legal frameworks that regulate surveillance activities differ greatly between nations and are frequently either antiquated or inadequate. Confusion and legal problems around data ownership, access rights, and reporting obligations for outbreaks might result from the absence of standardised international standards. The conflict between national sovereignty and international health security was exemplified by Indonesia's refusal to provide the WHO with samples of the H5N1 virus during the 2006–2007 avian influenza epidemic, claiming issues with intellectual property and fair access to vaccinations [133]. Discrimination and stigmatisation present another ethical dilemma. Affected people and communities may experience social exclusion or economic isolation during outbreaks of neglected diseases like Lassa fever or Ebola, especially if surveillance data are released to the public without proper context or protections. The efficacy of surveillance can be undermined by the misuse of such data, which can worsen already-existing disparities and discourage community involvement with health authorities [134, 135].

Furthermore, there are new ethical issues when using digital tools like AI, geolocation tracking, and big data analytics for disease surveillance. Despite their strength, these instruments have the potential to increase surveillance beyond what is necessary for public health, violating civil liberties and raising the possibility of abuse or overreach by the state. Therefore, in order to strike a balance between the protection of individual rights and the advantages to public health, transparent governance, ethical review boards, and respect to international human rights norms are crucial [136].

WHO and other international health organisations support ethical frameworks that prioritise openness, community involvement, informed consent, and fair benefit sharing in order to allay these worries. According to the WHO's 2017 guidance on ethical considerations in public health surveillance, all surveillance programmes must be planned and carried out in a way that upholds human dignity, guarantees equity, and reduces harm [137]. Without these moral and legal protections, international initiatives to combat neglected diseases risk losing credibility and encountering public opposition, which would lessen their efficacy.

10.7 CASE STUDIES AND SUCCESS STORIES

Numerous nations and organisations have effectively adopted monitoring techniques that have resulted in the efficient detection, management, and prevention of neglected viral infections in spite of ongoing obstacles. These case studies provide insightful lessons and emphasise the significance of early warning systems, international cooperation, and coordinated public health interventions.

10.7.1 Lassa Fever Surveillance in Nigeria

A viral haemorrhagic disease that is unique to West Africa, Lassa fever, has frequently erupted in Nigeria. The SORMAS a real-time digital platform that tracks Lassa fever cases and makes contact tracing and data reporting easier, was created by the Nigeria Centre for Disease Control (NCDC) in response. During the 2018 and 2019 outbreaks, this innovation resulted in better epidemiological data and faster outbreak response times [138]. Decentralisation to regional labs and integration with mobile technology improved early diagnosis and decreased case fatality rates in endemic areas [139].

10.7.2 Yellow Fever Vaccination and Surveillance in Angola and the Democratic Republic of Congo

Due to air travel and a shortage of vaccines, a significant yellow fever outbreak in 2016 that affected Angola, and the Democratic Republic of the Congo constituted a worldwide threat. Targeted vaccination programmes, enhanced vector surveillance, and cross-border cooperation were all part of the quick international response, which was spearheaded by WHO's EYE plan. Within a few months, the outbreak was successfully contained and almost 28 million people received vaccinations [140]. This achievement demonstrated how strong vaccine surveillance data, in conjunction with regional and international collaboration, may stop the transmission of viruses across national boundaries.

10.7.3 Nipah Virus Surveillance in Kerala, India

In 2018 and 2019, the state of Kerala in India provided a compelling illustration of early warning and containment of NiV outbreaks. Rapid diagnostic tests verified the virus when patients' unique neurological symptoms were identified. To stop the spread, public health professionals used inter-departmental collaboration, isolation procedures, and active contact tracking. The outbreak was contained to fewer than 20 cases thanks to community awareness initiatives and real-time updates provided by digital dashboards [141]. This case emphasises how crucial biosurveillance, quick reaction capabilities, and health system readiness are to zoonotic virus control (Figure 10.2).

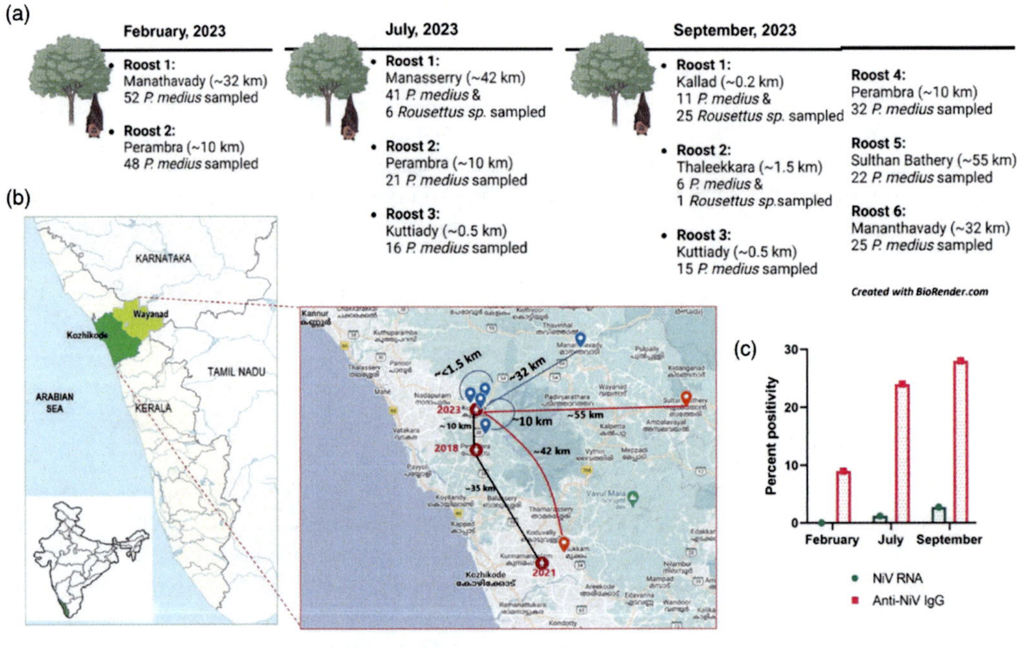

FIGURE 10.2 (a) The bat sampling sites in Kerala with details of the location of the roost (aerial distance from the NiV index case house of 2023) and the number of bats sampled in February, July, and September of 2023. (b) the geographical map of Kerala with the Kozhikode (green) and Wayanad (light green) districts from which sample collection was performed is highlighted, and in the inset, the locations of NiV outbreak reported areas (of the years 2018, 2019, and 2021 marked in red connected with black line showing the distance among each other) and the bat sampling roost locations (blue icon and red icon) with the aerial distance plotted from the 2023 NiV outbreak location as blue lines (to roosts which were NiV negative) and red lines (to roosts which showed Nipah viral RNA positivity). (c) the graphical representation of NiV RNA and antibody per cent positivity estimated during the months of February, July, and September from the outbreak region in 2023. (From Ref. [142].)

10.7.4 Zika Virus Monitoring in Brazil

Brazil's surveillance system was tested during the Zika virus outbreak in 2015–2016. The Brazilian Ministry of Health increased syndromic surveillance and expanded laboratory testing in several states in response to the onset of congenital Zika syndrome. Real-time spatial mapping of cases and genome sequencing were made possible by cooperation with international organisations like the WHO and CDC [143]. A reproducible model for future arboviral monitoring in Latin America was provided by the crisis, which resulted in the creation of birth defect registries and long-term improvements to vector control regulations.

10.7.5 Global Polio Surveillance Programme

The global surveillance network established by the Global Polio Eradication Initiative is a fundamental model for NV control, despite the fact that polio is not typically categorised as a NV. Since 1988, the initiative has decreased polio cases by more than 99% because to integrated laboratory networks and active surveillance in more than 70 countries. Transferable methodologies for neglected viral disease surveillance are provided by its focus on CBS, stool sample tracking, and early detection of circulating vaccine-derived polioviruses [144].

10.7.6 Effective Surveillance Programmes

10.7.6.1 Integrated Disease Surveillance and Response—Africa

One of the most extensive disease monitoring systems for Africa is the IDSR system, which was introduced by WHO-AFRO in 1998. It was created to combine disparate surveillance methods for a number of illnesses into a single, cohesive strategy, including neglected viral infections like Lassa fever, RVF, and yellow fever. This initiative encourages the prompt gathering, examination, and sharing of epidemiological data. Through the use of IDSR, nations such as Tanzania and Uganda have shown improved laboratory capacity, better outbreak response, and more effective case detection [145, 146].

10.7.6.2 Field Epidemiology Training Programmes (FETPs)

The development of national field epidemiological expertise has been greatly aided by FETPs. Local health workers are trained in public health reporting, case monitoring, and outbreak investigation through these programmes. During Lassa fever outbreaks in Nigeria, FETP-trained staff played a critical role in facilitating quicker case identification and control measure execution. FETPs provide quick, on-the-ground reactions by decentralising monitoring capabilities, particularly in underserved and challenging-to-reach areas [147].

10.7.6.3 Surveillance Outbreak Response Management and Analysis System

SORMAS is an open-source digital platform that was created in the wake of the West African Ebola pandemic and is currently utilised in Ghana, Nigeria, and other nations. It makes it possible to enter data in real time and visualise diseases that are prone to epidemics, such as NVs. The effectiveness and timeliness of surveillance have been greatly increased by features including automatic alarm creation, mobile case reporting, and coordination tools for healthcare professionals [148].

10.7.6.4 PAHO Arbovirus Surveillance—Latin America

In Latin America, the PAHO has created frameworks for surveillance of arboviral illnesses that are distinctive to each region. The Zika virus outbreak in 2015, for instance, enhanced Brazil's national response infrastructure. These systems, which can be expanded to include other NVs like Mayaro and Oropouche, featured vector monitoring, lab testing, and syndromic surveillance. These initiatives prioritise community involvement and intersectoral cooperation [149].

10.7.6.5 CBS Initiatives

CBS has been used in a number of nations to improve the identification of underreported virus epidemics. This method improves detection in isolated and under-resourced places by enabling local populations to report anomalous health incidents. During the post-Ebola era, CBS was crucial for tracking viral haemorrhagic fevers in Sierra Leone and Liberia. These approaches increase sensitivity to outbreaks by bridging the gap between rural populations and formal healthcare services [150].

10.7.7 Impact on Public Health Outcomes

10.7.7.1 Early Detection and Rapid Response

The early detection of outbreaks has been greatly enhanced by surveillance systems, especially when it comes to neglected viral illnesses like Lassa fever, dengue, and yellow fever. By promptly implementing control measures including isolation, vaccination campaigns, and vector control, early alarms help health authorities' lower transmission and morbidity. For example, during the 2017 monkeypox outbreak, Nigeria's early alarm via its IDSR system prompted a prompt containment effort that reduced the spread to less than 90 confirmed cases [151]. In neglected settings with inadequate access to healthcare, this early intervention is extremely important.

10.7.7.2 Reduced Mortality and Morbidity

Reducing mortality and morbidity is directly impacted by timely surveillance data availability. Hospitals and clinics can prepare resources such as antiviral medications, personal protective equipment, and community outreach initiatives when health systems are informed in advance. Areas with active surveillance exhibited up to 30% lower fatality rates than those without structured disease monitoring during the Lassa fever outbreaks in West Africa in 2018–2020 [152]. This illustrates how surveillance techniques have the ability to save lives.

10.7.7.3 Improved Vaccination Strategies and Preventive Measures

Evidence-based public health planning is supported by surveillance data, especially when it comes to focussing on immunisation and prevention initiatives. For instance, during the past ten years, outbreaks have decreased by over 40% thanks to the global yellow fever initiative, which is backed by data from WHO's Yellow Fever Surveillance and has made targeted vaccination possible in high-risk African nations [153]. This focused strategy guarantees effective resource utilisation in the most critical sectors.

10.7.7.4 Strengthened Health Systems

The infrastructure of public health is strengthened more broadly as a result of neglected viral surveillance. In addition to disease monitoring, initiatives like FETPs and IDSR enhance laboratory networks, educate medical staff, and expedite reporting procedures. Beyond only the viruses under observation, the health system as a whole gain from these long-lasting advancements. Following the Ebola outbreak, Sierra Leone was able to better handle the Lassa and Marburg virus outbreaks thanks to improvements made in surveillance capabilities [154].

10.7.7.5 Increased Community Trust and Engagement

Effective and transparent surveillance methods increase public confidence in public health officials. A feedback loop that is facilitated by open case reporting and prompt action motivates communities to take part in vaccination campaigns, prevent outbreaks, and heed public health advisories. Participatory surveillance for the NiV in Bangladesh has demonstrated that community involvement greatly enhances the calibre and velocity of data gathering during outbreaks [155].

10.8 FUTURE DIRECTIONS AND INNOVATIONS

10.8.1 Strengthening Integration of Surveillance Systems

Harmonising monitoring systems at the local, national, and international levels is a crucial future trend. Real-time response attempts are currently hampered by fragmented data silos, especially in underserved tropical regions. The One Health method, which combines environmental, animal, and human health surveillance, can improve the detection of zoonotic viruses such as NiV and RVF. In order to facilitate early warning across disciplines, WHO's Joint External Evaluation tool encourages nations to improve multisectoral collaboration [156].

10.8.2 Investment in Local Surveillance Infrastructure

It is imperative to move towards locally driven surveillance. Many nations continue to rely on unsustainable donor-driven vertical initiatives. A long-term basis for identifying new NVs can be established by developing in-country laboratory capacity, digital reporting systems, and skilled epidemiologists. In this sense, initiatives like Africa CDC's SURGE programme and the FETP are trailblazing models [157].

10.8.3 Genomic Surveillance and Pathogen Discovery

The identification and tracking of new infections is being revolutionised by genomic technologies like NGS. Real-time monitoring of viral evolution, antibiotic resistance, and changes that could compromise the effectiveness of vaccines is made possible by these technologies. Lassa, Zika, and CCHF viruses may now be sequenced at the regional level thanks to programmes like GISAID and the Africa Pathogen Genomics Initiative [158, 159]. Future priorities include expanding the use of genomics in low- and middle-income nations.

10.8.4 AI and Predictive Modelling

Predictive analytics, deep learning, and AI are being utilised more and more to spot outbreak trends before human specialists can. For example, days before the WHO formally confirmed the COVID-19 outbreak, the AI system BlueDot detected it. Outbreak preparedness could be revolutionised by using comparable methods for NVs [160]. Future work should concentrate on integrating syndromic surveillance, community reporting, and mHealth data to modify these systems for areas with little data.

10.8.5 Mobile Technologies and Citizen Participation

Rapid reporting of symptoms, suspected cases, and health behaviours is made possible by mobile phone-based data gathering and participatory surveillance platforms like Outbreaks Near Me and SORMAS. Future systems should use offline data access, local languages, and trusted community channels to empower residents in endemic areas [161]. In West Africa, this participatory method has already proven effective in enhancing response to outbreaks of yellow fever and Ebola.

10.8.6 Climate and Environmental Data Integration

Vector-borne NVs are significantly influenced by climate variables as temperature, precipitation, and patterns of deforestation. Future developments will probably include climate-informed risk forecasting, which predicts the likelihood of outbreaks by combining ecological factors, satellite data, and weather models. Climate-informed health decisions are already being promoted under the WHO- World Meteorological Organization (WMO) Global Framework for Climate Services [162].

10.8.7 Advancements in Predictive Modelling

10.8.7.1 Introduction to Predictive Modelling in Public Health

In public health, predictive modelling is the process of predicting the onset, progression, and possible consequences of diseases using statistical algorithms, computer simulations, and ML approaches. These models are essential for directing vaccination campaigns, resource allocation, and public health responses, especially for NVs for which there may be a lack of historical data. Predictive modelling allows authorities to shift from reactive to proactive public health measures as surveillance grows globally [163].

10.8.7.2 Application in Early Detection of Neglected Viruses

NVs frequently appear in underdeveloped areas with inadequate surveillance systems. By forecasting epidemics using available environmental, demographic, and movement data, predictive models assist in bridging this gap. Pre-emptive vaccination and vector control, for instance, are made possible by the modelling of RVF outbreaks based on rainfall patterns and livestock movement [164]. These models lessen reaction time, which is essential for illnesses that spread quickly.

10.8.7.3 Integration with Environmental and Climate Data

The combination of biological and climatic data has been one of the most significant developments in predictive modelling. Forecasts of vector population dynamics, which are frequently early markers of arbovirus epidemics, are made using variables like temperature, humidity, vegetation, and rainfall. In order to anticipate the possibility of outbreaks in Africa and Latin America, tools such as the epidemic prediction model for dengue and RVF use weather forecasts and satellite data [165].

10.8.7.4 Machine Learning and AI-Driven Forecasting

Neural networks and ML algorithms are used in recent advancements to increase prediction accuracy. AI-powered tools such as HealthMap and BlueDot forecast disease hotspots by combining standard surveillance data with data from unofficial sources (news articles, social media, and climatic data). After being used to detect Zika epidemics in Brazil, these methods are currently being explored for predicting outbreaks of lesser-known viruses such as Oropouche and Mayaro [166, 167].

10.8.7.5 Spatial and Spatio-Temporal Modelling Approaches

Researchers can now determine not just the potential locations of epidemics but also their timing and rate of spread thanks to developments in GIS and spatiotemporal modelling. The seasonal trends of arboviral illnesses, such as chikungunya, whose outbreaks are correlated with changes in land use, urbanisation, and human movement patterns, have been monitored with the aid of these techniques [168]. Targeted vector control and intervention tactics are supported by these models.

10.8.7.6 Limitations and the Need for Localised Data

Predictive models provide encouraging insights, but in many low-resource contexts, their accuracy is constrained by underreporting, poor data quality, and a lack of standardised inputs. Many models may not generalise to neglected viral hotspots since they are built using data from high-income nations. Building localised models that take into account area climatic, cultural, and epidemiological quirks is becoming more and more important [169].

10.8.8 Enhancing Early Detection Technologies

10.8.8.1 Introduction to Early Detection in Disease Control

In order to lessen the effect and spread of neglected viral infections, early detection is essential. Early detection of new infections enables prompt treatment, which lowers mortality, morbidity, and transmission. Innovations in detection technologies have become essential for enhancing global health outcomes in the setting of NVs, many of which have few diagnostic resources [170].

10.8.8.2 Development of Point-of-Care Diagnostic Tools

In environments with limited resources, point-of-care diagnostics have revolutionised the field of illness diagnosis. Real-time diagnosis is made possible by these portable, reasonably priced gadgets without requiring sophisticated laboratory infrastructure. Rapid diagnostic tests, for instance, have been created for the dengue, Lassa, and chikungunya viruses. These instruments' ease of use and quickness close the diagnostic gap in underserved and rural areas, greatly aiding in the early detection of outbreaks [171, 172].

10.8.8.3 Integration of Molecular Diagnostics

High sensitivity and specificity in virus identification have been made possible by contemporary molecular diagnostic techniques as PCR, loop-mediated isothermal amplification, and CRISPR-based tests. For example, CRISPR-Cas systems have demonstrated potential in identifying RNA viruses such as dengue and Zika in less than an hour with little equipment [173]. These techniques are currently being modified for NVs with little cross-reactivity into formats that may be used in the field.

10.8.8.4 Biosensor-Based Technologies

Another new technique for early detection is biosensors. These gadgets look for viral components like proteins or nucleic acids using biological substances. Electrochemical biosensors have been shown in recent research to be highly sensitive and inexpensive in detecting the antigens of the Japanese encephalitis and chikungunya viruses [174]. Wearable technology and mobile devices can incorporate biosensors, opening the door to community-level screening and real-time surveillance.

10.8.8.5 Smartphone-Based and Portable Diagnostics

Mobile health systems have been incorporated into diagnostic procedures because to the increasing use of smartphones. Healthcare professionals in remote areas can upload and analyse patient data promptly when using mobile apps in conjunction with diagnostic tools. Devices connected to smartphones for Zika virus detection and real-time case location mapping are two examples. Real-time alerts and analytics are made possible by these advancements, which are also transforming epidemic reporting [175].

10.8.8.6 Artificial Intelligence in Early Signal Detection

ProMED-mail and HealthMap have identified Ebola and Zika outbreaks before government notification, indicating that similar models can be tuned to monitor neglected viral threats [176, 177]. AI-powered systems can now scan digital data sources—such as social media, medical records, and internet search trends—to detect signals of emerging viral outbreaks, flagging unusual symptom clusters, spikes in keyword searches, or anomaly reports before official confirmation.

10.8.8.7 Challenges and Future Perspectives

Despite these developments, implementing detection methods in environments with limited resources still presents difficulties. These include supply chain problems, regulatory obstacles, excessive expenses, and a shortage of qualified workers. In the future, improving early warning systems for NVs will require scalable technologies that are affordable, reliable, and easy to use. The effectiveness of these technologies will also be improved by cross-border data sharing and integration with digital surveillance systems [178].

10.9 CONCLUSION

NVs continue to pose a threat, which emphasises how urgently strong, adaptable, and just global health systems are needed. As this chapter emphasises, these viruses mostly impact underprivileged and marginalised populations, flourishing in environments with inadequate surveillance, few diagnostic tools, and limited healthcare resources. Prioritising policies and implementing targeted interventions require an understanding of their epidemiology, regional prevalence, and case-specific symptoms. In the fight against NVs, the chapter has underlined that surveillance and early warning systems are crucial tools for disease control, outbreak prevention, and health system readiness and are not optional. The development of surveillance techniques, from conventional field epidemiology to cutting-edge technologies like AI, ML, and big data analytics, presents previously unheard-of possibilities for risk forecasting, real-time detection, and strategic intervention. However, to stop the global health gap from growing, these advances need to be distributed fairly and regulated ethically.

Global and regional networks, such as national public health organisations and WHO-led partnerships, are essential for organising surveillance operations, exchanging data, and sending out alerts. However, issues including ongoing underfunding, disjointed infrastructure, and moral and legal quandaries still restrict the efficacy of many monitoring projects. Furthermore, success stories demonstrate that even nations with little resources may put in place efficient early warning systems provided they have the necessary political will, community involvement, and scientific capability. Investing in predictive modelling, improving early detection technology, and cultivating an international culture of openness and cooperation are all critical to the future of NV surveillance. Prioritising these initiatives is essential to preserving the values of global solidarity and health fairness as well as preventing the next outbreak. In conclusion, if we decide to stop ignoring NVs, they will no longer be hidden. Our ability to identify, comprehend, and address these covert threats will be revolutionised by the development of robust, intelligent, and inclusive surveillance technologies, transforming a reactionary strategy into one that is predictive, preventive, and ultimately transformative for global health.

REFERENCES

1. Mayer SV, Tesh RB, Vasilakis N. The emergence of arthropod-borne viral diseases: A global prospective on dengue, chikungunya and Zika fevers. *Acta Trop.* 2017;166:155–63.
2. Leta S, Beyene TJ, De Clercq EM, Amenu K, Kraemer MU, Revie CW. Global risk mapping for major diseases transmitted by *Aedes aegypti* and *Aedes albopictus*. *Int J Infect Dis.* 2018;67:25–35.
3. Solomon T, Baylis M, Brown D. Zoonotic viral encephalitides: A view from the UK. *Curr Opin Infect Dis.* 2017;30(5):501–8.
4. Morens DM, Folkers GK, Fauci AS. The challenge of emerging and re-emerging infectious diseases. *Nature.* 2004;430(6996):242–9.
5. World Health Organization. Research and Development Blueprint: 2022 Annual Review of Diseases Prioritized Under the R&D Blueprint, 2022. WHO, Geneva. Available from: https://www.who.int/docs/default-source/blue-print/first-annual-review-of-diseases-prioritized-under-r-and-d-blueprint.pdf?sfvrsn=1f6b5da0_4&utm_source=chatgpt.com
6. Enchéry F, Horvat B. Understanding the interaction between henipaviruses and their natural host, fruit bats: Paving the way toward control of highly lethal infection in humans. *Int Rev Immunol.* 2017;36(2):108–21.

7. Chowdhury S, Chakraborty P. Universal health coverage – There is more to it than meets the eye. *J Fam Med Prim Care.* 2017;6(2):169–70.
8. Renault P, Solet JL, Sissoko D et al. A major epidemic of chikungunya virus infection on Réunion Island, France, 2005–2006. *Am J Trop Med Hyg.* 2007;77(4):727–31.
9. Gubler DJ. The global emergence/resurgence of arboviral diseases as public health problems. *Arch Med Res.* 2002;33(4):330–42.
10. Solomon T. Flavivirus encephalitis. *N Engl J Med.* 2004;351(4):370–8.
11. Spengler JR, Bergeron É, Rollin PE. Seroepidemiological studies of Crimean-Congo hemorrhagic fever virus in domestic and wild animals. *PLOS Negl Trop Dis.* 2016;10(1):e0004210.
12. Wilder-Smith A, Gubler DJ, Weaver SC et al. Epidemic arboviral diseases: Priorities for research and public health. *Lancet Infect Dis.* 2017;17(3):e101–6.
13. Purushotham JN, Lutz HL, Parker E, Andersen KG. Immunological drivers of zoonotic virus emergence, evolution, and endemicity. *Immunity.* 2025;58(4):784–96.
14. Ang BSP, Lim TCC, Wang L. Nipah virus infection. *J Clin Microbiol.* 2018;56(6):e01875–17.
15. Morens DM, Fauci AS. Emerging infectious diseases: Threats to human health and global stability. *PLOS Pathog.* 2013;9(7):e1003467.
16. Mandl JN, Schneider C, Schneider DS, Baker ML. Going to bat: Insights into host immune responses to neglected viruses from the animal reservoir. *Annu Rev Virol.* 2018;5(1):239–61.
17. Taylor L. Yellow fever outbreaks in Brazil worry experts. *The Lancet.* 2017;389(10070):613.
18. Hamer DH, Chen LH. Zika virus and microcephaly in Brazil: A call for action. *JAMA.* 2016;315(11):1104–5.
19. Young R, Bekele T, Gunn A, Chapman N, Chowdhary V, Corrigan K, Dahora L, Martinez S, Permar S, Persson J, Rodriguez B. Developing new health technologies for neglected diseases: A pipeline portfolio review and cost model. *Gates Open Res.* 2020;2:23.
20. World Health Organization. Ending the neglect to attain the sustainable development goals: a rationale for continued investment in tackling neglected tropical diseases 2021–2030. World Health Organization; September 26, 2022. https://www.who.int/publications/i/item/9789240052932?utm_source=chatgpt.com
21. Modjarrad K, Moorthy VS, Millett P et al. Developing global norms for sharing data and results during public health emergencies. *PLOS Med.* 2016;13(1):e1001935.
22. Tambo E, Ugwu CE, Ngogang JY. Need of surveillance response systems to combat Ebola outbreaks and other emerging infectious diseases in African countries. *Infect Dis Poverty.* 2014;3:29.
23. Morens DM, Daszak P, Taubenberger JK. Escaping Pandora's box—Another novel coronavirus. *N Engl J Med.* 2020;382(14):1293–5.
24. Thacker SB, Berkelman RL. Public health surveillance in the United States. *Epidemiol Rev.* 1988;10(1):164–90.
25. World Health Organization. Technical Guidelines for Integrated Disease Surveillance and Response in the African Region, 2010. 2nd ed. WHO Regional Office for Africa, Brazzaville.
26. Anyamba A, Linthicum KJ, Tucker CJ. Climate-disease connections: Rift Valley fever in Kenya. *Cadernos de Saúde Pública.* 2001;17(6):133–40.
27. Tom-Aba D, Olaleye A, Olayinka AT, Nguku P, Waziri N, Adewuyi P et al. Innovative technological approach to Ebola virus disease outbreak response in Nigeria using the open data kit and form hub technology. *PLOS One.* 2015;10(6):e0131000.
28. Munyua PM, Murithi RM, Wainwright S et al. Rift valley fever outbreak in livestock in Kenya, 2006–2007. *Am J Trop Med Hyg.* 2010;83(2 Suppl):58–64.
29. Yadav PD, Raut CG, Shete AM et al. Detection of Nipah virus RNA in fruit bat (Pteropus giganteus) from India. *Am J Trop Med Hyg.* 2012;87(3):576–8.
30. Hotez PJ, Savioli L, Fenwick A. Neglected tropical diseases of the Middle East and North Africa: Review of their prevalence, distribution, and opportunities for control. *PLOS Negl Trop Dis.* 2012;6(2):e1475.
31. Heymann DL. Control, elimination, eradication and re-emergence of infectious diseases: Getting the message right. *Bull World Health Organ.* 2006;84(2):82.
32. Ergönül O. Crimean-Congo haemorrhagic fever. *Lancet Infect Dis.* 2006;6(4):203–14.
33. Ferguson NM, Cucunubá ZM, Dorigatti I et al. Countering the Zika epidemic in Latin America. *Science.* 2016;353(6297):353–4.
34. World Health Organization. EYE Strategy: Eliminate Yellow Fever Epidemics, 2017. WHO, Geneva.
35. Malvy D, McElroy AK, de Clerck H, Günther S, van Griensven J. Ebola virus disease. *The Lancet.* 2019;393(10174):936–48.
36. Achee NL, Gould F, Perkins TA et al. A critical assessment of vector control for dengue prevention. *PLOS Negl Trop Dis.* 2015;9(5):e0003655.

37. Guillaume V, Lefeuvre A, Faure C, Marianneau P, Buckland R, Lam SK, Wild TF, Deubel V. Specific detection of Nipah virus using real-time RT-PCR (TaqMan). *J Virol Methods.* 2004;120(2):229–37.
38. Kasabi GS, Murhekar MV, Sandhya VK et al. Coverage and effectiveness of Kyasanur Forest disease (KFD) vaccine in Karnataka, South India. *Vaccine.* 2013;31(25):3171–7.
39. Peeling RW, Mabey D. Point-of-care tests for diagnosing infections in the developing world. *Clin Microbiol Infect.* 2010;16(8):1062–9.
40. Mackenzie JS, Jeggo M. The One Health approach—Why is it so important? *Trop Med Infect Dis.* 2019;4(2):88.
41. Heymann DL, Rodier GR. Surveillance of communicable diseases: A global perspective. *CMAJ.* 1998;158(4):493–5.
42. Huang YJ, Higgs S, Vanlandingham DL. Emergence and re-emergence of mosquito-borne arboviruses. *Curr Opin Virol.* 2019;34:104–9.
43. Ilori EA, Furuse Y, Ipadeola OB et al. Epidemiologic and clinical features of Lassa fever outbreak in Nigeria, 2019. *Emerg Infect Dis.* 2021;27(4):982–90.
44. Chen H, Liu K, Li Z, Wang P. Point of care testing for infectious diseases. *Clin Chim Acta.* 2019;493:138–47.
45. World Health Organization. Laboratory Strengthening for Communicable Disease Surveillance in the WHO African Region: a Strategic Framework 2016–2020, 2016. WHO, Brazzaville.
46. Kellner MJ, Koob JG, Gootenberg JS, Abudayyeh OO, Zhang F. Sherlock: Nucleic acid detection with CRISPR nucleases. *Nat Protoc.* 2019;14(10):2986–3012.
47. Nsubuga P, Eseko N, Wuhib T et al. Structure and performance of infectious disease surveillance and response systems in the WHO African region, 1998–2015. *BMC Public Health.* 2017;17(Suppl 4):801.
48. Brownstein JS, Freifeld CC, Madoff LC. Digital disease detection—Harnessing the web for public health surveillance. *N Engl J Med.* 2009;360(21):2153–5.
49. Ratnayake R, Tammaro M, Tiffany A, Kongelf A, Polonsky JA, McClelland A. People-centered surveillance: A systematic review of the implementation of community-based surveillance among crisis-affected populations. Available at SSRN 3577836. April 16, 2020. https://ssrn.com/abstract=3577836
50. Richards P. Ebola: How a People's Science Helped End an Epidemic, 2016. Zed Books Ltd.
51. Zinsstag J, Schelling E, Crump L, Whittaker M, Tanner M, Stephen C, editors. One Health: The theory and practice of integrated health approaches. 2nd ed. Cabi; 2020. ISBN: 978-1-78924-257-7, 472 pp.
52. Munyua PM, Njenga MK, Osoro E et al. Successes and challenges of the One Health approach in Kenya over the last decade. *BMC Public Health.* 2019;19(Suppl 3):465.
53. Paakkari L, Okan O. COVID-19: Health literacy is an underestimated problem. *Lancet Public Health.* 2020;5(5):e249–50.
54. Thacker SB, Qualters JR, Lee LM, Centers for Disease Control and Prevention. Public health surveillance in the United States: Evolution and challenges. *MMWR Suppl.* 2012;61(3):3–9.
55. LaBeaud AD, Bashir F, King CH. Measuring the burden of arboviral diseases: The spectrum of morbidity and mortality from four prevalent infections. *BMJ Glob Health.* 2011;2(2):e000412.
56. WHO Ebola Response Team. Ebola virus disease in West Africa—The first 9 months of the epidemic and forward projections. *N Engl J Med.* 2014;371(16):1481–95.
57. Brownstein JS, Freifeld CC, Reis BY, Mandl KD. Surveillance Sans Frontières: Internet-based emerging infectious disease intelligence and the HealthMap project. *PLOS Med.* 2008;5(7):e151.
58. World Health Organization. Early Detection, Assessment and Response to Acute Public Health Events: Implementation of Early Warning and Response With a Focus on Event-Based Surveillance, 2014. WHO, Geneva.
59. Watts N, Amann M, Ayeb-Karlsson S, Chambers J, Hamilton I, Lowe R, Pye S, Owfi F, Rabbaniha M, Tabatabaei M, Latifi AM. The Lancet Countdown on health and climate change: From 25 years of inaction to a global transformation for public health. *The Lancet.* 2018;391(10120):540.
60. Fall IS, Rajatonirina S, Yahaya AA, Zabulon Y, Nsubuga P, Nanyunja M, Wamala J, Njuguna C, Lukoya CO, Alemu W, Kasolo FC. Integrated disease surveillance and response (IDSR) strategy: Current status, challenges and perspectives for the future in Africa. *BMJ Global Health.* 2019;4(4):e001427.
61. Alemu W, Woldie M, Yitayal M, Mesfin K. Integrated disease surveillance and response implementation in Ethiopia: Successes and challenges. *BMC Public Health.* 2021;21(1):1–11.
62. Grubaugh ND, Faria NR, Andersen KG, Pybus OG. Genomic insights into Zika virus emergence and spread. *Cell.* 2018;172(6):1160–2.
63. Sazzad HM, Hossain MJ, Gurley ES et al. Nipah virus infection outbreak with nosocomial and corpse-to-human transmission, Bangladesh. *Emerg Infect Dis.* 2013;19(2):210–7.

64. Mehand MS, Al-Shorbaji F, Millett P, Murgue B. The WHO R&D blueprint: 2018 review of emerging infectious diseases requiring urgent research and development efforts. *Antiviral Res.* 2018;159:63–7.
65. Bhat P, Jagadeesha HS, Raju MK, Sooda S, Premanand K, Kumar R. Kyasanur Forest Disease, is our surveillance system healthy to prevent a larger outbreak? A mixed-method study, Shivamogga, Karnataka, India: 2019. *Int J Infect Dis.* 2021;110:S50–61.
66. Declich S, Carter AO. Public health surveillance: Historical origins, methods and evaluation. *Bull World Health Organization.* 1994;72(2):285.
67. Groseclose SL, Buckeridge DL. Public health surveillance systems: Recent advances in their use and evaluation. *Annual Rev Public Health.* 2017;38(1):57–79.
68. Anyamba A, Linthicum KJ, Small JL, Collins KM, Tucker CJ, Pak EW et al. Climate teleconnections and recent patterns of human and animal disease outbreaks. *PLOS Negl Trop Dis.* 2012;6(1):e1465.
69. Ofori B, Twum S, Yeboah SN, Ansah F, Sarpong KA. Towards the development of cost-effective point-of-care diagnostic tools for poverty-related infectious diseases in sub-Saharan Africa. *PeerJ.* 2024;12:e17198.
70. Nichols EK, Byass P, Chandramohan D, Clark SJ, Flaxman AD, Jakob R et al. The WHO 2016 verbal autopsy instrument: Revised for enhanced utility in facilitating mortality surveillance. *Global Health Action.* 2018;11(1):1527552.
71. Goldstone BJ, Brown B. The role of public knowledge, resources, and innovation in responding to the Ebola outbreak. *Disaster Med Public Health Prep.* 2015;9(5):595–7.
72. Shaffer JG, Grant DS, Schieffelin JS, Boisen ML, Goba A, Hartnett JN et al. Lassa fever in post-conflict Sierra Leone. *PLOS Negl Trop Dis.* 2014;8(3):e2748.
73. Olugasa BO, Dogba JB. Mapping of Lassa fever cases in post-conflict Liberia, 2008-2012: A descriptive and categorical analysis of age, gender and seasonal pattern. *Ann Afr Med.* 2015;14(2):120–2.
74. Lowe R, Gasparrini A, Van Meerbeeck CJ et al. Nonlinear and delayed impacts of climate on dengue risk in Barbados: A modelling study. *PLOS Med.* 2018;15(7):e1002613.
75. Gething PW, Casey DC, Weiss DJ et al. Mapping plasmodium falciparum mortality in Africa between 1990 and 2015. *N Engl J Med.* 2016;375(25):2435–45.
76. Jhutty SS, Boehme JD, Jeron A, Volckmar J, Schultz K, Schreiber J, Schughart K, Zhou K, Steinheimer J, Stöcker H, Stegemann-Koniszewski S. Predicting influenza A virus infection in the lung from hematological data with machine learning. *Msystems.* 2022;7(6):e00459–22.
77. Rangarajan P, Mody SK, Marathe M. Forecasting dengue and influenza incidences using a sparse representation of Google trends, electronic health records, and time series data. *PLOS Comput Bio.* 2019;15(11):e1007518.
78. Hansen F, Feldmann H, Jarvis MA. Targeting Ebola virus replication through pharmaceutical intervention. *Expert Opin Investig Drugs.* 2021;30(3):201–26.
79. Omoleke SA, Mohammed I, Saidu Y, Musa H. Ebola viral disease in West Africa: A threat to global health, economy and political stability. *J Public Health Afr.* 2016;7(1):534.
80. Dandekar R, Barbastathis G. Quantifying the effect of quarantine control in COVID-19 infectious spread using machine learning. *medRxiv.* 2020.
81. Grubaugh ND, Ladner JT, Lemey P et al. Tracking virus outbreaks in the twenty-first century. *Nat Microbiol.* 2019;4(1):10–9.
82. Faria NR, Sabino EC, Nunes MR et al. Mobile real-time surveillance of Zika virus in Brazil. *Genome Med.* 2016;8(1):97.
83. Cassidy R. Lives with others: Climate change and human-animal relations. *Annu Rev Anthropol.* 2012;41(1):21–36.
84. Talla C, Diallo D, Dia I, Ba Y, Ndione J-A, Sall AA et al. Statistical modeling of the abundance of vectors of West African Rift Valley fever in Barkédji, Senegal. *PLOS One.* 2014;9(12):e114047.
85. Samphutthanont R, Koyadun S. Spatial-temporal distribution pattern of dengue fever and spatial relationship of the Aedes mosquito larvae index in Phuket Province. *Int J Geoinformatics.* 2024;20(8):29–45.
86. Khan K, Arino J, Hu W et al. Spread of a novel influenza A (H1N1) virus via global airline transportation. *N Engl J Med.* 2009;361(2):212–4.
87. Udegbe FC, Nwankwo EI, Igwama GT, Olaboye JA. Real-time data integration in diagnostic devices for predictive modeling of infectious disease outbreaks. *Comp Sci IT Res J.* 2023;4(3):525–45.
88. Abaya ST, Ogoina D, Stow J, Abaye BB, Emeka C, Raimi MO. Beyond the epidemic: Effective public health strategies in response to Nigeria's first Lassa fever outbreak in a non-endemic region. *JMIR Preprints.* 2024;19(08):2024.

89. Kukkar A, Kumar Y, Sandhu JK, Kaur M, Walia TS, Amoon M. DengueFog: A fog computing-enabled weighted random forest-based smart health monitoring system for automatic dengue prediction. *Diagnostics*. 2024;14(6):624.
90. Liu WL, Wang Y, Chen YX, Chen BY, Lin AY, Dai ST, Chen CH, Liao LD. An IoT-based smart mosquito trap system embedded with real-time mosquito image processing by neural networks for mosquito surveillance. *Front Bioeng Biotechnol*. 2023;11:1100968.
91. King'ori T. Harnessing digital health technologies for improved healthcare delivery and disease surveillance in low-resource settings. *Glob Perspect Health, Med Nurs*. 2024;3(1):1–10.
92. Johansson MA, Apfeldorf KM, Dobson S et al. An open challenge to advance probabilistic forecasting for dengue epidemics. *Proc Natl Acad Sci USA*. 2019;116(48):24268–74.
93. Aldosery A, Musah A, Birjovanu G, Moreno G, Boscor A, Dutra L, Santos G, Nunes V, Oliveira R, Ambrizzi T, Massoni T. MEWAR: Development of a cross-platform mobile application and web dashboard system for real-time mosquito surveillance in Northeast Brazil. *Front Public Health*. 2021;9:754072.
94. Keshavamurthy R, Dixon S, Pazdernik KT, Charles LE. Predicting infectious disease for biopreparedness and response: A systematic review of machine learning and deep learning approaches. *One Health*. 2022;15:100439.
95. Beam AL, Kohane IS. Big data and machine learning in health care. *JAMA*. 2018;319(13):1317–8.
96. Anyamba A, Linthicum KJ, Small J et al. Climate teleconnections and recent patterns of human and animal disease outbreaks. *Philos Trans R Soc Lond B Biol Sci*. 2006;361(1469):1205–16.
97. Zainudin AN, Prawiragara FA. Leveraging technology for dengue fever management using mobile health GIS and Big Data to improve epidemic control in tropical regions: A literature review. *J Diverse Med Res: Medicosphere*. 2025;2(2):42–52.
98. Wesolowski A, Qureshi T, Boni MF et al. Impact of human mobility on the emergence of dengue epidemics in Pakistan. *Proc Natl Acad Sci USA*. 2015;112(38):11887–92.
99. D'Addiego J, Shah S, Pektaş AN, Bağci BK, Öz M, Sebastianelli S, Elaldı N, Allen DJ, Hewson R. Development of targeted whole genome sequencing approaches for Crimean-Congo haemorrhagic fever virus (CCHFV). *Virus Res*. 2024;350:199464.
100. Smith KF, Goldberg M, Rosenthal S, Carlson L, Chen J, Chen C, Ramachandran S. Global rise in human infectious disease outbreaks. *J Royal Soc Interf*. 2014;11(101):20140950.
101. Heymann DL, Rodier GR. Global surveillance, national surveillance, and SARS. *Emerg Infect Dis*. 2004;10(2):173–5.
102. Mackenzie JS, Drury P, Arthur RR, Ryan MJ, Grein T, Slattery R et al. The global outbreak alert and response network. *Glob Public Health*. 2014;9(9):1023–39.
103. Global Health Security Agenda. GHSA Framework and Action Packages [Internet]. 2022 [cited 2025 Jul 29]. Available from: https://ghsagenda.org/
104. Food and Agriculture Organization. GLEWS+ Global Early Warning System [Internet]. FAO/OIE/WHO; 2022 [cited 2025 Jul 29]. Available from: https://www.fao.org/one-health/areas-of-work/one-health-inteligence/en?utm_source=chatgpt.com
105. Freifeld CC, Mandl KD, Reis BY, Brownstein JS. HealthMap: Global infectious disease monitoring through automated classification and visualization of Internet media reports. *J Am Med Inform Assoc*. 2008;15(2):150–7.
106. Eysenbach G. Infodemiology and infoveillance: Tracking online health information and cyberbehavior for public health. *Am J Prev Med*. 2011;40(5):S154–8.
107. Elnaiem A, Mohamed-Ahmed O, Zumla A, Mecaskey J, Charron N, Abakar MF, Raji T, Bahalim A, Manikam L, Risk O, Okereke E. Global and regional governance of One Health and implications for global health security. *The Lancet*. 2023;401(10377):688–704.
108. World Health Organization. International Health Regulations (2005). World Health Organization; December 15, 2008.
109. World Health Organization. Global Outbreak Alert and Response Network (GOARN) [Internet]. 2023 [cited 2025 Jul 29]. Available from: https://www.who.int/initiatives/goarn
110. World Health Organization. R&D Blueprint: 2023 Annual Review of Priority Diseases [Internet]. 2023 [cited 2025 Jul 29]. Available from: https://cdn.who.int/media/docs/default-source/documents/r-d-blueprint-meetings/global-research-and-innovation-for-health-emergencies_report-2023.pdf?sfvrsn=9341366_1&utm_source=chatgpt.com
111. World Health Organization. Tripartite and UNEP support OHHLEP's definition of "One Health" [Internet]. WHO; 2021 [cited 2025 Jul 29]. Available from: https://www.who.int/news/item/01-12-2021-tripartite-and-unep-support-ohhlep-s-definition-of-one-health?utm_source=chatgpt.com

112. World Health Organization Regional Office for Africa. Integrated Disease Surveillance and Response: Technical Guidelines [Internet]. 3rd ed. 2019 [cited 2025 Jul 29]. Available from: https://www.who.int/publications/i/item/WHO-AF-WHE-CPI-05-2019?utm_source=chatgpt.com
113. National Centre for Disease Control (India). Integrated Health Information Platform (IHIP) [Internet]. 2024 [cited 2025 Jul 29]. Available from: https://ihip.nhp.gov.in
114. ICDDR, B - International Centre for Diarrhoeal Disease Research, Bangladesh. Nipah Virus Surveillance in Bangladesh [Internet]. 2023 [cited 2025 Jul 29]. Available from: https://www.icddrb.org/news-and-events/press-releases/nipah-virus
115. Pan American Health Organization (PAHO). PAHO Strategy for Arboviral Diseases [Internet]. 2023 [cited 2025 Jul 29]. Available from: https://www.paho.org/en/documents/strategy-arboviral-diseases
116. World Health Organization Eastern Mediterranean Regional Office. Regional Strategic Framework for Surveillance [Internet]. 2022 [cited 2025 Jul 29]. Available from: https://applications.emro.who.int/docs/Progress-report-IDS-eng.pdf?utm_source=chatgpt.com
117. Olayemi A, Cadar D, Magassouba N, Obadare A, Kourouma F, Oyeyiola A et al. New hosts of the Lassa virus. *Sci Rep.* 2016;6:25280.
118. Hossain MJ, Gurley ES, Montgomery JM, Bell M, Carroll DS, Hossain MS et al. Clinical presentation of Nipah virus infection in Bangladesh. *Clin Infect Dis.* 2008;46(7):977–84.
119. World Health Organization. Summary report on the WHO workshop on estimation of respiratory syncytial virus (RSV) disease burden based on RSV Surveillance of Global Influenza Surveillance and Response System, Geneva, Switzerland, 21–23 June 2023. World Health Organization; 2024.
120. Yu VL, Madoff LC. ProMED-mail: An early warning system for emerging diseases. *Clin Infect Dis.* 2004;39(2):227–32.
121. United Nations Office for the Coordination of Humanitarian Affairs. Humanitarian Needs Overview 2023: South Sudan. [Internet]. 2023 [cited 2025 Jul 29]. Available from: https://reliefweb.int
122. Musso D, Gubler DJ. Zika virus. *Clin Microbiol Rev.* 2016;29(3):487–524.
123. Fidler DP. global health surveillance reform and global health governance. *EMBO Rep.* 2010;11(1):7–11.
124. Fallatah DI, Adekola HA. Digital epidemiology: Harnessing big data for early detection and monitoring of viral outbreaks. *Infect Prev Pract.* 2024;6(3):100382.
125. Paixão ES, Teixeira MG, Rodrigues LC. Zika, chikungunya and dengue: The causes and threats of new and re-emerging arboviral diseases. *BMJ Global Health.* 2018;3(Suppl 1):e000530. doi:10.1136/bmjgh-2017-000530
126. Hotez PJ, Aksoy S. PLOS Neglected Tropical Diseases: Ten years of progress in neglected tropical disease control and elimination … more or less. *PLOS Negl Trop Dis.* 2017;11(4):e0005355.
127. Morse SS, Mazet JA, Woolhouse M, Parrish CR, Carroll D, Karesh WB et al. Prediction and prevention of the next pandemic zoonosis. *The Lancet.* 2012;380(9857):1956–65.
128. Dash AP, Bhatia R, Sunyoto T, Mourya DT. Emerging and re-emerging arboviral diseases in Southeast Asia. *J Vector Borne Dis.* 2013;50(2):77–84.
129. Moon S, Sridhar D, Pate MA, Jha AK, Clinton C, Delaunay S et al. Will Ebola change the game? Ten essential reforms before the next pandemic. *The Lancet.* 2015;386(10009):2204–21.
130. Piot P, Spencer J. From 1976 to 2018: Reflections on early investigations into the Ebola virus. *Trans Royal Soc Trop Med Hyg.* 2018;112(12):527–8.
131. World Health Organization. Taking a Multisectoral, One Health Approach: A Tripartite Guide to Addressing Zoonotic Diseases in Countries, 2019. WHO, Geneva. Available from: https://www.who.int/publications/i/item/9789241514934
132. Annas GJ. Your liberty or your life: Talking point on public health versus civil liberties. *EMBO Reports.* 2007;8(12):1093–8.
133. Fidler DP. Influenza virus samples, international law, and global health diplomacy. *Emerg Infect Dis.* 2008;14(1):88–94.
134. Abramowitz SA, Hipgrave DB, Witchard A, Heymann DL. Lessons from the West African Ebola epidemic: A systematic review of epidemiological and social and behavioral science research priorities. *J Infect Dis.* 2018;218(11):1730–8.
135. Pellecchia U, Crestani R, Decroo T, Van den Bergh R, Al-Kourdi Y. Social consequences of Ebola containment measures in Liberia. *PLOS One.* 2015;10(12):e0143036.
136. Morley J, Floridi L. The limits of empowerment: How to reframe the role of mHealth tools in the healthcare ecosystem. *Sci Eng Ethics.* 2020;26(3):1159–83.
137. World Health Organization. Guidance for Managing Ethical Issues in Infectious Disease Outbreaks, 2017. WHO, Geneva. Available from: https://www.who.int/publications/i/item/9789241549837

138. Cenciarelli O, Pietropaoli S, Malizia A, Carestia M, D'Amico F, Sassolini A, Di Giovanni D, Rea S, Gabbarini V, Tamburrini A, Palombi L. Ebola virus disease 2013-2014 outbreak in west Africa: An analysis of the epidemic spread and response. *Int J Microbiol.* 2015;2015(1):769121.
139. Nigeria Centre for Disease Control. 2019 Lassa Fever Outbreak Situation Report. Available from: https://ncdc.gov.ng
140. World Health Organization. Yellow fever – Angola and the Democratic Republic of the Congo. WHO; 2016. Available from: https://www.who.int/emergencies/disease-outbreak-news/item/13-april-2016-yellow-fever-angola-en?utm_source=chatgpt.com
141. Arunkumar G, Chandni R, Mourya DT, Singh SK, Sadanandan R, Sudan P et al. Outbreak investigation of Nipah virus disease in Kerala, India, 2018. *J Infect Dis.* 2019;219(12):1867–78.
142. Balasubramanian R, Mohandas S, Thankappan UP, Shete A, Patil D, Sabarinath K et al. Surveillance of Nipah virus in *Pteropus medius* of Kerala state, India, 2023. *Front Microbiol.* 2024;15:1342170.
143. Faria NR, Azevedo RS, Kraemer MU, Souza R, Cunha MS, Hill SC et al. Zika virus in the Americas: Early epidemiological and genetic findings. *Science.* 2016;352(6283):345–9.
144. Global Polio Eradication Initiative. Surveillance. Available from: https://polioeradication.org/polio-today/surveillance-indicators/
145. Nsubuga P, Eseko N, Wuhib T, Ndayimirije N, Chungong S, McNabb SJ et al. Structure and performance of infectious disease surveillance and response, United Republic of Tanzania, 1998. *Bull World Health Organ.* 2002;80(3):196–203.
146. Phalkey RK, Yamamoto S, Awate P, Marx M. Challenges with the implementation of an integrated disease surveillance and response (IDSR) system: Systematic review of the lessons learned. *Health Policy Plan.* 2015;30(1):131–43.
147. Nsubuga P, Johnson K, Tetteh C, Oundo J, Weathers A, Vaughan J, Elbon S, Tshimanga M, Ndugulile F, Ohuabunwo C. Evering-Watley M. Field epidemiology and laboratory training programs in sub-Saharan Africa from 2004 to 2010: Need, the process, and prospects. *Pan Afr Med J.* 2011;10.
148. Alonge O, Sonkarlay S, Gwaikolo W, Fahim C, Cooper JL, Peters DH. Understanding the role of community resilience in addressing the Ebola virus disease epidemic in Liberia: A qualitative study (community resilience in Liberia). *Global Health Action.* 2019;12(1):1662682.
149. Pan American Health Organization. Regional Zika Epidemiological Update. 2016. Available from: https://www.paho.org
150. Ratnayake R, Crowe SJ, Jasperse J, Privette G, Stone E, Miller L, Hertz D, Fu C, Maenner MJ, Jambai A, Morgan O. Assessment of community event–based surveillance for Ebola virus disease, Sierra Leone, 2015. *Emerg Infect Dis.* 2016;22(8):1431.
151. Yinka-Ogunleye A, Aruna O, Dalhat M, Ogoina D, McCollum A, Disu Y et al. Outbreak of human monkeypox in Nigeria in 2017–18: A clinical and epidemiological report. *Lancet Infect Dis.* 2019;19(8):872–9.
152. Ilori EA, Furuse Y, Ipadeola OB, Dan-Nwafor C, Abubakar A, Womi-Eteng OE et al. Epidemiologic and clinical features of Lassa fever outbreak in Nigeria, January 1–May 6, 2018. *Emerg Infect Dis.* 2019;25(6):1066–74.
153. World Health Organization. Eliminate Yellow Fever Epidemics (EYE) Strategy: A Global Initiative. 2021. Available from: https://www.who.int/initiatives/eye-strategy
154. Nyenswah TG. Strengthening public health integrated disease surveillance and response system post the 2014-2016 Ebola epidemic in Liberia: A case study (Doctoral dissertation, Johns Hopkins University).
155. Satter SM, Aquib WR, Sultana S, Sharif AR, Nazneen A, Alam MR, Siddika A, Akther Ema F, Chowdhury KI, Alam AN, Rahman M. Tackling a global epidemic threat: Nipah surveillance in Bangladesh, 2006–2021. *PLOS Negl Trop Dis.* 2023;17(9):e0011617.
156. World Health Organization. Joint External Evaluation Tool: International Health Regulations (2005) Monitoring and Evaluation Framework, 2018. 2nd ed. WHO, Geneva.
157. Africa Centres for Disease Control and Prevention. Surveillance and Epidemic Intelligence. 2022. Available from: https://africacdc.org/programmes/surveillance/
158. GISAID. Global Initiative on Sharing All Influenza Data. Available from: https://www.gisaid.org/
159. Africa CDC. Africa Pathogen Genomics Initiative (PGI). Available from: https://africacdc.org/institutes/africa-pathogen-genomics-initiative/
160. Bogoch II, Watts A, Thomas-Bachli A, Huber C, Kraemer MU, Khan K. Pneumonia of unknown aetiology in Wuhan, China: Potential for international spread via commercial air travel. *J Travel Med.* 2020;27(2):taaa008.

161. Hannatu MP. The use of technology in emerging disease prevention and control in Nigeria. *J Health Human Movement Studies*. 2025.
162. World Health Organization, World Meteorological Organization. Climate Services for Health: Improving Public Health Decision-Making in a New Climate, 2016. WHO and WMO, Geneva.
163. Chretien JP, George D, Shaman J, Chitale RA, McKenzie FE. Influenza forecasting in human populations: A scoping review. *PLOS One*. 2014;9(4):e94130.
164. Bridgman HA, Oliver JE. The Global Climate System: Patterns, Processes, and Teleconnections, 2014. Cambridge University Press.
165. Grover-Kopec EK, Kawano M, Klaver RW, Blumenthal MB, Ceccato P, Connor SJ. An online operational rainfall-monitoring resource for epidemic malaria early warning systems in Africa. *Malar J*. 2005;4:6.
166. Harish V. Digital Epidemiology for Resilient Health System Response to Pandemics (Doctoral dissertation, University of Toronto (Canada)), 2024.
167. Milinovich GJ, Williams GM, Clements ACA, Hu W. Internet-based surveillance systems for monitoring emerging infectious diseases. *Lancet Infect Dis*. 2014;14(2):160–8.
168. Khaki JJ, Acharya BK, Morita K, Pandey BD, Giorgi E. A regression based approach to modelling recurrent outbreaks: a case study of dengue fever in Nepal (2006 to 2022). *medRxiv*. 2025:2025-05.
169. Pigott DM, Golding N, Mylne A, Huang Z, Henry AJ, Weiss DJ et al. Mapping the zoonotic niche of Ebola virus disease in Africa. *Elife*. 2014;3:e04395.
170. [170, Peeling RW, Olliaro PL. Rolling out COVID-19 antigen rapid diagnostic tests: The time is now. *Lancet Infect Dis*. 2021;21(8):1052–3.
171. Broadhurst MJ, Brooks TJ, Pollock NR. Diagnosis of Ebola virus disease: Past, present, and future. *Clin Microbiol Rev*. 2016;29(4):773–93.
172. Waggoner JJ, Pinsky BA. Zika virus: Diagnostics for an emerging pandemic threat. *J Clin Microbiol*. 2016;54(4):860–7.
173. Myhrvold C, Freije CA, Gootenberg JS, Abudayyeh OO, Metsky HC, Durbin AF et al. Field-deployable viral diagnostics using CRISPR-Cas13. *Science*. 2018;360(6387):444–8.
174. Draz MS, Shafiee H. Applications of gold nanoparticles in virus detection. *Theranostics*. 2018;8(7):1985–2017.
175. Yang S, Rothman RE. PCR-based diagnostics for infectious diseases: Uses, limitations, and future applications in acute-care settings. *Lancet Infect Dis*. 2004;4(6):337–48.
176. Al-Surimi K, Khalifa M, Bahkali S, El-Metwally A, Househ M. The potential of social media and internet-based data in preventing and fighting infectious diseases: From internet to twitter. In Emerging and Re-Emerging Viral Infections: Advances in Microbiology, Infectious Diseases and Public Health, Vol 6, 2016 (pp. 131–9). Springer International Publishing, Cham.
177. Phalkey Keller M, Blench M, Tolentino H, Freifeld CC, Mandl KD, Mawudeku A, Eysenbach G, Brownstein JS. Use of unstructured event-based reports for global infectious disease surveillance. *Emerg Infect Dis*. 2009;15(5):689.
178. Okeke IN, Peeling RW, Goossens H, Auckenthaler R, Olmsted SS, de Lavison JF et al. Diagnostics as essential tools for containing antibacterial resistance. *Drug Resist Updat*. 2011;14(2):95–106.

11 Ethical Considerations in Virus Research

Almuthanna K. Alkaraki and Lina Al-Ebbini

11.1 INTRODUCTION

Recent literature has been full of work that documents the growing pressure on virologists' pressure that's coming from wider debates about transparency, biosecurity, and the ethics of research. Over the past several months, a large volume of notes, articles, blogs, and opinion pieces has been published, attracting interest from a wide and diverse audience [1, 2]. However, what could be particularly beneficial in this context is to shed light on the "how" of the ongoing debate by adopting a broad, bird's-eye perspective. Engaging in meta-ethical reflection may offer a more expansive view, enabling us to gain a clearer understanding of the dynamics at play "on the ground" amidst this complex discourse [3]. From a psychological perspective, moral judgments are shaped by underlying ethical intuitions and emotional responses. Conversely, an STS (Science and Technology Studies) scholar would typically set the stage by preparing a structured "script" of empirical research that precedes a broader thesis question, accompanied by several intriguing "hunches" and hypotheses that inform the inquiry [4, 5].

Yet, the task at hand requires us to take a step back, inviting us to reflect on the very fabric of these scripts that scholars and researchers weave: what if these empirical accounts constitute an integral part of the broader narrative that we seek to outline? The concepts of recursively and reflexively are essential at this juncture, especially given that the current virus emergency has significantly paralyzed fieldwork on the ongoing pandemic and created challenges for researchers. Our objective is to effectively adopt a broader analytical perspective and achieve a strategic overview of the ethical landscape surrounding virus research, particularly in an era when it seems that offering a strategic overview or a high-level perspective on ethical dimensions [6–8]. There is, one could argue, a common reluctance to attempt to "zoom out" of the academia-virology debate itself. However, that reluctance may itself reflect a deeper hesitation to reassess disciplinary boundaries, similar to exploring the conceptual isolation often applied to virology in ethical discourse. A meta-ethical investigation into research narratives could yield valuable insights into crisis ethics, particularly in light of the pressing historical juncture we are currently navigating [9]. This analysis adopts an STS lens to examine how institutions and technologies shape ethical discourse while framing the overall narrative within the prevailing trends and paradigms or even probing the crisis itself and the potential avenues for "repairing" the situation as it evolves. There is strong justification for including historical context when analyzing current bioethical challenges: if "multicentric" epidemiology is all the rage today. A similar nuanced and multi-faceted approach could also be applied to understanding the interconnected histories of viral diseases [10].

11.2 HISTORICAL CONTEXT OF VIRUS RESEARCH

The consequence of the "H5N1 affair": the first category includes "Regular" risks such as pathogen release, raising the question of whether badgers could serve as potential carriers for these pathogens. In contrast, there are "Unexpected" risks that emerge from the scenario, self-appointed enforcers of biosafety norms" or "individuals acting outside regulatory frameworks, regardless of the scientific consensus [11, 12]. Additionally, we observe a parallel development that introduces unique risks of "fuzziness," where the focus on biosecurity could Biosecurity paralysis is an invented or unclear term; that lacks definition, effectively diluting the priority and effectiveness of biosecurity measures. Theories of suspicion arise as a response, discussing cultural rituals that attempt to control societal fear, drawing on

DOI: 10.1201/9781003607816-11

the ideas of scholars like Boas and Douglas. This exploration goes beyond mere academic discourse, highlighting the need for a careful approach, or a "seize and freeze" mentality, in dealing with these risks. It insists on handling the situation with substantial care and consideration [13, 14].

We must embrace a more active stance to address these multifaceted challenges [15]. It prompts us to contemplate what is required beyond merely acknowledging "communicative incompetence." We must also confront methodological issues that arise in these discussions and consider what forms of strategic intervention may be necessary to effectively navigate the complexities involved [10].

11.3 ETHICAL FRAMEWORKS IN RESEARCH

This study argues that unified ethical principles with guidance for nonhuman animal research are prerequisites for multinational research projects with nonhuman animal models, similar to the standards of good ethical practice for research with human participants delineated in the Declaration of Helsinki. Sections outline several efforts to establish mutual international principles for the protection of humans and nonhuman animals. First, several sets of ethical principles are examined. Here the 3Rs principles are applied to ethical principles, critically assessing whether they can be Replaced, Reduced, or Refined. Then, international ethical principles are summarised, and how they are employed in various ethical frameworks. After describing several of these, it is concluded none of them are as comprehensive as an animal research "Helsinki" Declaration. However, the availability of widely shared ethical principles is found. Finally, a proposal that these principles can be consolidated and incorporated into research agreements is made [16].

The primary ethical framework relevant to this study concerns the treatment and use of animals in experimental investigations (animal research). The study utilizes nonhuman animal models in experimental psychology, neuroscience, and immunology. However, there is considerable international variability in international approaches to nonhuman animal research, and there is no consensus on best practices between researchers in different countries. One common notion is that animal research legislation and ethical oversight vary greatly between countries. Another is that legal and ethical standards in some countries, usually wealthier ones, are generally thought to be higher. As an example of this, recent reports critical to animal research practices in The Netherlands, Brazil, and the United Kingdom commonly imply that these practices are better regulated in the United States, Canada, India, and Australia. Thus, it seems there is an implicit hierarchy of legal and ethical sophistication vis-à-vis nonhuman animal research among different countries. Any large international collaboration involving researchers in these and other nations (as well as in developing nations) must necessarily deal with these concerns regarding the welfare and treatment of the nonhuman animals participating in their experiments [17]. To compare how distinct moral philosophies approach virus research dilemmas, Table 11.1 presents a structured comparison of utilitarian,

TABLE 11.1

Comparative Overview of Utilitarianism, Deontology, and Virtue Ethics in Virus Research Contexts

Ethical Framework	Defining Principle	Application in Virus Research	Strengths	Limitations
Utilitarianism	Maximize overall well-being	Supports cost-benefit analysis in vaccine trials	Balances outcomes	May justify harmful experiments
Deontology	Duty and adherence to rules	Emphasizes strict consent and animal rights	Ensures procedural integrity	May resist urgent public health responses
Virtue ethics	Moral character and integrity	Focuses on researcher intent and community trust	Promotes responsibility	Less prescriptive for policy design

This table summarizes the defining principles, strengths, limitations, and applications of the three major ethical theories as they pertain to virus research, particularly during outbreak and emergency response scenarios.

deontological, and virtue ethics perspectives, highlighting their implications for biosecurity, animal research, and informed consent.

11.3.1 Utilitarianism

On 7 December 1998 in Edinburgh, the Crown Office, the Office of the Procurator Fiscal, initiated a rare prosecution of five men for allegedly discussing the production of biological weapons via the Internet [18]. The evidence for this case is derived solely from the encrypted electronic mail that had passed between the defendants and Sarin Dum Binky. It is speculated that SDB is either a group of unknown Gallic pranksters or GCHQ [19].

This prosecution, alongside the recent trial of six senior scientists, has been interpreted as a symbolic response tied to transatlantic trade tensions, although its scientific impact remains ambiguous. Additionally, the prosecutors have sought to persuade a Scottish court that a biochemical arsenal can easily be manufactured by pouring various household liquids into a wheelie bin and mixing with a fine stick for 40 minutes [20].

The phylogeographic distribution of the Orf virus exemplifies the intersection between empirical virology and ethical oversight in pathogen handling, a species of the genus Parapoxvirus (PPV) that is endemic to Ireland and the United Kingdom. Scarification of the skin via mechanical or autoclaved rose thorns or splinters invaded with the virus appears to be necessary for disease manifestation. This was probably how it was historically passed on. Due to the lack of availability of non-viable DNA at the time, the well-validated methodology had failed to amplify the DNA of the 7th-century Kama oral vaccine [21, 22]. Bradlee demurred this explanation and felt it was sufficient only to pressure the administrative staff, who were already strained by the relentless late nights and outdated communication infrastructure hindered coordination [17, 23].

11.3.2 Deontology

One approach to ethics in public health is deontology, which holds that moral obligation comes, not from results or effectiveness, nor from the sum of individual rights or values, but from adherence to rules or principles that follow from the intrinsic nature of one's relationship to others. The parasitic relationship that is disease gives the medical profession its moral and epistemological authority so that authority comes with specific moral obligations [3, 24].

Deontology is a moral stance that has particular relevance for biosecurity. A central expression of deontological ethics is Immanuel Kant's moral theory, which holds that an action is morally right if it is performed out of duty and guided by a maxim that one could will to become a universal law. While science, public health, and clinical service would be impossible without the stability of such a rule, it is also so austere that Kant himself allows that in the face of a liar who demands the location or identity of innocents who will be murdered, the only moral act is to lie [25, 26].

Deontology bears some similarities and many differences with the Hippocratic Oath regarding the value of knowledge and intellectual property rights in medicine. The obligation of physicians to promote public health derives entirely from the final paragraph about the sanctity of medical knowledge [27]. This makes sense if the public right to benefit as much as possible from medical knowledge would check and balance a profession with an inherent conflict of interest. It makes less sense if one's primary concern about public health is with outbreaks of disease. While the deontological principles arising from the Hippocratic Oath are widely shared by public health practitioners, the division of roles and duties between public health principles and the medical profession is less clear [28].

11.3.3 Virtue Ethics

As experts work to better understand the SARS-CoV-2 virus and develop treatments and vaccines, ethical considerations will have to play a role in the ways that research is conducted and the development of treatments is prioritized. This criticism is not new and is partly on past ethical lapses in science and the continuity of inequitable treatment in healthcare treatment and viewing this issue through the perspective of virtue ethics [29]. Particularly, medical professionals are seen through this lens given their central role in addressing and containing the spread of this disease. Three considerations are focused on the latter. The first is an interest in the motives behind the relative restriction in revealing data from the influenza outbreak at a time when the alertness level had been raised to its highest point. This led to questions of how these motives and actions could be viewed from the standpoint of virtue ethics [30].

While several ethical frameworks have been used in considering actions related to the SARS-CoV-2 pandemic, so far physicians and other medical professionals have, as far as has been found, been considered only peripherally or incidentally within these frameworks. This consideration is critical, given the central role of frontline healthcare workers in pandemic responses to contain and treat the virus in addressing this type of concern [24]. This section explores how actions and motives related to healthcare professionals might be considered and criticized from the standpoint of the virtue of ethical tradition. Practitioners in the medical field possess expertise that can lead them to know many things about their patients and how to treat them effectively. Because this knowledge could be mishandled and misused, healthcare practitioners are typically expected to conform to codes of conduct that govern their decision-making. These codes tend to highlight things such as the need to respect the autonomy of patients, act in ways that are beneficial to them, refrain from actions that might cause harm, and be fair in the distribution of treatments [24].

11.4 BIOSAFETY AND BIOSECURITY

Vaccines have never been pursued for rapidly lethal, hemorrhagic viruses. Many researchers have experienced a "close call" in needle accidents and cannot afford vaccine development even in the face of an obvious need. Help is needed by the government, industry, and the medical profession. Concerning disulfide-linked oligomers, high-resolution electron microscope observations correlate well with spectrophotometric analyses for detecting soluble oligomers of the envelope glycoprotein of the Lassa secretary arenavirus and of the structural and nonstructural proteins of Coxiella burnetii [31].

Biosafety and biosecurity are defined and discussed as essential pillars of international health security since both involve activities and measures aimed at protecting human health, as well as animal and plant health, and the environment from accidental or deliberate release of biological agents and toxins. Biosecurity is a comprehensive, integrated risk-based multi-sectoral, and cross-cutting approach (at the national, regional, and global levels) to manage intentionally caused biological risks sustainably, securing the lowest possible level of detrimental impact on public health and safety, and the stability of economies, societies, and ecosystems. Biosecurity seeks to prevent, detect, contain, respond to, and mitigate the threat of non-state actors, and rogue states by the implementation of the appropriate multi-disciplinary, intersectoral, and integrated activities and measures. Biosecurity involves the protection of humans and animals against the malicious use of pathogens, toxins, and chemicals [32].

11.4.1 Biosafety Levels

Conducting experiments at Biosafety Level 2 (S2) was deemed necessary to effectively minimize the considerable risks associated with the potential spreading of laboratory-acquired infections, particularly following the critical recommendations issued by in-house consultants. These specific experiments resulted in the generation of high titers of an A/H1N1 pandemic influenza virus, among others, and, from multiple ethical points of view, it is important to thoroughly consider whether such actions were justified or not [33]. Biosafety Level-2 and laboratories designated for higher containment levels are specifically assigned for the work involving pathogens that carry a moderate risk of infection. This designation usually pertains to agents that could potentially cause human disease through various routes such as ingestion, inhalation, or exposure to the skin or mucous membranes [34]. Notable examples of these pathogens that fall under this classification include Staphylococcus aureus, Salmonella species, along with the A/H1N1 virus itself. The classification of biohazardous agents can furnish invaluable information to aid the informed consent process for voluntary donation to areas that are currently characterized by outbreaks. The informed consent process for a subject involving biocontainment measures and other protocols that fall outside routine clinical care must be clearly explained to ensure truly informed consent [35].

The issuance of critical and timely advice in response to the A/H1N1 influenza pandemic further underscored the strong emphasis on responsibility surrounding both animal and human transmission pathways of the flu virus. The A/H1N1 probe and cultured isolates were made available in consideration of potential mutational changes that might arise following the widespread use of the specific virus strain [35]. The work conducted in this context was largely generated by the national biosecurity regulations in both the United States and the European Union, whose passage was largely precipitated by significant events that took place just before the submission of the relevant documentation. Moreover, enabling free access to the draft for developing countries was viewed as a means of significantly benefiting global health, thereby contributing to efforts aimed at addressing the wealth inequalities that currently exist within the information age, and fostering a united response to health crises worldwide [36]. The varying levels of containment required for different viruses are summarized in Table 11.2, illustrating the alignment of bioethical obligations with biosafety regulations.

11.4.2 Laboratory Protocols

Safe and efficient manipulation of viral materials necessitates that an intricate combination of laboratory protocols, specialized facilities, and effective containment devices be employed meticulously to avert inadvertent infections. This multifaceted approach is critical in ensuring not just the safety of the laboratory personnel but also the surrounding environment [37]. In addition, various physical

TABLE 11.2
Classification of Biosafety Levels with Examples of Pathogens and Laboratory Requirements

Biosafety Level	Risk Group	Pathogen Examples	Laboratory Features	Ethical Considerations
BSL-1	Minimal	Adeno-associated viruses	Open bench work	Low ethical tension
BSL-2	Moderate	Influenza A (H1N1), *Staphylococcus aureus*	Biosafety cabinets	Requires informed staff
BSL-3	High	SARS-CoV-2, M. tuberculosis	Controlled access, negative pressure	Strong consent and containment
BSL-4	Extreme	Ebola, Marburg	Full-body suits, max containment	Dual-use and public safety conflicts

This table outlines the four biosafety levels (BSL-1 to BSL-4), detailing their associated risk groups, pathogen examples, containment requirements, and relevance to ethical oversight in virus research.

methods can inactivate viral materials, rendering them non-infective. This means that purified antigens or nucleic acids that are derived from viruses can be handled and manipulated with minimal risk to researchers and the public. Furthermore, it is imperative that the ethical considerations, as well as the biosafety clearance procedures that are mandated for the research institutions and the study initiators, are discussed in detail. Such considerations gain heightened importance particularly when the characteristics of the pathogen being studied, or the methodology being utilized, could potentially be utilized for the intentional or unintentional offensive development of biological warfare agents [38].

To underscore these essential points, the text presents the development and specific laboratory protocols of an ongoing study in which genetic recombinants of the smallpox vaccine have been modified to include green fluorescence protein, which is under the control of a viral promoter. This innovative research illustrates the potential advancements but also highlights the need for stringent oversight. Unfortunately, it is quite common for research protocols to be submitted within the context of an effort established or overseen by the sponsor [39]. Such arrangements may compromise reviewer objectivity, leading them to confront the challenge of writing a negative review if they find significant issues. These issues are presented to guide reviewers and policymakers intended for reviewers or policymakers at academic institutions or funding agencies. More positively, this serves as a comprehensive guide for researchers who are in the process of preparing submissions for research projects, ensuring that they navigate the complexities of ethical and biosafety considerations effectively [40].

11.5 INFORMED CONSENT IN RESEARCH

Virus research has peculiar ethical and legal aspects because an outbreak has important social and security impacts. Nevertheless, there are universal ethical principles concerning research on human subjects and patient care [41]. Among other things, a patient's biological samples and personal data should be protected. Such protection is enacted in a relevant document in the Italian system, at least, which occurs in every public and private medical facility. Moreover, patients should be made aware of each medical procedure that will concern him or her, and potential side effects and consequences of those procedures should be explained as well. In general, consent must be freed from any form of psychological conditioning [42].

There are difficulties in obtaining an individual's informed consent in the context of an outbreak concerning a novel virus, given the associated uncertainties. Difficulties related to virus transmission mechanisms suggest that researchers should protect themselves from blood and body fluid contamination of infected patients [43]. Although an infected patient's condition can easily be confirmed through clinical judgment, body fluids may be contaminated by air and bed sheets during treatment. Two confirmed cases were noted who soiled the bed and floors of a hospital restroom; some medical readers are now very anxious about isolated confirmed cases. An outbreak was similarly noted, and an organ transplant case died last week [44, 45]. The patient was not recognized as a confirmed case and the patient's excretion and body fluids were not properly disposed of, so all three medical staff in close contact were infected. Consequentially, a strong quarantine system was implemented within a week, and vigilance must be maintained [46].

In response to such a significant outbreak, it is ethically acceptable to take primary care or research actions, even with some limitations, such as a risk of infection. However, something should be done to prevent infections, and informed consent should be possibly obtained even if rapidly and in simple terms. This is not merely a problem in obtaining informed consent from an individual donor able to understand the situation and decide, but also an ethical issue of creating an informed and fair contract between donors and researchers, or, in broader terms, between the community and the scientific community [47].

11.5.1 Importance of Informed Consent

Recently, the COVID-19 pandemic has raised great concerns about the nature of virus research and the ethical rules that should govern it. An analysis of the question is offered of what researchers who conduct virus experiments must consider as regards their background [48]. Although the positions taken are normative, an attempt is made to provide defensible justifications for them, based on a wide variety of normative considerations (e.g., consequences and risk-benefit analysis, observational uses of the viruses, the use of vaccine strains and the delimitation of research), and the strength of these considerations are explored [49].

First, research proposals must be evaluated not only in terms of their intrinsic merits, but also concerning who the sponsors are, who the investigators are, where the work is to be done, and for what purpose the research is conducted. Second, certain consequences are anticipated, though they cannot be fully described in this context, but that by any common parlance would be catastrophic [50].

Considering these facts and possible consequences, in real-world research there is a risk of certain research objectives or venues being abused. So a decision has been made to impose certain limitations, and possibly a complete stop, on several research activities involving specific biological agents. This approach to the regulation of biological research involves Step One, the issuance of an interim final rule, which imposes certain special requirements on research involving "SELECT AGENTS AND TOXINS." About 40 viruses are listed as SELECT AGENTS, including many of interest for potential use with bioterrorist intent. Third, all laboratory work must necessarily be done safely and securely [51, 52]. This is violated in so many recent scientific reports, many of them even published in respectable scientific journals, where methods that consist of the overstress of experiments being performed violate morbidity/surgery rules [53].

Recent data regarding the natural infection of humans with a potential pandemic virus is a clear example. Mobile lab experiments hacked on the way to a military research center on bioweapons trafficking were not enough, as these experiments tried to produce variants that could resist harmful environmental conditions, creating an artificial body to create the virus allowing airborne transmission. If a proper experiment were ever conducted, the researchers would see that the highly resistant output has been produced, and the most feared pandemic would be established [54].

11.5.2 Challenges in Obtaining Consent

The rapid spread of the novel coronavirus around the world and the particular difficulties involved in prevention, public health measures, treatment, vaccine development, and a solid understanding of the effectiveness of the proposed interventions have led some countries to adopt radical restrictions on the spatial and social mobility of their populations. There are already many experimental treatments for COVID-19 and several vaccine candidates, but it is difficult to say in advance when they will be proven to be effective and how much time will be needed until the production, distribution, and global vaccination of a sufficient number of doses [55].

A few years back the world was not aware of the existence of the currently ongoing pandemic, therefore, the capacity to plan and anticipate the research is inadequate. However, it should be useful to record and conserve large quantities of viral biological samples, extractions of viral RNA, and all the clinical data available to deepen the understanding of the disease, the infection mechanism of the virus, and the effectiveness of various interventions. At the same time, the protection of privacy and the correctness of the treatments must also be guaranteed to the single infected individuals. To this end, national laws can be drafted that regulate the procedures for obtaining the informed consent of the donors of the biological samples [56].

11.6 ANIMAL WELFARE IN VIRUS RESEARCH

Scientists and ethicists have long debated the ethics of animal use in virus research due to the number of controversial issues it raises. Some of those issues include the ethics of using animals in biomedical research, ethical conduct in using animals in scientific research, and the boundaries of ethical animal research. Also, as science progresses, so does our understanding of evolutionary developments in species. This leads to the idea that, as a virus is a living being, its likelihood of interest in survival and reproduction would also apply to non-human animals, raising questions about the appropriateness of exposing them to the deliberate risk of extinction by purposefully causing their deaths – through the course of virus experimentation, for example. Concerns in this respect include the treatment of HIV in primates and the possibility of its extermination as a consequence of extinction [57].

11.6.1 Ethical Treatment of Animals

Any research that uses animals for investigational purposes must have a thought-out plan of which animal models what methods will be used and how the animals will be treated in their housing environment and during the experimental period. Researchers have a considerable moral obligation to treat these animals ethically. Ethical issues surrounding the treatment of animals are legion. The Federal Animal Welfare Law dictates that "reasonable efforts" should be made to minimize the pain, anguish, and suffering of experimental animals [58]. Researchers working with laboratory animals often face difficult choices in terms of poor experimental design leading to useless animal death, or extensive inefficacious use of the experimental animal. They have a moral obligation to treat these experimental animals properly. An institutionalized body exists to ensure ethical issues are addressed appropriately. The professional delegates appointed by the institution to review or conduct an experiment on live animals in scientific research perform the Institutional Animal Care and Use Committee (IACUC) in the United States [59]. An experiment may not be performed until the following obligations are observed. Before any experimental research project that may include animals can be considered, a written project proposal must be made to the institution, including the name of the responsible researcher, who will act as principal investigator to the IACUC members for their review and recommendations. Journals and supporting institutions won't consider or fund a study using animal models without clear ethical considerations [60].

11.6.2 Alternatives to Animal Testing

The use of in vitro models and the use of advanced testing methodologies, such as toxicogenomics, are increasing in this field, and in practice, these methods are replacing a substantial number of animal tests. Those that are not replaced have the potential to be refined. The use of HSE models for skin irritation and corrosion testing has been mandated in the European Union, effectively eliminating the need for such testing to be conducted on animals [61]. For transdermal delivery, similar models have gained acceptance as replacements for the traditional rodent soaking tests. The use of such models in risk assessment opened up a new field for the development of more complex in vitro models. As a result, in vitro models for skin sensitization (LLNA) and local coetaneous immune-response testing have come to be preferred over the use of animals [62]. Similarly, cell-based systems for the prediction of skin penetration and systemic toxicity have commonly been used, and alternatives to animal biocompatibility testing of medical devices have been developed. It is important to note that one model cannot mimic all aspects of human biology. Combining different models and approaches is important. An assessment of animal models may also be important, particularly in light of the increasing number of examples of drugs effective in animals that have gone on to fail in clinical trials [63].

11.7 HUMAN SUBJECT RESEARCH

Observation of the interaction of salivary compounds from specific insect vectors with the human immune response is necessary to more accurately predict human responses. A tool of volunteers to assess human responses to the salivary compounds of vector species is investigated. Understanding the human immune response to the introduction of pathogen-infected saliva through arthropod bites may also give insight into limiting the effectiveness of saliva or deriving intervention strategies that block this response. The assessment of protection from an anamnestic immune response to pathogen-infected saliva from infected arthropods is useful to guide experimental design [64]. Federal regulations lead investigators conducting vector biology research involving human participants through a series of recommendations and regulatory requirements when designing and implementing such studies. A general outline of what is necessary to conduct research involving human participants who are intentionally exposed to the bites of arthropod disease vectors is presented to provide a tool for both investigators and Institutional Review Boards [65].

Observations suggest that protective levels of immunity against natural exposure to bites by a disease vector might be against newly induced pathogen-infective salivary proteins. The basic principles of responsible conduct of research and research ethics must be followed. Ignorance of standards does not absolve scientists from infractions [66]. The efficacy and safety of drugs or vaccines capable of interfering with pathogen transmission by arthropods must eventually be tested in humans. Mosquito release field experiments will investigate the history and age structure of an insect population. Conformity with the appropriate Federal Regulations is required, including submission of a disclosure proposal to the policy official, the Department of Public Health [67].

11.7.1 Ethical Guidelines

Until recently, research ethics and public health ethics have gone separate ways. Rephrasing, "Public health practice and research operate as though they are separate, with distinct ethical frameworks guiding each." The recent outbreak of severe acute respiratory syndrome has revealed the critical importance of this clinical work of public health practice for public health research [68]. This chapter focuses neither on healthcare ethics nor on the ethics of public health practice but on the ethics of epidemiological research. The research SOP described here was based on experiences and insights gained in designing studies on SARS, which was a "crash" research funded and carried out as part of the public health response to the first emergence of SARS. However, the experiences and the arguments given turned out to be somewhat more general. This unsolicited dissemination of summaries of relevant literature scheduled early every morning combines two existing clinical research traditions: that of rapid dissemination of results for those involved in public health practice, and that of taking over unsuspecting subjects for caffeine delivery through dissemination of summaries of scientific publications with decided relevance for those writing while having breakfast. Since most cases of development of new syndromes are "one shot" events, understanding what happened during the SARS outbreak has general significance [69, 70].

Nothing like SARS had happened since AIDS, and comparing the emergence of SARS with that of AIDS misses mid- and long-term policy responses. It is not a matter of epidemiology but rather concerns the societal response to the outbreak, particularly the research put in place. Limited by internal policy and communications constraints, it has become possible to publicize most of the conclusions. Six elements are identified as crucial components of effective ethical or unethical practices of ethics of epidemiological research [71, 72].

11.7.2 Case Studies

Challenge studies involving human volunteers can indeed present a formidable risk-benefit profile, where the calculations behind the potential advantages are carefully weighed against the inherent risks. Nevertheless, these carefully controlled studies can play a crucial and useful role in the medical and scientific community when the disease in question is of significant public health importance and when there exists a proven and effective method of protection against it.

Currently, a novel mosquito-feeding technique is being researched for the administration of pre-enucleated Plasmodium falciparum sporozoites, an area that holds promise for advancing our understanding of malaria transmission. However, further thorough research is critically needed to refine the consistency of various parameters that will ultimately produce the maximum number of viable sporozoites needed for effective study outcomes. It is essential to have an effective and uncharged prophylactic method available for the prevention of mosquito-transmitted malaria to support both national and international efforts in combating this potentially devastating disease that affects millions of lives globally [73].

In regions where malaria is endemic, it is relatively rare for resident adults to experience episodes of severe, life-threatening cerebral falciparum malaria or recrudescent vivax malaria, or any other forms of complicated malaria when they travel to areas where malaria is not endemic, contributing to the complexity of transmission dynamics. However, notable episodes of malaria have been recorded among residents of Columbia, Maryland, and adults who are native to the Pacific coastal regions of Peru in the relatively recent past, highlighting that malaria remains a threat even in areas where it is not commonly perceived as a risk [74]. This chapter aims to discuss the various clinical manifestations associated with malaria and appropriate treatment measures that should be deployed based on the specific context of each case. Moreover, malaria, as well as infections from rickettsial and immune-complex arthropathies, are known to occur as complications in the context of post-transfusion scenarios, which necessitates increased awareness and preventive strategies within the healthcare system. Additionally, a maiden regional center has been indicated for the screening of blood and tissue donors to address and mitigate these risks, particularly regarding these possible complications related to transfusions. Furthermore, unrecognized viral infections remain a significant potential source of such iatrogenic infections, emphasizing the need for continued vigilance, research, and clinical education in this area to prevent further outbreaks and protect public health [75].

11.8 DUAL-USE RESEARCH OF CONCERN (DURC)

Both the revived virus and the information gained from the recovered frozen 1918 flu virus could be used for both beneficial and harmful purposes [76]. By definition, strategies for making vaccines for viruses depend on research, and the information gained from the research can be used for malevolent purposes. Furthermore, papers giving the complete DNA sequence of both very contagious and fatal viruses have been published [77]. However, the papers are in principle benevolence; publishing information can enable countermeasures and be used to harm. Currently, the line for the culling of millions of chickens and containing humans is the dateline of the revival of a Spanish flu virus; as the authors noted, the revival and study of the 1918 virus could save lives [24, 78].

On one hand, the research provides benefits; on the other hand, there is the risk of misuse by rogue states, terrorist groups, and the like. Legislation making the communication of findings to would-be bioterrorists illegal has recently been introduced in the United Kingdom. An influential article made clear that despite its objections, the editorial board of the journal was considering publishing a paper on how to make a new infectious and highly airborne form of the H5N1 bird flu virus [79].

Lack of easy access to the source material and complete experimental details would prevent someone possessing only the knowledge of this chapter from successfully recreating the virus. However, the newsweekly also reports that at least 40 of the spending on national research has denied access to their source material to the National Science Advisory Board for Biosecurity, which is trying to assess the paper. A journal on bioterrorism stresses the importance of ethical behavior in the scientific community in response to the paper controversy. The most obvious benefits of research in the biological sciences include the protection of human life and health against diseases, the protection of existing food sources, and the protection of human populations against biological weapons. The potential burdens of such research include death and sickness caused by the use of biological agents as weapons by malevolent actors [80].

11.8.1 Definition and Examples

The 8th Milestone emphasizes important materials that have been drawn from a variety of sources, including proscribed viral and dual-use research that raises significant concerns, gain-of-function research funding moratoria, classified research operations, lab security protocols, research involving offensive biological weapons, and critical ethical considerations that pertain to scientific determination amid contests of opinion and prevailing debates among stakeholders [37]. In this context, the ethics surrounding scientific determination are examined as they specifically relate to ongoing public debates about the origins of viral pathogens. There have been widely reported responses to four crucial questions that arguably establish, in a determinate manner, that there has been mistaken, misleading, or vague information circulating concerning SARS-CoV-2 and its origins. Such a possibility, when considered thoughtfully, clearly raises vital ethical questions that are particularly pertinent for virologists who are coming together to draft and author follow-up reports concerning the origins of the pandemic [37, 81].

In these complex cases, nuanced and well-considered strategies become necessary and imperative that, whenever the available evidence is found to be contestable or questionable, aim to critically engage with opponents. This engagement should actively seek constructive dialogue, rather than dismissing opponents as being inherently dishonest or deranged. The building models and framework controls are specifically intended to spare the journal from becoming enmeshed in partisan disputes that may obscure and divert attention from deeper scientific issues that require careful exploration and resolution [82, 83].

11.8.2 Regulatory Framework

Open virus research can provide important information in the field of diagnostics, prevention, and therapy of viral infections. Scientists' freedom of research in this field should be guaranteed. In research facilities, the pathogens dealt with must correspond to the risk levels. The development of strategies to intensify the antiviral immune response remains a key focus of current research. These activities can take place in the examination of established infection.

If vector-borne disease agents are released, technical measures must be taken to minimize the risk of infection for those involved in the research. Ethical considerations pertain primarily to the nature of the research and the possible consequences for the participants involved. Several federal regulations and institutional policies provide norms and standards for the ethical conduct of research. Topics that are necessary for ethical reflection when using human participants in VBR include procedures to obtain and manage consent, communication of risk and benefit, equity, justice, and the care for participants living in countries with limited resources or those who are socially marginalized. For certain types of inquiry, human observation is necessary [84].

One of the safest ways through which a thorough knowledge of the phenomenon can be drawn is through an understanding of insurance conditions. A more general understanding of how properties are evaluated in the insurance industry has developed along two divergent frameworks. These approaches either characterize the essential properties of the thing that is being insured or focus on the evaluation of risk, itself. Prosecutor and defense wave outputs, attending to cases of disputes or uncertainty about the nature of "the previously existing condition(s)" of an insured property. Where no external data can determine the outcome of litigation, the court must either decide which scientific interpretation is most credible or assess the case on the basis of legal principles [85].

11.9 PUBLIC HEALTH IMPLICATIONS

Pandemic influenza experts anticipate that a future influenza strain will easily spread between humans, will have a suitable minimum speed of human-to-human transmission, and will cause significant morbidity and mortality in those infected [23].

Measures that tend to slow the spread of contagious disease include isolation, quarantine, and "social distancing," which is coerced or otherwise enforced by the government. The grounds for their use must be strongest when those being considered for confinement are among the worst-off groups of society. The ethics of both isolation and quarantine must provide special protection for vulnerable members of society. Ethical criteria also indicate that quarantine must be minimally burdensome; those subjected to isolation and quarantine must be provided with necessities, such as food and water, shelter, and basic medical care; and those who are coerced must at least receive financial compensation for any inconvenience suffered, for lost wages, and for having their liberty unreasonably restricted. Such measures are legal, but also because they would generate distrust in the government and render the measures less effective [86, 87].

The potential of making available research resources and infrastructure for public health activities in response to pandemic influenza may extend beyond ethical requirements, as researchers may be bound by federal regulations surrounding bioethics, data privacy, and informed consent [88]. Because the implementation of public health activities inherently carries risks related to human subject research, extra considerations are now relevant for review. In situations where the primary risk to research participants posed by public health activities is dignitary harm, researchers should consider whether violations of safeguards on the privacy rights of participants will occur before proceeding with these activities. Refuse to provide other agencies with participants' private information that was collected for the research without appropriate consent and certification of confidentiality [89]. Regardless of the research nature, human subject protection remains essential to the relationship between the Public Health Service and those affected by the research. It is the realization that the human subjects research policies of the federal government are reciprocally endorsed by most states that necessitate the annual certification of compliance. Further to the human subject's regulations, the policies and practices of the Office of Human Subjects Research are also significant the ethical history within the federal government has naturally been examined within the context of what events and scandals prompted the creation of today's federal regulations [90].

11.9.1 Ethics in Vaccine Development

In today's world, the development, licensure, and use of vaccines have become an issue of great importance with increased concern about the potential use of viruses as bioweapons. The world currently faces a significant threat from pandemic influenza and bioterrorism. In addition, many emerging and re-emerging infectious diseases are caused by pathogens against which no vaccines are currently available or for which a vaccine is urgently needed [91]. This particular area of research is also plagued with accusations of scientific "blindness" as PhDs and doctors envision the renown and financial profit to be had from dangerous experimentation on those who have no true concept of what they are even agreeing to. However, legal professionals involved in the process of any clinical trial must take heed of the fact that there are several sets of rules governing the actual clinical trials themselves. This part of the vaccine development process is reviewed, as well as the process of submitting an investigational drug application (IDA) [92]. It is important to bear in mind, however, that human rights remain the primary and paramount consideration in any clinical trial. For, if such rights are not upheld and observed, then it renders all subsequent proceedings null and void as all information procured under such dubious circumstances must be dismissed as evidence. The IDA is a submission to both the Center for Drug Research and Evaluation and the Center for Biologics Evaluation and Research at the FDA, containing extensive information on all aspects of the vaccine. Of concern to lawyers are the informed consent documents and the site contracts. It is the role of the Client Representative to ensure that all site contracts are current, signed, and on file with the legal department. The clinical trial stage of a vaccine for any medical condition is the first step in its testing on humans. Before beginning a clinical trial, it is legally necessary to submit an IDA to the CBER. This application is also forwarded to the appropriate Institutional Review Boards for the clinical sites and to the Western IRB in the case of sites without their IRBs. The IDA must include a detailed Investigational Plan, detailing information on the design and monitoring of the study, safety, and effectiveness, etc. A master document contains all of this information and is available to sponsors and legal representatives [93].

11.9.2 Risk Communication

Introduction to Risk Communication Based in Partners Academy: risk communication in the context of the 2020 COVID-19 pandemic stands out as particularly complicated and fraught. This was no less the case in the first days following the outbreak of alarming viral pneumonia in Hubei province, China. In January 2020, the physical detection, genetic sequencing, and naming of a new β-coronavirus SARS-CoV-2 all occurred with almost unprecedented speed, in many ways marking a new chapter in the scientific and medical response to outbreaks of emerging zoonotic viruses [94, 95].

Despite the rapid pace of discovery, very little was known about the virus outside of increasingly harrowing public reports from China. Risks to global health and economies were extrapolated from the statistics of the sick and dead; travel bans were effected and recommended on the sparest of evidence. The shape and impact of the unfolding pandemic were largely elucidated through indirect and after-the-fact research. Released on 28 November 2019, just five weeks after the first detection of the novel virus, the publication was immediately widespread [56]. Outcomes ranged from a vast and largely unfettered global research effort to rapid development and widespread emergency use of novel vaccines, to emergent vaccines and treatments operated by profit-driven entities or imposed by global geopolitical power. On the heels of these events, the scientific and political community faced urgent and contentious questions about how to describe, quantify, and mitigate the risks of developing and publishing novel, laboratory-created pathogens [96]. As collaboration partners of the research Universities – the University of Virginia and the Universidade Nacional de Timor Lorosa'e (UNTL) also contemplated these controversies, which were contextually situated amidst long-standing grievances and power imbalances that hinder non-Western scientists in political efforts to prepare for and mediate rapidly emerging risks. As is often the case in low-income settings, key mechanisms for preventing and detecting new diseases exhibited structural weaknesses, including compromised biosecurity, poorly maintained equipment, and underfunded staffing levels. Compounding these difficulties, disparate public health and economic impacts of aggressive containment efforts – such as village lockdowns and imposed discord (forced physical distancing and cessation of most public transport) [97]. In this environment, even the most recent of the Bombali β-coronavirus discovered in May 2018 seemed less an object of curiosity than an affront.

11.10 GLOBAL PERSPECTIVES ON VIRUS RESEARCH ETHICS

11.10.1 International Guidelines

1.1. Ethical considerations in the selection of the topic and its implementation, care must be taken to minimize the risks of contamination and prevent the transmission of hazardous particles. This should be done following the laboratory biosafety guidelines [84, 88].

1.2. Appropriate procedures and techniques should be used to minimize the risks in the collection, storage, dispatch, and packaging of samples. The risks are minimized by performing appropriate manipulations under strict local conditions and implementing appropriate sterilization procedures. Health workers must be informed of the risks and follow the local laboratory safety guidelines [38].

1.3. Health workers are advised to consult with national health authorities to determine which tests require specific precautions. To minimize occupational risks, individuals exposed to hazardous particles must provide data on their vaccination status in the event of their illness and receive emergency care [67, 68, 88].

Ethical considerations also apply to the exchange of information. All parties must agree to exchange information in writing. Requests for information can only be addressed to the Ministry of Health, which has the power to decide whether or not to comply with them [20].

Aside from that, controversy arose over the publication of studies on the potential human-to-human transmission of the highly pathogenic avian influenza H5N1 virus, as well as on the attenuation of the same virus using genetic engineering. A troubled and profound international debate was triggered about the benefits and risks of permanently continuing virological research on amateur pathogens to reduce or possibly prevent their transmission. In this context, some significant steps have been undertaken at the national, international, and institutional levels [88]. Although there are reasonable chances of direct intentional use of genetically modified avian influenza viruses, the accident that occurred in China in August 2007 nevertheless highlights the importance of biosafety and biosecurity regulations in amateur virus research in the Member States. However, between the Member States there are imbalances in the application of these regulations, so the risk of leakage of potentially dangerous data and materials from the territory of one country to another has increased. It is considered expedient for the European Union to develop common criteria to guarantee adequate biosafety and biosecurity of individuals, laboratories, and the population of the Member States [86, 87]. To understand the global landscape of virus research governance, Table 11.3 summarizes leading ethical guidelines and the institutions tasked with ensuring compliance and transparency across jurisdictions.

11.10.2 Cultural Considerations

Scientific communities tend to be focused on their work to the exclusion of broader cultural considerations, but research into viruses and disease vulnerability cannot be pursued independently of specific social, economic, and political contexts where infectious diseases emerge and spread [10]. In particular, the uphill task of meeting expectations about private declarations of interest in the context of comparative poverty and a "gift relationship" economy

TABLE 11.3
International Ethical Standards and Oversight Frameworks for Virus Research

Organization/Framework	Scope	Focus Areas	Enforceability	Notable Provisions
Declaration of Helsinki	Global	Human subject research	Voluntary adherence	Consent, risk-benefit ratio
CIOMS Guidelines	WHO-affiliated	Developing countries' research	Guiding principles	Justice, community engagement
US NSABB	USA	Dual-use research	Strong advisory role	DURC, synthetic biology
EU Horizon Ethics Panel	EU	Biosecurity, data ethics	Mandatory for grants	Biorisk, consent, data transfer
Nagoya Protocol	UN CBD	Genetic resources and access	Treaty-level	Benefit-sharing obligations

This table maps key global ethics frameworks and regulatory bodies involved in virus research oversight, focusing on informed consent, dual-use research management, and transnational collaboration.

adds significantly to such considerations. In light of the destruction wrought by COVID-19, virologists, epidemiologists, and ecologists have the delicate task of defending grants based on disease emergence and vulnerability, in return for co-authorship, while also insisting on the priority of other interests.

Four authors got themselves into that very problem. For six months, they expressed interest in cooperating with that very problematic. Though that was perhaps the only serious expression of interest, it was made clear that acceptance would be subject to a more detailed discussion. That strand of the debate was abruptly halted by the acceptance of another research proposal – focusing on viral genetic analysis only. The other authors then hurriedly submitted a similar proposal, based on another strand of work. The experimental part of that strand was deemed to be the least original of the three proposals, but it came with a budget that was ten times greater than the pilot project. Extensive generous opportunities for joint publications and co-supervision were also generously offered [98].

Given the parasitic load in the first pilot region, there is plenty of scope for emerging disease. However, molecular biological techniques are neither used nor intended to be used, as a diagnostic tool. Instead, parasitological methods have been developed and are employed to monitor the success or failure of novel integrated approaches to combat "neglected" vector-borne diseases. To provide recognition of the innovative elements, and as a tardive response to queries, other pilot districts in Indonesia and Vietnam have undertaken to profile the environmentally and socio-economically determined susceptibility to tick-borne human infection. In light of the three Reviewer reports, it was thought best to adopt a broader analytical perspective in these times of coronavirus and anticipate the potential problems that might be generated by giving a political economy interpretation of the all-natural spread of the SARS-CoV-2 [29, 55, 94].

11.11 ETHICS OF EMERGING TECHNOLOGIES

On 30 November 2011, the independent US National Science Advisory Board for Biosecurity (NSABB) reviewed revised manuscripts submitted, for the first time in the history of the NSABB since its inception in 2004. In the following days, the US NSABB recommended the US Government to facilitate the publication of a redacted version of both manuscripts. In the meantime, the Dutch Government declared that researchers should have applied for an export permit before submission to the US journal, in the light of European Union directives on dual-use goods. These goods have legitimate commercial applications and can simultaneously be misused. In the rapidly evolving field of virology, with emerging viral threats and high mortality figures, the rapid microbiological diagnosis of the pathogen is crucial. An H5N1 avian influenza (bird flu) virus under certain conditions can acquire the capacity to be transmitted from mammals to humans. In the light of previous pandemics, enlightenment is crucial. Following recommendations of ad hoc review committees, various anthrax communication studies were conducted, improving knowledge on the mitigation of potential bioterrorist attacks. Since 15 years, worldwide an estimated number of two billion people died as a consequence of three influenza pandemics. In these processes, the actual spread of viral genetic information becomes an issue of growing concern [20]. Eighty years after the fact, the Spanish flu case still is an object of research efforts. On the one hand, its re-appearance is still an option, bearing in mind similar avian flu cases. On the other hand, remaining fully prepared for eventualities in these respects increases knowledge about the functioning and spread of a hitherto inscrutable disease. This communicational spread factor is novel to the rapidly evolving state-of-the-art technologies in virology [10].

11.11.1 Gene Editing

In the popular imagination, gene editing has the potential to cure genetic diseases (and much else besides), much as antibiotics can cure bacterial infections. For this to occur, germline gene-editing agents – tools capable of modifying the genome of an embryo that will develop into a person – would need to become routine, low-cost, and demonstrably safe. Among the main obstacles are the current ineffectiveness and undesirable side effects attendant to today's gene-editing technologies. It's worth, however, speaking more generally about two kinds of potential obstacles to gene editing. The first are technical issues, concerning safety, efficacy, or some other functional aspect of the editing process. These might, for instance, be about the ways by which a gene might be edited, or concerns about off-target effects, so that unintended mutations might be introduced as a side effect of an intended mutation. The second kind is non-technical issues, such as backlash, regulation, or other broader social and economic challenges that might face an emerging technology. These are important also, but here we will focus on the narrow technical concerns.

In 2018, a committee published a report reviewing scientific, ethical, and legal concerns about new genomic engineering technology. The report's key recommendations included that "clinical trials and potential use of heritable germline editing should be approached with caution," with the view to an initial focus on "severe monogenic diseases," and on "targeting early embryos." In addition, creating a heritable edition would necessarily require learning how to do precise and thetic edits, like base editing or prime editing, or about how to do unbiased edits. It's not clear yet, however, how to say in advance which edits would be desired [99].

11.11.2 Synthetic Biology

Regarding possible biosecurity issues, the application of the modular construction approach should be thoroughly examined in the context of Dual-Use Research of Concern. Despite modularization and biologists' lack of readiness for such research, targeted optimization is in principle feasible and allows a "bottom-up" approach to viral replication cycles, which could be legitimized by the framework for synthetic virus research approaches. It is imaginable that in the future, safe and self-limited replication of any double-stranded DNA or RNA virus can be realized from scratch [100]. Although so far only a few modular systems for synthetic biology are described and none for viral applications, viral replication inherently relies on a limited number of biomolecular processes and it simplifies the targeted modular approach [100]. Moreover, fundamentally, cell cultures and possibly model organisms are not necessary for research on the replication and the DNA recombinant systems of any virus. In conventional research on natural viruses, risks are connected to specific infectious agents, which are classified in biosecurity levels and focus more on predictable threats, such as the application of weapon agents, engineered bioweapon agents, or especially dangerous pathogens. On the contrary, the bottom-up approach relies on recently emerged technological knowledge involving serious potential risks, some of which are known but most of which are yet unknown to the involved biological community [101]. Additionally, emerging technologies in the field of virus research are only partially associated with catastrophic concerns and unconventional threats. In this light, risk assessment concerning the proposed research concepts on synthetic viral replication cycles should be specified. However, this is difficult due to the significant uncertainties about how the goals of the project will be achieved. A favorable opinion should be obtained on this research concept and it is also asked that a monitoring mechanism be established, which will evaluate biosafety and biosecurity issues of synthetic viruses while in progress [102].

11.12 CASE STUDIES OF ETHICAL DILEMMAS

Recently developed viruses have now become a significant economic, health, and security concern. Governments, research centers, and the virus recovery industry have made significant investments in establishing the biotechnologies required for the diagnosis and development of responses to these viruses. In recent years, within the field of science and safety, research and biotechnological-entrepreneurial centers focusing on the health sector have increasingly entered this system. This chapter examines ethical considerations in the development and application of these technologies and the relevant decisions made at the Guadiana launch center of the Fundación Ciencia y Salud, a non-profit health sector foundation in Buenos Aires. These viruses have certain characteristics of contagion and virulence, with significant consequences for individuals and various health systems around the world [10].

The transmission pathway for this specific new virus, as well as the development of a vaccine or the introduction of measures to avoid contagion, remains mostly unknown. The governments of the countries where it has arrived, including Peru, where the research to investigate and treat the virus has most developed at this point, responded by making significant investments to establish the biotechnologies necessary for diagnosis and construction. The opportunity that other governments, industries, and research centers, national or international, would be developing some other biotechnology to generate responses to the virus also began to be thought of in entrepreneurial sectors in the field of biomedicine. Since the coronavirus outbreak, a few health centers and private initiatives have begun to raise the possibility of conducting certain research lines [23].

11.12.1 COVID-19 Research

A proposal was therefore made for a "virtual challenge trial," in which volunteers would be infected with a specific SARS-CoV-2 strain under controlled, quarantined conditions. But the challenges this would involve are great, and any actual such trial would need to be carefully implemented: only a subset of putative candidates already producing information of potential benefit to others could be accepted [103]. Were a definitive efficacious therapy for COVID-19 to be established, sponsored, accredited, postdose unmasked, and widely available within the timeframe of a controlled challenge, then interaction with placebo and control groups while taking part in such seems to be ethically permissible. Whenever a group is infected, standard quarantine and infection management stratagems should apply; as such, it would not be possible to rely on unproven assumptions so that risks might thereby be safely courted. Beyond participant safety, selection and challenge-loss risks would also need to be very carefully and robustly managed. Given the SARS response back in 2003 and the strategic focus on a narrow set of potential pathogens that generated, there might well be criminatory vulnerabilities that are therefore overlooked as things stand [29, 55].

Welfare considerations could mean that a developed nation would be the only conceivable sponsor or host of a whole-virus challenge, should one ever be attempted. Further research was suggested into whether preparations might be thwarted or neutralized by patient factors or LIOB. It was important to emphasize that current infection with SARS-CoV-2 or any suspected exposure to an active case were declared trial exclusions [103]. Lastly, helpers were sought to track the transmission of a human coronavirus randomized controlled trial in the community. Finally, as the world struggles to make headway against the virus responsible, some are considering the controversial proposal of a controversial potential shortcut: human challenge experimentation with SARS-CoV-2 [103]. The United States has earmarked half a billion dollars to fund research into such. An advocacy group is not only coordinating volunteers but also pushing for the earliest possible implementation. Provide participants with specific debriefing or other relevant information after participation in a COVID-19 human infection study. Whole-virus live human challenge studies involve deliberately infecting healthy individuals with a pathogen to study host-pathogen interactions and vaccine formulation [29].

11.12.2 Ebola Virus Studies

Toward the final edition of WHO's "Ethical Considerations in Developing a Public Health Response to Pandemic Influenza," the identification of public engagement in all phases of research was brought out as a priority task. Emerging infectious diseases and risks released by them necessitated stakeholder consultation in advance of the event. Since the publication, several severe viral infections have caused global concern due to the failing control in the infected or high-risk areas and the threat of further spreading to countries disconnected from those affected. The recent outbreak of Ebola virus disease in West Africa has caused over 10,000 deaths [104].

The largest outbreak of filovirus infection since the discovery in 1976, it was reasoned, posed overwhelming challenges in the medical responses and research-intensive undertakings to find the curable efficacy of antiviral drugs in the fatal and highly infectious disease [105]. Responses by the international community, especially by the WHO, favored the experiments, known as ineffective methods based on experiences from the previous outbreaks of clinically manageable viruses. The evidence base for conducting public engagement in the research on experimental antiviral drugs with the possibility of adverse outcomes remains insufficient. That was because current studies on the clinical effects of the drugs are pursued only in hospitals located in resource-enriched countries. Early detection of the infected could effectively reduce the fatality of the virus-ridden EVD, according to speculative theories and observational findings. Such a common sense of the purported rationale called for establishing health monitoring and medication transferring systems in community settings where the poor or uneducated population in insufficiently treated countries becomes the major EVD victims. Confronted with widespread suspicion and dismissal by the expert community, experimental research for hypothetical novel treatment options was only a word of hope from the advanced scientists who never played on real scenes in ill-resourced settings [16]. The success unquestionably depended on public engagement, including the identification of good and important options and the trust and support needed to conduct the risky and costly experiments.

11.13 FUTURE DIRECTIONS IN VIRUS RESEARCH ETHICS

The COVID-19 pandemic should be examined within the broader context of ongoing research efforts. The novel COVID-19 has, at the time of writing, led to the confirmed infection of over 370,000 individuals and almost 6000 known fatalities across the Dutch borders. It is obvious that this virus and its multifarious implications invite a panoply of policy-relevant research interests of a multifarious nature; from the socio-psychological dynamics fueling the "online infodemic" (and its regulatory challenges), the epidemiological underpinnings of the rate of contagion (and the efficacy of policy measures poised to curb it), to and questions about the far-reaching economic ramifications [10]. Equally affected by the ongoing pandemic are the ways and spaces in which research can be performed and presented. Universities are relocating entire educational programs online, conference venues worldwide stand deserted, and laboratory experiments are put on hold. Consequently, the research infrastructure is put under significant stress. The outbreak has reinvigorated a debate on more fundamental issues, such as the range and depth of the public-private research collaboration in the Netherlands, the underlying dynamics driving the postdoc-as-"hydraulic" workforce, or the ethics of sharing viral samples (or not) with the labs from Wuhan. There are also concerns about the ways the academic community is affected by the current crisis, ranging from the potential long-term repercussions of disruptions of the research projects to the asymmetric capacity to conduct high-performance "lab-less" research projects (and the ways to cope with equity issues arising from it), or the horizons of the academic job market in a contracting economic landscape as the global recession looms on the horizon [106].

11.13.1 Evolving Ethical Standards

According to a recently published critical stance on the controversy, viro-scientist Ron Fouchier and a colleague behaved like the foolish alchemist who is hoisted by his own petard, being the inspiration of the character, Thomas Norland. Norland concludes from decreased mortality fertility rates that the global population will equalize vertically and that this has to be prevented to continue with the process of conspiracy not worth contemplating [10]. That notoriously results in enucleation bringing Norland to his demise. Such a destiny would take years of construction in confined environments and access to secret, specialized knowledge that even current R&D capabilities can hardly offer. So it is with an undisclosed array of arrangements and infrastructures that containment of the Courtier enunciation could be envisaged. Exemplified by fictional elaborations about the crossed re-shuffling of H5N1 strains, the just CU Stirling, published in the aftermath of the affair, was enough to trigger a Pandora's Box Art. 1 of that FL Booster Convention requires that parties must implement national regulations to control the export of dual-use goods and technologies, excepting those intended for organizations with adequate biosafety standards [107].

11.13.2 Role of Ethics Committees

The role of ethics committees is to ensure that research is conducted ethically. This includes ensuring that all research is conducted in compliance with national and international regulations and guidelines. For the safety of everyone involved in the trials, the devices or drugs that are being tested proceed through rigorous and lengthy research stages before being approved by the authorities. The scientific research must have passed through different levels of tests. There is no way to ensure the safety of a one-year vaccine for a pregnant woman. Many persons who participated in trials also tested positive for COVID-19, hence the herd immunity and the legal recourse in the event of serious adverse reactions [108].

11.14 CONCLUSION

As the world watches the spread of COVID-19, the importance of the rapidly evolving scientific field to have ethical guidelines comes into stark relief. Descriptive ethics describes the present principles by which people think rather than what principles they should think. Because ethics develops as knowledge accumulation, research ethics specifically could be descriptive in not only reporting present ethical guidelines but also historical, economic, and trade negotiations that shaped them [108].

It looks at the demanding research of the virologists, the ethical trade barrier raised by the secret session, and challenges to the scientific discussion of the Consell'arte e Mestiere of Genoa. Research in virology has a short time and is eminently concentrated on the characterization of new viruses. Since the typical research time was approximately six weeks, much necessary analysis is left unexplored in the urgent quest for the virus nature. Some funding institutions or peer reviewers judge research scholars or their research too harshly, award or recommend low competitive scores, and describe the research as incompetent, not feasible or innovative, unimportant, and other later reasons.

But there is still evidence of the desiderata, joined with a substantial increase in popular demand, emerge a strong ethical concern among researchers themselves and a bio-political agenda by supranational bodies. National delegations raised an ethical trade concern about the European Community's proposal to ease the exchange of accidental pathogens to protect public health. Because of ethical reasons this possibility emerges in an informal discussion, is never mentioned to policy renovators or presented in the final document and is not shared by other responsible delegates. The issue of the trade implications of the regulation of biohazards and biotechnologies is publicly raised by an official of the pharmaceutical industry during a closed session.

The discussion on ethics had been first raised by the industry lobby less than one year earlier and at the beginning focused on law and safety. He accused the European Community obligates restrictive measures on easily exporting viruses, and the United States that only visitors may know by telechirurgical means what is done in the laboratories. In the same session, a European Community official revealed the planned organization of secret consultations among Western delegations to harmonize their medical and scientific strategies. This elicited another bioethical objection: "Given the secret nature of this type of meeting, any strategy the European Community decides in this area will be effectively deprived of scientific discussion in the public domain of democratic societies."

REFERENCES

1. Fraumann, G.; Colavizza, G. The role of blogs and news sites in science communication during the COVID-19 pandemic. *Frontiers in Research Metrics and Analytics* 2022, *7*. https://doi.org/10.3389/frma.2022.824538.
2. López-Baucells, A.; Rocha, R.; Fernández-Llamazares, Á. When bats go viral: Negative framings in virological research imperil bat conservation. *Mammal Review* 2017, *48*, 62–66. https://doi.org/10.1111/mam.12110.
3. Empathic patterns in complex discourse. *Journal of Organizational Psychology* 2022, *22*. https://doi.org/10.33423/jop.v22i1.5109.
4. Skitka, L.J.; Hanson, B.E.; Morgan, G.S.; Wisneski, D.C. The psychology of moral conviction. *Annual Review of Psychology* 2020, *72*, 347–366. https://doi.org/10.1146/annurev-psych-063020-030612.
5. Landmann, H.; Heß, U. Testing moral foundation theory: Are specific moral emotions elicited by specific moral transgressions? *Journal of Moral Education* 2017, *47*, 34–47. https://doi.org/10.1080/03057240.2017.1350569.
6. Howlett, M. Looking at the 'field' through a Zoom lens: Methodological reflections on conducting online research during a global pandemic. *Qualitative Research* 2021, *22*, 387–402. https://doi.org/10.1177/1468794120985691.
7. Roy, R.; Uekusa, S. Collaborative autoethnography: "self-reflection" as a timely alternative research approach during the global pandemic. *Qualitative Research Journal* 2020, *20*, 383–392. https://doi.org/10.1108/qrj-06-2020-0054.
8. Rodríguez, J.K.; Ridgway, M. Intersectional reflexivity: Fieldwork experiences of ethnic minority women researchers. *Gender Work and Organization* 2023, *30*, 1273–1295. https://doi.org/10.1111/gwao.12977.
9. Warrick, D.D. Revisiting resistance to change and how to manage it: What has been learned and what organizations need to do. *Business Horizons* 2023, *66*, 433–441. https://doi.org/10.1016/j.bushor.2022.09.001.
10. Zwart, H. Emerging viral threats and the simultaneity of the non-simultaneous: Zooming out in times of Corona. *Medicine, Health Care and Philosophy* 2020, *23*, 589–602.
11. Falkow, S. The lessons of Asilomar and the H5N1 "affair." *mBio* 2012, *3*. https://doi.org/10.1128/mBio.00354-12.
12. Casadevall, A.; Shenk, T. The H5N1 moratorium controversy and debate. *mBio* 2012, *3*. https://doi.org/10.1128/mBio.00379-12.
13. Shibata, T. Building parallel development for technological transition. *Organizational Science* 2012, *46*, 53–63. https://doi.org/10.11207/soshikikagaku.46.2_53.
14. Sun, T.; Song, J.; Wang, M.; Zhao, C.; Zhang, W. Challenges and recent progress in the governance of biosecurity risks in the era of synthetic biology. *Journal of Biosafety and Biosecurity* 2022, *4*, 59–67. https://doi.org/10.1016/j.jobb.2022.02.002.
15. Jeste, D.V.; Lee, E.; Cassidy, C.; Caspari, R.; Gagneux, P.; Glorioso, D.; Miller, B.L.; Semendeferi, K.; Vogler, C.; Nusbaum, H.C. et al. The new science of practical wisdom. *Perspectives in Biology and Medicine* 2019, *62*, 216–236. https://doi.org/10.1353/pbm.2019.0011.
16. Calain, P. The Ebola clinical trials: A precedent for research ethics in disasters. *Journal of Medical Ethics* 2018, *44*, 3–8.
17. Petkov, C.I.; Flecknell, P.; Murphy, K.; Basso, M.A.; Mitchell, A.S.; Hartig, R.; Thompson-Iritani, S. Unified ethical principles and an animal research 'Helsinki' declaration as foundations for international collaboration. *Current Research in Neurobiology* 2022, *3*, 100060.
18. Crown Office and Procurator Fiscal Service. Prosecution Code. 2001. https://www.copfs.gov.uk/publications/prosecution-code/html/ (accessed on 14/05/2025).
19. Tombs, J. Prosecution-in the public interest? *Proceedings Prosecutorial* 1984, 11.
20. Keuleyan, E. Liberty to decide on dual use biomedical research: An acknowledged necessity. *Science and Engineering Ethics* 2010, *16*, 43–58.
21. Fleming, S.B.; Wise, L.M.; Mercer, A.A. Molecular genetic analysis of ORF virus: A poxvirus that has adapted to skin. *Viruses* 2015, *7*, 1505–1539. https://doi.org/10.3390/v7031505.

22. Coradduzza, E.; Sanna, D.; Scarpa, F.; Azzena, I.; Fiori, M.S.; Scivoli, R.; Rocchigiani, A.M.; Bechere, R.; Dettori, M.A.; Pintus, D. et al. A deeper insight into evolutionary patterns and phylogenetic history of ORF virus through the whole genome sequencing of the first Italian strains. *Viruses* 2022, *14*. https://doi.org/10.3390/v14071473.
23. Selgelid, M.J. Pandethics. *Public Health* 2009, *123*, 255–259.
24. Varkey, B. Principles of clinical ethics and their application to practice. *Medical Principles and Practice* 2021, *30*, 17–28. https://doi.org/10.1159/000509119.
25. Alexander, L.; Moore, M. Deontological ethics. In Stanford Encyclopedia of Philosophy, Zalta, E.N., Ed.; Stanford University: 2007; Fall 2007 Edition. Available from: https://plato.stanford.edu/entries/ethics-deontological.
26. Mandal, J.; Ponnambath, D.K.; Parija, S.C. Utilitarian and deontological ethics in medicine. *Tropical Parasitology* 2016, *6*, 5–7. https://doi.org/10.4103/2229-5070.175024.
27. Marcucci, L.; Fourel, D. Understanding the modern relevance of the Hippocratic Oath. *Ethics, Medicine and Public Health* 2024, *32*, 101005. https://doi.org/10.1016/j.jemep.2024.101005.
28. Kopel, J. The Hippocratic Oath across the interfaith spectrum. *Proceedings (Baylor University Medical Center)* 2022, *35*, 266–269. https://doi.org/10.1080/08998280.2021.1993113.
29. Angius, F.; Puxeddu, S.; Zaimi, S.; Canton, S.; Nematollahzadeh, S.; Pibiri, A.; Delogu, I.; Alvisi, G.; Moi, M.L.; Manzin, A. SARS-CoV-2 evolution: Implications for diagnosis, treatment, vaccine effectiveness and development. *Vaccines (Basel)* 2024, *13*. https://doi.org/10.3390/vaccines13010017.
30. Henzler, D.; Schmidt, S.; Koçar, A.; Herdegen, S.; Lindinger, G.L.; Maris, M.T.; Bak, M.A.R.; Willems, D.L.; Tan, H.L.; Lauerer, M. et al. Healthcare professionals' perspectives on artificial intelligence in patient care: A systematic review of hindering and facilitating factors on different levels. *BMC Health Services Research* 2025, *25*, 633. https://doi.org/10.1186/s12913-025-12664-2.
31. Lipkin, W.I. Biocontainment in gain-of-function infectious disease research. *mBio* 2012, *3*, 10–1128.
32. Bakanidze, L.; Imnadze, P.; Perkins, D. Biosafety and biosecurity as essential pillars of international health security and cross-cutting elements of biological nonproliferation. *BMC Public Health* 2010, *10*, 1–8.
33. Ta, L.; Gosa, L.; Nathanson, D.A. Biosafety and biohazards: Understanding biosafety levels and meeting safety requirements of a biobank. *Methods in Molecular Biology* 2019, *1897*, 213–225. https://doi.org/10.1007/978-1-4939-8935-5_19.
34. Burnett, L.C.; Lunn, G.; Coico, R. Biosafety: Guidelines for working with pathogenic and infectious microorganisms. *Current Protocols in Microbiology* 2009, *Chapter 1*, Unit 1A.1. https://doi.org/10.1002/9780471729259.mc01a01s13.
35. de la Cal, M.A.; Cerdà, E.; Abella, A.; Garcia-Hierro, P. Classification of microorganisms according to their pathogenicity. In *Infection Control in the Intensive Care Unit*, van Saene, H.K.F., Silvestri, L., de la Cal, M.A., Gullo, A., Eds.; Springer Milan: 2012; pp. 29–40.
36. Nambisan, P. Laboratory biosafety and good laboratory practices. In *An Introduction to Ethical, Safety and Intellectual Property Rights Issues in Biotechnology*, Nambisan, P., Ed.; Elsevier, Academic Press: 2017, pp. 253–273.
37. Ruppach, H. Viral safety for biotherapeutics and biosimilar. *Drug Discovery Today: Technologies* 2020, *37*, 23–29. https://doi.org/10.1016/j.ddtec.2020.08.001.
38. Kaushik, N.; Mitra, S.; Baek, E.J.; Nguyen, L.N.; Bhartiya, P.; Kim, J.H.; Choi, E.H.; Kaushik, N.K. The inactivation and destruction of viruses by reactive oxygen species generated through physical and cold atmospheric plasma techniques: Current status and perspectives. *Journal of Advanced Research* 2023, *43*, 59–71. https://doi.org/10.1016/j.jare.2022.03.002.
39. Rashid, A.B.; Kausik, M.D.A.K. AI revolutionizing industries worldwide: A comprehensive overview of its diverse applications. *Hybrid Advances* 2024, *7*, 100277. https://doi.org/10.1016/j.hybadv.2024.100277.
40. Earl, J. Innovative practice, clinical research, and the ethical advancement of medicine. *The American Journal of Bioethics* 2019, *19*, 7–18. https://doi.org/10.1080/15265161.2019.1602175.
41. Voit, K.; Skuban-Eiseler, T.; Orzechowski, M.; Steger, F. Informed consent in COVID-19-research: An ethical analysis of clinical studies performed during the pandemic. *Healthcare (Basel)* 2023, *11*(12), 1793. https://doi.org/10.3390/healthcare11121793.
42. Al Habsi, S.S. Ethical considerations in obtaining informed consent in research participation. *International Journal of Educational Contemporary Explorations* 2024, *1*(1), 22–32. https://doi.org/10.69481/DIVL1152
43. Jepson, R.G.; Robertson, R. Difficulties in giving fully informed consent. *BMJ* 2003, *326*, 1039. https://doi.org/10.1136/bmj.326.7397.1039/a.

44. Chalmers, C.; Gourlay, H. Infection prevention and control. *Essentials of Nursing Adults* 2018, 231.
45. Coli, E. Infection control. 1979.
46. Centers for Disease Control and Prevention. *Guideline for Isolation Precautions: Preventing Transmission of Infectious Agents in Healthcare Settings.* Updated November 22, 2023. Available from: https://www.cdc.gov/infection-control/hcp/isolation-precautions/precautions.html
47. World Health Organization. *Guidance for Managing Ethical Issues in Infectious Disease Outbreaks.* World Health Organization: 2016. ISBN: 978-92-4-154983-7; 68 pp. Available from: https://apps.who.int/iris/handle/10665/250580
48. Xafis, V.; Schaefer, G.O.; Labude, M.K.; Zhu, Y.; Hsu, L.Y. The perfect moral storm: Diverse ethical considerations in the COVID-19 pandemic. *Asian Bioethics Review* 2020, *12*, 65–83. https://doi.org/10.1007/s41649-020-00125-3.
49. Raus, K.; Mortier, E.; Eeckloo, K. Ethical issues in managing the COVID-19 pandemic. *Bioethics* 2021, *35*, 581–588. https://doi.org/10.1111/bioe.12877.
50. Gross, R.; Karyadi, D.; Sastroamidjojo, S.; Schultink, W. Guidelines for the development of research proposals following a structured, holistic approach for a research proposal (SHARP). *Food and Nutrition Bulletin* 1998, *19*, 268–282.
51. Ware, J.J.; Munafò, M.R. Significance chasing in research practice: Causes, consequences and possible solutions. *Addiction* 2015, *110*, 4–8. https://doi.org/10.1111/add.12673.
52. Galvan, M.C.; Pyrczak, F. *Writing Empirical Research Reports: A Basic Guide for Students of the Social and Behavioral Sciences*; Routledge: 2023.
53. Fedeli, P.; Scendoni, R.; Cingolani, M.; Corrales Compagnucci, M.; Cirocchi, R.; Cannovo, N. Informed consent and protection of personal data in genetic research on COVID-19. *Healthcare (Basel)* 2022, *10*, 349.
54. Zwerwer, L.R.; Peto, T.E.A.; Pouwels, K.B.; Walker, A.S.; Eyre, D.W.; Stroesser, N.; Matthews, P.C.; Wei, J.; Diamond, I.; Studley, R. et al. Identification of undetected SARS-CoV-2 infections by clustering of nucleocapsid antibody trajectories. *Nature Communications* 2025, *16*, 4466. https://doi.org/10.1038/s41467-025-57370-z.
55. Alizon, S.; Sofonea, M.T. SARS-CoV-2 epidemiology, kinetics, and evolution: A narrative review. *Virulence* 2025, *16*, 2480633. https://doi.org/10.1080/21505594.2025.2480633.
56. Singh, S.; McNab, C.; Olson, R.M.; Bristol, N.; Nolan, C.; Bergstrøm, E.; Bartos, M.; Mabuchi, S.; Panjabi, R.; Karan, A. et al. How an outbreak became a pandemic: A chronological analysis of crucial junctures and international obligations in the early months of the COVID-19 pandemic. *The Lancet* 2021, *398*, 2109–2124. https://doi.org/10.1016/S0140-6736(21)01897-3.
57. Ruiz, S.I.; Zumbrun, E.E.; Nalca, A. Animal models of human viral diseases. In *Animal Models for the Study of Human Disease*; Elsevier: 2017; pp. 853–901.
58. Kiani, A.K.; Pheby, D.; Henehan, G.; Brown, R.; Sieving, P.; Sykora, P.; Marks, R.; Falsini, B.; Capodicasa, N.; Miertus, S. Ethical considerations regarding animal experimentation. *Journal of Preventive Medicine and Hygiene* 2022, *63*, E255.
59. Naderi, M.M.; Sarvari, A.; Milanifar, A.; Boroujeni, S.B.; Akhondi, M.M. Regulations and ethical considerations in animal experiments: International laws and Islamic perspectives. *Avicenna Journal of Medical Biotechnology* 2012, *4*, 114.
60. Kiani, A.K.; Pheby, D.; Henehan, G.; Brown, R.; Sieving, P.; Sykora, P.; Marks, R.; Falsini, B.; Capodicasa, N.; Miertus, S. et al. Ethical considerations regarding animal experimentation. *Journal of Preventive Medicine and Hygiene* 2022, *63*, E255–E266. https://doi.org/10.15167/2421-4248/jpmh2022.63.2S3.2768.
61. Cheluvappa, R.; Scowen, P.; Eri, R. Ethics of animal research in human disease remediation, its institutional teaching; and alternatives to animal experimentation. *Pharmacology Research & Perspectives* 2017, *5*, e00332.
62. Alkilani, A.Z.; Nasereddin, J.; Hamed, R.; Nimrawi, S.; Hussein, G.; Abo-Zour, H.; Donnelly, R.F. Beneath the skin: A review of current trends and future prospects of transdermal drug delivery systems. *Pharmaceutics* 2022, *14*. https://doi.org/10.3390/pharmaceutics14061152.
63. Suthar, D.; Raut, R.; Bajaj, A. Advances in skin-mimetic platforms: A comprehensive review of drug permeation models. *Journal of Drug Delivery Science and Technology* 2024, *98*, 105887. https://doi.org/10.1016/j.jddst.2024.105887.
64. Achee, N.L.; Youngblood, L.; Bangs, M.J.; Lavery, J.V.; James, S. Considerations for the use of human participants in vector biology research: A tool for investigators and regulators. *Vector-Borne Zoonotic Diseases* 2015, *15*, 89–102. https://doi.org/10.1089/vbz.2014.1628.
65. Tirloni, L.; Calvo, E.; Konnai, S.; da Silva Vaz, I. Jr. Editorial: The role of saliva in arthropod-host-pathogen relationships. *Frontiers in Cellular and Infection Microbiology* 2020, *10*, 630626. https://doi.org/10.3389/fcimb.2020.630626.

66. Schneider, D.S. Battling the bite: Tradeoffs in immunity to insect-borne pathogens. *Immunity* 2016, *44*, 1251–1252. https://doi.org/10.1016/j.immuni.2016.06.008.
67. Saadatian-Elahi, M.; Horstick, O.; Breiman, R.F.; Gessner, B.D.; Gubler, D.J.; Louis, J.; Parashar, U.D.; Tapia, R.; Picot, V.; Zinsou, J.A. et al. Beyond efficacy: The full public health impact of vaccines. *Vaccine* 2016, *34*, 1139–1147. https://doi.org/10.1016/j.vaccine.2016.01.021.
68. Klingler, C.; Barrett, D.H.; Ondrusek, N.; Johnson, B.R. Jr.; Saxena, A.; Reis, A.A. Beyond research ethics: Novel approaches of 3 major public health institutions to provide ethics input on public health practice activities. *Journal of Public Health Management and Practice* 2020, *26*, E12–e22. https://doi.org/10.1097/phh.0000000000000734.
69. Brännmark, J. The independence of medical ethics. *Medicine, Health Care and Philosophy* 2019, *22*, 5–15. https://doi.org/10.1007/s11019-018-9842-1.
70. How can we draw the line between clinical care and medical research. *PLOS Medicine* 2007, *4*, e340. https://doi.org/10.1371/journal.pmed.0040340.
71. Valdiserri, R.O.; Holtgrave, D.R. Responding to pandemics: What we've learned from HIV/AIDS. *AIDS and Behavior* 2020, *24*, 1980–1982. https://doi.org/10.1007/s10461-020-02859-5.
72. Fauci, A.S.; Folkers, G.K. HIV/AIDS and COVID-19: Shared lessons from two pandemics. *Clinical Infectious Diseases* 2024. https://doi.org/10.1093/cid/ciae585.
73. Dilani, P.V.D.; Wickramasinghe, P.D.S.U.; Lakshman, G.V.C.P.; Ranathunge, T.; Dassanayake, R.S.; Silva Gunawardene, Y.I.N. Hemocup blood feeder: An affordable and simplified blood-feeding device for maintenance of Aedes aegypti mosquito colonies in Sri Lanka. *Experimental Parasitology* 2023, *249*, 108518. https://doi.org/10.1016/j.exppara.2023.108518.
74. Autino, B.; Noris, A.; Russo, R.; Castelli, F. Epidemiology of malaria in endemic areas. *The Mediterranean Journal of Hematology and Infectious Diseases* 2012, *4*, e2012060. https://doi.org/10.4084/mjhid.2012.060.
75. Dholakia, S. Conducting controlled human infection model studies in India is an ethical obligation. *Indian Journal of Medical Ethics* 2018, *3*(4), 263–266. https://doi.org/10.20529/IJME.2018.083
76. Miller, S.; Selgelid, M.J. Ethical and philosophical consideration of the dual-use dilemma in the biological sciences. *Science and Engineering Ethics* 2007, *13*, 523–580.
77. Knobler, S.L.; Mahmoud, A.A.F.; Pray, L.A. Vaccines: Research, development, production, and procurement issues. In *Biological Threats and Terrorism: Assessing the Science and Response Capabilities – Workshop Summary*. National Academies Press (US): 2002; Chapter 3, pp. 85–112. doi:10.17226/10290
78. Snelling, P.; Quick, O. Confidentiality and public interest disclosure: A framework to evaluate UK healthcare professional regulatory guidance. *Medical Law International* 2022, *22*, 3–32. https://doi.org/10.1177/09685332221079124.
79. Larcher, V. Consent, competence, and confidentiality. *BMJ* 2005, *330*, 353–356.
80. Burki, T. Warnings over the future of UK medical research. *The Lancet* 2023, *401*, 1641. https://doi.org/10.1016/S0140-6736(23)01002-4.
81. Saalbach, K.P. Chapter three - gain-of-function research. In *Advances in Applied Microbiology*, Gadd, G.M., Sariaslani, S., Eds.; Academic Press: 2022; Volume 120, pp. 79–111.
82. Hudon, C.; Kessler, R. Integrating case management for patients with complex needs in the ground practice: The importance of context in evaluative designs. *Health Research Policy and Systems* 2023, *21*, 9. https://doi.org/10.1186/s12961-023-00960-4.
83. Anvari, F.; Lakens, D. The replicability crisis and public trust in psychological science. *Comprehensive Results in Social Psychology* 2018, *3*, 266–286. https://doi.org/10.1080/23743603.2019.1684822.
84. Achee, N.L.; Youngblood, L.; Bangs, M.J.; Lavery, J.V.; James, S. Considerations for the use of human participants in vector biology research: A tool for investigators and regulators. *Vector-Borne and Zoonotic Diseases* 2015, *15*, 89–102.
85. Rodriguez, A.; Smith, J. Phenomenology as a healthcare research method. *Evid Based Nurs* 2018, *21*, 96–98.
86. Evans, N.G. Covid-19: The ethics of clinical research in quarantine. *BMJ* 2020, *369*.
87. Spitale, G. COVID-19 and the ethics of quarantine: A lesson from the Eyam plague. *Medicine, Health Care and Philosophy* 2020, *23*, 603–609. https://doi.org/10.1007/s11019-020-09971-2.
88. Doerr, M.; Wagner, J.K. Research ethics in a pandemic: Considerations for the use of research infrastructure and resources for public health activities. *Journal of Law and the Biosciences* 2020, *7*, lsaa028.
89. Schofield, B. Informed consent in research. *European Health Psychologist* 2014, *16*, 101–106.
90. Walker, S.; Potts, J.; Martos, L.; Barrera, A.; Hancock, M.; Bell, S.; Geddes, J.; Cipriani, A.; Henshall, C. Consent to discuss participation in research: A pilot study. *BMJ Mental Health* 2020, *23*, 77–82.

91. Keitel, W.A. Clinical trials of vaccines for biodefense and emerging and neglected diseases. In *Vaccines for Biodefense and Emerging and Neglected Diseases*, Barrett, A.D.T. & Stanberry, L.R., Eds.; Academic Press (Elsevier): 2009; pp. 1451–1478. ISBN: 978-0-12-369408-9.
92. Bülow, W.; Helgesson, G. Criminalization of scientific misconduct. *Medicine, Health Care and Philosophy* 2019, *22*, 245–252. https://doi.org/10.1007/s11019-018-9865-7.
93. Pallocci, M.; Treglia, M.; Passalacqua, P.; Tittarelli, R.; Zanovello, C.; De Luca, L.; Caparrelli, V.; De Luna, V.; Cisterna, A.M.; Quintavalle, G. et al. Informed consent: Legal obligation or cornerstone of the care relationship? *International Journal of Environmental Research and Public Health* 2023, *20*. https://doi.org/10.3390/ijerph20032118.
94. Feng, Y.; Zhao, X.; Chen, Z.; Nie, K.; Yin, Z.; Xia, Y.; Wang, J.; Niu, P.; A, R.; Li, L. et al. Genomic surveillance for SARS-CoV-2 variants of concern from imported COVID-19 cases - the Mainland of China, 2021. *China CDC Weekly* 2022, *4*, 680–684. https://doi.org/10.46234/ccdcw2022.144.
95. Feng, Z.; Cui, S.; Lyu, B.; Liang, Z.; Li, F.; Shen, L.; Xu, H.; Yang, P.; Wang, Q.; Zhang, D.; et al. Genomic characteristics of SARS-CoV-2 in Beijing, China, 2021. *Biosafety and Health* 2022, *4*, 253–257. https://doi.org/10.1016/j.bsheal.2022.04.006.
96. Milanovic, B. The three eras of global inequality, 1820–2020 with the focus on the past thirty years. *World Development* 2024, *177*, 106516. https://doi.org/10.1016/j.worlddev.2023.106516.
97. Peprah, D.; Bangura, J.; Vandi, M.; Thomas, H.; Dea, M.; Schneider, A.; Chittenden, K. Social and political dimensions of disseminating research findings on emerging zoonotic viruses: Our experience in Sierra Leone. *Global Health: Science and Practice* 2021, *9*, 459–466.
98. Crowley, S. Reflections on an argument that won't go away: Or, a turn of the ideological screw. *Quarterly Journal of Speech* 1992, *78*, 450–465. https://doi.org/10.1080/00335639209384010.
99. Brokowski, C. Do CRISPR germline ethics statements cut it? *CRISPR Journal* 2018, *1*, 115–125. https://doi.org/10.1089/crispr.2017.0024.
100. Staufer, O.; Gantner, G.; Platzman, I.; Tanner, K.; Berger, I.; Spatz, J.P. Bottom-up assembly of viral replication cycles. *Nature Communications* 2022, *13*, 6530. https://doi.org/10.1038/s41467-022-33661-7.
101. Xhexhi, K.; Aliaj, B. Modular construction and bioclimatic strategies: A sustainable approach to building design. In 4th International Conference on Scientific and Academic Research (ICSAR 2024), Konya, Turkey; 2024 Jul 19–20. Vol. *4*(1): February; p. 282–292.
102. Hoffmann, S.A.; Diggans, J.; Densmore, D.; Dai, J.; Knight, T.; Leproust, E.; Boeke, J.D.; Wheeler, N.; Cai, Y. Safety by design: Biosafety and biosecurity in the age of synthetic genomics. *iScience* 2023, *26*, 106165. https://doi.org/10.1016/j.isci.2023.106165.
103. Bull, S.; Jamrozik, E.; Binik, A.; Parker, M.J. SARS-CoV-2 challenge studies: Ethics and risk minimisation. *Journal of medical ethics* 2021, *47*, e79–e79.
104. Centers for Disease Control and Prevention; National Centre for Infectious Diseases. *Emerging Infectious Diseases*; Centers for Disease Control and Prevention (CDC): 2002; Volume 8.
105. Folayan, M.O.; Haire, B.G. Ethical considerations in the conduct of research on therapies for the prevention and treatment of Ebola virus disease in developing countries. *The Pan African Medical Journal* 2015, *22*, 8.
106. Lalin, S.A.A.; Ahmed, M.N.Q.; Haq, S.M.A. The effects of the COVID-19 pandemic on students' academic performance and mental health: An overview. *Regional Science Policy & Practice* 2024, *16*, 100046. https://doi.org/10.1016/j.rspp.2024.100046.
107. Zhang, Y.X.; Perry, K.; Vinci, V.A.; Powell, K.; Stemmer, W.P.C.; del Cardayré, S.B. Genome shuffling leads to rapid phenotypic improvement in bacteria. *Nature* 2002, *415*, 644–646. https://doi.org/10.1038/415644a.
108. Sheehy, A.; James, J.R.; Horgan, M. Implementing a national approach to research ethics review during a pandemic–the Irish experience. *HRB Open Research* 2020, *3*, 63.

12 Community Engagement in Public Health

Strengthening Responses to Neglected Viruses

Abdelrahim Alqudah, Esam Qnais, Omar Gammoh, Alaa A. A. Aljabali, and Mohamed El-Tanani

12.1 INTRODUCTION

12.1.1 Overview of Community Engagement

Community engagement (CE) refers to the active involvement of local populations in decision-making processes and initiatives that influence their health outcomes. CE is key to building partnerships among researchers, health workers, and communities. Collaborative efforts have demonstrated improved health outcomes, including in the context of neglected and emerging viral infections [1, 2]. Such partnerships ensure that community voices are meaningfully integrated into public health decision-making. Community contributions are incorporated into both research design and intervention strategies. This approach can result in more effective, sustainable, and locally accepted health solutions. Table 12.1 illustrates key strategies for community participation in the context of neglected viral diseases, including participatory models, capacity-building approaches, and culturally appropriate communication frameworks [3].

Different models of engagement, like participatory research and capacity development initiatives, have been successful in strengthening community resilience to outbreaks. For example, Southeast Asian models of CBPR have enhanced vaccine coverage and dissemination of information in the case of viral epidemics [10].

12.2 IMPORTANCE OF COMMUNITY PARTICIPATION FOR THE CONTROL OF NEGLECTED VIRUSES

12.2.1 Challenges in Addressing Neglected Viruses

Neglected and lesser-known viruses pose significant challenges to global health systems. They suffer from chronic underfunding, limited public awareness, and inadequate health infrastructure [11, 12]. Addressing these issues requires robust resource mobilization strategies and improved access to health system support.

Neglected viruses are also frequently associated with stigma and widespread misconceptions. Such stigma compounds public health challenges by reducing engagement with prevention and care services [13, 14]. Engaging local communities is a key strategy to counteract these barriers. Community involvement helps dispel myths and misinformation surrounding these diseases. It also facilitates the dissemination of accurate information on risk factors and preventive measures [15, 16].

DOI: 10.1201/9781003607816-12

TABLE 12.1
Community Engagement Strategies for Enhanced Prevention, Surveillance, and Response to Neglected Viruses

Strategy	Description	Illustrative Examples	Expected Outcome	References
Participatory and Transdisciplinary Research	• *Power-centric co-governance*: Explicitly addresses intra-community hierarchies; co-develops data sovereignty protocols • *Decolonial praxis*: Equitably integrates academic, indigenous, and local knowledge systems • *Ethics of burden*: Compensates community labor; manages participation fatigue	• CBPR with co-signed governance charters and authorship agreements • PAR projects using community-led power mapping + Reflexive diaries • TDR teams co-modeling systems with traditional healers and epidemiologists	• Measurable decision-power shifts to communities • Co-created interventions addressing structural drivers • Documented resolution of knowledge conflicts	[4]
Capacity Building and Strengthening	• *Critical pedagogy*: Freirean approaches fostering consciousness of inequity • *Infrastructure focus*: Co-establishing community-owned research units/labs • *Reciprocity*: Community-to-academy mentorship on ethics/context	• Workshops on political economy of health disparities • Community-led data hubs for ongoing surveillance • Traditional birth attendants training clinicians in cultural safety	• Communities as knowledge producers (e.g., local journals) • Sustainable community institutions • Policy co-creation with marginalized groups	[5]
Culturally Radical Communication	• *Structural co-creation*: Messages address land rights, medical racism • *Counter-narrative development*: Community media challenging stigma • *Accountability*: Community-led cultural safety audits of health systems	• Radio dramas co-produced by sex workers dismantling stigma • Cholera prevention linked to water sovereignty campaigns • Indigenous elders evaluating hospital discrimination policies	• Critical health literacy enabling collective action • Amplified counter-narratives reducing structural stigma • Institutional reforms from audit findings	[6]
Power-Sensitive Network Engagement	• *Inclusion safeguards*: Power-network overlap mapping • Quotas for marginalized groups • *Accountability mechanisms*: Co-designed oversight of leaders • *Anti-extraction*: Rejecting tokenistic partnerships	• Grassroots disability collectives leading outreach design • Youth councils holding chiefs accountable for health budgets • Feminist groups monitoring gender-based violence programs	• Redistributed influence to marginalized voices • Cross-movement solidarity (e.g., health-climate justice) • Transparent resource allocation	[7]

(*Continued*)

TABLE 12.1 (*Continued*)

Community Engagement Strategies for Enhanced Prevention, Surveillance, and Response to Neglected Viruses

Strategy	Description	Illustrative Examples	Expected Outcome	References
Bidirectional Accountability Systems	• *Centering marginality*: Feedback channels designed with/for oppressed groups • *Institutional metrics*: Public dashboards tracking responsiveness • *Syndromic rumor surveillance*: Early detection of mistrust	• Migrant-led committees with veto power over research protocols • Real-time platforms showing how feedback changed policies • Rumor heatmaps triggering rapid-response dialogues	• Health systems demonstrably adapt to community priorities • Mistrust early-warning systems reducing crises • Power asymmetry reductions (validated metrics)	[8]
Anti-Colonial Systems Integration	• *Root-cause focus*: Modeling land dispossession/climate drivers of disease • *Trade-off transparency*: Participatory analysis of conservation-livelihood tensions • *Knowledge decolonization*: Indigenous ecological knowledge as science	• Co-developed systems map of mining-disease-poverty feedback loops • Scenario planning with pastoralists for zoonosis control • Indigenous-led protected areas as health interventions	• Policy coherence across health/environment/justice sectors • Equitable multi-species outcomes • Decolonial resource governance (e.g., land repatriation)	[9]

This table outlines community-driven approaches using Community-Based Participatory Research (CBPR; collaborative research equitably involving communities in all phases) and Participatory Action Research (PAR; empowering communities through iterative reflection and action) alongside transdisciplinary methods, emphasizing co-governance, structural accountability, and decolonial praxis to achieve sustainable health outcomes for neglected viral diseases.

12.2.2 Methods Sensitive to Culture

Engaging local populations is essential for ensuring the cultural relevance and effectiveness of public health interventions. Such engagement allows interventions to be tailored to local cultural norms, thereby increasing their acceptability and impact [17, 18]. Cultural beliefs and practices are key determinants in shaping the design and implementation of health campaigns. Incorporating cultural context facilitates the adoption of interventions by local communities [19].

For example, traditional funerals and mourning rituals played a critical role in the transmission of Ebola during past outbreaks. This underscores the necessity of culturally specific communication strategies to ensure effective containment and behavioral change [20, 21].

12.3 CONTRIBUTION BY COMMUNITY ENGAGEMENT TOWARD PREVENTION, SURVEILLANCE, AND RESPONSE FOR DISEASES

12.3.1 Disease Prevention

CE is also key to the prevention of disease through mobilizing communities for health-promoting activities. Community groups can support and reinforce preventive measures such as vaccination campaigns and the promotion of hygiene practices—both critical in controlling the spread of endemic viruses [22, 23]. Public participation in health programs enhances surveillance by enabling earlier symptom reporting and increasing healthcare-seeking behavior, thereby facilitating timely medical intervention [17].

12.3.2 Oversee and Response

Also bolstering efficacy is the possibility for monitoring health trends through local knowledge systems and community-based networks [24, 25]. In the aftermath of outbreaks, mobilized communities can rapidly implement epidemic control measures targeting often-overlooked viral threats. These include public sensitization campaigns and the execution of contact-tracing protocols [26, 27]. CE represents a foundational strategy for addressing the challenges posed by neglected viral diseases. By mobilizing partnerships, raising awareness, and enabling culturally sensitive interventions, CE maximizes its impact on strengthening public health systems and empowering communities to take ownership of their health outcomes [28].

12.4 IMPORTANCE OF CE FOR NEGLECTED VIRUSES

12.4.1 Importance of Community Involvement for Neglected Viruses

CE is also an essential component in addressing the public health challenges posed by neglected viral diseases, given its capacity to bridge knowledge gaps, combat misinformation, and reinforce prevention strategies. These neglected viruses often spread in contexts marked by limited health literacy and widespread misinformation, leading to harmful behaviors and social stigma. CE helps overcome these barriers by disseminating accurate information and fostering a culture of trust and collaboration between public health authorities and local communities. For instance, community-led groups can act as trusted conduits for delivering evidence-based health information, correcting misconceptions, and promoting preventive measures [29, 30]. This not only boosts knowledge about neglected viruses, but also empowers individuals to take proactive roles in safeguarding both personal and community health [31].

Thus, the necessity for inclusive, culturally sensitive CE cannot be overestimated. Every community has its own culture, beliefs, and practices, all which shape perceptions of illness, wellness, and acceptable health interventions. Therefore, interventions by the public sector must be customized to appeal to the cultural contexts and social dynamics of the targeted populations [32]. The involvement of local members and leaders during the intervention design and execution ensures not only the cultural appropriateness of the intervention but also the regard for local culture and norms [18].

For example, Ethiopia's CHAMPS initiative demonstrated increased community understanding and participation by engaging local leaders during both the design and implementation phases, which significantly improved acceptance and support from the community [33]. Culturally tailored interventions foster trust, strengthen relationships between public health authorities and local communities, and contribute to sustainable improvements in health outcomes [34].

Numerous real-world examples demonstrate how publicly funded, community-led initiatives significantly improve public health outcomes. One good illustration is the Jackson Heart Study, where the effectiveness of community mobilization was shown to be effective in addressing health disparities experienced by African American populations [18]. The study is representative of the potential for improving health literacy, increasing participation in health screening, and managing non-communicable diseases through active CE. Equally, urban communities from disadvantaged backgrounds have successfully contributed to the prevention and management of non-communicable diseases through local mobilization efforts and support from community health workers (CHWs) [35]. These cases illustrate the high acceptability and long-term sustainability of health interventions when local populations are actively involved.

Chapter Goals:

1. *CE Function*: CE functions as a critical mechanism for raising awareness, empowering local populations, and enhancing communication capacity in public health initiatives. It serves to build trust between health authorities and communities, disseminate accurate health information, and secure community support for interventions targeting previously

neglected viral threats. By involving community members in dialogue and decision-making, CE fosters local ownership of health initiatives, enhances the cultural relevance of interventions, and improves the effectiveness of prevention and response strategies [36, 37].

2. *Scope for Discussion*: This section outlines the key thematic pillars of the discussion, including capacity building, participatory methodologies, and the convergence of policy frameworks. The chapter emphasizes the importance of executing participatory processes to foster citizen engagement in strengthening health-related decision-making systems [17, 38]. It further underscores the significance of capacity building for local professionals and community leaders, while examining the policy convergence required to ensure the sustainability of engagement initiatives. At its core, citizen engagement contributes to mitigating neglected viral threats by enabling the exchange of knowledge, addressing misinformation, and initiating culturally appropriate intervention strategies. Moreover, by fostering collaborative partnerships between public health authorities and local knowledge holders, citizen engagement enhances overall population health and empowers individuals to take proactive responsibility for their own well-being [39].

12.4.2 Building Awareness

12.4.2.1 Understanding Local Context

Understanding the distinctive specific social, economic, and cultural determinants underpinning the perceptions and behaviors of local populations toward neglected viruses is a foundation for the effectiveness of interventions aimed at improving public health outcomes. Communities are not homogenous; rather, they are shaped by complex histories, cultural frameworks, and socioeconomic conditions that influence how health information is received and acted upon. For instance, where traditional beliefs about sickness and health prevail public health messages that are not aligned with these belief systems may be rejected or misinterpreted, leading to ineffective interventions [40, 41]. Therefore, understanding the local context is essential for adapting health messages in a way that resonates with local populations and fosters sustainable health behaviors.

Formative research is essential for identifying existing knowledge gaps, dominant beliefs, and practice levels in the community. This is achieved through the execution of various tools such as interviews, surveys, and focus group discussions focus groups, where the actors actively engage in their perceptions of neglected viruses, the source of their knowledge, and their health-seeking activities [41]. An exemplary instance is demonstrated by a study conducted in Ethiopia, where widespread misconceptions about the mode of transmission for some viruses were identified among rural populations, thus preventing the adoption of preventive measures by the community [42]. Additionally, formative research aids public health experts in identifying design-targeted interventions by addressing specific misconceptions and tailoring health education strategies to close identified knowledge gaps related to neglected viral diseases [43].

Challenges to sensitization and knowledge transmission are complex and deeply rooted. Socioeconomic inequalities, linguistic heterogeneity, and access limitations hamper the knowledge transmission of neglected viruses [3]. Traditional modes of transmission in most low-resource settings often exclude marginalized or hard-to-reach segments of the population, while digital communication is constrained by poor internet quality and limited digital literacy [44]. Compounding the issue is the widespread dissemination of health-related misinformation online, which further complicates accurate knowledge transfer. To counter the limitations of this form of constraint, innovative knowledge transmission models have to combine various transmission modes in trained community-based facilitators to ensure accurate and inclusive communication [45].

Intervention	Description	Primary Objective	Key Outcomes	Methodologies	Challenges	References
Multi-Sectoral Collaboration	Facilitating coordinated efforts among health, education, environment, and social welfare sectors to address neglected viral diseases through shared resources and expertise.	Foster a holistic, integrated response to viral outbreaks by leveraging diverse sectoral strengths.	Improved outbreak response efficiency, enhanced policy coherence, and broader community impact.	Inter-sectoral task forces, joint planning workshops, and cross-ministerial data-sharing platforms.	Coordination barriers, conflicting priorities, and resource silos.	[46]
Incentivizing Community Participation	Implementing financial (e.g., stipends) and non-financial (e.g., recognition, training) incentives to engage community health volunteers in outbreak response and prevention.	Increase sustained volunteer engagement and improve response efficiency in community health programs.	Higher volunteer retention, increased community trust, and faster outbreak response times.	Microfinance programs, community recognition events, and performance-based incentives.	Sustainability of funding, equitable incentive distribution, and volunteer burnout.	[47]
Integrating Health Education in School Curricula	Embedding education on infectious disease prevention, transmission, and response into primary and secondary school curricula, tailored to local disease contexts.	Build long-term community knowledge and resilience against neglected viral diseases from an early age.	Increased health literacy, reduced disease stigma, and proactive preventive behaviors among youth.	Curriculum development workshops, teacher training, and integration with STEM education.	Limited school resources, curriculum overload, and varying educational standards.	[48]
Ensuring Equitable Resource Distribution	Prioritizing allocation of health resources (e.g., vaccines, diagnostics, personnel) to marginalized, rural, and underserved populations to address health inequities.	Reduce disparities in health access and enhance epidemic preparedness in vulnerable communities.	Improved health equity, increased access to care, and stronger community-level preparedness.	Needs-based resource mapping, community-led distribution networks, and equity-focused funding models.	Logistical barriers, political resistance, and data gaps on marginalized groups.	[49, 50]
Monitoring and Evaluation Frameworks	Establishing robust systems to assess the effectiveness, reach, and impact of community health interventions through quantitative and qualitative metrics.	Enhance accountability, measure program success, and drive continuous improvement in health initiatives.	Evidence-based program adjustments, increased transparency, and improved intervention outcomes.	Logic models, key performance indicators (KPIs), and participatory evaluation methods.	Data collection challenges, resource constraints, and stakeholder alignment.	[51]

12.4.2.2 Tailoring Information

The translation of scientific information regarding neglected viruses into linguistically acceptable forms for the local environment is a measure that leads to good communication. A best practice includes the use of infographics and visual aids to convert complex healthcare information into simplified, visually engaging formats that resonate with local audiences [52]. The best practice includes the use of infographics and visual aids to convert complex healthcare information into simplified, visually engaging formats that resonate with local audiences [52]. Visual storytelling anchored in locally relevant environmental imagery—has also proven effective in conveying essential messages about disease prevention in ways that are meaningful and culturally aligned [52].

Effective awareness campaigns also incorporate indigenous languages, pictograms, and narrative tools to enhance the effectiveness of the message. One notable example is a state-led vector-borne disease awareness campaign that integrated indigenous languages and culturally resonant visual aids to strengthen the delivery of prevention messages [52, 53]. Moreover, the addition of indigenous narration and dramatizations further enhanced health knowledge transmission by increasing psychological resonance and cultural relevance among target communities [52, 54]. Apart from increase indigenous community participation in health-related initiatives.

The discipline of behavioral science, particularly behavioral economics, can be leveraged to increase immunization uptake through targeted intervention strategies [55]. For instance, the practice of the principles of the science of behavioral economics, the emphasis of the stress of loss frame instead of gain frame affects the practice of the improvement of the levels of the measures of immunizations [55, 56]. Additionally, tailoring communication strategies to address the specific psychological and contextual challenges of vulnerable populations enhances the effectiveness of immunization campaigns [55].

12.4.2.3 Utilizing Community Networks

Effective dialogue about stigmatized viruses requires the involvement of key societal institutions, including educational systems, faith-based organizations (FBOs), and local government bodies. These institutions have unique capacity to deliver health communication with both speed and perceived credibility due to their established community trust and outreach networks [44]. For instance, schools raise awareness among youth and their caregivers about stigmatized viruses, while religious groups can mobilize their followers to engage in health promotion and disease prevention efforts [45]. Participation by institutions is key to the formation of a sense of belonging and the legitimacy of health communication efforts.

CHWs also have a major role to play by focusing on the vulnerable groups and disseminating information about the viral diseases that are hitting the vulnerable groups disproportionally. CHWs often share cultural, linguistic, and socioeconomic backgrounds with the communities they serve, which enhances their credibility and facilitates trust-building [57]. For instance, the ability of CHWs to provide culturally sensitive information and support is best utilized to tackle specific health challenges while overcoming the challenges faced by group members [58].

In addition to raising awareness, their contribution to health promotion empowers individuals toward proactive measures for their health. Establishing trust and achieving community buy-in are essential components of effective engagement. Trust is built through consistent, transparent communication and demonstrated commitment to community well-being [44]. Involving community members in the planning and implementation of health initiatives fosters a sense of ownership and increases participation levels [44, 59]. Examples include the involvement of the community in the planning of the awareness campaign to facilitate right and effective information communication, hence the amplification of public health outcomes that are linked to underprioritized viruses. To sum up, awareness for underprioritized viruses necessitates knowledge of the local environment, customized information communication, and community networks' mobilization. Addressing knowledge gaps, utilizing the influence of behavioral science, along with the construction of the community's intervention credibility, can amplify awareness and add effectiveness to prevention [60].

12.4.2.4 Empowering Local Communities

12.4.2.4.1 Participatory Approaches

Advancement of practice is essential for the meaningful engagement of community members across all phases of the research continuum from priority setting to intervention design and impact evaluation. Participatory approaches, such as Community-Based Participatory Research (CBPR), empower communities by creating inclusive environments where their concerns are articulated and their priorities guide the research agenda. This interaction fosters a sense of ownership among participants, which supports the development of context-appropriate and sustainable health interventions particularly those addressing neglected viral threats [61, 62]. Effectiveness is further enhanced when local and indigenous knowledge systems are integrated throughout the research process, as guided by CBPR frameworks. This has improved the community acceptability level and potential effectiveness [63].

It is essential to analyze the ethical implications and power dynamics involved community-led intervention processes. Traditional research designs risk reinforcing hierarchical power structures, thereby marginalizing community voices. Important ethical considerations need to involve transparency throughout the process, equitable distribution of benefits, and explicit recognition of community authority and autonomy [64]. Involving community members not only increases the relevance and validity of the data but also helps build relationships and allows rapport building between the researchers and the community [57, 63]. Additionally, it can assist in resolving ethical dilemmas and ensure that interventions are aligned with local cultural values and social norms. To establish sustainable CE, capacity building is essential. Table 12.2 outlines key strategies such as training in epidemiology, leadership development through mentorship, and the application of digital tools, all of which are widely recognized in global health research as essential for empowering local actors and enhancing intervention sustainability.

12.4.2.4.2 Capacity Building

The improvement of local capacities for preventing neglected viruses requires the empowerment of local citizens through the attainment of core skills and knowledge, thus boosting the effectiveness of disease prevention and health-promoting activities. This capacity building can involve the establishment of local expertise through the attainment of skills through training on the core principles of the study of diseases (epidemiology), the prevention of diseases, and health promotion skills [79]. For instance, training sessions and seminars can empower individuals with the skills to conduct health education, screen activities, and mobilize resources [80]. Ongoing support for both local populations and frontline healthcare workers is essential to ensure the consistent delivery of community-level health services. The incorporation of regular training procedures and the available resources can also assist healthcare activities at the local level to keep the public aware and interested [79].

Strengthening workforce capacity, along with the equitable allocation of financial and infrastructural resources, is crucial to address capacity-building activities. Many communities are hampered by limited funding and resource shortages, which inhibit their ability to train and recruit skilled personnel [81]. To overcome these constraints, innovative strategies involving digital technology adoption championed by local leaders and CHWs should be pursued [82]. For example, the deployment of cellular phones is capable of imparting unplanned capacity building, enhancing communication in the health workforce, and facilitating the acquisition of information to support tracking and evaluation [83]. Strategic investments in workforce development and improvements in the quality of training resources significantly strengthen community resilience in managing neglected viral diseases.

TABLE 12.2
Capacity-Building Strategies for Enhancing Community Engagement in Combating Neglected Viral Diseases

Intervention	Description	Primary Objective	Key Outcomes	Methodologies	Challenges	References
Training in Basic Epidemiology	Comprehensive education on disease transmission dynamics, surveillance techniques, and outbreak investigation, including case identification, contact tracing, and data reporting.	Equip community members with skills to monitor and respond to infectious disease outbreaks, particularly neglected viruses.	Enhanced early detection of outbreaks, improved reporting accuracy, and increased community participation in surveillance systems.	Workshops, simulation-based training, and community-led surveillance programs.	Limited literacy, resource constraints, and sustaining engagement.	[47, 65, 66]
Health Promotion Skills Development	Structured programs to educate communities on hygiene practices (e.g., handwashing, sanitation), vaccination campaigns, and preventive behaviors tailored to local contexts.	Empower individuals to design and lead public health campaigns to reduce disease transmission.	Increased vaccination uptake, improved hygiene adherence, and reduced incidence of preventable diseases.	Interactive training sessions, peer-led education, and culturally adapted communication materials.	Misinformation, cultural resistance, and logistical barriers.	[46, 67, 68]
Mentorship Programs	Structured pairing of community leaders with public health experts for ongoing training, knowledge transfer, and leadership development.	Build sustainable leadership capacity to drive long-term community health initiatives.	Strengthened community leadership, improved program sustainability, and enhanced trust in health systems.	One-on-one mentoring, virtual coaching, and leadership workshops.	Scalability, mentor availability, and maintaining long-term engagement.	[69–71]
Use of Digital Tools	Deployment of mobile apps, SMS-based platforms, and telehealth systems for health education, real-time data collection, and communication.	Enhance accessibility of health education and support continuous learning and monitoring in remote areas.	Increased reach of health campaigns, improved data collection, and higher engagement in health programs.	Mobile app development, SMS alerts, and integration with existing health information systems.	Digital literacy gaps, infrastructure limitations, and data privacy concerns.	[47, 72]

(Continued)

TABLE 12.2 (*Continued*)
Capacity-Building Strategies for Enhancing Community Engagement in Combating Neglected Viral Diseases

Intervention	Description	Primary Objective	Key Outcomes	Methodologies	Challenges	References
Resource Allocation and Workforce Development	Strategic investment in healthcare infrastructure (e.g., clinics, diagnostic tools) and training of community health workers to address neglected viruses.	Build resilient health systems capable of sustained response to infectious diseases in resource-limited settings.	Improved healthcare access, increased workforce capacity, and reduced disease burden.	Infrastructure grants, task-shifting models, and vocational training programs.	Funding constraints, staff retention, and equitable resource distribution.	[50, 73]
Community-Based Participatory Research (CBPR)	Engaging communities in designing and conducting research to address local health priorities, focusing on neglected viruses.	Foster community ownership of health research and interventions to ensure relevance and sustainability.	Locally relevant solutions, increased trust in research, and enhanced community-researcher collaboration.	Co-design workshops, participatory data collection, and community feedback loops.	Time-intensive processes, ethical considerations, and varying community capacity.	[74, 75]
Psychosocial Support Training	Training community members to provide mental health and psychosocial support during outbreaks to address stigma and fear.	Mitigate psychological impacts of outbreaks and promote community resilience.	Reduced stigma, improved mental health outcomes, and stronger community cohesion.	Peer support groups, counseling training, and stigma reduction campaigns.	Limited mental health expertise and cultural stigma around mental health.	[76–78]

This table details evidence-based interventions, including epidemiological training, health promotion skills, mentorship programs, digital health tools, resource allocation, participatory research, and psychosocial support, designed to empower communities, strengthen local resilience, and bridge gaps between public health systems and communities in managing neglected viral diseases.

12.4.2.4.3 Delivery of Healthcare

CE can also tackle the underlying social, economic, and environmental determinants that contribute to the transmission of neglected viral diseases. Poverty, lack of sanitation and access to clean water, and sub-standard housing have an effect on health and the likelihood of the transmission of diseases [84]. CE aimed at addressing these determinants can enhance the effectiveness and sustainability of broader public health interventions. For example, community-driven initiatives to improve access to sanitation infrastructure and clean water can substantially reduce the incidence of waterborne infections linked to neglected viruses [85].

There are numerous examples of successful community-initiated projects that have addressed the root causes of disease transmission on a global scale. India's "Swachh Bharat Abhiyan" (Clean India Mission) mobilized communities to improve sanitation and hygiene practices, resulting in measurable reductions in the spread of communicable diseases [86]. Likewise, intersectoral coordination—linking health, education, and economic development can amplify the impact of community-based interventions. As an example, coordination among the health ministry and schools can promote health awareness and education among students, who can become agents for the health of their families and the community itself [87]. Multisectoral collaboration fosters collective commitment by recognizing the complexity and multidimensional nature of neglected viral diseases and promoting inclusive, context-sensitive approaches to prevention and treatment.

Strengthening local population capacity through targeted capacity development, participatory methodologies, and efforts to alleviate social determinants represents a critical strategy for both the prevention and treatment of neglected viral diseases. Placing local communities at the center of intervention design, research, and implementation ensures that solutions are contextually appropriate, sustainable, and empowering ultimately fostering stronger, more resilient populations [88].

12.4.3 Effective Communication Strategies

12.4.3.1 Two-Way Communication

Effective communication is a cornerstone of public health, and when implemented through multimedia platforms, two-way dialogue, and crisis-responsive messaging, it fosters engagement, builds trust, and enhances responsiveness among researchers, health authorities, and community members [89]. Establishing robust channels for two-way communication during public health campaigns is essential. This not only enables the effective dissemination of health information, but also allows for the collection of community feedback, which is essential for the management of concerns and refutation of misinformation. Wegwarth et al. [90] reported that communication toward the improvement of the public's health is heavily supported when institutions demonstrate openness and reliability, particularly during emergencies. Incorporating feedback mechanisms—such as community forums, surveys, or participatory dialogue—enables health authorities to assess public sentiment and refine communication strategies accordingly. For instance, the need for outreach through local leaders and volunteer groups for the treatment of child and maternal health has been identified by Cofie et al. [91], which can easily address citizens' concerns and bring about changes in the health sector.

Moreover, communication plays an indispensable role in advancing public health outcomes. Evidence from the study by Lee and Li indicates that transparent communication fosters public trust and facilitates adherence to protective health behaviors, such as social distancing during pandemics [92]. This was the case during the COVID-19 pandemic, when consistent and transparent messaging significantly influenced compliance with public health regulations [93]. Two-way communication enables health authorities to better understand community concerns and needs, thereby enhancing the effectiveness and relevance of public health interventions.

12.4.3.2 Multimedia Approaches

Employing a combination of online and offline communication platforms is essential for ensuring broad and inclusive public outreach. Zhang et al. demonstrated that designating authoritative medical professionals as public spokespersons significantly enhances message credibility and increases public receptivity [94]. This is particularly fundamental when dealing with complicated health issues where clear, culturally sensitive messaging can help overcome informational disparities across diverse population segments.

Digital campaigns have demonstrated significant influence on public attitudes and levels of CE. For example, Lee et al. addressed critical disparities in internet access among marginalized populations such as migrant farmworkers and recommended collaborative efforts to implement digital tools for improved outreach [92]. This is also evidenced by a study by Joly et al., where the mental health services' need for outreach was stressed [95]. By employing various communication platforms, health authorities can make their messages available and understandable to audiences.

Multimedia health messages must be crafted with clarity and cultural sensitivity to effectively reach target audiences. Effective communication strategies should be grounded in the specific cultural contexts of the populations being addressed. For example, the incorporation of culture-specific terms and imagery can strengthen the impact of the health campaign through campaign work being undertaken under the barber parlor environment for populations being targeted by the campaign [96]. Such culturally embedded approaches not only promote greater audience engagement but also foster a sense of belonging and lay the groundwork for long-term community-health system relationships.

12.4.3.3 Crisis Communication

Effective communication is essential during public health crises and outbreaks. Transparency, empathy, and honesty must guide communication efforts to establish credibility and encourage public adherence to health guidelines. Previous public health emergencies have underscored the importance of transparent communication strategies to address uncertainty and public concern. Wegwarth et al. argued that lack of transparency can foster vaccine hesitancy and erode trust in public health authorities [90]. Therefore, open and proactive communication is critical to preventing misinformation and securing public support during health emergencies.

Crisis communication should be informed by lessons learned from previous public health emergencies. For instance, the COVID-19 pandemic demonstrated the necessity of adaptable communication strategies capable of responding to rapidly changing information and public sentiment [93]. To cultivate and maintain high levels of public trust, health authorities must incorporate feedback mechanisms and communicate changes in guidance with transparency and clarity. In addition, increased public engagement timeliness, relevance, and public acceptance of official guidance, as demonstrated by qualitative analysis during the pandemic [97]. In summary, effective crisis communication relies on two-way communication platforms, inclusive messaging frameworks, and the establishment of resilient communication infrastructures. Governments must prioritize transparency, cultural and contextual relevance, and active citizen engagement to ensure that public health interventions are perceived as legitimate and trustworthy.

12.4.4 Promising Research Areas

12.4.4.1 Evaluating CE Interventions

Exploring pathways for translating CE interventions into practice is essential for enhancing their effectiveness in improving public health, particularly in addressing neglected viral infections. This chapter outlines three key focus areas: empirical measurement of CE interventions, dissemination of best-practice models for replication, and targeted efforts to reduce health disparities. Together, these streams underscore the importance of evidence-based evaluation, scalable intervention models, and the critical role of CE in advancing health equity and social justice [98, 99].

Comprehensive measurement frameworks are necessary to evaluate the effectiveness of interventions in disease prevention, population health outcomes, and overall well-being. Intervention strategies designed to strengthen population engagement are considered cost-effective when their broader health impacts are systematically assessed and interpreted in relation to overall population health outcomes [95]. Systematic evaluation requires the development of metrics that assess both the level of population engagement and the corresponding health impacts of engagement-driven interventions. Mixed-methods approaches combining quantitative data collection with qualitative insights are essential for capturing the full spectrum of intervention outcomes and understanding the contextual factors driving success.

It has been argued that the mixed-method approach allows for the explanation of engagement outcomes by integrating quantitative data with individual accounts and lived experiences. This approach is suitable for capturing the complexity of engagement activities, which may be inadequately represented through quantitative methods alone [100]. For example, it has been demonstrated how transformational engagement methods can tailor interventions to population-specific needs [101]. Mixed-method designs can demonstrate the long-term outcomes of engagement activities and their potential to sustainably improve public health conditions.

To assess the long-term impact of CE projects, measurement must extend beyond immediate outcome evaluation. Assessment should include changes in attitudes toward healthcare services, the sustainability of engagement activities over time, and their influence on key health determinants. Memon et al. indicated that perinatal and neonatal mortality rates could decline significantly when interventions are implemented through community-level engagement, particularly when designed for sustained participation. Such findings support the establishment of valid long-term impact metrics for CE project evaluation by researchers [102].

12.4.4.2 Scaling Up Effective Models

An important area of research is the identification of adaptable models that can be implemented across diverse settings to guide the tailoring of interventions. Successful models commonly involve the development of local partnerships, the presence of local leadership, and communication strategies sensitive to the cultural and environmental context. The importance of tailoring engagement processes to local settings in order to strengthen intervention outcomes and improve the scalability and generalizability of such models. Flexibility in model adaptation is particularly critical for scaling interventions in contexts with significant health disparities, such as low- and middle-income countries [101].

Sustainable CE involves grassroots community participation, long-term financial commitment, and the integration of CE into the routine operations of the public health sector. Establishing sustained relationships between public health institutions and local communities is necessary to build trust and enable long-term collaboration [18]. Embedding CE within the institutional culture of the health sector requires supportive policy frameworks to ensure long-term continuity. This may involve mobilizing local resources for healthcare delivery and allocating funding to build capacity among frontline public-sector workers.

It has been emphasized the importance of involving communities to empower vulnerable populations. Participatory policymaking serves as a foundational approach for advancing efforts to improve health equity. Integrating participatory models into health systems may contribute to dismantling structural mechanisms that reinforce health inequities, thereby improving overall sector performance. Robust policy frameworks are essential for institutionalizing CE and strengthening preparedness for neglected viral diseases [103]. Table 12.3 outlines integrative policy measures, including cross-sector coordination, equitable resource allocation, and long-term community participation as critical pillars.

12.4.4.3 Addressing Health Inequities

A comprehensive examination of the community's potential to reduce health disparities resulting from poorly managed viral infections is needed to promote social equity. Involving the community can empower vulnerable groups through their inclusion in decision-making about their health. In addition [17], emphasized the importance of leveraging community involvement in designing interventions tailored to the specific needs of target groups. Empowerment can increase participation in activities that support health and contribute to improved outcomes, particularly among groups disproportionately affected by neglected viral infections.

TABLE 12.3

Policy Recommendations for Strengthening Community Responses to Neglected Viral Diseases

Intervention	Description	Primary Objective	Key Outcomes	Methodologies	Challenges	References
Multi-Sectoral Collaboration	Facilitating coordinated efforts among health, education, environment, and social welfare sectors to address neglected viral diseases through shared resources and expertise.	Foster a holistic, integrated response to viral outbreaks by leveraging diverse sectoral strengths.	Improved outbreak response efficiency, enhanced policy coherence, and broader community impact.	Inter-sectoral task forces, joint planning workshops, and cross-ministerial data-sharing platforms.	Coordination barriers, conflicting priorities, and resource silos.	[104]
Incentivizing Community Participation	Implementing financial (e.g., stipends) and non-financial (e.g., recognition, training) incentives to engage community health volunteers in outbreak response and prevention.	Increase sustained volunteer engagement and improve response efficiency in community health programs.	Higher volunteer retention, increased community trust, and faster outbreak response times.	Microfinance programs, community recognition events, and performance-based incentives.	Sustainability of funding, equitable incentive distribution, and volunteer burnout.	[46, 50]
Integrating Health Education in School Curricula	Embedding education on infectious disease prevention, transmission, and response into primary and secondary school curricula, tailored to local disease contexts.	Build long-term community knowledge and resilience against neglected viral diseases from an early age.	Increased health literacy, reduced disease stigma, and proactive preventive behaviors among youth.	Curriculum development workshops, teacher training, and integration with STEM education.	Limited school resources, curriculum overload, and varying educational standards.	[105]
Ensuring Equitable Resource Distribution	Prioritizing allocation of health resources (e.g., vaccines, diagnostics, personnel) to marginalized, rural, and underserved populations to address health inequities.	Reduce disparities in health access and enhance epidemic preparedness in vulnerable communities.	Improved health equity, increased access to care, and stronger community-level preparedness.	Needs-based resource mapping, community-led distribution networks, and equity-focused funding models.	Logistical barriers, political resistance, and data gaps on marginalized groups.	[106]
Monitoring and Evaluation Frameworks	Establishing robust systems to assess the effectiveness, reach, and impact of community health interventions through quantitative and qualitative metrics.	Enhance accountability, measure program success, and drive continuous improvement in health initiatives.	Evidence-based program adjustments, increased transparency, and improved intervention outcomes.	Logic models, key performance indicators (KPIs), and participatory evaluation methods.	Data collection challenges, resource constraints, and stakeholder alignment.	[70, 107]

This table outlines evidence-based strategies to enhance community resilience, promote equity, and foster cross-sectoral collaboration. Key actions include incentivizing participation, integrating health education, ensuring equitable resource allocation, and implementing robust monitoring and evaluation frameworks for sustained impact.

Investigating the intersection between infectious disease management and social determinants is central to understanding the broader landscape of health disparities [108]. It has been highlighted the need for engagement tools at the environmental level, where support from caregivers and the public is essential to improving health outcomes among low- and middle-income populations. Addressing social determinants including poverty, education, and access to care can amplify the impact of CE in controlling neglected viruses among at-risk populations. Activities by FBOs in building vaccine confidence and reducing disparities underscore the potential of CE to promote health equity. Syed et al. [109] demonstrated that FBOs can act as intermediaries between public health systems and communities by facilitating communication and trust-building. Leveraging existing environmental and community structures can enhance intervention reach and promote equitable health environments. Promising directions for CE interventions include rigorous evaluation methods, scaling of effective models, and addressing health inequities. Focusing on these elements will enhance the effectiveness of CE interventions and support health equity in marginalized communities. The use of mixed methods, model replication, and emphasis on social determinants are critical to advancing CE as a core component of public health strategy [110].

12.5 CONCLUSION

Neglected viruses continue to represent a major global health concern, worsened by structural limitations such as inadequate health infrastructure, widespread misinformation, and limited resources. This chapter highlights the essential role of CE in addressing these challenges by fostering credibility, improving disease surveillance, and supporting preventive interventions. Through participatory models such as CBPR and capacity-building activities, CE empowers the community by closing the knowledge gaps between local populations and the authorities responsible for their health. Additionally, culturally sensitive communication has been shown to support strong public health responses and improve the effectiveness of intervention programs.

To strengthen the effectiveness of CE in preventing neglected viruses, researchers must adopt interdisciplinary approaches that integrate the perspectives of affected populations throughout all stages of health intervention design and implementation. Policymakers should embed CE models into national and international public health frameworks to ensure structural support and dedicated funding for grassroots-level health activities. Practitioners must implement evidence-based CE interventions that incorporate multimedia tools, two-way communication, and crisis response strategies to support credibility and enable knowledge exchange at the community level. Despite these efforts, gaps persist in the systematic evaluation and scalability of CE interventions. Future research should focus on establishing standardized metrics for evaluating CE effectiveness, exploring tools such as digital health technologies and social network-based interventions, and fostering interdisciplinary collaborations across epidemiology, social science, and behavioral research.

Embedding CE into the broader public health agenda is essential for building global capacity to address emerging infectious diseases, reducing health disparities, and empowering populations as primary stakeholders in disease prevention and treatment. Capacity building for CE is not peripheral to public health, it is a core component necessary to achieving long-term global health security and equity.

REFERENCES

1. Roy, B. Combination of social determinants and improvements in population health outcomes. *JAMA Network Open* 2024, *7*, e2425957–e2425957. https://doi.org/10.1001/jamanetworkopen.2024.25957.
2. Saputra, R.; Hayati, I.R.; Lidyawati, Y.; Pohan, R.A.; Ramadhani, E.; Fau, S. Improving the quality of community mental health services for schizophrenia patients in China. *Asian Journal of Psychiatry* 2025, 104378–104378. https://doi.org/10.1016/j.ajp.2025.104378.

3. Yuan, M.; Lin, H.; Wu, H.; Yu, M.; Tu, J.; Lü, Y. Community engagement in public health: A bibliometric mapping of global research. *Archives of Public Health* 2021, *79*, 1–17.
4. Collins, S.E.; Clifasefi, S.L.; Stanton, J.; The Leap Advisory, B.; Straits, K.J.E.; Gil-Kashiwabara, E.; Rodriguez Espinosa, P.; Nicasio, A.V.; Andrasik, M.P.; Hawes, S.M. et al. Community-based participatory research (CBPR): Towards equitable involvement of community in psychology research. *American Psychologist* 2018, *73*, 884–898. https://doi.org/10.1037/amp0000167.
5. Birgel, V.; Decker, L.; Röding, D.; Walter, U. Community capacity for prevention and health promotion: A scoping review on underlying domains and assessment methods. *Systematic Reviews* 2023, *12*, 147. https://doi.org/10.1186/s13643-023-02314-1.
6. Brooks, L.A.; Manias, E.; Bloomer, M.J. Culturally sensitive communication in healthcare: A concept analysis. *Collegian* 2019, *26*, 383–391. https://doi.org/10.1016/j.colegn.2018.09.007.
7. Lewis, V.A.; MacGregor, C.A.; Putnam, R.D. Religion, networks, and neighborliness: The impact of religious social networks on civic engagement. *Social Science Research* 2013, *42*, 331–346. https://doi.org/10.1016/j.ssresearch.2012.09.011.
8. Andreu Perez, L.; Kim, N.; Martino, V.; Lee, S. Constructive roles of organizational two-way symmetrical communication: Workplace pseudo-information gatekeeping. *American Behavioral Scientist* 2025, *69*, 206–218.
9. Hopkins, A.G. Rethinking decolonization. *Past and Present* 2008, *200*, 211–247.
10. Kadariya, S.; Ball, L.; Chua, D.; Ryding, H.; Hobby, J.; Marsh, J.; Bartrim, K.; Mitchell, L.; Parkinson, J. Community organising frameworks, models, and processes to improve health: A systematic scoping review. *International Journal of Environmental Research and Public Health* 2023, *20*. https://doi.org/10.3390/ijerph20075341.
11. Dodd, L.E.; Lane, H.C.; Muyembé-Tamfum, J.J. Vaccines for Mpox—An unmet global need. *NEJM Evidence* 2024, *3*. https://doi.org/10.1056/evide2300348.
12. Jiang, J.; Huang, L. Special issue "Molecular advances and perspectives of lung disease." *International Journal of Molecular Sciences* 2025, *26*, 946. https://doi.org/10.3390/ijms26030946.
13. Pinardi, T.; Prasetyo, A. Community behavior in prevention of dengue hemorrhagic fever through activities of eradication of dengue fever mosquito nests. *Health Notions* 2017, *1*, 296–299. https://doi.org/10.33846/hn10329.
14. Adnan, H.; Ahmed, M.; Salomon, I. Chandipura virus outbreak: A growing public health crisis that demandsimmediate action. *Annals of Medicine and Surgery* 2024. https://doi.org/10.1097/ms9.0000000000002841.
15. Mittal, P.; Bansal, R. Issues and challenges in community engagement. *Community Engagement for Sustainable Practices in Higher Education: From Awareness to Action* 2024, 211–224.
16. Rheumatology, T.L. Patient and public involvement: Better research for patients. *The Lancet Rheumatology* 2025, *7*, e73–e73. https://doi.org/10.1016/s2665-9913(25)00004-9.
17. Morales-Garzón, S.; Parker, L.A.; Hernández-Aguado, I.; González-Moro Tolosana, M.; Pastor-Valero, M.; Chilet-Rosell, E. Addressing health disparities through community participation: A scoping review of co-creation in public health. *Healthcare (Basel)* 2023, *11*. https://doi.org/10.3390/healthcare11071034.
18. Haldane, V.; Chuah, F.L.H.; Srivastava, A.; Singh, S.R.; Koh, G.C.H.; Seng, C.K.; Legido-Quigley, H. Community participation in health services development, implementation, and evaluation: A systematic review of empowerment, health, community, and process outcomes. *PLOS One* 2019, *14*, e0216112. https://doi.org/10.1371/journal.pone.0216112.
19. Abdel-Azim, G.G.; Osman, K.A.A. The importance of cultural dimensions in the design process of the vernacular societies. *Ain Shams Engineering Journal* 2018, *9*, 2755–2765. https://doi.org/10.1016/j.asej.2017.09.005.
20. Park, C. Traditional funeral and burial rituals and Ebola outbreaks in West Africa: A narrative review of causes and strategy interventions. *Journal of Health and Social Sciences* 2020, *5*, 073.
21. Kagawa-Singer, M. The cultural context of death rituals and mourning practices. *Oncology Nursing Forum* 1998, *25*, 1752–1756.
22. Erku, D.; Khatri, R.; Endalamaw, A.; Wolka, E.; Nigatu, F.; Zewdie, A.; Assefa, Y. Community engagement initiatives in primary health care to achieve universal health coverage: A realist synthesis of scoping review. *PLOS One* 2023, *18*, e0285222.
23. Odugleh-Kolev, A.; Parrish-Sprowl, J. Universal health coverage and community engagement. *Bulletin of the World Health Organization* 2018, *96*, 660.
24. Paganelli, A.I.; Mondéjar, A.G.; da Silva, A.C.; Silva-Calpa, G.; Teixeira, M.F.; Carvalho, F.; Raposo, A.; Endler, M. Real-time data analysis in health monitoring systems: A comprehensive

systematic literature review. *Journal of Biomedical Informatics* 2022, *127*, 104009. https://doi.org/10.1016/j.jbi.2022.104009.

25. World Health Organization; International Health Partnership+ (IHP+). *Monitoring, Evaluation and Review of National Health Strategies: A Country-Led Platform for Information and Accountability.* WHO: 2011; 132 pp. Available from: https://apps.who.int/iris/handle/10665/104857
26. Abramowitz, S.; Bedson, J. Community engagement in disease outbreak preparedness and response: Lessons from recent outbreaks, key concepts, and quality standards for practice. In *Communication and Community Engagement in Disease Outbreaks: Dealing With Rights, Culture, Complexity and Context*, Manoncourt, E., Obregon, R., Chitnis, K., Eds.; Springer International Publishing: 2022; pp. 43–72.
27. Shannon, C. Understanding community-level disaster and emergency response preparedness. *Disaster Medicine and Public Health Preparedness* 2015, *9*, 239–244.
28. O'Mara-Eves, A.; Brunton, G.; Oliver, S.; Kavanagh, J.; Jamal, F.; Thomas, J. The effectiveness of community engagement in public health interventions for disadvantaged groups: A meta-analysis. *BMC Public Health* 2015, *15*, 129. https://doi.org/10.1186/s12889-015-1352-y.
29. Corbin, J.H.; Oyene, U.E.; Manoncourt, E.; Onya, H.; Kwamboka, M.; Amuyunzu-Nyamongo, M.; Sørensen, K.; Mweemba, O.; Barry, M.M.; Munodawafa, D. A health promotion approach to emergency management: Effective community engagement strategies from five cases. *Health Promotion International* 2021, *36*, i24–i38.
30. Afolabi, A.A.; Ilesanmi, O.S. Community engagement for COVID-19 prevention and control: A systematic review. *Public Health Toxicology* 2022, *2*, 1–17.
31. Tripathi, T. Advances in vaccines: Revolutionizing disease prevention. *Scientific Reports* 2023, *13*, 11748. https://doi.org/10.1038/s41598-023-38798-z.
32. Turkson-Ocran, R.N.; Nkimbeng, M.; Erol, D.; Hwang, D.A.; Aryitey, A.A.; Hughes, V. Strategies for providing culturally sensitive care to diverse populations. *Journal of Christian Nursing* 2022, *39*, 16–21. https://doi.org/10.1097/cnj.0000000000000900.
33. Drown, L.; Amberbir, A.; Teklu, A.M.; Zelalem, M.; Tariku, A.; Tadesse, Y.; Gebeyehu, S.; Semu, Y.; Ntawukuriryayo, J.T.; VanderZanden, A. et al. Reducing the equity gap in under-5 mortality through an innovative community health program in Ethiopia: An implementation research study. *BMC Pediatrics* 2024, *23*, 647. https://doi.org/10.1186/s12887-023-04388-1.
34. Joo, J.Y.; Liu, M.F. Culturally tailored interventions for ethnic minorities: A scoping review. *Nursing Open* 2021, *8*, 2078–2090. https://doi.org/10.1002/nop2.733.
35. Nutbeam, D.; McGill, B.; Premkumar, P. Improving health literacy in community populations: A review of progress. *Health Promotion International* 2018, *33*, 901–911. https://doi.org/10.1093/heapro/dax015.
36. Christens, B.D.; Zeldin, S. Community engagement. In *Encyclopedia of Adolescence*, Levesque, R.J.R., Ed.; Springer: 2016.
37. Sørensen, K.; Levin-Zamir, D.; Duong, T.V.; Okan, O.; Brasil, V.V.; Nutbeam, D. Building health literacy system capacity: A framework for health literate systems. *Health Promotion International* 2021, *36*, i13–i23. https://doi.org/10.1093/heapro/daab153.
38. Ricciardelli, A. Governance, local communities, and citizens participation. In *Global Encyclopedia of Public Administration, Public Policy, and Governance*, Farazmand, A., Ed.; Springer International Publishing: 2018; pp. 2919–2932.
39. Gilmore, B.; Ndejjo, R.; Tchetchia, A.; de Claro, V.; Mago, E.; Diallo, A.A.; Lopes, C.; Bhattacharyya, S. Community engagement for COVID-19 prevention and control: A rapid evidence synthesis. *BMJ Global Health* 2020, *5*. https://doi.org/10.1136/bmjgh-2020-003188.
40. Bedford, J.; Farrar, J.; Ihekweazu, C.; Kang, G.; Koopmans, M.; Nkengasong, J. A new twenty-first century science for effective epidemic response. *Nature* 2019, *575*, 130–136.
41. Van den Broucke, S. Why health promotion matters to the COVID-19 pandemic, and vice versa. *Health Promotion International* 2020, *35*, 181–186.
42. Leulseged, H.; Abdi, F.; Mohamed, A.; Wakoya, G.; Aliyi, M.; Tesfaye, A.; Yali, S.; Mohamed, K.; Mohamed, A.; Yilma, E. Building bridges: Enhancing cultural understanding and community engagement in the CHAMPS program—experience sought from Gobe Challa Village, Ethiopia. *Discover Social Science and Health* 2024, *4*, 51.
43. Pleasant, A.; O'Leary, C.; Carmona, R.H. Using formative research to tailor a community intervention focused on the prevention of chronic disease. *Evaluation and Program Planning* 2020, *78*, 101716. https://doi.org/10.1016/j.evalprogplan.2019.101716.

44. Nickel, P.M.; Hunt, C. Community engagement, equity, and cross-sector public health practice. *Public Health Reports* 2024, *139*, 669–674.
45. Aghakhani, N.; Ewalds-Kvist, B.M.; Azami, M.; Rarani, S.A. Community involvement in COVID-19 management: A central component of comprehensive public health responses to COVID-19 and future health crises. *Disaster Medicine and Public Health Preparedness* 2024, *18*, e327.
46. Ackley, C.; Elsheikh, M.; Zaman, S. Scoping review of neglected tropical disease interventions and health promotion: A framework for successful NTD interventions as evidenced by the literature. *PLOS Neglected Tropical Diseases* 2021, *15*, e0009278.
47. King'ori, T. Harnessing digital health technologies for improved healthcare delivery and disease surveillance in low-resource settings. *Global Perspectives in Health, Medicine, and Nursing* 2024, *3*, 1–10.
48. Yue, J.L.; Yan, W.; Sun, Y.K.; Yuan, K.; Su, S.Z.; Han, Y.; Ravindran, A.V.; Kosten, T.; Everall, I.; Davey, C.G. Mental health services for infectious disease outbreaks including COVID-19: A rapid systematic review. *Psychological Medicine* 2020, *50*, 2498–2513.
49. Joshi, R.; Alim, M.; Kengne, A.P.; Jan, S.; Maulik, P.K.; Peiris, D.; Patel, A.A. Task shifting for non-communicable disease management in low and middle income countries – A systematic review. *PLOS One* 2014, *9*, e103754.
50. Okoroafor, S.C.; Christmals, C.D. Task shifting and task sharing implementation in Africa: A scoping review on rationale and scope. *Healthcare (Basel)* 2023, *11*, 1200.
51. Oladeji, O.; Oladeji, A.; Halima, A.; Baitwabusa, A.E. Financial incentive for community health workers, a necessity for achieving universal health coverage: Integrated literature review. *International Journal of Community Medicine and Public Health* 2023, *11*, 545–555.
52. Kulkarni, A.; Joshi, K.; Jasani, K. Empowering public health: The impact of infographics in raising awareness, informing, and educating India. *Journal of the Epidemiology Foundation of India* 2024, *2*, 83–85.
53. McGuire, J.; Evans, E.; Kane, E.; McGuire, J.; Evans, E.; Kane, E. What works in public awareness campaigns? A scoping review. In *Evidence-based Policing and Community Crime Prevention*, Knutsson, J., Tilley, N., Eds.; Routledge: 2021; pp. 417–433.
54. Schipper, J. Awareness campaigns are only as effective as the planning that goes into them. *Canadian Pharmacists Journal* 2006, *139*, 42.
55. Ford, K.L.; West, A.B.; Bucher, A.; Osborn, C.Y. Personalized digital health communications to Increase COVID-19 vaccination in underserved populations: A double diamond approach to behavioral design. *Frontiers in Digital Health* 2022, *4*, 831093.
56. Beatty, A.; Moffitt, R.; Buttenheim, A.; National Academies of Sciences, Engineering and Medicine. The behavioral economics toolkit: Policy levers and intervention strategies. In *Behavioral Economics: Policy Impact and Future Directions*; National Academies Press (US): 2023.
57. Astale, T.; Abebe, T.; Mitike, G. Workload and emerging challenges of community health workers in low- and middle-income countries: A mixed-methods systematic review. *PLOS One* 2023, *18*, e0282717. https://doi.org/10.1371/journal.pone.0282717.
58. Pinto, D.; Carroll-Scott, A.; Christmas, T.; Heidig, M.; Turchi, R. Community health workers: Improving population health through integration into healthcare systems. *Current Opinion in Pediatrics* 2020, *32*, 674–682. https://doi.org/10.1097/mop.0000000000000940.
59. Bera, O.P.; Mondal, H.; Bhattacharya, S. Empowering communities: A review of community-based outreach programs in controlling hypertension in India. *Cureus* 2023, *15*, e50722.
60. Bugshan, W.M.; Qahtani, S.J.; Alwagdani, N.A.; Alharthi, M.S.; Alqarni, A.M.; Alsuat, H.M.; Almotairi, A.H. Role of health awareness campaigns in improving public health: A systematic review. *International Journal of Life Science and Pharma Research* 2022, *12*, 129–135.
61. Meredith, G.R.; Leong, D.; Frost, C.; Travis, A.J. Facilitated asynchronous online learning to build public health strategic skills. *Journal of Public Health Management and Practice* 2024, *30*, 56–65.
62. Manafò, E.; Petermann, L.; Vandall-Walker, V.; Mason-Lai, P. Patient and public engagement in priority setting: A systematic rapid review of the literature. *PLOS One* 2018, *13*, e0193579.
63. Astale, T.; Mesele, H.; Pasquino, S.L.; Zewdie, A.; Wolka, E.; Endalamaw, A.; Assefa, Y.; Mitike, G. Capacity-building interventions for health extension workers in Ethiopia: A scoping review. *PLOS One* 2025, *20*, e0317198.
64. Chhabra, D.; Jahangiri, K.; Sohrabizadeh, S.; Ghomian, Z.; Shahsavani, A.; Wahi, D.C. Exploring strategies to mitigate the adverse health impacts of air pollution on children in India: A qualitative study. *Cureus* 2024, *16*.

65. Were, M.C. Challenges in digital medicine applications in under-resourced settings. *Nature Communications* 2022, *2022*, 3020.
66. McGowan, C.R.; Takahashi, E.; Romig, L.; Bertram, K.; Kadir, A.; Cummings, R.; Cardinal, L.J. Community-based surveillance of infectious diseases: A systematic review of drivers of success. *BMJ Global Health* 2022, *7*, e009934.
67. Baatiema, L.; Sumah, A.M.; Tang, P.N.; Ganle, J.K. Community health workers in Ghana: The need for greater policy attention. *BMJ Global Health* 2016, *1*, e000141.
68. Naghiloo, Z. Global health interventions in resource-limited settings: Successes and lessons learned. *Health Nexus* 2023, *1*, 56–60.
69. Lescano, A.G.; Cohen, C.R.; Raj, T.; Rispel, L.; Garcia, P.J.; Zunt, J.R.; Hamer, D.H.; Heimburger, D.C.; Chi, B.H.; Ko, A.I. Strengthening mentoring in low-and middle-income countries to advance global health research: An overview. *The American Journal of Tropical Medicine and Hygiene* 2018, *100*, 3.
70. Hamid, M.; Rasheed, M.A. A new path to mentorship for emerging global health leaders in low-income and middle-income countries. *The Lancet Global Health* 2022, *10*, e946–e948.
71. Rodríguez, D.C.; Jessani, N.S.; Zunt, J.; Ardila-Gómez, S.; Muwanguzi, P.A.; Atanga, S.N.; Sunguya, B.; Farquhar, C.; Nasuuna, E. Experiential learning and mentorship in global health leadership programs: Capturing lessons from across the globe. *Annals of Global Health* 2021, *87*, 61.
72. Latif, S.; Rana, R.; Qadir, J.; Ali, A.; Imran, M.A.; Younis, M.S. Mobile health in the developing world: Review of literature and lessons from a case study. *IEEE Access* 2017, *5*, 11540–11556.
73. Das, S.; Grant, L.; Fernandes, G. Task shifting healthcare services in the post-COVID world: A scoping review. *PLOS Global Public Health* 2023, *3*, e0001712.
74. Wallerstein, N.; Duran, B.; Oetzel, J.G.; Minkler, M. *Community-Based Participatory Research for Health: Advancing Social and Health Equity*; John Wiley & Sons: 2017.
75. Ochola, E.A.; Karanja, D.M.S.; Elliott, S.J. Local tips, global impact: Community-driven measures as avenues of promoting inclusion in the control of neglected tropical diseases: A case study in Kenya. *Infectious Diseases of Poverty* 2022, *11*, 81–90.
76. Patel, V.; Saxena, S.; Lund, C.; Thornicroft, G.; Baingana, F.; Bolton, P.; Chisholm, D.; Collins, P.Y.; Cooper, J.L.; Eaton, J. The Lancet Commission on global mental health and sustainable development. *The Lancet* 2018, *392*, 1553–1598.
77. Parsekar, S.; Dhyani, V.S.; Rajwar, E.; Krishnan, J.B.; Venkatesh, B.T.; Saravu, K.; Brand, H. An overview of preventive and control strategies for common mental health disorders during infectious disease epidemics. *F1000Research* 2022, *11*, 297.
78. Anindhita, M.; Haniifah, M.; Putri, A.M.N.; Karnasih, A.; Agiananda, F.; Yani, F.F.; Haya, M.A.N.; Pakasi, T.A.; Widyahening, I.S.; Fuady, A. Community-based psychosocial support interventions to reduce stigma and improve mental health of people with infectious diseases: A scoping review. *Infectious Diseases of Poverty* 2024, *13*, 1–17.
79. Naslund, J.A.; Shidhaye, R.; Patel, V. Digital technology for building capacity of nonspecialist health workers for task sharing and scaling up mental health care globally. *Harvard Review of Psychiatry* 2019, *27*, 181–192.
80. Atinga, R.A.; Dery, S.; Katongole, S.P.; Aikins, M. Capacity for optimal performance of healthcare supply chain functions: Competency, structural and resource gaps in the Northern Region of Ghana. *Journal of Health Organization and Management* 2020, *34*, 899–914.
81. Mohajer, N.; Singh, D. Factors enabling community health workers and volunteers to overcome socio-cultural barriers to behaviour change: Meta-synthesis using the concept of social capital. *Human Resources for Health* 2018, *16*, 1–9.
82. Brown, A.; Philipson, A.; Dunstan, K.; Frazer-Ryan, S. "Healthy Kids"—A capacity building approach for the early childhood education and care sector. *Health Promotion Journal of Australia* 2023, *34*, 765–774.
83. Pani, S.R.; Nallala, S.; Rout, S.K.; Sundari, S.; Chokshi, M.; Mokashi, T.; Nair, A.; Kadam, S.M. Effects of various financial and non-financial incentives on the performance of accredited social health activist: Evidence from two selected districts of Odisha. *Journal of Health Management* 2022, *24*, 74–86.
84. Chosewood, L.C.; Schill, A.L.; Chang, C.C.; Childress, A.M.; Hudson, H.L.; Tamers, S.L.; Howard, J. The National Institute for Occupational Safety and Health Total Worker Health® Program: The third decade. *Journal of Occupational and Environmental Medicine* 2024, *66*, 6–8.
85. Wilhelm, A.K.; Schwedhelm, M.; Bigelow, M.; Bates, N.; Hang, M.; Ortega, L.; Pergament, S.; Allen, M.L. Evaluation of a school-based participatory intervention to improve school

environments using the consolidated framework for implementation research. *BMC Public Health* 2021, *21*, 1–14.

86. Wei, K.; Chopra, P.; Strehlow, S.; Stow, M.; Kaplan, I.; Szwarc, J.; Minas, H. The capacity-building role of community liaison workers with refugee communities in Victoria, Australia. *International Journal of Mental Health Systems* 2021, *15*, 1–13.
87. Saputra, D.; Assaf, R.A.; Achmad, A.Z. Enhancing community participation in public services through participatory innovation. *Jurnal Ilmiah Ilmu Administrasi Publik* 2022, *12*, 677.
88. Alhassan, J.A.K.; Gauvin, L.; Judge, A.; Fuller, D.; Engler-Stringer, R.; Muhajarine, N. Improving health through multisectoral collaboration: Enablers and barriers. *Canadian Journal of Public Health* 2021, *112*, 1059–1068. https://doi.org/10.17269/s41997-021-00534-3.
89. Bernhardt, J.M. Communication at the core of effective public health. *American Journal of Public Health* 2004, *94*, 2051–2053. https://doi.org/10.2105/ajph.94.12.2051.
90. Wegwarth, O.; Hertwig, R.; Giese, H.; Fineberg, H.V. The impact of nontransparent health communication during the COVID-19 pandemic on vaccine-hesitant people's perception of vaccines. *Frontiers in Public Health* 2024, *11*, 1256829.
91. Cofie, L.E.; Barrington, C.; Akaligaung, A.; Reid, A.; Fried, B.; Singh, K.; Sodzi-Tettey, S.; Barker, P.M. Integrating community outreach into a quality improvement project to promote maternal and child health in Ghana. *Global Public Health* 2014, *9*, 1184–1197.
92. Lee, J.G.L.; LePrevost, C.E.; Harwell, E.L.; Bloss, J.E.; Cofie, L.E.; Wiggins, M.F.; Firnhaber, G.C. Coronavirus pandemic highlights critical gaps in rural Internet access for migrant and seasonal farmworkers: A call for partnership with medical libraries. *Journal of the Medical Library Association* 2020, *108*, 651.
93. Lowe, M.; Harmon, S.H.E.; Kholina, K.; Parker, R.; Graham, J.E. Public health communication in Canada during the COVID-19 pandemic. *Canadian Journal of Public Health* 2022, *113*, 34–45.
94. Zhang, Y.S.D.; Young Leslie, H.; Sharafaddin-Zadeh, Y.; Noels, K.; Lou, N.M. Public health messages about face masks early in the COVID-19 pandemic: Perceptions of and impacts on Canadians. *Journal of Community Health* 2021, *46*, 903–912.
95. Joly, B.M.; Williamson, M.E.; Bernard, K.P.; Mittal, P.; Pratt, J. Evaluating community outreach efforts: A framework and approach based on a national mental health demonstration project. *Journal of MultiDisciplinary Evaluation* 2012, *8*, 46–56.
96. Luque, J.S.; Ross, L.; Gwede, C.K. Qualitative systematic review of barber-administered health education, promotion, screening and outreach programs in African-American communities. *Journal of Community Health* 2014, *39*, 181–190.
97. LePrevost, C.E.; Cofie, L.E.; Nieuwsma, J.; Harwell, E.L.; Rivera, N.D.; Acevedo, P.A.; Lee, J.G.L. Community health worker outreach to farmworkers in rural North Carolina: Learning from adaptations to the SARS-CoV-2 pandemic. *Health Expectations* 2024, *27*, e14047.
98. Woolf, S.H.; Purnell, J.Q.; Simon, S.M.; Zimmerman, E.B.; Camberos, G.J.; Haley, A.; Fields, R.P. Translating evidence into population health improvement: Strategies and barriers. *Annual Review of Public Health* 2015, *36*, 463–482. https://doi.org/10.1146/annurev-publhealth-082214-110901.
99. Guerin, R.J.; Glasgow, R.E.; Tyler, A.; Rabin, B.A.; Huebschmann, A.G. Methods to improve the translation of evidence-based interventions: A primer on dissemination and implementation science for occupational safety and health researchers and practitioners. *Safety Science* 2022, *152*, 105763. https://doi.org/10.1016/j.ssci.2022.105763.
100. Stern, C.; Lizarondo, L.; Carrier, J.; Godfrey, C.; Rieger, K.; Salmond, S.; Apóstolo, J.; Kirkpatrick, P.; Loveday, H. Methodological guidance for the conduct of mixed methods systematic reviews. *JBI Evidence Synthesis* 2020, *18*, 2108–2118. https://doi.org/10.11124/jbisrir-d-19-00169.
101. McHugh, S.M.; Riordan, F.; Kerins, C.; Curran, G.; Lewis, C.C.; Presseau, J.; Wolfenden, L.; Powell, B.J. Understanding tailoring to support the implementation of evidence-based interventions in healthcare: The CUSTOMISE research programme protocol. *HRB Open Research* 2023, *6*, 7. https://doi.org/10.12688/hrbopenres.13675.1.
102. Memon, Z.A.; Khan, G.N.; Soofi, S.B.; Baig, I.Y.; Bhutta, Z.A. Impact of a community-based perinatal and newborn preventive care package on perinatal and neonatal mortality in a remote mountainous district in Northern Pakistan. *BMC Pregnancy Childbirth* 2015, *15*, 106. https://doi.org/10.1186/s12884-015-0538-8.
103. Rong, T.; Ristevski, E.; Carroll, M. Exploring community engagement in place-based approaches in areas of poor health and disadvantage: A scoping review. *Health & Place* 2023, *81*, 103026. https://doi.org/10.1016/j.healthplace.2023.103026.

104. Van Panhuis, W.G.; Paul, P.; Emerson, C.; Grefenstette, J.; Wilder, R.; Herbst, A.J.; Heymann, D.; Burke, D.S. A systematic review of barriers to data sharing in public health. *BMC Public Health* 2014, *14*, 1–9.
105. Perry, H.B.; Zulliger, R.; Rogers, M.M. Community health workers in low-, middle-, and high-income countries: An overview of their history, recent evolution, and current effectiveness. *Annual Review of Public Health* 2014, *35*, 399–421.
106. Déglise, C.; Suggs, L.S.; Odermatt, P. SMS for disease control in developing countries: A systematic review of mobile health applications. *Journal of Telemedicine and Telecare* 2012, *18*, 273–281.
107. Pillai, G.; Chibale, K.; Constable, E.C.; Keller, A.N.; Gutierrez, M.M.; Mirza, F.; Sengstag, C.; Masimirembwa, C.; Denti, P.; Maartens, G. The next generation scientist program: Capacity-building for future scientific leaders in low-and middle-income countries. *BMC Medical Education* 2018, *18*, 1–11.
108. Tripathi, S.; Jain, M.; Bagai, A.; Rao, K.V. Designing appropriate, acceptable and feasible community-engagement approaches to improve routine immunisation outcomes in low-and middle-income countries: A synthesis of 3ie-supported formative evaluations. *PLOS One* 2022, *17*, e0275278.
109. Syed, U.; Kapera, O.; Chandrasekhar, A.; Baylor, B.T.; Hassan, A.; Magalhães, M.; Meidany, F.; Schenker, I.; Messiah, S.E.; Bhatti, A. The role of faith-based organizations in improving vaccination confidence & addressing vaccination disparities to help improve vaccine uptake: A systematic review. *Vaccines* 2023, *11*, 449.
110. King, A.C.; Odunitan-Wayas, F.A.; Chaudhury, M.; Rubio, M.A.; Baiocchi, M.; Kolbe-Alexander, T.; Montes, F.; Banchoff, A.; Sarmiento, O.L.; Bälter, K. et al. Community-based approaches to reducing health inequities and fostering environmental justice through global youth-engaged citizen science. *International Journal of Environmental Research and Public Health* 2021, *18*. https://doi.org/10.3390/ijerph18030892.

13 Case Studies in Virus Elimination and Control

Muhamed Adilović

13.1 SUCCESS STORIES

Successes in virus elimination and control demonstrate the potential of well-coordinated public health strategies. From the global eradication of smallpox to regional achievements in eliminating diseases such as polio, measles, and canine rabies, these case studies demonstrate the importance of sustained vaccination campaigns, strong political leadership, community engagement, and international collaboration. Each success offers valuable insights into the mechanisms and factors that enable disease control, while also revealing the importance of adaptability and long-term commitment. These examples can serve as models for future efforts, showing what is possible when science, policies, and society strive toward a common goal.

13.1.1 Eradication of Smallpox

Smallpox is caused by the Variola virus, and its mortality rate was close to 30%, frequently resulting in long-term complications, including disfiguring scars and blindness among survivors [1]. Smallpox represents the only human infectious disease that has been completely eradicated, which represents a significant milestone in global public health. This was achieved through a coordinated international campaign led by the World Health Organization (WHO), employing vaccination, epidemiological surveillance, and containment methodologies [2].

The basis for smallpox eradication was made in 1796, when Edward Jenner introduced a vaccine based on the Vaccinia virus [3]. Routine immunization led to a significant decline in smallpox incidence across industrialized nations by the mid-20th century, while it persisted in several regions of Asia and Africa [4]. Due to this, the WHO launched the Intensified Smallpox Eradication Program in 1967, which consisted of mass immunization prioritizing the rapid identification and isolation of cases, and the vaccination of close contacts, which disrupted chains of transmission [5].

The last case of naturally occurring smallpox was documented in Somalia in October 1977 [6]. This was followed by a two-year period of global surveillance confirming the absence of new infections, after which smallpox was officially declared eradicated in 1980 [7]. Several factors were crucial for the eradication, including:

- the absence of an animal carrier;
- the availability of an effective thermostable vaccine;
- the distinct clinical presentation of the disease allowing early diagnosis;
- as well as the international collaboration led by the WHO [8, 9].

Despite its eradication, potential risks, including accidental release from laboratory stocks or bioterrorism, have resulted in continued research aimed at developing novel antiviral therapeutics and enhancing vaccines [10–12]. As a result, the smallpox eradication campaign represents a model for successful global health intervention, based on sustained immunization strategies and global cooperation [13].

DOI: 10.1201/9781003607816-13

13.1.2 Polio Elimination in Africa, the Americas, and Beyond

Poliomyelitis is an acute viral disease caused by the poliovirus, primarily affecting children under the age of five and may result in irreversible paralysis or fatal outcomes [14]. The global initiative to eradicate polio had a great success, mainly due to large immunization programs, disease surveillance, and coordinated outbreak response [15, 16].

13.1.2.1 Polio Elimination in the Americas

The Americas became the first region to achieve the interruption of poliovirus transmission, with the last case reported in Peru in 1991 and official eradication declared in 1994. This was accomplished by extensive vaccination campaigns utilizing both the oral polio vaccine and the inactivated polio vaccine, coordinated through the efforts of the Pan American Health Organization (PAHO) and national ministries of health [17, 18]. Additional factors contributing to this included strong political support, community engagement, and high vaccination rates [18, 19].

13.1.2.2 Polio Elimination in Africa

Africa was declared free of wild poliovirus in 2020, with the last case documented in Nigeria in 2016 [20]. The regional eradication was accomplished through the deployment of mobile health teams, enhanced surveillance systems, and collaboration with community leaders to combat both logistical challenges in remote areas and misinformation [21]. Persistent immunization activities are still important in order to prevent the re-establishment of transmission [22, 23].

13.1.2.3 Tajikistan's Success in Halting Polio Spread

Tajikistan was the first region experiencing a Polio outbreak in 2010, after being declared polio-free by the WHO. It resulted in 463 laboratory-confirmed cases and 29 associated fatalities [24]. The disease spread rapidly among children, with Tajikistan's Ministry of Health initiating immunization campaigns for people of all ages to quickly ensure herd immunity. This was followed by surveillance measures, including environmental monitoring and sewage sampling, resulting in the last case in Tajikistan being reported in July of 2010 [25]. The virus was still able to spread neighboring countries, including Kazakhstan, Russia, Turkmenistan, and Uzbekistan, which have improved their immunization campaigns and surveillance frameworks to decrease further transmission [24].

13.1.3 Measles and Rubella Elimination and Control in the Americas and Other Regions

Measles and rubella are two highly transmissible viral diseases accompanied by severe complications, including encephalitis, miscarriage, and death.

13.1.3.1 Elimination Milestones in the Americas

The Americas became the first region to eliminate rubella and measles in 2015 and 2016, respectively. This was achieved through mass immunization strategies, surveillance, political support, and coordination by PAHO [26]. New measles outbreaks were still reported between 2019 and 2023 in 18 countries; however, by the end of 2023, no country maintained endemic measles transmission. This signifies the importance of continuous surveillance and the maintenance of high vaccine coverage [27]. Figure 13.1 shows the number of reported Measles cases in the United States from 1985 to 2025, showcasing a sharp drop in recent history, after the introduction of the second MMR dose [28].

13.1.3.2 Global Status and Setbacks

Global progress in eliminating measles and rubella has shown variable success. As of 2020, 81 countries had verified the elimination of measles, while rubella elimination had been confirmed

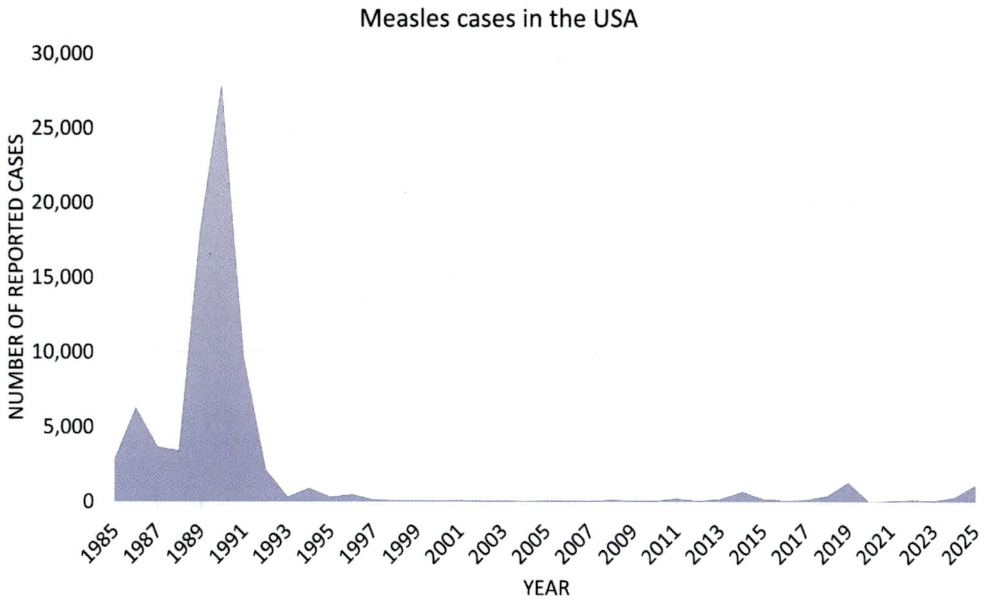

FIGURE 13.1 Reported measles cases in the United States between 1985 and 2025.

in 93 countries. However, the COVID-19 pandemic significantly disrupted routine immunization services, resulting in disease resurgence. For instance, the United Kingdom recorded very high numbers of measles cases in 2024 [29].

13.1.4 Egypt's Rapid Hepatitis C Elimination

Egypt's national response to hepatitis C virus (HCV) infection represents an exemplary public health achievement, transitioning from one of the highest global HCV rates to a near-elimination status within a single decade. During the early 2000s, Egypt exhibited an HCV prevalence of approximately 10% [30]. The Egyptian government launched a comprehensive HCV elimination strategy comprising four core components:

- *Nationwide Mass Screening*: more than 60 million individuals aged 18–59 years were screened for HCV between 2018 and 2019, representing one of the largest public health campaigns globally [31].
- *Universal Treatment Access*: no-cost access to highly effective direct-acting antiviral therapies was ensured, which demonstrated cure rates exceeding 95% among treated individuals [32].
- *Strengthened Healthcare Delivery*: efficient case management and follow-up were facilitated through the improvements in healthcare infrastructure and the establishment of a network of specialized treatment centers [33].
- *Political and International Support*: strong political support, together with collaboration from international organizations, enabled rapid mobilization of resources and technical expertise [34, 35].

As a result of these concerted efforts, Egypt became the first country to attain the WHO's "gold tier" status in progress toward HCV elimination [36].

13.1.5 Nigeria's Effective Containment of Ebola

In 2014, Nigeria faced a potentially devastating outbreak of Ebola virus disease (EVD) following the introduction of the virus into Lagos – one of Africa's most densely populated cities. Despite this challenge, Nigeria managed to rapidly contain the outbreak, becoming a global model in public health emergency response and outbreak containment strategies.

Nigeria's successful containment of EVD was facilitated by the rapid implementation of several key strategies:

- *Activation of the Emergency Operations Center (EOC)*: EOCs were developed during the polio eradication campaign, allowing for their quick reactivation and deployment of a coordinated incident management system, enabling real-time decision-making, resource allocation, and coordination [37].
- *Comprehensive Contact Tracing*: public health authorities have implemented intensive contact tracing, identifying and monitoring 894 individuals who had potential exposure to the patient zero, enabling prompt identification of secondary infections and the interruption of transmission chains [38].
- *Risk Communication and Public Awareness*: a large-scale public education campaign disseminated accurate information on EVD symptoms, modes of transmission, and prevention, enhancing community engagement and early reporting of suspected cases [39, 40].
- *Healthcare Worker Training and Infection Control*: targeted training programs equipped healthcare personnel with knowledge of infection prevention and control protocols [41].
- *International Technical Collaboration*: Nigeria worked closely with international agencies, including the WHO and Centers for Disease Control and Prevention (CDC), receiving technical expertise and resources, strengthening the national response to the disease [42].

These interventions successfully restricted the outbreak to 20 confirmed cases and 8 mortalities. On October 20, 2014, the WHO officially declared Nigeria Ebola-free [43].

13.1.6 Rabies Control around the World

Rabies continues to pose a global health challenge, especially in regions where canine-mediated transmission remains prevalent. However, numerous countries represent successful stories in rabies elimination or control through strategic vaccination, sustained surveillance, and rigorous public health policies.

13.1.6.1 Japan: Rabies-Free Status Maintained Since 1957

Japan achieved rabies elimination through a sophisticated approach including strict dog registration laws, mass vaccination campaigns, and aggressive control of stray dog populations. Continuous surveillance and strict animal import quarantine protocols have preserved the country's rabies-free status for over six decades [44].

13.1.6.2 Taiwan: Rapid Containment Following Re-emergence

Taiwan's rabies control efforts intensified following the reappearance of the virus in wildlife populations in 2013. The national response included mass vaccination campaigns targeting both domestic animals and wildlife reservoirs [45]. Enhanced border controls and public education further reduced the risk of transmission [46].

13.1.6.3 United States: Elimination of Canine Rabies and Wildlife Surveillance

The United States eliminated canine rabies transmission by 2007 through widespread use of rabies vaccination in dogs, as well as oral rabies vaccination (ORV) targeting raccoons, coyotes, and foxes.

The national Rabies Management Program integrates annual ORV baiting with thorough surveillance [47, 48].

13.1.6.4 South Korea: Sustained Reduction through Dual Vaccination Strategy

South Korea's control of rabies includes mandatory vaccination of domestic dogs and ORV campaigns directed at raccoon dogs. These measures, together with public awareness campaigns and government-coordinated surveillance, have drastically reduced human and animal rabies cases since the early 2010s [49–51].

13.1.6.5 European Union: Regional Coordination and ORV Success

Several European countries, including France, Germany, and Switzerland, eliminated fox-mediated rabies during the 1980s and 1990s through organized ORV programs. The European Union continues to invest in regional strategies to maintain rabies-free zones, including the support to Eastern European countries, where isolated cases still persist [52, 53].

13.1.6.6 Australia: Rabies Prevention via Biosecurity

Australia has maintained its rabies-free status through strict biosecurity protocols, focusing on animal import controls and continuous surveillance. The country has implemented a risk-based approach targeting potential incursions from nearby endemic regions [54].

13.1.6.7 Latin America: A Regional Public Health Success

In the early 1980s, Latin America reported on average 300 human rabies cases annually, with domestic dogs representing the majority of transmissions. PAHO launched a region-wide initiative in 1983 targeting the elimination of canine rabies, focusing on three strategies: mass dog vaccination, widespread public health education, and timely administration of post-exposure prophylaxis to bite victims. This has resulted in significant improvements, with human rabies cases decreasing to 13 in 2018 [55, 56].

13.1.7 SARS-CoV-1 Containment: Lessons from 2003

The 2003 outbreak of Severe Acute Respiratory Syndrome (SARS), caused by the SARS-CoV-1 virus, emerged as a global health crisis, affecting 29 countries with more than 8,000 confirmed infections and 774 reported deaths [57]. However, the outbreak was contained within six months, offering valuable insights into the effective epidemic response.

The outbreak was initially identified in Guangdong Province, China, in late 2002, with early containment being impeded due to delays in case reporting. By February 2003, international spread had occurred, prompting the WHO to issue a global health alert in March [58, 59]. An important milestone in the response was the rapid sequencing of the SARS-CoV-1 genome in April 2003, enabling the development of reliable diagnostic assays and improved epidemiological tracking [60].

On a local level, contact tracing and isolation measures were important to the disease containment efforts, with countries like Singapore utilizing electronic monitoring and financial incentives to enforce home quarantine, limiting virus transmission [61, 62]. Internationally, affected countries imposed travel restrictions and started airport screening, including temperature checks for departing passengers [63, 64]. Hospital-based transmission was also a major driver of the disease spread, requiring infection prevention and control practices, including the provision of personal protective equipment, staff training, and the implementation of negative-pressure isolation wards [65].

The rapid containment of SARS-CoV-1 signified the importance of early case detection, transparent data sharing, and coordinated international response. These lessons were crucial in responding to subsequent outbreaks, including Middle East Respiratory Syndrome in 2012 and the COVID-19 pandemic in 2019 [66].

13.1.8 New Zealand's COVID-19 Elimination Strategy

New Zealand's approach to managing the COVID-19 pandemic has been internationally recognized as one of the most effective [67]. Figure 13.2 shows the number of COVID-19 deaths per 100,000 people in G7 countries and New Zealand, according to the Johns Hopkins University, with the data up to March 10, 2023 [68]. In March 2020, there were only 102 confirmed COVID-19 cases and no recorded deaths. At the time, the New Zealand government adopted an aggressive elimination strategy aimed at completely preventing the disease transmission. A nationwide lockdown was announced, which came into effect on March 26 [69].

Overall, the elimination strategy consisted of a range of interventions underpinned by a robust suite of non-pharmaceutical interventions, including:

- Comprehensive lockdowns limiting physical movement, gatherings, and non-essential economic activities.
- Strict border controls consisting of mandatory quarantine and testing for all incoming travelers.
- Testing and contact tracing infrastructure allowing rapid identification, isolation, and management of cases [70].

These interventions had a synergistic effect, resulting in the quick suppression of disease spread and enabling precise epidemiological control. By late July 2020, New Zealand had reported no community-based transmission for over 80 consecutive days [70]. This success allowed social and economic activities with minimal restrictions. As more transmissible variants emerged and global case numbers surged, New Zealand implemented a hybrid strategy emphasizing minimization and protection, balancing public health with economic and social needs [67].

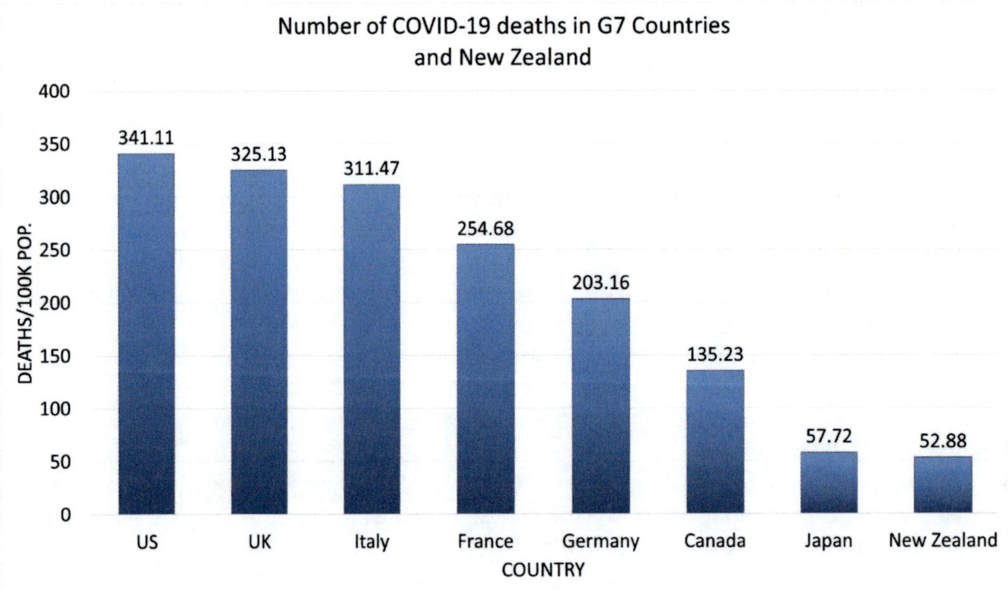

FIGURE 13.2 Number of COVID-19 deaths per 100,000 people in G7 countries and New Zealand.

TABLE 13.1
Number of Deaths Worldwide during Influenza Pandemics

Year	1918	1957	1968	2009
Name	Spanish Flu	Asian Flu	Hong Kong Flu	Swine Flu
Global Deaths	20 million	1.1 million	1 million	12,400

13.2 ONGOING CHALLENGES

13.2.1 Influenza Pandemics: A Recurring Threat

Influenza pandemics pose repeating challenges to global health security, with notable occurrences in 1918, 1957, 1968, and 2009. Table 13.1 shows the number of deaths worldwide during each of the major influenza pandemics. It is worth noting that although the Swine Flu had a significantly lower number of deaths compared to the previous ones, the number of reported illnesses was around 61 million [71]. These events emphasize the influenza virus's capacity for genetic variation, including antigenic shift and interspecies transmission, which together contribute to the emergence of novel strains increasing global morbidity and mortality [72]. The dynamic nature of influenza viruses requires constant global surveillance, research efforts, as well as the infrastructure in place to combat the impact of future pandemics.

The 1918 H1N1 pandemic, also known as the "Spanish flu," resulted in an estimated 50 million deaths worldwide. It remains the influenza pandemic with the largest impact in recorded history [73]. Later pandemics in 1957 (H2N2), 1968 (H3N2), and 2009 (H1N1) were characterized by antigenic shifts, which led to the emergence of novel viral strains against which humans had limited preexisting immunity. These recurring influenza pandemics emphasize the cyclical and unpredictable nature of flu [74].

Rapid detection of emerging influenza strains is instrumental for effective containment; however, global surveillance capabilities are highly unbalanced, with specific regions unable to rapidly respond to novel threats due to lower resources to fight the disease, inadequate laboratory capacity, delayed case reporting, and insufficient funding [75]. The 2009 H1N1 pandemic further showed gaps in the global surveillance approach and emphasized the need for collaborative international monitoring networks and coordinated response [76].

Currently, one of the major threats is the highly pathogenic avian influenza virus H5N1, due to its high mortality rate in humans and ongoing transmission in bird populations. Although human infections are relatively rare, the potential for mutation raises concerns regarding future pandemic risk [77]. Different countries develop specific policies to prepare for potential outbreaks, including stockpiling of existing vaccines and the development of novel ones [78, 79].

13.2.2 HIV/AIDS: Progress and Persistent Barriers in Prevention and Treatment

Over the past four decades, the global response to HIV/AIDS has moved from crisis management to chronic disease control, largely due to the introduction of antiretroviral therapy (ART), sophisticated prevention strategies and campaigns, as well as global policy frameworks. ART has transformed HIV into a manageable chronic condition, which has also resulted in a significant decline in AIDS-related mortality [80]. Nevertheless, the goal of ending the HIV epidemic is constantly impeded by structural, social, and economic barriers.

ART has revolutionized HIV treatment through the suppression of viral replication, which has significantly lowered the transmission risk, and has consequently reduced morbidity and mortality [81]. Pre-exposure prophylaxis, when taken consistently, can decrease the risk of acquisition by up to 99% in high-risk populations [82]. Additionally, global initiatives have contributed to the HIV prevention and treatment, including the Joint United Nations Programme on HIV/AIDS (UNAIDS) with "95-95-95" targets, aiming to diagnose 95% of people positive on HIV, provide ART to 95% of those diagnosed, and achieve viral suppression in 95% of those treated by 2025 [83]. As of 2023, 86% knew their HIV status, 89% were on ART, and 93% had achieved viral suppression [84].

Despite these advancements, several ongoing challenges obstruct the full realization of HIV prevention and treatment objectives, including:

- Stigma and discrimination, resulting in a barrier to testing, treatment, and are, particularly in marginalized communities [85].
- Structural and economic inequities, specifically access to healthcare, education, and funding, which disproportionately increase the HIV burden in low- and middle-income countries [86].
- Comorbid behavioral health conditions, with mental health disorders and substance abuse being prevalent among people living with HIV and further complicating successful treatment. Integrated care models addressing both HIV and behavioral health prove to be essential for improving overall HIV/AIDS prevention and treatment [87].

13.2.3 Hepatitis B Control through Vaccination

The introduction of the hepatitis B vaccine in 1981 represents a major milestone in the prevention of the hepatitis B virus (HBV), offering a safe, immunogenic, and efficient intervention that significantly altered the global epidemiology of the disease [88].

The international integration of HBV vaccination into immunization schedules has contributed to substantial reductions in disease burden. By 2016, all countries in the Americas had incorporated HBV vaccination into routine infant immunization programs. These measures led to an estimated three-dose coverage of 89% and a birth dose coverage of 75% by 2015, correlating with a decline in HBV surface antigen (HBsAg) prevalence among children [89].

In the African region, intensive efforts were made to eliminate mother-to-child transmission of HBV, together with the development of regional elimination frameworks. This has resulted in a considerable decrease in disease incidence between 2016 and 2021, however, challenges still persist. HBV infection remains a significant concern for global health, affecting an estimated almost 300 million individuals worldwide and contributing to approximately 820,000 deaths annually [90].

13.2.4 The Struggle against Measles Resurgence: Causes and Countermeasures

Although specific countries have had great success in fighting measles, due to its highly transmissible nature, it remains a serious global health threat, despite the availability of a safe and effective vaccine. The recent resurgence of measles in multiple regions is concerning and emphasizes the fragile nature of herd immunity in the presence of vaccination gaps and sociopolitical challenges [91]. Figure 13.3 shows the top 10 countries with measles outbreaks, together with the reported numbers between October 2024 and April 2025, according to the CDC [92].

Several underlying factors contributing to the measles resurgence, including:

- A decline in vaccination rates; sustained high vaccination rates are essential for maintaining herd immunity against measles. However, several countries have reported very low coverage rates, creating immunity gaps that facilitate viral transmission [93].
- Vaccine hesitancy and misinformation – undermines public confidence, partly driven by the dissemination of a now-debunked study falsely linking the MMR vaccine to autism [94, 95].

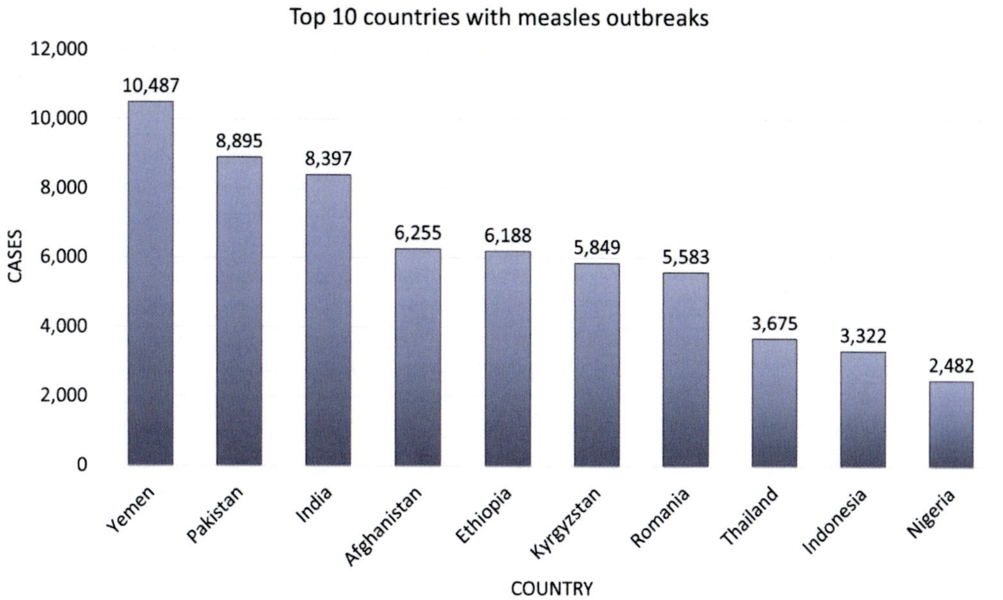

FIGURE 13.3 Countries with the largest measles outbreaks between October 2024 and April 2025.

- International travel and viral importation – global mobility enables the reintroduction of measles into regions with weakened immunization coverage, which could trigger local outbreaks [96].
- Healthcare access inequities, with large groups of people being susceptible to measles outbreaks, including those with limited access to healthcare services, often due to ongoing conflicts in the area [97, 98].

13.2.5 Vector-Borne Diseases: The Fight against Malaria, Dengue, and Other Arboviruses

Vector-borne diseases – including malaria, dengue, and other arboviral infections are a significant global health concern, especially in tropical and subtropical regions. Despite numerous advancements in control strategies, various environmental, biological, and sociopolitical factors interfere with eradication and containment efforts [99].

Climate change plays an important role in altering the distribution and activity of disease vectors. Rising global temperatures, changes in precipitation patterns, and increased frequency of extreme weather events have expanded the geographic range of mosquitoes. This has enabled the transmission of dengue and Zika viruses in previously non-endemic regions, including parts of Europe and North America, further complicating vector control and requiring adaptation of surveillance strategies [100]. Moreover, prolonged use of insecticides has led to the emergence of resistance among vector populations, including *Anopheles funestus* mosquitoes in Tanzania, developing insecticide-resistant mutations [101]. These developments emphasize the importance of novel insecticide research and integrated pest management techniques.

Technological breakthroughs such as gene-drive systems represent new possibilities for vector population suppression and the interruption of disease transmission; however, these innovations raise ethical and ecological questions [102].

Integrated Vector Management represents the foundation of vector control, combining various approaches, including chemical, biological, environmental, and community-based interventions, specifically designed for local epidemiological conditions. However, the success of such strategies is affected by behavioral variability among vectors, environmental dynamics, and inconsistent resource allocation [103]. Different tools, including molecular diagnostics, Geographic Information Systems, and remote sensing technologies, could enhance epidemic forecasting and support data-driven health interventions [100, 104].

13.2.6 Emerging and Re-emerging Viruses

The constant threat posed by emerging and re-emerging viruses, including Ebola virus and SARS-CoV-2, underscores the need for proactive surveillance systems and global preparedness strategies. Even though these pathogens exhibit diverse transmission pathways and clinical presentations, they are unified by common underlying drivers that facilitate their emergence.

13.2.6.1 Ebola Virus Disease (EVD)

Since its identification in 1976, EVD has re-emerged in multiple outbreaks, mainly in sub-Saharan Africa. The disease is characterized by high mortality rates, rapid person-to-person transmission, and potential for nosocomial amplification, increasing the burden on the healthcare system. The Democratic Republic of Congo (DRC) remains a center of recurrent outbreaks. Over the past decade, the DRC has implemented several containment strategies, including epidemiological surveillance, contact tracing, community engagement, and targeted vaccination campaigns [105]. During the Kivu Ebola epidemic (2018–2020), health authorities prioritized early detection, quarantine, infection prevention in healthcare settings, and the enforcement of safe burial practices. However, these efforts were frequently hindered by armed conflict, population displacement, and mistrust toward public health authorities [106, 107].

The detection of a new Ebola cluster in Uganda in March 2025 signifies the ongoing risk of resurgence, especially in regions with widespread human-wildlife interaction and insufficient health system resilience [108].

Vaccination remains an important response to Ebola, with the rVSV-ZEBOV vaccine significantly contributing to the transmission control. In Sierra Leone, a nationwide vaccination campaign was launched in December 2024, with the goal to immunize over 20,000 frontline workers. This is an important step, considering the country's high mortality rate during the 2014–2016 West African epidemic [109, 110].

13.2.6.2 COVID-19 Pandemic

The emergence of SARS-CoV-2 in late 2019 led to the most significant pandemic of the 21st century to date, exposing systemic vulnerabilities in global healthcare systems. Despite the rapid development of vaccines, several factors pose barriers to successful control, including limited vaccine access, the emergence of novel variants able to evade the immune system, and widespread pandemic fatigue [111–114].

Factors contributing to the viral emergence and re-emergence include:

- Zoonotic spillover, with a high proportion of emerging infectious diseases having zoonotic origin, being contributed to by activities such as deforestation and wildlife trade [115–117]
- Globalization and urbanization [118]
- Climate change [118]
- Armed conflict and insecurity, which restricts access to public healthcare and diagnostic laboratories [119, 120]
- Community mistrust and misinformation [121, 122].

13.3 LESSONS FOR OTHER REGIONS

13.3.1 Combating Vaccine Hesitancy: Strategies for Public Health Campaigns, Community Engagement, and Public Trust

Vaccine hesitancy has emerged as a significant problem for successful global immunization, contributing to the resurgence of vaccine-preventable diseases in multiple regions worldwide. Combatting this phenomenon requires an integrated strategy that combines evidence-based public health campaigns, local community engagement, and sustained efforts to build institutional trust.

Effective public health campaigns are fundamental to countering vaccine misinformation and teaching people the benefits of immunization. The "19 to Zero" initiative in Canada represents a data-driven approach to vaccine promotion done by a coalition of healthcare professionals, academics, and marketing experts [123–125]. In the DRC, there was a lot of misinformation and conspiracy theories during the mpox vaccination campaign. UNICEF, in collaboration with local leaders, countered this by establishing community action cells responsible for delivering accurate information and addressing vaccine-related concerns [126, 127].

Reinforcing public trust in vaccination is of high importance, especially since it can be influenced by political, cultural, and historical factors, as evidenced during the COVID-19 pandemic [128, 129]. In the United States, barbershops and salons have been used as effective platforms for disseminating health information and promoting clinical trial participation through partnerships with pharmacy chains [130]. Additionally, testimonies from individuals who were formerly hesitant about vaccination can persuade others to participate in the immunization process [131, 132].

13.3.2 Building Pandemic-Ready Health Systems: Key Components

The COVID-19 pandemic revealed critical vulnerabilities in global health infrastructure, and has emphasized the need for health systems capable of mounting rapid, coordinated responses. Various components play a key role in the successful response to an emerging crisis.

- *Surveillance Infrastructure and Early Warning Systems*: timely pathogen detection is crucial to effective outbreak control. The WHO's pandemic hub helps with global pathogen monitoring using advanced data analytics and genomic surveillance [133], coupled with real-time data sharing by the International Pathogen Surveillance Network [134, 135]. The WHO's Global Influenza Surveillance and Response System, established in 1952, is an example of a long-standing, coordinated effort in tracking the evolution of the influenza virus. It comprises of 142 centers across 115 countries, promptly identifying novel strains, and decreasing the risk of widespread transmission [136].
- *Wastewater-based Epidemiology Monitoring*: a non-invasive method enabling the detection of viral RNA and other biomarkers shed in human waste, enabling the identification of outbreaks weeks before clinical cases are reported. Recent studies have shown the predictive capabilities of wastewater analysis, including the detection of avian influenza six weeks before outbreaks in poultry populations [137, 138].
- *Scalable Healthcare Infrastructure*: health systems should be able to adapt to sudden increases in demand without compromising the service. Virtual hospitals, such as Saudi Arabia's Seha Virtual Hospital, show how telemedicine can scale healthcare access and minimize the infection risk during pandemics, especially in regions with geographical and logistical barriers to healthcare access [139].
- *Integrated Data Systems*: the COVID-19 pandemic highlighted the fragmentation of national data infrastructures, prompting the developed countries to establish a unified digital health framework that would enhance coordination, surveillance, and evidence-based policymaking [140]. During the 2024 Marburg virus outbreak in Rwanda, rapid

implementation of isolation, screening, and contact tracing led to successful containment, highlighting the role of transparent communication, genomic data sharing, and cross-border collaboration [141].

- *Global Partnerships*: global health threats require collaborative responses. The mission of Coalition for Epidemic Preparedness Innovations (CEPI) is to develop vaccines for emerging threats, aiming to significantly reduce the time required for deployment in future pandemics [142].
- *Equitable Access to Medical Countermeasures*: equity remains an important part of an effective global health response. The Access to COVID-19 Tools (ACT) Accelerator, launched at the beginning of COVID-19 pandemic in 2020, helped with the equitable distribution of vaccines, diagnostics, and therapeutics. The COVAX initiative, a part of ACT framework, aimed to deliver COVID-19 vaccines to low- and middle-income countries, helping decrease the disparities in access during global health emergencies [143].
- *Zoonotic Surveillance*: The Global Virome Project (GVP) represents an effort to catalog unknown viruses present in wildlife species with high zoonotic potential. By characterizing these pathogens before they transfer to human populations, GVP enhances predictive modeling and early warning capacities [144].

13.3.3 Adapting Strategies to Local Contexts

The effectiveness of virus control and elimination is significantly influenced by their level of adaptation to the specific sociocultural, economic, and infrastructural conditions of target populations, being able to promote community ownership, trust, and long-term sustainability.

Paying attention to cultural norms and engaging local actors is fundamental to the success of public health strategies. During the West African Ebola outbreak, early containment efforts struggled due to public mistrust; however, community cooperation significantly improved when public health authorities engaged religious leaders [145].

Economic constraints influence the scope and sustainability of public health interventions, particularly in low-income settings, with cost-effective, community-based models demonstrating success in bridging resource gaps. In Uganda, Village Health Teams consisting of trained community health workers played a vital role in malaria control by providing both physical supplies and health education [146].

The adaptability of viral control strategies is tied to the existing healthcare infrastructure, with technologically advanced countries integrating digital health systems to enhance response efficiency. South Korea's COVID-19 digital contact tracing was successful in rapidly identifying and notifying possible exposures to the virus [147].

Political commitment and governance are instrumental in the efficient coordination of public health responses. Vietnam's early response to COVID-19 benefited from centralized governance, resulting in prompt border closures, mass quarantines, and efficient resource allocation, slowing down the spread of the virus [148].

Taiwan is another example of an effective strategy that drew on experience from prior epidemics. The government started early border screening, real-time health data integration, and widespread testing. This enabled proactive containment without initial nationwide lockdowns [149].

13.3.4 One Health Approach: Integrating Human, Animal, and Environmental Health

The One Health approach recognizes the interdependence between human, animal, and environmental health systems, as shown in Figure 13.4. The emergence of zoonotic pathogens increasingly threatens global health security, which is the reason why an interdisciplinary approach is essential for effective viral disease prevention, control, and elimination.

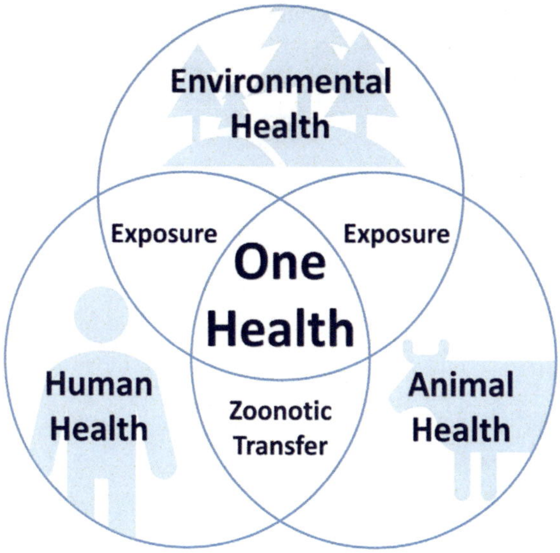

FIGURE 13.4 The One Health concept.

The One Health approach emphasizes that the health of humans is closely linked to that of animals and the ecosystems they share. The data indicates that approximately 75% of emerging or re-emerging infectious diseases are zoonotic in origin [150]. Moreover, intrusion into animal habitats, wildlife exploitation, and climate change has intensified human-wildlife contact, elevating the probability of viral spillover events [151]. This was emphasized by the West Nile virus outbreak in New York City in 1999 – a mosquito-borne pathogen affecting both avian species and humans [152].

Additionally, research reveals that birds in urban environments harbor antibiotic-resistant bacteria, with human-derived contaminants contributing to this [153]. The COVID-19 pandemic, likely originating from wildlife such as bats, transmitted to humans and caused a global crisis, further emphasizing the significance of One Health frameworks to prevent future pandemics. This could be achieved through enhanced biosurveillance, interdisciplinary collaboration, and the implementation of appropriate policies and legislations [154–156].

13.4 CONCLUSION

The global experience with viral epidemics and pandemics offers valuable insights into what drives success in disease control and elimination. As illustrated through the case studies covered, effective viral containment is rarely the result of a single strategy. Instead, it comes from the dynamic interplay of various factors, including proactive surveillance, political commitment, community participation, technological innovation, and sustained public health infrastructure investment.

One recurring theme is the need for adapting interventions to local contexts. Strategies that were successful in one setting may fail in another unless they are tailored to local cultural norms, taking into account economic and health system capacities. The success of Uganda's Village Health Teams, Taiwan's integrated data systems, and South Korea's digital contact tracing demonstrate how context-specific responses contribute to effectiveness and public compliance.

The One Health framework, integrating human, animal, and environmental health, has received increasing popularity, especially with the rise of zoonotic threats such as COVID-19, Ebola, and avian influenza. Emerging and re-emerging viruses continue to find the gaps in fragmented health systems unless a more integrated and interdisciplinary approach is taken.

Political will and funding remain essential for disease control, as witnessed in various cases, from Rwanda's quick action during the Marburg outbreak to the delayed Ebola vaccine development. The COVID-19 pandemic has further emphasized the importance of transparency, equity, and international work. Initiatives such as COVAX and CEPI reveal the potential of global collaboration when contributed to with appropriate political and financial support.

Finally, public trust is the key component that underlies all successful strategies. Whether through risk communication that is culturally sensitive, ethical use of digital technologies, or transparent governance, trust-building remains a top priority for all interventions. Without it, even technically advanced systems that have enough resources are susceptible to public disengagement and resistance.

Viral threats are inevitable with increasing ecological disruption, urbanization, and global mobility. However, large-scale pandemics can be avoided, minimized, and better managed. With the right combination of preparedness, surveillance, adaptability, and cooperation, virus elimination and control are achievable goals. The key components include readiness, research, cooperation, commitment to equity, and a sense of collective responsibility.

REFERENCES

1. Henderson DA. The eradication of smallpox – an overview of the past, present, and future. Vaccine 2011;29:D7–D9.
2. Fenner F (Ed.). Smallpox and Its Eradication, 1988. World Health Organization, Geneva.
3. Riedel S. Edward Jenner and the history of smallpox and vaccination. Proc (Bayl Univ Med Cent) 2005;18(1):21–25.
4. Poland GA. Smallpox vaccines: From first to second to third generation. Lancet 2005;365(9457):362–363.
5. Foege WH, Millar JD, Lane JM. Selective epidemiologic control in smallpox eradication. Am J Epidemiol 1971;94(4):311–315.
6. Breman JG, Arita I. The confirmation and maintenance of smallpox eradication. N Engl J Med 1980;303(22):1263–1273.
7. Strassburg MA. The global eradication of smallpox. Am J Infect Control 1982;10(2):53–59.
8. Henderson DA. Smallpox: The Inside Story of Eradicating a Worldwide Killer, 2009. Prometheus, Amherst.
9. Tomori O. From smallpox eradication to the future of global health: Innovations, application and lessons for future eradication and control initiatives. Vaccine 2011;29:D145–148.
10. Henderson DA, Inglesby TV, Bartlett JG et al. Smallpox as a biological weapon: Medical and public health management. Working Group on Civilian Biodefense. JAMA 1999;281(22):2127–2137.
11. Masiello M, Brown L, Gostin LO (Eds.). Future State of Smallpox Medical Countermeasures [Internet], 2024. National Academies Press (US), Washington (DC) [cited 2025 May 11]. Available from: http://www.ncbi.nlm.nih.gov/books/NBK605458/.
12. Nishiyama Y, Matsukuma S, Matsumura T et al. Preparedness for a smallpox pandemic in Japan: Public health perspectives. Disaster Med Public Health Prep 2015;9(2):220–223.
13. Metzger WG, Köhler C, Mordmüller B. Lessons from a modern review of the smallpox eradication files. J R Soc Med 2015;108(12):473–477.
14. Aylward B, Tangermann R. The global polio eradication initiative: Lessons learned and prospects for success. Vaccine 2011;29:D80–85.
15. Makoni M. Africa eradicates wild polio. Lancet Microbe 2020;1(6):e243.
16. Adebisi YA, Eliseo-Lucero Prisno D, Nuga BB. Last fight of wild polio in Africa: Nigeria's battle. Public Health Pract (Oxf) 2020;1:100043.
17. Robbins FC. Eradication of polio in the Americas. JAMA 1993;270(15):1857–1859.
18. Emanuele CA, Jean Baptiste AE, Chévez AE et al. Maintaining the region of the Americas free of polio: Best practices for incident management support teams. Rev Panam Salud Publica 2024;48:e23.
19. Hovi T, Shulman LM, van der Avoort H et al. Role of environmental poliovirus surveillance in global polio eradication and beyond. Epidemiol Infect 2012;140(1):1–13.
20. Africa Regional Commission for the Certification of Poliomyelitis Eradication. Certifying the interruption of wild poliovirus transmission in the WHO African region on the turbulent journey to a polio-free world. Lancet Glob Health 2020;8(10):e1345–e1351.

21. Mohammed A, Tomori O, Nkengasong JN. Lessons from the elimination of poliomyelitis in Africa. Nat Rev Immunol 2021;21(12):823–828.
22. Cooper LV, Bandyopadhyay AS, Gumede N et al. Risk factors for the spread of vaccine-derived type 2 polioviruses after global withdrawal of trivalent oral poliovirus vaccine and the effects of outbreak responses with monovalent vaccine: A retrospective analysis of surveillance data for 51 countries in Africa. Lancet Infect Dis 2022;22(2):284–294.
23. Alleman MM, Jorba J, Riziki Y et al. Vaccine-derived poliovirus serotype 2 outbreaks and response in the Democratic Republic of the Congo, 2017-2021. Vaccine 2023;41 Suppl 1(Suppl 1):A35–A47.
24. Yakovenko ML, Gmyl AP, Ivanova OE et al. The 2010 outbreak of poliomyelitis in Tajikistan: Epidemiology and lessons learnt. Euro Surveill 2014;19(7):20706.
25. Khetsuriani N, Pallansch MA, Jabirov S et al. Population immunity to polioviruses in the context of a large-scale wild poliovirus type 1 outbreak in Tajikistan, 2010. Vaccine 2013;31(42):4911–4916.
26. Ropero AM, Kurtis H, Vulanovic L et al. Vaccination week in the Americas: An ongoing initiative to strengthen and sustain measles and rubella elimination in the region. Vaccines (Basel) 2024;12(7):812.
27. Rey-Benito G, Pastor D, Whittembury A et al. Sustaining the elimination of measles, rubella and congenital Rubella syndrome in the Americas, 2019-2023: From challenges to opportunities. Vaccines (Basel) 2024;12(6):690.
28. CDC. Measles Cases and Outbreaks [Internet]. Measles (Rubeola). 2025 [cited 2025 May 31]. Available from: https://www.cdc.gov/measles/data-research/index.html.
29. De Francesco MA. Measles resurgence in Europe: An open breakthrough in the field of vaccine-preventable diseases. Pathogens 2023;12(10):1192.
30. El-Akel W, El-Sayed MH, El Kassas M et al. National treatment programme of hepatitis C in Egypt: Hepatitis C virus model of care. J Viral Hepat 2017;24(4):262–267.
31. Kandeel A, Genedy M, El-Refai S et al. The prevalence of hepatitis C virus infection in Egypt 2015: Implications for future policy on prevention and treatment. Liver Int 2017;37(1):45–53.
32. Waked I, Esmat G, Elsharkawy A et al. Screening and treatment program to eliminate hepatitis C in Egypt. N Engl J Med 2020;382(12):1166–1174.
33. Shiha G, Soliman R, Mikhail NNH et al. An educate, test and treat model towards elimination of hepatitis C infection in Egypt: Feasibility and effectiveness in 73 villages. J Hepatol 2020;72(4):658–669.
34. Mohamoud YA, Mumtaz GR, Riome S et al. The epidemiology of hepatitis C virus in Egypt: A systematic review and data synthesis. BMC Infect Dis 2013;13:288.
35. Gomaa A, Gomaa M, Allam N et al. Hepatitis C elimination in Egypt: Story of success. Pathogens 2024;13(8):681.
36. WHO commends Egypt for its progress on the path to eliminate hepatitis C [Internet]. World Health Organization. [cited 2025 May 30]. Available from: https://www.who.int/news/item/09-10-2023-who-commends-egypt-for-its-progress-on-the-path-to-eliminate-hepatitis-c.
37. Ebola Virus Disease Outbreak — Nigeria, July–September 2014 [Internet]. Centers for Disease Control and Prevention. [cited 2025 May 17]. Available from: https://www.cdc.gov/mmwr/preview/mmwrhtml/mm6339a5.htm.
38. Fasina FO, Shittu A, Lazarus D et al. Transmission dynamics and control of Ebola virus disease outbreak in Nigeria, July to September 2014. Euro Surveill 2014;19(40):20920.
39. Otu A, Ameh S, Osifo-Dawodu E et al. An account of the Ebola virus disease outbreak in Nigeria: Implications and lessons learnt. BMC Public Health 2017;18(1):3.
40. Olatunji RW. Health crisis communication and post-Ebola virus disease containment era in Lagos State, Nigeria. Int J Crisis Commun 2020;4(1):23–32.
41. Shuaib F, Gunnala R, Musa EO et al. Ebola virus disease outbreak - Nigeria, July-September 2014. MMWR Morb Mortal Wkly Rep 2014;63(39):867–872.
42. Breakwell L, Gerber AR, Greiner AL et al. Early identification and prevention of the spread of Ebola in high-risk African countries. MMWR Suppl 2016;65(3):21–27.
43. WHO Declares Nigeria Ebola-Free [Internet]. World Health Organization. 2025 [cited 2025 May 30]. Available from: https://www.afro.who.int/news/who-declares-nigeria-ebola-free.
44. Takayama N. Rabies control in Japan. Jpn J Infect Dis 2000;53(3):93–97.
45. Huang ASE, Chen WC, Huang WT et al. Public health responses to reemergence of animal rabies, Taiwan, July 16-December 28, 2013. PLOS One 2015;10(7):e0132160.
46. Chang SS, Tsai HJ, Chang FY et al. Government response to the discovery of a rabies virus reservoir species on a previously designated rabies-free island, Taiwan, 1999-2014. Zoonoses Public Health 2016;63(5):396–402.

47. Sterner RT, Meltzer MI, Shwiff SA et al. Tactics and economics of wildlife oral rabies vaccination, Canada and the United States. Emerg Infect Dis 2009;15(8):1176–1184.
48. Kirby JD, Chipman RB, Nelson KM et al. Enhanced rabies surveillance to support effective oral rabies vaccination of raccoons in the Eastern United States. Trop Med Infect Dis 2017;2(3):34.
49. Yang DK, Kim HH, Cho I.S. Strategies to maintain Korea's animal rabies non-occurrence status. Clin Exp Vaccine Res 2018;7(2):87–92.
50. Yang DK, Kim HH, Lee KK et al. Mass vaccination has led to the elimination of rabies since 2014 in South Korea. Clin Exp Vaccine Res 2017;6(2):111–119.
51. Cheong Y, Kim B, Lee KJ et al. Strategic model of national rabies control in Korea. Clin Exp Vaccine Res 2014;3(1):78–90.
52. Freuling CM, Hampson K, Selhorst T et al. The elimination of fox rabies from Europe: Determinants of success and lessons for the future. Philos Trans R Soc Lond B Biol Sci 2013;368(1623):20120142.
53. Müller FT, Freuling CM. Rabies control in Europe: An overview of past, current and future strategies. Rev Sci Tech 2018;37(2):409–419.
54. Sparkes J, Fleming PJS, Ballard G et al. Canine rabies in Australia: A review of preparedness and research needs. Zoonoses Public Health 2015;62(4):237–253.
55. Schneider MC, Min KD, Romijn PC et al. Fifty years of the national rabies control program in Brazil under the one health perspective. Pathogens 2023;12(11):1342.
56. Rabies persists in only four countries of Latin America and the Caribbean [Internet]. Pan American Health Organization. 2018 [cited 2025 May 30]. Available from: https://www.paho.org/en/news/21-9-2018-rabies-persists-only-four-countries-latin-america-and-caribbean
57. Summary of probable SARS cases with onset of illness from 1 November 2002 to 31 July 2003 [Internet]. World Health Organization. [cited 2025 May 30]. Available from: https://www.who.int/publications/m/item/summary-of-probable-sars-cases-with-onset-of-illness-from-1-november-2002-to-31-july-2003.
58. Heymann DL. The international response to the outbreak of SARS in 2003. Philos Trans R Soc Lond B Biol Sci 2004;359(1447):1127–1129.
59. Heymann DL, Rodier G. Global surveillance, national surveillance, and SARS. Emerg Infect Dis 2004;10(2):173–175.
60. Rota PA, Oberste MS, Monroe SS et al. Characterization of a novel coronavirus associated with severe acute respiratory syndrome. Science 2003;300(5624):1394–1399.
61. Peiris JSM, Yuen KY, Osterhaus ADME et al. The severe acute respiratory syndrome. N Engl J Med 2003;349(25):2431–2441.
62. Tan CC. SARS in Singapore–key lessons from an epidemic. Ann Acad Med Singap 2006;35(5):345–349.
63. Chan GCT, Koh D. Reviewing lessons learnt of SARS in Singapore during planning for influenza pandemic. Int Marit Health 2006;57(1–4):163–176.
64. Singh K, Hsu LY, Villacian JS et al. Severe acute respiratory syndrome: Lessons from Singapore. Emerg Infect Dis 2003;9(10):1294–1298.
65. Seto WH, Tsang D, Yung RWH et al. Effectiveness of precautions against droplets and contact in prevention of nosocomial transmission of severe acute respiratory syndrome (SARS). Lancet 2003;361(9368):1519–1520.
66. Prompetchara E, Ketloy C, Palaga T. Immune responses in COVID-19 and potential vaccines: Lessons learned from SARS and MERS epidemic. Asian Pac J Allergy Immunol 2020;38(1):1–9.
67. Managing COVID-19 in New Zealand [Internet]. Ministry of Health NZ. 2024 [cited 2025 May 14]. Available from: https://www.health.govt.nz/strategies-initiatives/programmes-and-initiatives/covid-19/managing-covid-19-in-new-zealand.
68. Mortality Analyses [Internet]. Johns Hopkins Coronavirus Resource Center. [cited 2025 May 31]. Available from: https://coronavirus.jhu.edu/data/mortality.
69. Baker M, Kvalsvig A, Verrall AJ et al. New Zealand's elimination strategy for the COVID-19 pandemic and what is required to make it work. N Z Med J 2020;133(1512):10–14.
70. Jefferies S, French N, Gilkison C et al. COVID-19 in New Zealand and the impact of the national response: A descriptive epidemiological study. Lancet Public Health 2020;5(11):e612–e623.
71. History of flu (influenza): Outbreaks and vaccine timeline [Internet]. Mayo Clinic. [cited 2025 May 18]. Available from: https://www.mayoclinic.org/diseases-conditions/history-disease-outbreaks-vaccine-timeline/flu.
72. Morens DM, Taubenberger JK. The mother of all pandemics is 100 years old (and going strong)! Am J Public Health 2018;108(11):1449–1454.
73. Taubenberger JK, Morens DM. 1918 influenza: The mother of all pandemics. Emerg Infect Dis 2006;12(1):15–22.

74. Kilbourne ED. Influenza pandemics of the 20th century. Emerg Infect Dis 2006;12(1):9–14.
75. Oshitani H, Kamigaki T, Suzuki A. Major issues and challenges of influenza pandemic preparedness in developing countries. Emerg Infect Dis 2008;14(6):875–880.
76. Ziegler T, Mamahit A, Cox NJ. 65 years of influenza surveillance by a World Health Organization-coordinated global network. Influenza Other Respir Viruses 2018;12(5):558–565.
77. Adlhoch C, Baldinelli F. Avian influenza, new aspects of an old threat. Euro Surveill 2023;28(19):2300227.
78. Jennings LC, Monto AS, Chan PKS et al. Stockpiling prepandemic influenza vaccines: A new cornerstone of pandemic preparedness plans. Lancet Infect Dis 2008;8(10):650–658.
79. Cargnin Faccin F, Perez DR. Pandemic preparedness through vaccine development for avian influenza viruses. Hum Vaccin Immunother 2024;20(1):2347019.
80. Phanuphak N, Gulick RM. HIV treatment and prevention 2019: Current standards of care. Curr Opin HIV AIDS 2020;15(1):4–12.
81. Menéndez-Arias L, Delgado R. Update and latest advances in antiretroviral therapy. Trends Pharmacol Sci 2022;43(1):16–29.
82. Khalili J, Landovitz RJ. HIV preexposure prophylaxis-the role of primary care clinicians in ending the HIV epidemic. JAMA Intern Med 2020;180(1):126–130.
83. Swinkels HM, Nguyen AD, Gulick PG. HIV and AIDS. StatPearls [Internet], 2025. StatPearls Publishing, Treasure Island (FL) [cited 2025 May 31]. Available from: http://www.ncbi.nlm.nih.gov/books/NBK534860/.
84. Global HIV & AIDS statistics — Fact sheet [Internet]. UNAIDS. [cited 2025 May 30]. Available from: https://www.unaids.org/en/resources/fact-sheet.
85. Mahajan AP, Sayles JN, Patel VA et al. Stigma in the HIV/AIDS epidemic: A review of the literature and recommendations for the way forward. AIDS 2008;22 Suppl 2(Suppl 2):S67–79.
86. Gupta GR, Parkhurst JO, Ogden JA et al. Structural approaches to HIV prevention. Lancet 2008;372(9640):764–775.
87. Junaid K, Afzal S, Daood M et al. Substance abuse and mental health issues among HIV/AIDS patients. J Coll Physicians Surg Pak 2023;33(3):325–334.
88. Zanetti AR, Van Damme P, Shouval D. The global impact of vaccination against hepatitis B: A historical overview. Vaccine 2008;26(49):6266–6273.
89. Ropero Álvarez AM, Pérez-Vilar S, Pacis-Tirso C et al. Progress in vaccination towards hepatitis B control and elimination in the region of the Americas. BMC Public Health 2017;17(1):325.
90. Al-Busafi SA, Alwassief A. Global perspectives on the hepatitis B vaccination: Challenges, achievements, and the road to elimination by 2030. Vaccines (Basel) 2024;12(3):288.
91. Gooch MD. The resurgence of measles: A rash from the past. Adv Emerg Nurs J 2024;46(3):217–227.
92. CDC. Global Measles Outbreaks [Internet]. Global Measles Vaccination. 2025 [cited 2025 May 18]. Available from: https://www.cdc.gov/global-measles-vaccination/data-research/global-measles-outbreaks/index.html.
93. Wang R, Jing W, Liu M et al. Trends of the global, regional, and national incidence of measles, vaccine coverage, and risk factors in 204 countries from 1990 to 2019. Front Med (Lausanne) 2021;8:798031.
94. Gabis LV, Attia OL, Goldman M et al. The myth of vaccination and autism spectrum. Eur J Paediatr Neurol 2022;36:151–158.
95. Novilla MLB, Goates MC, Redelfs AH et al. Why parents say no to having their children vaccinated against measles: A systematic review of the social determinants of parental perceptions on MMR vaccine hesitancy. Vaccines (Basel) 2023;11(5):926.
96. Jost M, Luzi D, Metzler S et al. Measles associated with international travel in the region of the Americas, Australia and Europe, 2001-2013: A systematic review. Travel Med Infect Dis 2015;13(1):10–18.
97. Guha-Sapir D, Moitinho de Almeida M, Scales SE et al. Containing measles in conflict-driven humanitarian settings. BMJ Glob Health 2020;5(9):e003515.
98. Mehtar S, AlMhawish N, Shobak K et al. Measles in conflict-affected northern Syria: Results from an ongoing outbreak surveillance program. Confl Health 2021;15(1):95.
99. Chala B, Hamde F. Emerging and re-emerging vector-borne infectious diseases and the challenges for control: A review. Front Public Health 2021;9:715759.
100. Zhang Y, Wang M, Huang M et al. Innovative strategies and challenges mosquito-borne disease control amidst climate change. Front Microbiol 2024;15:1488106.
101. Matowo NS, Martin J, Kulkarni MA et al. An increasing role of pyrethroid-resistant Anopheles funestus in malaria transmission in the Lake Zone, Tanzania. Sci Rep 2021;11(1):13457.
102. Monawwer SA, Alzubaidi AOI, Yasmin F et al. Gene-drive mosquitoes: A prospect for future malaria control. Pan Afr Med J 2022;41:109.

103. Benelli G, Beier JC. Current vector control challenges in the fight against malaria. Acta Trop 2017;174:91–96.
104. McMahon A, Mihretie A, Ahmed AA et al. Remote sensing of environmental risk factors for malaria in different geographic contexts. Int J Health Geogr 2021;20(1):28.
105. Rosello A, Mossoko M, Flasche S et al. Ebola virus disease in the democratic Republic of the Congo, 1976-2014. Elife 2015;4:e09015.
106. Juma CA, Mushabaa NK, Abdu Salam F et al. COVID-19: The current situation in the democratic Republic of Congo. Am J Trop Med Hyg 2020;103(6):2168–2170.
107. Chapwanya M, Lubuma J, Terefe Y et al. Analysis of war and conflict effect on the transmission dynamics of the tenth Ebola outbreak in the democratic Republic of Congo. Bull Math Biol 2022;84(12):136.
108. Rojek A, Fieggen J, Apiyo P et al. Ebola disease: Bridging scientific discoveries and clinical application. Lancet Infect Dis 2025;25(3):e165–e176.
109. Alabi A, Kokou K, Mahmoudou S et al. Replication, safety and immunogenicity of the vectored Ebola vaccine rVSV-ΔG-ZEBOV-GP in a sub-Saharan African paediatric population: A randomised controlled, open-label trial in children aged 1-12 years living in Lambaréné, Gabon. J Infect 2024;89(4):106237.
110. Watson CH, Gsell PS, Hall Y, et al. rVSV-ZEBOV vaccination in people with pre-existing immunity to Ebolavirus: An open-label safety and immunogenicity study in Guinean communities affected by Ebola virus disease (l'essai proches). BMC Med. 2024;22(1):523.
111. Zavras D. Access to the COVID-19 vaccine. Int J Environ Res Public Health 2022;19(17):11054.
112. Aleem A, Akbar Samad AB, Vaqar S. Emerging variants of SARS-CoV-2 and novel therapeutics against coronavirus (COVID-19). StatPearls [Internet], 2025. StatPearls Publishing, Treasure Island (FL) [cited 2025 May 17]. Available from: http://www.ncbi.nlm.nih.gov/books/NBK570580/.
113. Beyer S, Haufe S, Dirks M et al. Post-COVID-19 syndrome: Physical capacity, fatigue and quality of life. PLOS One 2023;18(10):e0292928.
114. Morgul E, Bener A, Atak M et al. COVID-19 pandemic and psychological fatigue in Turkey. Int J Soc Psychiatry 2021;67(2):128–135.
115. Plowright RK, Parrish CR, McCallum H et al. Pathways to zoonotic spillover. Nat Rev Microbiol 2017;15(8):502–510.
116. Ellwanger JH, Fearnside PM, Ziliotto M et al. Synthesizing the connections between environmental disturbances and zoonotic spillover. An Acad Bras Cienc 2022;94:e20211530.
117. Tajudeen YA, Oladunjoye IO, Bajinka O et al. Zoonotic spillover in an era of rapid deforestation of tropical areas and unprecedented wildlife trafficking: Into the wild. Challenges 2022;13(2):41.
118. Marie V, Gordon ML. The (re-)emergence and spread of viral zoonotic disease: A perfect storm of human ingenuity and stupidity. Viruses 2023;15(8):1638.
119. Gayer M, Legros D, Formenty P et al. Conflict and emerging infectious diseases. Emerg Infect Dis 2007;13(11):1625–1631.
120. Goniewicz K, Burkle FM, Horne S et al. The influence of war and conflict on infectious disease: A rapid review of historical lessons we have yet to learn. Sustainability 2021;13(19):10783.
121. Törnberg P. Echo chambers and viral misinformation: Modeling fake news as complex contagion. PLOS One 2018;13(9):e0203958.
122. Nsoesie EO, Cesare N, Müller M et al. COVID-19 misinformation spread in eight countries: Exponential growth modeling study. J Med Internet Res 2020;22(12):e24425.
123. Baker MN, Merkley E. Dynamic role of personality in explaining COVID-19 vaccine hesitancy and refusal. Front Psychol 2023;14:1163570.
124. Lunsky Y, Kithulegoda N, Thai K et al. Beliefs regarding COVID-19 vaccines among Canadian workers in the intellectual disability sector prior to vaccine implementation. J Intellect Disabil Res 2021;65(7):617–625.
125. Kandasamy S, Manoharan B, Khan Z et al. Perceptions of COVID-19 risk, vaccine access and confidence: A qualitative description of South Asians in Canada. BMJ Open 2023;13(4):e070433.
126. Ackah BBB, Woo M, Stallwood L et al. COVID-19 vaccine hesitancy in Africa: A scoping review. Glob Health Res Policy 2022;7(1):21.
127. Hopkins KL, Underwood T, Iddrisu I et al. Community-based approaches to increase COVID-19 vaccine uptake and demand: Lessons learned from four UNICEF-supported interventions. Vaccines (Basel) 2023;11(7):1180.
128. Man SS, Wen H, Zhao L et al. Role of trust, risk perception, and perceived benefit in COVID-19 vaccination intention of the public. Healthcare (Basel) 2023;11(18):2589.
129. Vergara RJD, Sarmiento PJD, Lagman JDN. Building public trust: A response to COVID-19 vaccine hesitancy predicament. J Public Health (Oxf) 2021;43(2):e291–e292.

130. Nuss H, Privor-Dumm L, Ukachukwu C et al. Building and sustaining flu vaccine acceptance and trust in the black community through partnerships with churches, salons, and barbershops. J Racial Ethn Health Disparities 2024. doi: 10.1007/s40615-024-02141-7.
131. Opel DJ, Salmon DA, Marcuse EK. Building trust to achieve confidence in COVID-19 vaccines. JAMA Netw Open 2020;3(10):e2025672.
132. Purvis RS, Hallgren E, Moore RA et al. Trusted sources of COVID-19 vaccine information among hesitant adopters in the United States. Vaccines (Basel) 2021;9(12):1418.
133. Morgan O, Pebody R. The WHO hub for pandemic and epidemic intelligence; supporting better preparedness for future health emergencies. Euro Surveill 2022;27(20):2200385.
134. Getchell M, Wulandari S, de Alwis R et al. Pathogen genomic surveillance status among lower resource settings in Asia. Nat Microbiol 2024;9(10):2738–2747.
135. Koopmans MPG. Fast-forwarding collaborative surveillance. Front Sci 2024;2:1415415.
136. Hay AJ, McCauley JW. The WHO global influenza surveillance and response system (GISRS)-A future perspective. Influenza Other Respir Viruses 2018;12(5):551–557.
137. Kilaru P, Hill D, Anderson K et al. Wastewater surveillance for infectious disease: A systematic review. Am J Epidemiol 2023;192(2):305–322.
138. Hassard F, Bajón-Fernández Y, Castro-Gutierrez V. Wastewater-based epidemiology for surveillance of infectious diseases in healthcare settings. Curr Opin Infect Dis 2023;36(4):288–295.
139. Sheerah HA, AlSalamah S, Alsalamah SA et al. The rise of virtual health care: Transforming the health care landscape in the Kingdom of Saudi Arabia: A review article. Telemed J E Health 2024;30(10):2545–2554.
140. Latorre F, Hawks CE, Colmenares B et al. Unified identification protocol for cross-border healthcare. Stud Health Technol Inform 2023;309:111–115.
141. Ahmed MM, Okesanya OJ, Gilbert A. Comprehensive public health response and preparedness: Key lessons from Rwanda's effective containment of the 2024 Marburg virus outbreak. Ethics, Medicine and Public Health 2025;33:101037.
142. Halabi S, Gostin LO, Aneja K et al. The coalition for epidemic preparedness innovations (CEPI) and the partnerships of equitable vaccine access. J Law Med Ethics 2023;51(2):234–246.
143. Eccleston-Turner M, Upton H. International collaboration to ensure equitable access to vaccines for COVID-19: The ACT-accelerator and the COVAX facility. Milbank Q 2021;99(2):426–449.
144. Carroll D, Daszak P, Wolfe ND et al. The Global Virome Project. Science 2018;359(6378):872–874.
145. Richards P, Amara J, Ferme MC et al. Social pathways for Ebola virus disease in rural Sierra Leone, and some implications for containment. PLOS Negl Trop Dis 2015;9(4):e0003567.
146. Nanyonjo A, Nakirunda M, Makumbi F et al. Community acceptability and adoption of integrated community case management in Uganda. Am J Trop Med Hyg 2012;87(5 Suppl):97–104.
147. Park YJ, Choe YJ, Park O et al. Contact tracing during coronavirus disease outbreak, South Korea, 2020. Emerg Infect Dis 2020;26(10):2465–2468.
148. Dinh L, Dinh P, Nguyen PDM et al. Vietnam's response to COVID-19: Prompt and proactive actions. J Travel Med 2020;27(3):taaa047.
149. Wang CJ, Ng CY, Brook RH. Response to COVID-19 in Taiwan: Big data analytics, new technology, and proactive testing. JAMA 2020;323(14):1341–1342.
150. Tazerji SS, Nardini R, Safdar M et al. An overview of anthropogenic actions as drivers for emerging and re-emerging zoonotic diseases. Pathogens 2022;11(11):1376.
151. Aguirre AA, Catherina R, Frye H et al. Illicit wildlife trade, wet markets, and COVID-19: Preventing future pandemics. World Med Health Policy 2020;12(3):256–265.
152. Nash D, Mostashari F, Fine A et al. The outbreak of West Nile virus infection in the New York City area in 1999. N Engl J Med 2001;344(24):1807–1814.
153. Martín-Vélez V, Navarro J, Figuerola J et al. A spatial analysis of urban gulls contribution to the potential spread of zoonotic and antibiotic-resistant bacteria. Sci Total Environ 2024;912:168762.
154. Shaheen MNF. The concept of one health applied to the problem of zoonotic diseases. Rev Med Virol 2022;32(4):e2326.
155. Keusch GT, Amuasi JH, Anderson DE et al. Pandemic origins and a one health approach to preparedness and prevention: Solutions based on SARS-CoV-2 and other RNA viruses. Proc Natl Acad Sci U S A 2022;119(42):e2202871119.
156. Degeling C, Johnson J, Kerridge I et al. Implementing a One Health approach to emerging infectious disease: Reflections on the socio-political, ethical and legal dimensions. BMC Public Health 2015;15:1307.

14 Next-Generation Immunotherapies and Vaccines

Kenneth Lundstrom

14.1 INTRODUCTION

Since the first demonstration of functional vaccines against smallpox in 1796 [1] vaccine development has seen a tremendous progress not the least during the last few years due to the accelerated vaccine development in the context of the COVID-19 pandemic [2]. Originally, whole virus vaccines were developed against a number of infectious diseases [3]. Most commonly vaccines based on live attenuated [4] or inactivated [5] viruses have been used. The advantage of this approach is that these vaccines target several epitopes on viral surface proteins inducing strong and pan-viral immune responses. However, the production procedure is time consuming and the flexibility to re-engineer vaccines to face emerging novel strains and variants of viruses is limited. Since the advent of recombinant protein technology, protein- and subunit-based vaccines have expanded the range of vaccine development approaches from protein isolation to production in heterologous expression systems [6]. Furthermore, peptide vaccines have proven successful in eliciting neutralizing antibodies against the foot-and-mouth disease virus (FMDV) in several animal models [7]. Nucleic acid vaccines, either based on DNA [8] or RNA [9] have received enormous attention recently not least due to the unpresidential vaccine development occurring during the COVID-19 pandemic [10].

Finally, vaccine development has frequently utilized viral vectors as delivery vehicles [11]. In this context, vectors based on both DNA and RNA viruses have been evaluated in preclinical animal models, in clinical trials and even some vaccines have received market approval (Table 14.1). For example, double-stranded DNA (dsDNA) adenovirus (Ad) has generated high levels of episomal heterologous gene expression in a broad range of host cells [12]. As the original Ad vectors induced strong anti-Ad immune responses in vivo, the introduction of genomic deletions in second- and third-generation Ad vectors has eliminated this problem [13]. Furthermore, engineering of packaging cell lines has facilitated large-scale production of GMP-grade Ad particles suitable for clinical trials [14].

Self-replicating RNA viruses possess the unique feature of self-amplification of their genomic and subgenomic single-stranded RNA (ssRNA) in infected host cells resulting in an estimated 200,000-fold RNA amplification [15]. For this reason and due to their broad host range [16], alphaviruses have become attractive delivery vehicles for immunotherapy and vaccine development [17, 18]. Moreover, alphavirus vectors can be utilized as viral particles, RNA replicon vectors and layered DNA plasmid vectors [19]. Flaviviruses are also self-replicating ssRNA viruses, which have been frequently used for heterologous gene expression related to cancer immunotherapy and vaccine development [20]. Typically, Kunjin virus (KUN) [21] and yellow fever virus (YFV) [22] vector systems have been utilized.

In contrast to alphaviruses and flaviviruses with an ssRNA genome of positive polarity, both measles viruses (MV) [23] and rhabdoviruses [26] carry a negative sense ssRNA genome. For this reason, MV particles are rescued by reverse genetics from helper cell lines [24]. Oncolytic MV vectors have been evaluated for cancer immunotherapy [25]. Reverse genetics have also been used in vaccine development of rhabdoviruses [27]. Vesicular stomatitis virus (VSV) vectors have been successfully used in vaccine development against Ebola virus disease (EVD) [28].

 DOI: 10.1201/9781003607816-14

TABLE 14.1
Viral Expression Vectors

Virus	Genome	Insert	Features	References
Adenoviruses	dsDNA	7.5 kb	• Broad host range, pre-existing immunity	[12]
			• Reduced immunity of second- and third-generation vectors	[13]
			• Packaging cell line for GMP production	[14]
Alphaviruses	+ssRNA	8 kb	• Broad host range	[16]
			• saRNA vectors provide extreme transgene expression	[17, 18]
			• Flexible use of particles, RNA and DNA delivery	[19]
Flaviviruses	+ssRNA	6 kb	• Relatively broad host range	[20]
			• Frequently used for immunotherapy and vaccines	[21, 22]
Measles viruses	–ssRNA	6 kb	• Relatively broad host range	[23]
			• Reverse genetics and engineered packaging cell lines	[24]
			• Oncolytic MV vectors used for cancer therapy	[25]
Rhabdoviruses	–ssRNA	6 kb	• Relatively broad host range	[26]
			• Reverse genetics used for rhabdovirus vectors	[27]
			• Successful application of VSV in vaccine development	[28]
Herpes viruses	dsDNA	>30 kb	• Broad host range, excellent packaging capacity	[29]
			• Latent infection, long-term expression	[30]
			• Oncolytic HSV applied for cancer therapy	[31]
Retroviruses	ssRNA	8 kb	• Broad host range	[32]
			• Long-term expression, chromosomal integration	[32]
			• Successful gene therapy in SCID-X1 patients	[33]
Lentiviruses	ssRNA	8 kb	• Infection of dividing and non-dividing cells	[34]
			• Non-random chromosomal integration	[35]
Newcastle disease virus	–ssRNA	4 kb	• Specific replication of tumor cells	[36]
			• Targeted killing of tumor cells	[37]
			• Efficient in preclinical and clinical cancer trials	[38]
Reoviruses	dsRNA	unknown	• Oncolytic reoviruses specifically kill cancer cells	[39]
			• Induced immune responses, antitumor activity	[40]
Poxviruses	dsDNA	>30 kb	• Excellent packaging capacity	[41]
			• Cancer immunotherapy and vaccine development	[42]
			• No prevention of infection by systemic immunity	[43]
Picornaviruses	ssRNA	6 kb	• Oncolytic activity in cancer cells	[44]
			• Replication in certain non-dividing cells	
Polyoma viruses	dsDNA	17.7 kb	• Large packaging capacity, broad host range	[45]
			• SV40-free VLPs for delivery of antiviral agents and DNA vaccines	[46]

Abbreviations: dsDNA, double-stranded DNA; dsRNA, double-stranded RNA, GMP, good manufacturing procedure; HSV, herpes simplex virus; MV, measles virus; SCID, severe combined immunodeficiency; +ssRNA, positive sense single-stranded RNA; –ssRNA, negative sense single-stranded RNA; VLPs, virus-like particles; VSV, vesicular stomatitis virus.

Herpes simplex viruses (HSVs) are large dsDNA viruses with a broad host range and a packaging capacity of more than 30 kb of foreign DNA [29]. HSV vectors are characterized by their latent infection mode and long-term transgene expression [30], making it a priority to remove non-essential HSV genes to assure reduced cytotoxicity and enhanced safety [29]. Oncolytic HSV vectors have proven efficient in cancer therapy [31].

Viral vectors based on the ssRNA retroviruses produce intermediate dsDNA in infected cells allowing integration into the host genome [32]. Retroviruses have been used for long-term transgene expression, particularly applied for successful treatment of Severe Combined Immunodeficiency-X1 (SCID-X1) patients [33]. Random chromosomal integration has, however, resulted in serious adverse events culminating in leukemia in some patients [47], which has been addressed by establishing controlled genome integration [48]. Alternatively, lentiviruses, a subgroup of retroviruses which can infect both dividing and non-dividing cells [34] and demonstrate non-random chromosomal integration have been used [35]. Several lentiviruses such as human immunodeficiency virus (HIV) [34], simian immunodeficiency virus (SIV) [49], feline immunodeficiency virus (FIV) [50], and equine infectious anemia virus (EIAV) [51] have been subjected to gene therapy applications.

The Newcastle disease virus (NDV), belonging to the family of Ortho RNA viruses, is a negative sense ssRNA virus, which can specifically replicate in tumor cells [36]. For this reason, NDV is considered an attractive alternative for cancer therapy. Although NDV has a limited packaging capacity of 4 kb of foreign genetic material [37], it has proven efficient in both preclinical animal models and clinical trials [38]. Reoviruses are double-stranded RNA (dsRNA) viruses which have oncolytic properties and the capacity to specifically kill cancer cells [39]. Moreover, reoviruses induce immune responses leading to antitumor immune responses [40]. Although the packaging capacity of reoviruses is unknown, their oncolytic properties as such contribute to efficient tumor cell killing.

Poxviruses are large dsDNA viruses with foreign DNA packaging capacity of more than 30 kb [41]. Typically, vaccinia virus (VV) strains have been used for cancer immunotherapy and vaccine development [42]. It has also been demonstrated that systemic immunity against VV in adult cancer patients did not prevent infection of tumors in vivo [43]. Picornaviruses are small ssRNA viruses, which have been used both in gene therapy and vaccine applications [44]. The dsRNA simian virus 40 (SV40), a polyoma virus, has a broad host range and an astonishing packaging capacity of 17.7 kb despite its small genome of 5 kb [45]. SV40 wild type-free virus-like particles (VLPs) have been engineered [46]. Moreover, SV40 VLPs have been used for the delivery of antiviral and DNA vaccines.

14.2 VACCINE DEVELOPMENT CHALLENGES

Viral vectors have proven safe and efficacious as delivery vehicles for vaccines. As vaccines against both infectious diseases and different types of cancers have been reviewed thoroughly elsewhere [52, 53], the focus is here on the several challenges that vaccine development is facing. For example, generation of strong immune responses is essential for the development of efficient vaccines. Target selection and optimization can be supported by molecular modeling [54] and computational design [55]. Moreover, efficient transgene expression is mandatory to elicit strong immune responses. In this context, different types of vectors have proven excellent [56] (Table 14.2). For example, Ad-vectors expressing the influenza virus A hemagglutinin (HA) have demonstrated strong immune responses in mice and chicken [57]. Similarly, robust immune responses against HIV were detected in BALB/c mice immunized with a VV vector expressing the HIV Env-GM-CSF fusion protein [58]. Moreover, immunization of mice with a Semliki Forest virus (SFV) vector, an alphavirus, expressing the HIV Gag-Env-PolRT proteins elicited antigen-specific T-cell responses [59]. Another important issue concerns the type of immune response that is elicited and how it influences vaccine efficacy. Humoral immunity comprises adaptive immunity involving the production of antibodies to neutralize pathogens with the key players being B lymphocytes, plasma cells, and memory B cells [60]. Their function relates to recognizing extracellular bacteria and viruses, neutralizing them by binding to their surface, and activating the complement system. In contrast, cellular immunity presents an adaptive immune response, which does not involve antibodies relying on the destruction of infected or abnormal cells by T cells [61]. The key players of cellular immunity are T cells, helper T (Th) cells (CD4+), cytotoxic T cells (CD8+), and memory T cells. Their functions are to target intracellular bacteria and viruses, to destroy infected and cancer cells via cytotoxic mechanisms, and to release cytokines for modulation of the immune response. Moreover, the Th type bias is important for the potency of vaccines as the Th1 subset of T cells secrets such cytokines as interferon-γ (IFN-γ) and tumor necrosis factor (TNF), generally associated with inflammation whereas the Th2 subset

TABLE 14.2
Viral Vectors as Delivery Vectors for Vaccines against Infectious Diseases

Vector	Target	Features	References
Ad	HA	Strong immune response in mice and chicken	[57]
VV	HIV Env-GM-CSF	Strong immune response in BALB/c mice	[58]
SFV	HIV Env-Gag-PolRT	Antigen-specific T-cell responses in mice	[59]
VEE	EBOV-NP + EBOV-GP	Protection against EBOV in mice and guinea pigs	[63]
	EBOV-NP	Protection only in mice	[63]
VSV	EBOV-GP	Protection against EBOV in non-human primates	[64]
	MARV-GP	Protection against MARV in non-human primates	[64, 65]
	ZEBOV-GP	Excellent vaccine efficacy in phase III	[66]
		Excellent vaccine efficacy in phase III Vaccine approval in 2019 by the FDA	[28]
VEE	CMV gB or PP65/IE1	Safe, Ab responses in phase I	[67]
		Safe, not sufficient vaccine efficacy in phase IIb	[68]
VEE	HIV-Gag	Safe, well tolerated, only modest Ab responses in phase I	[69]
ChAdOx1	SARS-CoV-2 S	Humoral and cellular responses in mice	[70]
		Prevention of pneumonia in primates	[71]
		Good safety and efficacy in phase III	[72]
		EUA in the United Kingdom in January 2021	[73]
Ad5	SARS-CoV-2 S	Protection against SARS-CoV-2 in macaques	[74]
		Robust immunogenicity of pooled clinical trials	[75]
		EUA in China	[76]
Ad26	SARS-CoV-2 S	Protection against SARS-CoV-2 in hamsters	[77]
		Protection against SARS-CoV-2 in macaques	[78]
		Vaccine efficacy in clinical trials	[79]
		EUA by the FDA	[80]
Ad26 + Ad5	SARS-CoV-2 S	Protection against SARS-CoV-2 in phase I/II	[81]
		91.6% vaccine efficacy in phase III	[82]
		EUA in Russia	[83]
MVA	SARS-CoV-2 S	Complete protection against SARS-CoV-2 in hamsters	[84]
	SARS-CoV-2 S and ST	Humoral and cellular immune responses in phase I	[85]
VSV	SARS-CoV-2 S/EBOV-GP	Protection against SARS-CoV-2 in hamsters	[86]
		Protection against SARS-CoV-2 in primates	[87]
	SARS-CoV-2 S	Phase I discontinued due to weak immune responses	[88]
	SARS-CoV-2 S	Positive responses in phase I, phase II in progress	[89]
MV	SARS-CoV-2	Protection against SARS-CoV-2 in hamsters	[90]
		Termination after weak immune responses in phase I	[91]
VEE	LNP-SARS-CoV-2 S saRNA	Robust immune responses in mice	[92]
		SARS-CoV-2 neutralization in 15–48% cases in phase I	[93]
		Prime-boost regimen enhanced seroconversion in phase II	[94]
		95% protection against severe COVID-19	[95]
		Safety and immunogenicity in phase II/III	[96]
		EUA in India	[97]
		Long-term immunity, transmission prevention in hamsters	[98]

Abbreviations: Ab, antibody; Ad, adenovirus; CMV gB, cytomegalovirus glycoprotein; EBOV, Ebola virus; Env, envelope protein; EUA, emergency use authorization; GM-CSF, granulocyte macrophage-colony stimulating factor; GP, glycoprotein; HA, hemagglutinin; HIV, human immunodeficiency virus; LNP, liposomal nanoparticle; MARV, Marburg virus; MV, measles virus; MVA, modified vaccinia virus Ankara strain; NP, nucleoprotein; S, SARS-CoV-2 spike protein; saRNA, self-amplifying RNA; SFV, Semliki Forest virus; ST, prefusion-stabilized SARS-CoV-2 spike protein; VEE, Venezuelan equine encephalitis virus; VSV, vesicular stomatitis virus; VV, vaccinia virus.

produces IL-4 and Il-5 cytokines, which contribute to the proliferation and differentiation of B cells and is associated with humoral immunity [62]. Among cancer vaccines, numerous studies have demonstrated strong immune responses and therapeutic activity in animal models and clinical trials as described below.

14.3 PROTECTION AND THERAPEUTIC ACTIVITY

Obviously, the goal of vaccine development is to provide protection against challenges with infectious agents. In this context, Venezuelan equine encephalitis virus (VEE), an alphavirus, has been applied for the expression of the Ebola virus (EBOV) glycoprotein (GP) and nucleoprotein (NP) in BALB/c mice and guinea pigs [63]. Administration of VEE-EBOV-GP particles alone or in combination with VEE-EBOV-NP particles protected mice against lethal doses of EBOV [63]. In contrast, only mice were protected against EBOV when VEE-EBOV-NP particles were administered alone. Moreover, VSV, a rhabdovirus, expressing the EBOV-GP and Marburg virus GP (MARV-GP) protected primates against EBOV and MARV strains, respectively [64]. Most importantly, VSV vectors expressing the EBOV-GP from the Zaire strain showed excellent vaccine efficacy in a phase III study in Guinea [65]. Moreover, similar results were observed in another phase III study in Guinea and Sierra Leone [66]. The VSV-ZEBOV vaccine was approved for the treatment of Ebola virus disease (EVD) by the FDA in 2019 [28]. In the context of clinical trials, VEE particles expressing the cytomegalovirus (CMV) glycoprotein B (gB) or the PP65/IE1 fusion protein were subjected to a phase I study in 40 CMV-seronegative volunteers [67]. The vaccination was well tolerated showing mild to moderate local and minimal systemic reactogenicity and elicited neutralizing antibody and multifunctional T-cell responses essential for protective immunity. Furthermore, the CMV vaccine was evaluated in a phase IIb trial [68]. Although safe the three-dose administration showed 44.6% efficacy, which was not sufficient for reduction of the incidence of primary CMV infection compared to placebo. A phase I study in healthy HIV-negative individuals of subcutaneous administration of VEE particles expressing HIV-Gag demonstrated good safety and tolerability but only modest local immunogenicity and T-cell responses [69].

Development of vaccines against COVID-19 has received a special attention due to the recent pandemic. For example, the chimpanzee adenovirus ChAdOx1 vector carrying the SARS-CoV-2 S spike protein gene elicited strong immune responses in mice [70] and prevented pneumonia development in macaques [71]. Furthermore, the ChAdOx1-nCoV-19 vaccine has been evaluated in several clinical trials. In phase III, the safety profile was acceptable and the efficacy against symptomatic COVID-19 was good [72]. The ChAdOx1-nCoV-19 vaccine was granted emergency use authorization (EUA) in the United Kingdom in January 2021 [73]. Moreover, vaccines based on the Ad types 5 and 26 have been engineered. The Ad5-nb2 vaccine based on the expression of the SARS-CoV-2 S protein, protected immunized macaques against SARS-CoV-2 challenges [74]. Based on a number of clinical trials, the Ad5-nb2 vaccine elicited strong immune responses, which was superior after aerosol inhalation compared to intramuscular administration [75]. The Ad5-nb2 vaccine was granted EUA in China [76]. Another approach has been to develop a COVID-19 vaccine based on a single injection of Ad26 expressing the SARS-CoV-2 S gene, which demonstrated protection against challenges with SARS-CoV-2 in hamsters [77] and rhesus macaques [78]. The Ad26-CoV-2.S vaccine have proven efficient in clinical trials [79], resulting in EUA in the United States [80]. The Russian Sputnik V vaccine, which is based on a prime vaccination with the Ad26 serotype vector expressing the SARS-CoV-2 S gene followed by a booster administration of Ad5-SARS-CoV-2 S, showed protection of hamsters challenged with SARS-CoV-2 [81]. In clinical trials, 91.6% vaccine efficacy was claimed [82] and the vaccine has received EUA in Russia [83].

In addition to Ad vectors, MVA (modified vaccinia virus Ankara strain) poxvirus vectors expressing the SARS-CoV-2 S protein showed complete protection against SARS-CoV-2 challenges in hamsters [84]. In a clinical setting, humoral and cellular immune responses were detected after vaccination with an MVA vector expressing the native SARS-CoV-2 S gene (MVA-SARS-S), but superior immunogenicity was obtained with the prefusion-stabilized SARS-CoV-2 protein (MVA-SARS-2-ST) [85]. In the case of rhabdovirus-based COVID-19 vaccines, VSV vectors expressing either the SARS-CoV-2 S protein (VSV-SARS2) or a combination with the EBOV-GP (VSV-SARS2-EBOV) protected intranasally vaccinated hamsters against pneumonia [86]. Moreover, non-human primates immunized with the VSV-SARS2-EBOV vaccine were protected against SARS-CoV-2 challenges [87]. Related to clinical evaluation, the VSV-SARS-CoV-2 S V590 phase I study revealed weaker immune responses compared to those observed in convalescent COVID-19 patients leading to the discontinuation of the trial [88]. However, in a phase I/II study on the VSVΔG-SARS-CoV-2 vaccine, the favorable results from the phase I part allowed expansion into phase II [89]. In the case of MV vectors, Syrian hamsters immunized with MV vectors expressing the full-length SARS-CoV-2 S protein were protected against challenges with SARS-CoV-2 [90]. However, although the prefusion-stabilized MV-SARS-CoV-2 S TMV-083 vaccine showed good safety and tolerability in a phase I study, the insufficient immune responses resulted in the early termination of the study [91].

RNA molecules of self-replicating RNA viruses, mainly VEE, have been utilized for COVID-19 vaccine development [2, 11]. For example, VEE replicon RNA expressing the prefusion-stabilized SARS-CoV-2 S protein was encapsulated in liposomal nanoparticles (LNPs) (LNP-nCoV saRNA), which resulted in strong immune responses in BALB/c mice [92]. Moreover, LNP-nCoV saRNA demonstrated good safety and tolerability, and neutralized SARS-CoV-2 in 15–48% of vaccinated volunteers in phase I [93]. In phase II, a prime-boost regimen with 1 μg and 10 μg LNP-nCoV saRNA, respectively, enhanced the seroconversion rates significantly [94]. Another LNP formulation carrying the VEE SARS-CoV-2 S replicon RNA (LUNAR-COV19 RNA) showed 95% protection against severe COVID-19 and 55% overall protection against symptomatic disease [95]. Additionally, vaccination with the lipid inorganic nanoparticle (LION) encapsulated VEE-SARS-CoV-2 S RNA was deemed safe and immunogenic in phase II/III [96], which was granted EUA in India [97]. Recently, the intranasal thermostable LNP-encapsulated VEE-SARS CoV-2 S RNA vaccine (AAHI-SC2) has demonstrated long-term immunity, improved protection, and prevention of transmission in hamsters [98].

Cancer therapy, including immunotherapy and vaccine development, has been frequently targeted by viral vectors as described below and summarized in Table 14.3. Alternative approaches include the overexpression of tumor associated antigens (TAAs), cytotoxic and apoptotic proteins, immunostimulatory proteins such as cytokines and the application of oncolytic viruses, which specifically replicate in tumor cells [99]. For example, oncolytic Ad carrying the cancer cell targeting ligand SYENFSA (SYE) have shown specific targeting of pancreatic cancer cells leading to efficient oncolysis of pancreatic ductal adenocarcinoma (PDAC) cells [100]. In another study, the engineering of the survivin (Sur) promoter in the Ad-SYE vector (AdSur-SYE) showed high transduction efficiency of pancreatic neuroendocrine tumors (PNETs) and complete tumor regression in mice [101]. Moreover, the chimeric Ad serotype 5 and 3 vector carrying the melanoma differentiation-associated gene-7 (MDA-7) and interleukin-24 (IL-24) selectively killed tumor cells and additionally generated a bystander activity in distant tumors after intratumoral injection in mice [102].

Alphavirus vectors have proven efficient for immunotherapy and cancer vaccines. For example, SFV particles expressing IL-12 administered via an implanted cannula reduced RG2 gliomas by 87% in rats [103]. In another approach, replication-proficient SFV particles expressing GFP prolonged the survival of mice with A549 lung tumor xenografts [104]. Interestingly, immunization of mice with 0.1 μg of SFV-LacZ replicon RNA provided protection against colon tumor challenges [105]. Additionally, therapeutic effects and prolonged survival were observed in mice with pre-existing tumors after SFV-LacZ RNA administration. The combination of VEE particles expressing the tyrosine-related protein-2 (TRP-2) and the antagonist anti-CTL antigen-4 (CTLA-4) monoclonal antibody (mAb) resulted in complete B16F10 melanoma tumor regression in 50% of immunized mice [106]. Furthermore, complete regression in 90% of mice was observed for the combination of VEE-TRP-2 with the agonist anti-glucocorticoid-induced tumor necrosis factor

TABLE 14.3
Viral Vectors as Delivery Vectors for Cancer Immunotherapy and Vaccines

Vector	Target	Features	References
Ad	SYE	Targeting and killing of PDAC cells	[100]
AdSur	SYE	Complete regression of PNETs in mice	[101]
Ad5-Ad3	MDA-7, IL-24	Selective tumor cell killing	[102]
SFV	IL-12	87% regression of RG2 gliomas in rats	[103]
	GFP	Prolonged survival in mice with A549 lung tumors	[104]
SFV RNA	LacZ RNA	Protection and therapeutic efficacy in mice	[105]
SFV + CTLA-4 mAb	TRP-2	Complete tumor regression in 50% of mice	[106]
SFV + GITR mAb	TPR-2	Complete tumor regression in 90% of mice	[106]
SFV DNA	VEGFR-2/IL-12+ Surv/β-hCG	Superior tumor growth and survival after combination therapy	[107]
	HPV E6/E7	Antitumor responses up to 340 days in mice	[108]
	HPV-E6/E7	Strong Ab responses in all 12 pts in phase I	[109]
	HPV-E6/E/	Complete responses in 50% of CIN3 pts in phase II	[110]
VEE	CEA	Prolonged survival in colorectal cancer pts in phase I	[111]
	PSMA	Weak immunogenicity in prostate cancer pts	[112]
M1	Oncolytic M1	Tumor regression in mice with 4T1 and B16 tumors	[113]
KUN	GM-CSF	>50% (CT26) and 67% (B16-OVA) of mice cured	[21]
MV	NIS	Enhanced ATC cell killing in phase in mice	[114]
	SCD	Killing of tumor cell lines and primary tumors	[115]
	SCD	Killing of hepatocellular carcinoma in mice	[116]
	SCD	huCCT1 tumor regression, prolonged survival in mice	[25]
MV-Edm	Oncolytic MV	Complete tumor regression in CTCL pts in phase I	[117]
MV	CEA	Double overall survival in pts with ovarian cancer	[118]
		No dose-limiting toxicity in glioma pts in phase I	[119]
VSV	IFNβ-NIS	Dose dependent tumor regression in mice	[120]
	IFNβ-NIS	Good safety and tolerability in dogs	[121]
	MUC1	Reduction in tumor growth in mice	]122]
HSV	Oncolytic HSV	Tumor reduction, prolonged survival in mice	[123]
	Oncolytic HSV-2	Inhibition of tumor growth in mice	[124]
HSV HF10	Oncolytic HSV	Good safety, efficacy in breast, H&N, pancreatic cancers	[125]
	Oncolytic HSV	Phase I: 3 pts with SD, 4 pts with PR, 9 pts with PD	[126]
T-VEC + ipilimumab	Oncolytic HSV	Phase II: objective responses in melanoma pts	[127]
	Oncolytic HSV	Higher objective response rates after combination	[128]
RRV Toca511	CD	Long-term survival in murine glioma model	[129]
	CD	Superior survival in HGG pts in phase I	[130]
NDV	IL-15, IL-12	Superior survival of NDV-IL15 in mice with melanoma	[131]
NDV D90	GFP	Suppression of tumor growth in mice with lung tumors	[132]
NDV PV101	Oncolytic NDV	Progression-free survival in pts with solid tumors	[133]
NDV La Sota	Oncolytic NDV	Prolonged survival in colorectal cancer pts in phase III	[134]
Reovirus	Oncolytic reovirus	Prolonged survival in mice with breast cancer	[135]
Reovirus	Reolysin + paclitaxel/ carboplatin	Phase II: Tolerability, benefits in NSCLC pts	[136]
Reovirus	Reolysin + GEM	Phase II: successful combination in pts with PDAC	[137]
VV	PANVAC-VF	CEA and MUC-1 CTL-specific responses in mice	[138]
VV	GLV-2b372	50% reduction of HCC tumors in mice	[139]
VV + Radio-iodide	GLV-1h153-NIS	Synergistic effect of VV and radio-iodide in mice	[140]
VV	vvDD	Resolution of metastasis, tumor regression in pts with colorectal cancer in phase I	[141]
VV	PANVAC-VF	Phase III: 50% tumor regression in 67% of melanoma pts	[142]

(Continued)

TABLE 14.3 (*Continued*)
Viral Vectors as Delivery Vectors for Cancer Immunotherapy and Vaccines

Vector	Target	Features	References
CAV21	ICAM-1-DAF	Tumor regression in mice with melanoma	[143]
	ICAM-1-DAF	Regression, elimination of metastases in mice	[144]
	ICAM-1-DAF+DOX	Enhanced tumor regression murine breast tumor model	[145]
	ICAM-1	Phase I: Good safety, tumor cell death in NMIBC pts	[146]
SV40	hTR-SST2	Inhibition of pancreatic tumor progression in mice	[147]

Abbreviations: Ab, antibody; Ad, adenovirus; ATC, anaplastic thyroid cancer; CD, cytosine deaminase; CEA, carcinoembryonic antigen; CIN3, cervical intraepithelial neoplasia, grade 3; CTCL, cutaneous T-cell lymphoma; CTLA-4, CTL antigen-4; CAV21, coxsackievirus A21; DAF, decay-accelerating factor; DOX, doxorubicin hydrochloride; GEM, gemcitabine; GITR, glucocorticoid-induced tumor necrosis factor receptor; GM-CSF, granulocyte macrophage-colony stimulating factor; HCC, hepatocellular cancer; HPV, human papillomavirus; HSV, herpes simplex virus; huCCT1, human cholangiocarcinoma; ICAM-1, intercellular adhesion molecule-1; IFN, interferon; IL-24, interleukin-24; KUN, Kunjin virus; mAb, monoclonal antibody; MDA-7, melanoma differentiation-associated gene-7; MUC-1, mucin-1; MV, measles virus; NDV, Newcastle disease virus; NIS, sodium iodine symporter; NMIBC, non-muscle invasive bladder cancer; NSCLC, non-small cell lung cancer; PD, progressive disease; PDAC, pancreatic ductal adenocarcinoma; PNET; pancreatic neuroendocrine tumor; PR, partial response; PSMA, prostate-specific membrane antigen; pts, patients; SCD, super cytosine deaminase; SD, stable disease; SFV, Semliki Forest virus; Sur, survivin promoter; SYE, SYENSFA cancer cell targeting ligand; TRP-2, tyrosine-related protein-2; VEE, Venezuelan equine encephalitis virus; VEGFR-2, vascular endothelial growth factor receptor-2; VSV, vesicular stomatitis virus; VV, vaccinia virus.

receptor (GITR) mAb [106]. Self-replicating DNA replicons have also been evaluated such as the SFV DNA replicon co-expressing the vascular endothelial growth factor receptor-2 (VEGFR-2) and interleukin-12 (IL-12) and the SFV DNA replicon co-expressing survivin and the β-hCG antigen [107]. Co-administration of the two SFV DNA replicons resulted in superior tumor growth inhibition and survival compared to treatment with either SFV DNA replicon alone in a B16 mouse melanoma model. In another approach SFV vectors expressing the human papillomavirus (HPV) fusion protein E6-E7 was administered to tumor-bearing mice leading to regression and complete elimination of established tumors [108]. Long-term cytotoxic T lymphocyte (CTL) and antitumor responses lasting up to 340 days were also obtained. In phase I, the SFV-HPV E6/E7 vaccine (Vvax001) showed good safety and tolerability profiles in patients with HPV-induced cancers with strong HPV16 specific antibody responses in all 12 evaluated patients [109]. In phase II, 18 patents with cervical intraepithelial neoplasia grade 3 (CIN3) were fully vaccinated, of which reduction in CIN3 lesions were detected in 94% of the patients already three weeks after vaccine administration [110]. Moreover, histopathologic complete responses were seen in 50% of the patients and HPV16 clearance was achieved in 63% of the patients. In phase I, VEE particles expressing the carcinoembryonic antigen (CEA) elicited antigen-specific immunogenicity and showed prolonged survival in patients with stages III and IV colorectal cancer [111]. In contrast, VEE particles expressing the prostate-specific membrane antigen (PSMA) induced only weak PSMA-specific immunogenicity in castration-resistant metastatic prostate cancer (CRPC) patients in phase I [112] despite previously eliciting robust immune responses in mice [148]. The oncolytic M1 alphavirus has shown high tumor tropism and efficient tumor killing in mice with implanted 4T1 mammary carcinoma or B16 melanoma cells [113].

Among flaviviruses, intratumoral injection of KUN vectors expressing GM-CSF into mice with subcutaneous CT26 colon tumors resulted in cure of more than 50% of animals [21]. Likewise, significant tumor regression and a 67% cure rate were observed in mice with B16-OVA melanoma tumors [21].

Oncolytic MV strains such as MV-Edm expressing the sodium iodide symporter (NIS) enhanced tumor cell killing in a murine anaplastic thyroid cancer (ATC) model [114]. Moreover, expression of the bifunctional suicide gene super cytosine deaminase (SCD) comprising cytosine deaminase (CD) and uracil phosphoribosyl transferase from MV vectors efficiently killed ovarian tumor cell lines and primary tumors [115]. MV-SCD also provided long-term replication in tumor tissue and induced cell death in mice with hepatocellular carcinoma (HCC) xenografts [116]. In another study, MV-SCD delivery to mice with human cholangiocarcinoma (huCCT1) resulted in tumor regression and significant prolonged survival [25]. In phase I, the MV-Edm-Zagreb strain showed complete tumor regression in one patient with cutaneous T-cell lymphoma (CTCL) [117]. In a phase I/II study in patients with recurrent ovarian cancer disease, stabilization was achieved in all nine patients with twice as long as expected overall survival of 12.5 months [118]. Additionally, MV-CEA showed no dose-limiting toxicity in patients with recurrent glioblastoma multiforme in phase I [119].

In the context of rhabdoviruses, expression of interferon-β (IFN-β) and NIS from an oncolytic VSV vector generated a dose-dependent regression of syngeneic 5TGM1 plasmacytoma tumors in mice [120]. Moreover, intravenous administration of VSV-IFNβ-NIS showed good safety and tolerability in dogs with naturally occurring tumors [121]. Expression of the human mucin 1 (MUC1) gene from VSV vectors resulted in significant reduction of tumor growth in a PDAC mouse model [122].

Oncolytic HSV vectors have demonstrated efficient replication in tumor cells, which induces T cells and natural tumor killer cells leading to a significant reduction in tumor growth and prolonged survival of immunized mice [123]. Likewise, immunization with an oncolytic HSV-2 vector caused a significant inhibition of tumor growth [124]. In phase I, the oncolytic HSV HF10 was evaluated in patients with recurrent breast cancer, head and neck cancer, and pancreatic cancer showing good safety and some therapeutic efficacy [125]. In another phase I study, in patients with nonresectable locally advanced pancreatic cancer, intratumoral HSV HF10 administration resulted in partial responses (PR) in three patients, stable disease (SD) in four patients, and progressive disease (PD) in nine patients [126]. The HSV based Talimogene Laherprarepvec (T-VEC) vector was evaluated in combination with ipilimumab in a phase II study in comparison to ipilimumab monotherapy in patients with advanced unresectable melanoma [127]. The combination treatment did not show differential benefit in progression-free survival although objective responses were seen in patients. In a five-year follow-up phase II study on patients with unresectable stage IIIB–IV melanoma, the combination of T-VEC and ipilimumab showed higher objective response rates compared to ipilimumab monotherapy [128].

The nonlytic amphitropic retroviral replicating vector (RRV) Toca511 expressing yeast CD showed prolonged long-term survival in an orthotopic glioma model [129]. In the context of clinical trials, Toca511 demonstrated statistically superior survival of 13.6 months in patients with recurrent or progressive high-grade glioma (HGG) in phase I [130].

Oncolytic NDV vectors expression IL-12 or IL-15 suppressed tumor growth in a mouse melanoma model [131]. However, the administration of NDV-IL15 generated superior survival rates in comparison to NDV-IL-12 and tumor re-challenging indicated a 26.6% better survival for NDV-IL-15 compared to NDV-IL-12 [131]. Moreover, the engineered oncolytic NDV D90 strain expressing GFP showed suppression of tumor growth and loss of body weight in athymic mice with implanted lung tumors [132]. In phase II, the NDV PV101 strain provided progression-free survival of 4–31 months in patients with solid tumors [133]. In phase III, NDV La Sota-based immunotherapy resulted in prolonged survival and short-term improvement in quality of life of colorectal cancer patients [134].

Oncolytic reoviruses have also been subjected to cancer therapy. For example, various pancreatic cancer cell lines are susceptible to reoviruses [149]. Moreover, oncolytic reoviruses prolonged survival in a murine breast cancer model, which was further enhanced by the combination with the checkpoint inhibitor anti-PD1 [135]. Reoviruses have been subjected to a large number of clinical trials [150]. For example, the oncolytic type 3 Dearing reovirus (Reolysin) was combined with paclitaxel and carboplatin in a phase II study showing good tolerability and response benefits in patients with non-small cell lung cancer (NSCLC) [136]. Moreover, the combination of Reolysin and Gemcitabine was favorable in PDAC patients in phase II [137].

Among poxviruses, VV vectors have been frequently applied for cancer therapy [151]. For example, The PANVAC-VF system consists of recombinant VV and fowl poxvirus expressing the MUC1 and CEA genes, respectively [138]. The PANVAC-VF system also contains three costimulatory molecules: cluster of differentiation B7.1, intercellular adhesion molecule-1 (ICAM-1) and leukocyte function associated antigen-3 (LFA-3), collectively named TRICOM which has elicited MUC-1 and CEA CTL responses in mice [138]. Moreover, the light-emitting GLV-2b372 VV vector generated a 50% reduction in HCC tumor volumes in mice compared to the 400% increased tumor growth in control mice [139]. Furthermore, the GLV-1h153 VV vector expressing NIS showed synergistic effect of VV-NIS and radio-iodide treatment in the transgenic adenocarcinoma of the mouse prostate (TRAMP) model related to tumor growth and survival [140]. In clinical evaluation, intravenous administration of the vvDD vector to patients with advanced colorectal cancer and other solid tumors induced potent Th1-mediated immune responses and resolution of liver metastasis and tumor regression were observed [141]. In phase III, the PANVAC-HF vaccine provided a 50% reduction in tumor size in 67% of patients with unresectable stage IIIB, IIIC, and IV melanoma [142].

Among picornaviruses, subcutaneously injected coxsackievirus A21 (CAV21) expressing the ICAM-1 and decay-accelerating factor (DAF) genes reduced tumor burden in a melanoma mouse model [143]. Moreover, CAV21-ICAM-1-DAF significantly reduced pre-existing tumors and eliminated metastases in mice with T47D and MDA-MB-231-luc breast tumor xenografts [144]. Tumor regression was further enhanced by combination therapy with doxorubicin hydrochloride [145]. Moreover, the CAV21-ICAM-1 vector has been subjected to a phase I trial in patients with a non-muscle-invasive bladder cancer (NMIBC) showing an acceptable safety profile, viral targeting, replication, and tumor cell death [146].

Among polyoma viruses, a replication-deficient SV40 vector carrying the human telomerase RNA (hTR) tumor-specific promoter and the somatostatin receptor tumor-suppressor 2 (SST2) gene has been used [147]. Intratumoral injection resulted in inhibition of tumor cell progression in mice with Capan-1 pancreatic tumors.

14.4 INNOVATIVE APPROACHES TO IMMUNIZATION

Despite the numerous examples presented above on immunotherapeutic applications, novel approaches for improvements have recently been described. For example, the COVID-19 pandemic has highlighted the problems with waning SARS-CoV-2 vaccine potency with time and the continuous emerging of novel vaccine variants with potentially increased infectivity and pathogenicity [152]. The natural waning of COVID-19 vaccine potency with time has been addressed by booster vaccinations, which has to a certain extent enhanced and prolonged the vaccine efficacy [153]. However, the emerging SARS-CoV-2 variants such as the alpha (B.1.1.7) variant with 17 mutations, beta (B.1.351) with 9 mutations, gamma (P.1) with 10 mutations, delta (B.1.617.2) with 13 mutations, lambda (C.13) with 8 mutations, and omicron (B.1.1.529) with 50 mutations have generated concern of vaccine potency [154]. The potency of first-generation vaccines against SARS-CoV-2 variants has been reviewed elsewhere [155] and is briefly summarized below. For example, homologous booster vaccinations with the whole-virus vaccine VLA2001 and the heterologous ChAdOx1 nCoV-19 vaccine, respectively, enhanced vaccine efficacy against the delta and omicron variants [156]. However, in another study it was demonstrated in a phase I/II study that the ChAdOx1 nCoV-19 vaccine failed to protect against mild-to-moderate COVID-19 caused by the beta variant [157]. In a comparative study, the ChAdOx1 nCoV-19 and BNT162b2 mRNA vaccines showed superior reduction in SARS-CoV-2 transmission of the alpha variant compared to the delta variant [158]. In another study, a clear decrease in neutralizing antibody titers against the beta and gamma variants compared to the Wuhan strain were observed after immunization with the Ad26.COV2.S vaccine

[159]. Flu VLPs expressing the SARS-CoV-2 S receptor-binding domain (RBD) induced strong neutralizing antibodies against the Wuhan strain and the delta variant [160].

Significant attention has been paid to the re-engineering of first-generation COVID-19 vaccines to enhance the immunogenicity against SARS-CoV-2 variants. For example, the VSV-ΔG-S vector containing a chimeric minispike of the globular RBD was engineered and evaluated for immunogenicity against several variants of concern (VoC) in mice [161]. Comparable neutralizing antibodies were obtained for the Wuhan strain and the alpha, gamma and delta variants, in contrast to the beta and omicron variants showing a 3-fold reduction in titers. Moreover, engineering of the bivalent mRNA-1273.214 vaccine, comprising the original mRNA-1273 and the omicron BA.1 corresponding RNA, elicited significantly higher neutralizing antibody titers against the omicron variants than the original mRNA-1273 vaccine [162]. It also induced strong immunogenicity against the alpha, beta, gamma, and delta variants. In another study, the XBB.1.5 mRNA vaccine induced 27.0-fold and 27.6-fold higher neutralizing antibody titers against the XBB.1.5 and EG5.1 variants, respectively, compared to first-generation mRNA vaccines in uninfected individuals [163]. Furthermore, a 13.3- to 27.4-fold increase was obtained against the emerging variants HV.1, HK.3, JD.1.1, and JN.1.

Among self-amplifying RNA (saRNA) vaccines, the LNP-nCoVsaRNa vaccine elicited strong immune responses and provided protection in hamsters against the D614G strain and the alpha variant [164]. In another approach, the LNP-formulated dual saRNA vaccine expressing both the SARS-CoV-2 S RBD and N RNA (ZIP1642) elicited high neutralizing antibody titers against the beta and delta variants [165]. Moreover, the saRNA/NLC SARS-CoV-2-S vaccine, a thermostable nanolipid carrier (NLC)-encapsulated saRNA vaccine, induced robust neutralizing antibody titers against the alpha, beta, and delta variants [166]. In a systematic review and meta-analysis COVID-19 vaccines such as BBIBP-CorV, BBV152, CoronaVac, SCB-2019, ChAdOx1 nCOV-19, Ad26.COV2.S, NVX-CoV2373, mRNA-1273, BNT162b2, and CVnCoV showed vaccine efficacies for SARS-CoV-2 variants of 55.9–88.0% [167]. The efficacy against the alpha variant was high, while only moderate efficacy was seen for the beta, gamma, and delta variants. Booster vaccinations enhanced the efficacy against the delta and omicron variants.

An interesting approach for cancer immunotherapy has been to target endogenous retroviruses (ERVs) with adenovirus vectors [168]. In humans, ERVs contribute to 8% of the genome and have been detected in different types of cancers and due to their low immunogenicity and immunosuppression they contribute to cancer escapeing immunosurveillance. The combination of adenovirus vectors and VLPs can ideally target ERVs and induce B-cell responses. For example, the expression of ERV Env-packaged into Gag-based VLPs has been demonstrated to target colorectal carcinomas eliminating small tumors in mice [169].

Appropriate viral vector delivery has proven crucial for obtaining strong immune responses in vivo resulting in protection against challenges against infectious agents and tumor cells and also generating therapeutic antitumor responses. In the context of COVID-19, most of the vaccine development has focused on intramuscular delivery despite the respiratory nature of SARS-CoV-2 leading to persistent viral replication in the nasal mucosa and viral transmission despite vaccinations. Despite the success seen for intramuscularly administered COVID-19 vaccines, limited development of intranasal vaccines has been conducted [170]. Recently, a vaccine based on liposomal nanoparticle (LNP)-encapsulated VEE replicon RNA expressing the prefusion-stabilized SARS-CoV-2 spike protein gene (AAH1-SC2) was evaluated for immunogenicity, protection, and transmission in rodents after intranasal administration [98]. Intranasal delivery can stimulate critical immune cells within the mucosa and respiratory tract resulting in prevention of early-stage viral infection and reduced viral transmission [171]. The study demonstrated that the AAH1-SC2 vaccine elicited robust systemic humoral and cellular immune responses and induced long-lasting respiratory-specific immunity in mice. Moreover, intranasal administration of AAH1-SC2 conferred protection against SARS-CoV-2 challenge and prevented viral transmission in hamsters.

14.5 CONCLUSIONS

Viral vectors have demonstrated good safety and efficacy as delivery vehicles for immunotherapy and vaccine development. Numerous studies have demonstrated success in preclinical and clinical settings for infectious diseases (Table 14.2) and cancers (Table 14.3). In addition to the approved mRNA-based [172, 173] and EUA granted viral vector-based [73, 76, 80, 83, 97] COVID-19 vaccines, also viral vector-based cancer drugs have received approval. For example, Gendicine™, based on an oncolytic Ad virus vector expressing the p53 gene, was approved in China in 2003 [174]. Good safety and superior responses compared to standard therapies have been obtained in more than 30,000 head and neck cancer patients with Gendicine™ [175]. The oncolytic HSV vector expressing GM-CSF has been approved in the United States and Europe for the treatment of melanoma patients [176, 177]. In the case of vaccine development, the VSV-ZEBOV vaccine has been approved by the FDA [28].

Although progress has been made, there is still room for improvement in several areas. For example, delivery of therapeutics and antigens is crucial for achieving success. As described above, targeting the intranasal delivery route improved the quality of immunogenicity, protection against SARS-Cov-2 challenges and prevented transmission of disease in hamsters. The choice of viral vector is also important. Although both DNA and RNA viruses have proven successful for immunotherapy and vaccine development, long-term expression and chromosomal integration of retroviruses and HSV have caused some concern for targeting infectious diseases and cancers. In this context, vectors providing high-level transient heterologous gene expression might be favorable. Traditionally, Ad vectors have proven efficient although there have been some issues with pre-existing immunity against Ad in humans. Self-replicating RNA viruses present an attractive alternative due to their efficient RNA replication in host cells. It has been confirmed that 100- to 1000-fold lower doses are needed to achieve clinically relevant immune responses and tumor regression compared to other viral vectors, potentially resulting in significantly reduced adverse events in patients. Oncolytic viruses also present important assets in cancer therapy as their specific replication in tumor cells results in efficient tumor killing without causing damage to normal tissues. Moreover, combination therapy with antiviral and cancer drugs and mAbs has proven useful and often superior to monotherapy with either viral vectors or drugs. Overall, viral vectors will play an important role in future activities in the areas of immunotherapy and vaccine development.

REFERENCES

1. Jenner's Breakthrough. The History of Vaccines. Philadelphia: The College of Physicians of Philadelphia. 2020. Archived from the original on 6 June 2017 (accessed on 19 May 2025).
2. Lundstrom K. COVID-19 Vaccines: Where did we stand at the end of 2023? Viruses 2024; 16: 20.
3. Amanna IJ, Slifka MK. Successful vaccines. Curr Top Microbiol Immunol 2020; 428: 1–30.
4. Delrue I, Verzele D, Madder A, Nauwynck HJ. Inactivated virus vaccines from chemistry to prophylaxis: Merits, risks and challenges. Expert Rev Vaccines 2012; 11: 695–719.
5. Minor PD. Live attenuated vaccines: Historical successes and current challenges. Virology 2015; 479-480: 379–392.
6. Chen WH, Du L, Chag SM, Ma C, Tricoche N, Tao X et al. Yeast expressed recombinant protein of the receptor-binding domain in SARS-CoV spike protein with deglycosylated forms as a SARS vaccine candidate. Hum Vaccin Immunother 2014; 10: 648–658.
7. Bittle JL, Houghton RA, Alexander H, Shinnick TM, Sutcliffe JG, Lerner RA et al. Protection against foot and mouth disease by immunisation with a chemically synthesised peptide predicted from the viral nucleotide sequence. Nature 1982; 298: 30–33.
8. Lee J, Kumar SA, Jhan YY, Bishop CJ. Engineering DNA vaccines against infectious diseases. Acta Biomater 2018; 80: 31–47.
9. Gote V, Bolla PK, Kommineni N, Butreddy A, Nukala PK, Palakurthi SS et al. A comprehensive review of mRNA vaccines. Int J Mol Sci 2023; 24: 2700.

10. Lamb YN. BNT162b2 mRNA COVID-19 vaccine: First approval. Drugs 2021; 81: 495–501.
11. Lundstrom K. Viral vectors in gene therapy. Diseases 2018; 6: 42.
12. Schiedner G, Morral N, Parks RS, Wu Y, Koopmans SC, Langston C et al. Genomic DNA transfer with a high-capacity 1890 adenovirus vector results in improved in vivo gene expression and decreased toxicity. Nat Genet 1998; 18: 180–183.
13. Wang F, Wang Z, Tian H, Qi M, Zhai Z, Li S et al. Biodistribution and safety assessment of bladder cancer specific oncolytic adenovirus in subcutaneous xenografts tumor model in nude mice. Curr Gene Ther 2012; 12: 67–76.
14. Wei Q, Fan J, Liao J, Zou Y, Song D, Liu J et al. Engineering the rapid adenovirus production and amplification (RAPA) cell line to expedite the generation of recombinant adenoviruses. Cell Physiol Biochem 2017; 41: 2383–2398.
15. Frolov I, Hoffman TA, Pragal BM, Dryga SA, Huang H, Schlesinger S et al. Alphavirus-based expression vectors: Strategies and applications. Proc Natl Acad Sci USA 1996; 93: 11371–11377.
16. Strauss JH, Strauss EG. The alphaviruses; gene expression, replication and evolution. Microbiol Rev 1994; 58: 491–562.
17. Liljeström P, Garoff H. A new generation of animal cell expression vectors based on the Semliki Forest virus replicon. Biotechnology 1991; 9: 1356–1361.
18. Davis NL, Willis LV, Smith JF, Johnston RF. In vitro synthesis of infectious Venezuelan equine encephalitis virus RNA from a cDNA clone: Analysis of a viable deletion mutant. Virology 1989; 171: 189–204.
19. Lundstrom K. Self-amplifying RNA viruses as RNA vaccines. Int J Mol Sci 2020; 21: 5130.
20. Pijlman GP, Suhrbier A. Khromykh AA Kunjin virus replicons: An RNA-based, non-cytopathic viral vector system for protein production, vaccine and gene therapy applications. Exp Opin Biol Ther 2006; 6: 134–145.
21. Hoang-Le D, Smeenk L, Anraku I, Pijlman GP, Wang XJ, de Vrij J et al. A Kunjin replicon vector encoding granulocyte macrophage colony-stimulating factor for intra-tumoral gene therapy. Gene Ther 2009; 16: 190–191.
22. Jones CT, Patkar CG, Kuhn RJ. Construction and applications of yellow fever virus replicons. Virology 2005; 331: 247–259.
23. Apostolopoulos V. Vaccine delivery methods into the future. Vaccine 2016; 4: 9.
24. Singh M, Cattaneo R, Billeter MA. A recombinant measles virus expressing hepatitis B surface antigen induces humoral responses in genetically modified mice. J Virol 1999; 73: 4823–4828.
25. Lange S, Lampe J, Bossow S, Zimmermann M, Bitzer M, Lauer UM. A novel armed oncolytic measles vaccine virus for the treatment of cholangiocarcinoma. Hum Gene Ther 2013; 24: 554–556.
26. Finke S, Conzelmann KK. Recombinant rhabdoviruses: Vectors for vaccine development and gene therapy. Curr Top Microbiol Immunol 2005; 292: 165–200.
27. An H, Kim GN, Kang CY. Genetically modified VSV(NJ) vector is capable of accommodating a large foreign gene insert and allows high level gene expression. Virus Res 2013; 171: 168–177.
28. Ollmann Saphire E. A vaccine against Ebola virus. Cell 2020; 181: 6.
29. Holmes KD, Cassam AK, Chan B, Peters AA, Weaver LC, Dekaban GA. A multi-mutant herpes simplex virus vector has minimal cytotoxic effects on the distribution of filamentous actin, alpha-actinin and a glutamate receptor in differentiated PC-12 cells. J Neurovirol 2000; 6: 33–34.
30. Epstein AL, Marconi P, Argnani R, Manservigi R. HSV-1 derived recombinant and amplicon vectors for gene transfer and gene therapy. Curr Gene Ther 2005; 5: 445–458.
31. Li JM, Kao KC, Li LF. Micro-RNA-145 regulates oncolytic herpes simplex virus-1 for selective killing of human non-small lung cancer cells. Virol J 2013; 10: 241.
32. Schambach A, Morgan M. Retroviral vectors for cancer gene therapy. Recent Results Cancer Res 2016; 209: 17–35.
33. Hacein-Bey-Abina S, Pai SY, Gaspar HB, Armant M, Berry CC, Blanche S et al. A modified γ-retrovirus vector for X-linked severe combined immunodeficiency. N Engl J Med 2014; 371: 1407–1417.
34. Vigna E, Naldini L. Lentiviral vectors: Excellent tools for experimental Gene transfer and promising candidates for gene therapy. J Gen Med 2000; 2: 308–316.
35. Ciuffi A. Mechanisms governing lentivirus integration site selection. Curr Gene Ther 2008; 8: 419–429.
36. Ganar K, Das M, Sinha S, Kumar S. Newcastle disease virus: Current status and our understanding. Virus Res 2014; 184: 71–78.
37. Reichard K, Lorence RM, Cascino CJ et al. Newcastle disease virus selectively kills human tumor cells. J Surg 1992; 52: 448–453.

38. Schirrmacher V, Griesbach A, Ahlert T. Antitumor effects of Newcastle disease virus in vivo: Local versus systemic effects. Int J Oncol 2001; 18: 945–952.
39. Clements D, Helson E, Gujar SA, Lee PW. Reovirus in cancer therapy: An evidence-based review. Oncol Virother 2014; 3: 69–82.
40. Gujar SA, Marcato P, Pan D, Lee PW. Reovirus virotherapy overrides tumor antigen presentation evasion and promotes protective antitumor immunity. Mol Cancer Ther 2010; 9: 2924–2933.
41. Kwak H, Honig H, Kaufman HL. Poxviruses as vectors for cancer immunotherapy. Curr Opin Drug Discov Devel 2003; 6: 161–168.
42. Xu L, Sun H, Lemoine NR, Xuan Y, Wang P. Oncolytic vaccinia virus and cancer immunotherapy. Front Immunol 2024; 14: 1324744.
43. Mastrangelo MJ, Lattime EC. Virotherapy clinical trials for regional disease: In situ immune modulation using recombinant poxvirus vectors. Cancer Gene Ther 2002; 9: 1013–1021.
44. Bradley S, Jakes AD, Harrington K, Pandha H, Melcher A, Errington-Mais F. Applications of coxsackievirus A21 in oncology. Oncol Virother 2014; 3: 47–55.
45. Kimchi-Sarfaty C, Gottesman MM. SV40 pseudovirions as highly efficient vectors for gene transfer and their potential application in cancer therapy. Curr Pharm Biotechnol 2004; 5: 451–458.
46. Kawano M, Matsui M, Handa H. SV40 virus-like particles as an effective delivery system and its application to a vaccine carrier. Expert Rev Vaccines 2013; 12: 199–210.
47. McCormack MP, Rabbitts TH. Activation of the T-cell oncogene LMO_2 after gene therapy for X-linked severe combined immunodeficiency. N Engl J Med 2004; 350: 913–922.
48. Hu WS, Pathak VK. Design of retroviral vectors and helper cells for gene therapy. Pharmacol 2000; 52: 493–511.
49. Nakajima T, Nakamaru K, Ido E, Terao K, Hayami M, Hasegawa M. Development of novel simian immunodeficiency virus vectors carrying a dual gene expression system. Hum Gene Ther 2000; 11: 1863–1874.
50. Hartmann K. Clinical aspects of feline retroviruses: A review. Viruses 2012; 4: 2684–2710.
51. Olsen JC. Gene transfer vectors derived from equine infectious anemia virus. Gene Ther 1998; 5: 1481–1487.
52. Wang S, Liang B, Wang W, Li L, Feng N, Zhao Y et al. Viral vectored vaccines: Design, development, preventive and therapeutic applications in human diseases. Signal Transduct Target Ther 2023; 8: 149.
53. Lundstrom K. Viral vector-based cancer vaccines. Methods Mol Biol 2025; 2926: 101–127.
54. De Groot AS, Moise L, Terry F, Gutierrez AH, Hindocha P, Richard G et al. Better epitope discovery, precision immune engineering and accelerated vaccine design using immunoinformatics tools. Front Immunol 2020; 11: 442.
55. Kim YA, Mousavi K, Yazdi A, Zwierzyna M, Cardinali M, Fox D et al. Computational design pf mRNA vaccines. Vaccine 2024; 42: 1831–1840.
56. Lundstrom K. Application of viruses for gene therapy and vaccine development. In The Biological Role of a Virus. Advances in Environmental Microbiology, vol. 9. Hurst CJ (Ed.). 2022: 285–341. Springer Nature, Switzerland.
57. Gao W, Soloff AC, Lu X, Montecalvo A, Nguyen DC, Matusoka Y et al. Protection of mice and poultry from lethal H5N1 avian influenza virus through adenovirus-based immunization. J Virol 2006; 80: 1959–1964.
58. Rodr Guez D, Rodr Guez JR, Llorente M, V Zquez I, Lucas P, Esteban M et al. A human immunodeficiency virus type 1 Env-granulocyte-macrophage-colony-stimulating factor fusion protein enhances the cellular immune response to Env in a vaccinia virus-based vaccine. J Gen Virol 1999; 80: 217–223.
59. Ajbani SP, Velhal SM, Kadam RB, Patal VV, Lundstrom K, Bandivdekar AH. Immunogenicity of virus-like Semliki Forest virus replicon particles expressing Indian HIV-1C gag, env and pol RT genes. Immunol Lett 2017; 190: 221–232.
60. Shishido SN, Varahan S, Yuan K, Li X, Fleming SD. Humoral innate immune response and disease. Clin Immunol 2012; 144: 142–158.
61. Attfield DC. Cellular immunity: The final paradigm? Immunol Cell Biol 1997; 75: 96–101.
62. Romagnani S. T-cell subsets (Th1 versus Th2). Ann Allergy Asthma Immunol 2000; 85: 9–18.
63. Pushko P, Bray M, Ludwig GV, Parker M, Schmaljohn A, Sanchez A et al. Recombinant RNA replicons derived from attenuated Venezuelan equine encephalitis virus protect guinea pigs and mice from Ebola hemorrhagic fever virus. Vaccine 2000; 19: 142–153.
64. Geisbert TW, Feldmann H. Recombinant vesicular stomatitis virus-based vaccines against Ebola and Marburg infections. J Infect Dis 2011; 204: S1075–S1081.

65. Henao-Restrepo AM, Longini IM, Egger M, Dean NE, Edmunds WJ, Camacho A et al. Efficacy and effectiveness of an rVSV vectored vaccine expressing Ebola surface glycoprotein: Interim results from the Guinea ring vaccination cluster-randomised trial. Lancet 2015; 386: 857–866.
66. Henao-Restrepo AM, Camacho A, Longini IM, Watson CH, Edmunds WJ, Egger M et al. Efficacy and effectiveness of an rVSV vectored vaccine in preventing Ebola virus disease: Final results from the Guinea ring vaccination, open-label, cluster-randomised trial (Ebola Ca Suffit!). Lancet 2017; 389: 505–518.
67. Bernstein DI, Reap EA, Katen K, Watson A, Smith K, Norberg P et al. Randomized, double-blind, phase I trial on an alphavirus replicon vaccine for cytomegalovirus in CMV negative volunteers. Vaccine 2010; 28: 484–493.
68. Das R, Blázquez-Gamero D, Bernstein DI, Gantt S, Bautista O, Beck K et al. Safety, efficacy, and immunogenicity of A replication-defective human cytomegalovirus vaccine, V160, in cytomegalovirus-seronegative women: A double-blind, randomised, placebo-controlled, phase 2b trial. Lancet Infect Dis 2023; 23: 1383–1394. doi: 10.1016/S1473-3099(23)00343-2.
69. Wecker M, Gilbert P, Russell N, Hural J, Allen M, Pensiero M et al. Phase I safety and immunogenicity evaluations of an alphavirus replicon HIV-1 subtype C gag vaccine in healthy HIV-1-uninfected adults. Clin Vaccine Immunol 2012; 19: 1651–1660.
70. Folegatti PM, Bellamy D, Roberts R, Powlson J, Edwards NJ, Mair CF et al. Safety and immunogenicity of a novel recombinant simian adenovirus ChAdOx2 as a vectored vaccine. Vaccines 2019; 7: 40.
71. Van Doremalen N, Lambe T, Spencer A, Belij-Rammerstorfer S, Purushotham JN, Port JR et al. ChAdOx1 nCov-19 vaccination prevents SARS-CoV-2 pneumonia in rhesus macaques. bioRxiv 2020; 586: 578–582.
72. Voysey M, Costa Clemens SA, Madhi SA, Weckx LY, Folegatti PM, Aley PK et al. Safety and efficacy of the ChAdOx1 nCoV-19 vaccine (AZD1222) against SARS-CoV-2: An interim analysis of four randomized controlled trials in Brazil, South Africa, and the UK. Lancet 2021; 397: 99–111.
73. GOV.UK. Regulatory Approval of COVID-19 Vaccine AstraZeneca. 2021. Available online: https://www.gov.uk/government/publications/regulatory-approval-of-covid-19-vaccine-astrazeneca (accessed on July 2, 2025).
74. Feng L, Wang Q, Shan C, Yang C, Feng Y, Wu J et al. An adenovirus-vectored COVID-19 vaccine confers protection from SARS-CoV-2 challenge in rhesus macaques. Nat Commun 2020; 11: 4207.
75. Liu W, Li Y, Li X, Wang F, Qi R, Zhu T, Li J. Pooled analysis of the effect of pre-existing Ad5 neutralizing antibodies on the immunogenicity of adenovirus type 5 vector-based COVID-19 vaccine from eight clinical trials. Vaccines 2025; 13: 333.
76. China Approves Two More Domestic COVID-19 Vaccines for Public Use. Available online: https://www.reuters.com/article/idUSKBN2AP1V8/ (accessed on July 2, 2025).
77. Tostanoski LH, Wegmann F, Martinot AJ, Loos C, McMahan K, Mercado NB et al. Ad26 vaccine protects against SARS-CoV-2 severe clinical disease in hamsters. Nat Med 2020; 26: 1694–1700.
78. Mercado NNB, Zahn R, Wegmann F, Loos C, Chandrashekar A, Yu J et al. Single-shot Ad26 vaccine protects against SARS-CoV-2 in rhesus macaques. Nature 2020; 586: 583–588.
79. Sadoff J, Le Gars M, Shukarev G, Heerwegh D, Truyers C, de Groot AM et. al. Interim results of a phase 1-2 a trial of Ad26.COV2.S covid-19 vaccine. N Engl J Med 2021; 384: 1824–1835.
80. Food and Drug Administration. Fact Sheet for Healthcare Providers Administrating Vaccine (Vaccination Providers), Emergency Use Authorization (EUA) of the Janssen COVID-19 Vaccine to Prevent Coronavirus Disease (COVID-19); Food and Drug Administration: Silver Spring, MD, USA, 2019. Available online: https://www.fda.gov/media/146304/download (accessed on July 2, 2025).
81. Logunov DY, Dolzhikova IV, Zubkova OV, Tukhvatulin AI, Shcheblyakov DV, Dzharullaeva AS et al. Safety and immunogenicity of an rAd26 and rAd5 vector-based heterologous prime-boost COVID-19 vaccine in two formulations: Two open, non-randomised phase 1/2 studies from Russia. Lancet 2020; 6736: 31866–31863.
82. Logunov DY, Dolzhikova IV, Shcheblyakov D, Tukhvatulin AI, Zubkova OV, Dzharullaeva AS et al. Safety and efficacy of an rAd26 and rAd5 vector-based heterologous prime-boost COVID-19 vaccine: An interim analysis of a randomised controlled phase 3 trial in Russia. Lancet 2021; 397: 671–681.
83. Callaway E. Russia's fast-track coronavirus vaccine draws outrage over safety. Nature 2020; 584: 334–335.
84. Abdelnabi R, Perez P, Astorgano D, Albericio G, Kerstens W, Thibaut HJ et al. Optimized vaccine candidate MVA-S(3P) fully protects against SARS-CoV-2 infection in hamsters. Fron Immunol 2023; 14: 1163159.

85. Mayer L, Weskamm LM, Fathi A, Kono M, Heidepriem J, Krähling V et al. MVA-based vaccine candidates encoding the native or prefusion-stabilized SARS-CoV-2 spike reveal differential immunogenicity in humans. NPJ Vaccines 2024; 9: 20.
86. O'Donnell KL, Clancy CS, Griffin AJ, Shifflett K, Gourdine T, Thomas T et al. Optimization of single-dose VSV-based COVID-19 vaccination in hamsters. Front Immunol 2022; 12: 788235.
87. Furuyama W, Shifflett K, Pinski AN, Griffin AJ, Feldmann F, Okumura A et al. Rapid protection from COVID-19 in nonhuman primates vaccinated intramuscularly but not intranasally with a single dose of a vesicular stomatitis virus-based vaccine. mBio 2022; 13: e0337921.
88. Robbins JA, Tait D, Huang Q, Dubey S, Crumley T, Cote J et al. Safety and immunogenicity of intramuscular, single-dose V590 (rVSV-SARS-CoV-2 vaccine) in healthy adults: Results from a phase 1 randomised, double-blind, placebo-controlled, dose-ranging trial. EBioMedicine 2022; 82: 104138.
89. ModernaTX, Inc. *A Phase 3 Study to Evaluate the Efficacy, Safety, and Immunogenicity of mRNA-1273 SARS-CoV-2 Vaccine (NCT04470427)*. ClinicalTrials.gov; 2020. Available from: https://clinicaltrials.gov/study/NCT04470427 (accessed on June 11, 2025.
90. Hörner C, Schürmann C, Auste A, Ebenig A, Muraleedharan S, Dinnon KH 3rd et al. A highly immunogenic and effective measles virus-based Th1-biased COVID-19 vaccine. Proc Natl Acad Sci U S A 2020; 117: 32657–32666.
91. Launay O, Artaud C, Lachatre M, Ait-Ahmed M, Klein J, Luong Nguyen LB et al. Safety and immunogenicity of a measles-vectored SARS-CoV-2 vaccine candidate, V591/TMV-083, in healthy adults: Results of a randomized, placebo-controlled phase I study. EBioMedicine 2022; 75: 103810.
92. McKay PF, Hu K, Blakney AK, Samnuan K, Brown JC, Penn R et al. Self-amplifying RNA SARS-CoV-2 lipid nanoparticle vaccine candidate induces high neutralizing antibodies in mice. Nat Commun 2020; 11: 3523.
93. Pollock KM, Cheeseman HM, Szubert AJ, Libri V, Boffito M, Owen D et al. Safety and immunogenicity of a self-amplifying RNA vaccine against COVID-19: COVAC1, a phase I, dose-ranging trial. eClinicalMedicine 2022; 44: 101262.
94. Szubert AJ, Pollock KM, Cheeseman HM, Alagaratnam J, Bern H, Bird O et al. COVAC1 phase 2a expanded safety and immunogenicity study of a self-amplifying RNA vaccine against SARS-CoV-2. EClinicalMedicine 2023; 56: 101823.
95. Arthur R. Arcturus' self-amplifying COVID-19 mRNA vaccine meets phase 3 primary end points. 2022. Available from: www.biopharma-reporter.com (accessed on June 11, 2025).
96. Safety, tolerability and immunogenicity of the candidate GEMCOVAC-19 (COVID-19) vaccine in healthy pediatric subjects of 5 to less than 18 years. Available from: www.ctri.nic.in/Clinicaltrials/pmaindet2.php?trialid=66638 (accessed on June 11, 2025).
97. HDT bio receives emergency use approval for covid-19 vaccine in India. Available from: www.pharmaceutical-technology.com/news/hdt-bio-vaccine-india-approval (accessed on June 11, 2025).
98. Jennewein MF, Schultz MD, Beaver S, Battisti P, Bakken J, Hanson D et al. Intranasal replicon SARS-CoV-2 vaccine produces protective respiratory and systemic immunity and prevents viral transmission. Mol Ther 2025; 33: 3286–3306. doi:10.1016/j.ymthe.2025.04.007.
99. Zajakina A, Spunde K, Lundstrom K. Application of alphaviral vectors for immunomodulation in cancer therapy. Curr Pharm Des 2017; 23: 4906–4932.
100. Nagasato M, Rin Y, Yamamoto Y, Henmi M, Hiraoka N, Chiwaki F et al. A tumor-targeting adenovirus with high gene transduction efficiency for primary pancreatic cancer and ascites cells. Anticancer Res 2017; 37: 3599–3605.
101. Yamamoto Y, Nagasato M, Rin Y, Henmi M, Ino Y, Yachida S et al. Strong antitumor efficacy of a pancreatic tumor targeting oncolytic adenovirus for neuroendocrine tumors. Cancer Med 2017; 6: 2385–2397.
102. Emdad L, Das SK, Wang XY, Sarkar D, Fisher PB. Cancer terminator viruses (CVT): A better solution for viral-based therapy of cancer. J Cell Physiol 2018; 233: 5684–5695.
103. Roche FP, Sheahan BJ, O'Mara SM, Atkins GJ. Semliki Forest virus-mediated gene therapy of the RG2 rat glioma. Neuropathol Appl Neurobiol 2010; 36: 64.
104. Määttä AM, Mäkinen K, Ketola A, Liimatainen T, Yongabi FN, Vähä-Koskela M et al. Replication competent Semliki Forest virus prolongs survival in experimental lung cancer. Int J Cancer 2008; 123: 1704–1711.
105. Ying H, Zaks TZ, Wang R-F, Irvine KR, Kammula US, Marincola FM et al. Cancer therapy using a self-replicating RNA vaccine. Nat Med 1999; 5: 823–827.

106. Avogadri F, Zappasodi R, Yang A, Budhu S, Malandro N, Hirschhorn-Cymerman D et al. Combination of alphavirus replicon particle-based vaccination with immunomodulatory antibodies: Therapeutic activity in the B16 melanoma mouse model and immune correlates. Cancer Immunol Res 2014; 2: 448–458.
107. Yin X, Wang W, Zhu X, Wang Y, Wu S, Wang Z et al. Synergistic antitumor efficacy of combined DNA vaccines targeting tumor cells and angiogenesis. Biochem Biophys Res Comm 2015; 465: 239–244.
108. Daemen T, Riezebos-Brilman A, Bungener L, Regts J, Dontje B, Wilschut J. Eradication of established HPV16-transformed tumours after immu nisation with recombinant Semliki Forest virus expressing a fusion protein of E6 and E7. Vaccine 2003; 21: 1082–1088.
109. Komdeur FL, Singh A, van de Wall S, Meulenberg JJM, Boerma A, Hoogeboom BN et al. First-in-human phase I clinical trial of an SFV based RNA replicon cancer vaccine against HPV induced cancers. Mol Ther 2021; 29: 611–625.
110. Eerkens AL, Esajas MD, Brummel K, Vledder A, van Rooij N, Plat A et al. Vvax001, a therapeutic vaccine, for patients with HPV16-positive high-grade cervical intraepithelial neoplasia: A phase II trial. Clin Cancer Res 2025; 31: 1016–1026.
111. Crosby EJ, Hobeika AC, Niedzweicki D, Rushing C, Hsu D, Berglund P et al. Long-term survival of patients with stage III colon cancer treated with VRP-CEA(6D), an alphavirus vector that increases the CD8+ effector memory T cell to Treg ratio. J Immunother Cancer 2020; 8: e001662.
112. Slovin SF, Kehoe M, Durso R, Fernandez C, Olson W, Gao JP et al. A phase I dose escalation trial of vaccine replicon particles (VRP) expressing prostate-specific membrane antigen (PSMA) in subjects with prostate cancer. Vaccine 2013; 31: 943–949.
113. Lin Y, Zhang H, Liang J, Li K, Zhu W, Fu L et al. Identification and characterization of alphavirus M1 as a selective oncolytic virus targeting ZAP-defective human cancers. Proc Natl Acad Sci USA 2014; 111: E4504–E4512.
114. Reddi HV, Madde P, McDonough SJ, Rujillo MA, Morris JC, 3rd, Myers RM et al. Preclinical efficacy of the oncolytic measles virus expressing the sodium iodide symporter in iodine non-avid anaplastic thyroid cancer: A novel therapeutic agent allowing non-invasive imaging and radio iodine therapy. Cancer Gene Ther 2012; 19: 659–665.
115. Hartkopf AD, Bossow S, Lampe J, Zimmermann M, Taran F-A, Wallwiener D et al. Enhanced killing of ovarian carcinoma using oncolytic measles vaccine virus armed with a yeast cytosine deaminase and uracil phosphoribosyl transferase. Gynecol Oncol 2013; 130: 362–368.
116. Lampe J, Bossow S, Weiland T, Smirnow I, Lehmann R, Neubert W et al. An armed oncolytic measles vaccine virus eliminates human hepatoma cells independently of apoptosis. Gene Ther 2013; 20: 1033–1041.
117. Heinzerling L, Kunzi V, Oberholzer PA, Kündig T, Naim H, Dummer R. Oncolytic measles virus in cutaneous T-cell lymphomas mounts antitumor immune responses in vivo and targets interferon resistant tumor cells. Blood 2005; 106: 2287–2294.
118. Galanis E, Hartmann LC, Cliby WA, Long HJ, Peethambaram PP, Barrette BA et al. Phase I trial of intraperitoneal administration of an oncolytic measles virus strain engineered to express carcinoembryonic antigen for recurrent ovarian cancer. Cancer Res 2010; 70: 875–882.
119. Myers R, Harvey M, Kaufmann TJ, Greiner SM, Krempski JW, Raffel C et al. Toxicology study of repeat intracerebral administration of a measles virus derivative producing carcinoembryonic antigen in rhesus monkeys in support of a phase I/II clinical trial for patients with recurrent gliomas. Hum Gene Ther 2008; 19: 690–698.
120. Zhang L, Steele MB, Jenks N, Grell J, Suksanpaisan L, Naik S et al. Safety studies in tumor and non-tumor-bearing mice in support of clinical trials using oncolytic VSV-IFNβ-NIS. Hum Gene Ther Clin Dev 2016; 27: 111–122.
121. LeBlanc AK, Naik S, Galyon GD, Jenks N, Steele M, Peng KW et al. Safety studies on intravenous administration of oncolytic recombinant vesicular stomatitis virus in purpose-bred beagle dogs. Hum Gene Ther Clin Dev 2013; 24: 174–181.
122. Hastie E, Grdzelishvili VZ. Vesicular stomatitis virus as a flexible platform for oncolytic virotherapy against cancer. J Gen Virol 2012; 93: 2529–2545.
123. Eissa IR, Naoe Y, Bustos-Villalobos I, Ichinose T, Tanaka M, Zhiwen W et al. Genomic signature of the natural oncolytic herpes simplex virus HF10 and its therapeutic role in preclinical and clinical trials. Front Oncol 2017; 7: 149.
124. Yang H, Peng T, Li J, Wang Y, Zhang W, Zhang P et al. Treatment of colon cancer with oncolytic herpes simplex virus in preclinical models. Gene Ther 2016; 23: 450–459.

125. Kasuya H, Kodera Y, Nakao A, Yamamura K, Gewen T, Zhiwen T et al. Phase I dose-escalation clinical trial of HF10 oncolytic herpes virus in 17 Japanese patients with advanced cancer. Hepato-Gastroenterology 2014; 61: 599–605.
126. Hirooka Y, Kasuya H, Ishikawa T, Kawashima H, Ohno E, Villalobos IB et al. A phase I clinical trial of EUS-guided intratumoral injection of the oncolytic virus, HF10 for unresectable locally advanced pancreatic cancer. BMC Cancer 2018; 18: 596.
127. Almutairi AR, Alkhatib NS, Oh M, Curiel-Lewandrowski, Babiker HM, Cranmer LD et al. Economic evaluation of talimogene laherparepvec plus ipilimumab combination therapy vs ipilimumab monotherapy in patients with advanced unresectable melanoma. JAMA Dermatol 2019; 155: 22–28.
128. Chesney JA, Puzanov I, Collichio FA, Singh P, Milhem MM, Glaspy J et al. Talimogene laherparepvec in combination with ipilimumab versus ipilimumab alone for advanced melanoma: 5-year final analysis of a multicenter, randomized, open-label, phase II trial. J Immunother Cancer 2023; 11: e006270.
129. Huang TT, Parab S, Burnett R, Diago O, Ostertag D, Hofman FM et al. Intravenous administration of retroviral replicating vector, Toca511, demonstrates efficacy in orthotopic immune-competent mouse glioma model. Hum Gene Ther 2015; 26: 82–93.
130. Cloughesy TF, Landolfi J, Hogan DJ, Bloomfield S, Carter B, Chen CC et al. Phase I trial of vocimagine amiroretrorepvec and 5-fluoro cytosine for recurrent high-grade glioma. Sci Transl Med 2016; 8: 341–375.
131. Niu Z, Bai F, Sun T, Tian H, Yu D, Yin J et al. Recombinant Newcastle disease virus expressing IL15 demonstrates promising antitumor efficiency in melanoma model. Technol Cancer Res Treat 2015; 14: 607–615.
132. Chai Z, Zhang P, Fu F, Zhang X, Liu Y, Hu L et al. Oncolytic therapy of a recombinant Newcastle disease virus D90 strain for lung cancer. Virol J 2014; 11: 84.
133. Pecora AL, Rizvi N, Cohen GI, Meropol NJ, Sterman D, Marshall JL et al. Phase I trial of intravenous administration of PV701, an oncolytic virus, in patients with advanced solid cancers. J Clin Oncol 2002; 20: 2251–2266.
134. Liang W, Wang H, Sun TM, Yao WQ, Chen LL, Jin Y et al. Application of autologous tumor cell vaccine and NDV vaccine in treatment of tumors of digestive tract. World J Gastroenterol 2003; 9: 495–498.
135. Mostafa AA, Meyers DE, Thirukkumaran CM, Liu PJ, Gratton K, Spurrell J et al. Oncolytic Reovirus and immune checkpoint inhibitor as a novel immunotherapeutic strategy for breast cancer. Cancers 2018; 10: 205.
136. Villalona-Calero MA, Lam E, Otterson G, Zhao WQ, Timmons M, Subramaniam D et al. Oncolytic reovirus in combination with chemotherapy in metastatic or recurrent non-small cell lung cancer patients with KRAS-activated tumors. Cancer 2016; 122: 875–883.
137. Mahalingam D, Goel S, Aparo S, Patel Arora S, Noronha N, Tran H et al. A phase II study of pelareorep (REOLYSIN®) in combination with gemcitabine for patients with advanced pancreatic adenocarcinoma. Cancers 2018; 10: 160.
138. Madan RA, Arlen PM, Gullye JL. PANVAC-VF: Poxviral-based vaccine therapy targeting CEA and MUC1 in carcinoma. Expert Opin Biol Ther 2007; 7: 543–554.
139. Ady JW, Johnsen C, Mojica K, Heffner J, Love D, Pugalenthi A et al. Oncolytic gene therapy with recombinant vaccinia virus strain GLV-2b372 efficiently kills hepatocellular carcinoma. Surgery 2015; 158: 331–338.
140. Mansfield DC, Kyula JN, Rosenfelder N, Chao-Chu J, Kramer-Marek G, Khan AA et al. Oncolytic vaccinia virus as a vector for therapeutic sodium iodide symporter gene therapy in prostate cancer. Gene Ther 2016; 23: 357–368.
141. Downs-Canner S, Guo ZS, Ravindranathan R, Breitbach CJ, O'Malley ME, Jones HL et al. Phase I study of intravenous oncolytic poxvirus (vvDD) in patients with advanced solid cancers. Mol Ther 2016; 24: 1492–1501.
142. Chi J, Patel R, Rehman H, Goyal S, Wasif Saif M. Recent advances in immunotherapy for pancreatic cancer. J Cancer Metastasis Treat 2020; 6: 43.
143. Shafren DR, Au GG, Nguyen T, Newcombe NG, Haley ES, Beagley L et al. Systemic therapy of malignant human melanoma tumors by a common cold-producing enterovirus, coxsackievirus a21. Clin Cancer Res 2014; 10: 53–60.
144. Skelding KA, Barry RD, Shafren DR. Systemic targeting of metastatic human breast xenografts by Coxsackievirus A21. Breast Cancer Res Treat 2009; 113: 21–30.
145. Skelding KA, Barry RD, Shafren DR. Enhanced oncolysis mediated by Coxsackievirus A21 in combination with doxorubicin hydrochloride. Investig New Drugs 2012; 30: 568–581.

146. Annels NE, Mansfield D, Arif M, Ballesteros-Merino C, Simpson GR, Denyer M et al. Phase I trial of an ICAM-1-targeted immunotherapeutic-coxsackievirus A 21 (CVA21) as an oncolytic agent against non-muscle-invasive bladder cancer. Clin Cancer Res 2019; 25: 5818–5831.
147. Cordelier P, Bienvenu C, Lulka H, Marrache F, Bouisson M, Openheim A et al. Replication-deficient rSV40 mediate pancreatic gene transfer and long-term inhibition of tumor growth. Cancer Gene Ther 2007; 14: 19–29.
148. Durso RJ, Andjelic S, Gardner JP, Margitich DJ, Donovan GP, Arrigale RR et al. A novel alphavirus vaccine encoding prostate-specific membrane antigen elicits potent cellular and humoral immune responses. Clin Cancer Res 2007; 13: 3999–4008.
149. Etoh T, Himenio Y, Matusmoto T, Aramaki M, Kawano K, Nishizono A et al. Oncolytic viral therapy for human pancreatic cancer cells by reovirus. Clin Cancer Res 2003; 9: 1218–1223.
150. Gong J, Sachdev E, Mita AC, Mita MM. Clinical development of reovirus for cancer therapy: An oncolytic virus with immune-mediated antitumor activity. World J Methodol 2016; 6: 25–42.
151. Al Yaghchi C, Zhang Z, Alusi G, Lemoine NR, Wang Y. Vaccinia virus, a promising new therapeutic agent for pancreatic cancer. Immunotherapy 2015; 7: 1249–1258.
152. Choi JY, Smith DM. SARS-CoV-2 variants of concern. Yonsei Med J 2021; 62: 961.
153. Nashwan A, Yassin M, Soliman A, De Sanctis V, Ibrahim M. mRNA-based COVID-19 vaccine booster dose: Benefits, risks, and coverage. Acta Biomed 2022; 93: e2022236.
154. Lundstrom K. Role of nucleic acid vaccines for the management of emerging variants of SARS-CoV-2. In SARS-CoV-2 Variants and Global Population: Diagnostic Strategies, Vaccine Development, and Therapeutic Management. Chavda VP, Uversky VN (Eds.). 2024; 285–316. Apple Academic Press, Inc.
155. Lundstrom K. Immunobiology and immunotherapy of COVID-19. Prog Mol Biol Transl Sci 2025; 213: 73–133.
156. Taucher C, Lazarus R, Dellago H, Maurer G, Weisova P, Corbic-Ramljak I et al. Safety and immunogenicity against ancestral, delta and omicron virus variants following a booster dose of an inactivated whole-virus COVID-19 vaccine VLA2001: A randomized, controlled, phase 3 COV-COMPARE trial. J Infect 2003; 87: 242–254.
157. Madhi SA, Baillie V, Cutland CL, Voysey M, Koen AL, Fairlie L et al. Efficacy of the ChAdOx1 nCoV-19 COVID-19 vaccine against the B.1.351 variant. N Engl J Med 2021; 384: 1885–1898.
158. Eyre DW, Taylor D, Purver M, Chapman D, Fowler T, Pouwels KB et al. Effect of COVID-19 vaccination on transmission of alpha and delta variants. N Engl J Med 2022; 386: 744–756.
159. Alter G, Yu J, Liu J, Chandrashekar A, Borducchi EN, Tostanoski LH et al. Immunogenicity of Ad26. COV2.S vaccine against SARS-CoV-2 variants in humans. Nature 2021; 596: 268–272.
160. Wang Z, Li Z, Shi W, Zhu D, Hu S, Dinh PU-C et al. SARS-CoV-2 and influenza double hit vaccine based on RBD-conjugated inactivated influenza A virus. Sci Adv 2023; 9: eabo4100.
161. Hennrich AA, Sawatsky B, Santos-Mandujano R, Banda DH, Oberhuber M, Schopf A et al. For COVID-19: Protection of mice after a single immunization. PLOS Pathog 2021; 17: e1009064.
162. Chalkias S, Harper C, Vrbicky K, Walsh SR, Essink B, Brosz A et al. A bivalent omicron-containing booster vaccine against COVID-19. N. Engl. J. Med 2022; 387: 1279–1291.
163. Wang Q, Guo Y, Bowen A, Mellis IA, Valdez R, Gherasim C et al. XBB.1.5 monovalent mRNA vaccine booster elicits robust neutralizing antibodies against XBB subvariants and JN.1. Cell Host Microbe 2024; 32: 315–321.e3.
164. Frise R, Baillin L, Zhou J, Kugathasan R, Peacock TP, Brown JC et al. A self-amplifying RNA vaccine protects against SARS-CoV-2 (D614G) and alpha variant of concern (B.1.1.7) in a transmission-challenge hamster model. Vaccine 2022; 40: 2848–2855.
165. McCafferty S, Ashiqui Haque AKM, Vandierendonck A, Weidensee B, Plovyt M, Stuchlikova M et al. A dual-antigen self-amplifying RNA SARS-CoV-2 vaccine induces potent humoral and cellular immune responses and protects against SARS-CoV-2 variants through T cell-mediated immunity. Mol Ther 2022; 30: 2968–2983.
166. Voigt EA, Gerhardt A, Hanson D, Jennewein MF, Battisti P, Reed S et al. Author correction: A self-amplifying RNA vaccine against COVID-19 with long-term room-temperature stability. NPJ Vaccines 2022; 7: 150.
167. Zeng B, Gao L, Zhou Q, Yu K, Sun F. Effectiveness of COVID-19 vaccines against SARS-CoV-2 variants of concern: A systematic review and meta-analysis. BMC Med 2022; 20: 200.
168. Vergara Bermejo A, Ragonnaud E, Daradoumis J, Holst P. Cancer associated endogenous retroviruses: Ideal immune targets for adenovirus-based immunotherapy. Int J Mol Sci 2020; 21: 4843.

169. Neukirch L, Nielsen TK, Laursen H, Daradoumis J, Thirion C, Holst RJ. Adenovirus based virus-like-vaccines targeting endogenous retroviruses can eliminate growing colorectal cancers in mice. Oncotarget 2019; 10: 1458–1472.
170. Lundstrom K. Improved SARS-CoV-2 vaccine based on intranasally administered replicon RNA. Mol Ther 2025; 33: 2970–2971.
171. Tiboni M, Casettari L, Illum L. Nasal vaccination against SARS-CoV-2: Synergistic or alternative to intramuscular vaccines? Int J Pharm 2021; 603: 120686.
172. FDA Approves First COVID-19 Vaccine. Available online: https://www.fda.gov/news-events/press-announcements/fda-approves-first-covid-19-vaccine (accessed June 26, 2025).
173. FDA Takes Additional Action in Fight against COVID-19 by Issuing Emergency Use Authorization for Second COVID-19 Vaccine. Available online: https://www.fda.gov/news-events/press-announcements/fda-takes-additional-action-fight-against-covid-19-issuing-emergency-use-authorization-second-covid-19 (accessed on June 26, 2025).
174. Räty JK, Pikkarainen JT, Wirth T, Ylä-Herttuala S. Gene therapy: The first approved gene-based medicines, molecular mechanisms and clinical indications. Curr Mol Pharmacol 2008; 1: 13–23.
175. Zhang WW, Li L, Li D, Liu J, Li W, Xu X et al. The first approved gene therapy product for cancer ad-p53 (gendicine): 12 years in the clinic. Hum Gene Ther 2018; 29: 160–179.
176. Fukuhara H, Ino Y, Todo T. Oncolytic virus therapy: A new era of cancer treatment at dawn. Cancer Sci 2016; 107: 1373–1379.
177. Kaufman HL, Kim DW, De Raffele G, Mitcham J, Coffin RS, Kim-Schulze S et al. Local and distant immunity induced by intralesional vaccination with an oncolytic herpes virus encoding GM-CSF in patients with stage IIIc and IV melanoma. Ann Surg Oncol 2010; 17: 718–730.

15 Conclusions
A Call to Action

Alaa A. A. Aljabali

15.1 INTRODUCTION

The study of viruses has traditionally been driven by public health crises and pandemic responses, and in spite of the vast reservoir of virus like pathogens that remain uncharacterized, many are insufficiently studied despite their potential public health threat [1, 2]. This book has aimed to expose neglected viruses—pathogens that remain understudied despite posing clear risks to public health. Through the synthesis of expert perspectives from various fields, we have emphasized the urgent need to strengthen monitoring systems, advance research agendas, and mobilize policy level responses to these overlooked viral threats.

Neglected viruses are often trapped in a state of scientific neglect, most commonly because they are geographically constrained, cause sporadic epidemics, or fail to attract enough funds due to economic and political constraints [3]. However, as in the case of Chikungunya, Nipah, and Lassa viruses, these pathogens have demonstrated zoonotic potential, rapid transmissibility, and the capacity to cause significant morbidity and mortality [4, 5].

Understanding and combating neglected viruses is not solely a scientific pursuit, but a global health imperative. The COVID-19 pandemic exposed the risks of failing to recognize and act on emerging viral threats. Neglected viruses represent the next frontier in infectious disease research and demand proactive strategies to prevent their escalation into future pandemics [6–8].

This final chapter synthesizes the book's central findings, highlighting the similarities among the viruses, the latest scientific developments, and the public health implications of inaction. It concludes with a direct call to action for scientists, decision-makers, and global health officials to prioritize neglected viral threats in order to prevent an avoidable global health crisis.

15.2 KEY FINDINGS AND LESSONS LEARNED

15.2.1 Commonalities in Neglected Viruses

Despite their varying genetic makeup and epidemiologic patterns, the so-called neglected viruses all share common characteristics that contribute to their persistent neglect and long-term public health threat [9–11].

Underreporting and global deprioritization: These viruses primarily affect impoverished or geographically isolated populations, resulting in limited surveillance and sparse reporting. Compared to high-profile viruses like influenza and HIV, neglected viruses receive significantly less research funding and attention, and remain poorly understood in terms of pathogenesis, treatment options, and control strategies [12, 13]. Many neglected viruses share critical features that exacerbate their public health impact, including zoonotic reservoirs, underdeveloped countermeasures, and disproportionate effects on resource-limited regions. For instance, Nipah, Lassa, and Chikungunya viruses—despite taxonomic and ecological differences—exemplify these shared challenges through their capacity for cross-species transmission, high outbreak potential, and severe clinical manifestations (Table 15.1).

Zoonotic origins and transmission mechanisms: A substantial proportion of neglected viruses originate from zoonotic reservoirs, with transmission occurring between wildlife or domesticated

 DOI: 10.1201/9781003607816-15

TABLE 15.1
Comparative Overview of Key Neglected Viruses

Feature	Nipah Virus	Lassa Virus	Chikungunya Virus	Hantaviruses
Virus Family/ Order	*Paramyxoviridae*	Arenaviridae	Togaviridae	*Hantaviridae* (Order: *Bunyavirales*)
Primary Reservoir(s)	Fruit bats (genus *Pteropus*)	Rodents (*Mastomys* spp.)	Humans (amplifying host)	Rodents (species-specific)
Transmission Route(s)	Direct contact (bats/ pigs) Human-to-human	Rodent excreta Human-to-human (body fluids)	Mosquitoes (*Aedes* spp.)	Inhalation of aerosolized rodent excreta
Key Endemic Regions	Southeast Asia, South Asia	West Africa	Africa, Asia, Americas, Europe	Worldwide (syndrome-dependent)
Primary Clinical Syndrome(s)	Encephalitis, respiratory illness	Hemorrhagic fever, deafness	Fever, chronic arthralgia	HFRS: Hemorrhagic fever with renal syndrome HPS: Hantavirus pulmonary syndrome
Case Fatality Rate (CFR)	40–75%	~1% (community); ≥15% (hospitalized)	<0.1%	HFRS: 1–15%; HPS: >30%
Key Challenges	Sporadic outbreaks, high CFR, diagnostics	Underreporting, lack of vaccines/treatment	Co-circulation with Dengue/Zika	Zoonotic spillover, environmental drivers

This table summarizes the epidemiological characteristics, transmission routes, clinical features, and specific challenges associated with Nipah virus, Lassa virus, Chikungunya virus, and Hantaviruses, which are discussed as representative examples of neglected viral threats. Data synthesized from chapter content and established virological knowledge relevant to the discussion. CFR, case fatality rate.

animals and humans. Bats, rodents, and mosquitoes are the main vectors for many such viruses, including Hantavirus, Nipah virus, and Lassa virus. The expanding wildlife-human interface, resulting from climate change, urbanization, and deforestation, has accelerated spillover events and contributed to the emergence of localized epidemics [14–16].

Challenges in diagnosis, treatment, and control: Limited clinical awareness and inadequate diagnostic infrastructure often result in delayed or inaccurate diagnoses, impeding timely outbreak control. Most of these diseases remain in early-stage drug or vaccine development pipelines, with some relying solely on supportive care. Their sporadic occurrence renders them economically unattractive for pharmaceutical investment, as the industry typically prioritizes diseases with stable, high-demand markets [17–19].

15.2.1.1 Scientific Advancements

Emerging diagnostic technologies and surveillance: Despite persistent challenges, significant progress has been made in the study of neglected viruses, particularly in genomics, immunotherapy, and diagnostics, offering new opportunities for improved disease detection, control, and management [20–22]. Advances in metagenomic and next-generation sequencing (NGS) have enabled rapid characterization of novel viral threats, including their genomic patterns and adaptability. Advances in monoclonal antibody treatments and RNA-based vaccines have also created new intervention opportunities. Recent innovations in mRNA vaccine platforms demonstrated during the COVID-19 pandemic show promise for adaptation against other viral threats such as Lassa and Nipah viruses [23, 24].

Emerging diagnostic tools and enhancing surveillance: Traditional diagnostic methods such as serology and PCR are now complemented by CRISPR-based diagnostics and point-of-care technologies, enabling faster and more accurate detection even in remote settings. The integration of AI and big data in disease monitoring networks globally has improved the forecasting and real-time tracking of epidemics, allowing for more timely and effective public health responses [25–27].

15.2.1.2 Public Health and Policy Implications

While scientific advances are crucial, global health strategies and policy frameworks are also essential for addressing the overlooked threats posed by neglected viruses. A multitude of systemic problems must be addressed to ensure equitable access to diagnostics, therapeutics, and preventive healthcare [28, 29]. Despite the significant impact of neglected viral diseases, research also remains disproportionately focused on a small number of high-profile pathogens, leaving many others chronically underfunded. Many epidemics occur in nations with fragile healthcare systems, where limited resources hinder timely containment and response. A shift to long-term, sustainable investments in pandemic preparedness is urgently needed to prevent future global health crises [30, 31].

15.2.1.3 Importance of Global Cooperation and Exchange of Data

The interconnected nature of viral emergence calls for multilateral cooperation among governments, academic research, and global health organizations. The Global Virome Project and WHO's Emerging Diseases Surveillance Network are examples of how open access data sharing has become critical for enabling real-time tracking of viral mutations and emerging outbreaks by the global scientific community. Enhancing such initiatives is essential to strengthening global health security and outbreak preparedness [32–34].

The overarching takeaway from the study is that proactive intervention yields better outcomes than delayed, reactive control. As shown by past pandemics, delaying action until a forgotten virus reaches pandemic scale is a high-risk and unsustainable approach. Investing in research, building up surveillance, and promoting global partnerships are necessary to prevent emerging viral threats from escalating into global pandemics [35, 36].

15.3 URGENT NEEDS IN NEGLECTED VIRUS RESEARCH

The urgent need to enhance the monitoring and integration of research on understudied viruses is underscored by the call to provide real-time monitoring mechanisms to offer early outbreak detection. Robust surveillance systems are essential for identifying emerging viral threats before they escalate into widespread epidemics. The incorporation of newer tools, such as metagenomic sequencing and machine learning algorithms, could significantly enhance real-time surveillance capabilities [37]. The tools enable rapid identification of pathogens across diverse environments, allowing public health authorities to implement timely responses. Improving cross-border data exchange and international cooperation is vital, as viral outbreaks often transcend national boundaries. Collaborative mechanisms, such as the One Health strategy, can facilitate epidemiologic data sharing and support coordinated responses to neglected viral threats [37–39].

15.3.1 Development of Targeted Diagnostics and Therapeutics

Advancing targeted diagnostics and treatments is a top research priority for addressing neglected viruses. The development of rapid, low-cost diagnostic tools suitable for use in low resource settings is urgently needed. Current diagnostic methods in most of the neglected viruses are slow and require highly specialized laboratory infrastructure, which is often unavailable in the regions most affected [40]. Improved point-of-care diagnostics and multiplex assays can enhance diagnostic capacity, enabling earlier detection and more effective clinical intervention. Expediting the development of antiviral drugs and vaccines is also critical for controlling neglected viral infections. The COVID-19 pandemic has demonstrated the value of accelerated research and development, and similar approaches must be applied to neglected viruses to enable timely responses to future outbreaks [41].

15.3.2 Investments in Public Health Infrastructure

Investments in public health infrastructure are required to strengthen healthcare systems in regions at high risk for neglected viral outbreaks. Many countries affected by neglected viruses, healthcare systems face severe challenges due to inadequate facilities, limited access to essential equipment and medications, and shortages of trained healthcare personnel [42]. Public health infrastructure improvement is essential for improving outbreak preparedness and ensuring timely care for affected populations. Integrating vector control into national public health strategies is critical for preventing the spread of vector-borne viruses. Integrating vector control into national public health strategies is critical for preventing the spread of vector-borne viruses. Vector control methods, particularly insecticide-treated nets and by implementing community-based educational interventions, have worked well proven effective in controlling the spread of infectious diseases such as dengue and Zika [37].

15.3.3 Closing the Gap between Funding and Policy

Bridging the gaps in funding to support research on neglected viruses is necessary. Government and private investments in public health and in the conduct of research are necessary in bridging the funding gaps to contain the neglected diseases [43]. Public–private partnerships (PPPs) are also crucial for fostering innovation and supporting neglected virus research through sustainable, integrated funding mechanisms. Leveraging global health policies and inclusion in the global health strategy are key to securing long-term investments in public health infrastructure and research capacity. The policymaker is necessary to take note of the necessity for containment of the neglected diseases and to provide funding in support of healthcare delivery, monitoring, and also in conducting research [42]. In summary, addressing the immediate demands of underfunded virus research requires a multipronged approach that includes enhanced surveillance, targeted diagnostics, investments in public health infrastructure, and bridging gaps in funding and policymaking. By prioritizing these efforts, stakeholders can strengthen global health security and proactively reduce the risk of future viral epidemics.

15.4 THE FUTURE OF NEGLECTED VIRUS RESEARCH

15.4.1 Harnessing Technological Advances

The future research on the so-called neglected viruses depends heavily on the application of technological advancements, foremost in the realms of artificial intelligence (AI), big data analytics, and genomics. These tools are central to predicting, tracking, and mitigating viral threats. For instance, AI-based algorithms can rapidly process large volumes of genomic and epidemiological data to identify early indicators of emerging outbreaks involving neglected viruses [44]. Through machine learning and the application of genomics, researchers can enhance predictive capacity, supporting proactive public health preparedness and response. In addition, the use of big data facilitates the integration of diverse datasets to generate a comprehensive understanding of viral transmission dynamics and outbreak behavior [45].

Aside from the use of nanotechnology and new drug delivery systems, both of which hold promises for accelerating research progress on neglected viruses, utilization of big data and AI also contributes to expediting discovery and development. Nanotechnology enhances the efficacy, bioavailability, and stability of vaccines and therapeutic agents. For example, vaccine platforms using nanoparticles enable targeted delivery to specific cells, thereby improving immune response

TABLE 15.2
Comparative Analysis of Persistent Challenges and Innovative Approaches in Neglected Virus Management

Challenge Area	Key Obstacles	Emerging Solutions
Surveillance and Detection	• Underreporting in remote areas • Delayed outbreak recognition	• Portable NGS for field surveillance • AI-driven predictive analytics (e.g., healthmap.org) • One Health-integrated reporting systems
Diagnostics	• Lab-dependent assays • Limited POC options in LMICs	• CRISPR-based tools (e.g., SHERLOCK) • Paper-based multiplex assays • Cloud-connected mobile diagnostics
Therapeutics/ Vaccines	• Few approved countermeasures • Low commercial viability	• mRNA platforms (e.g., Nipah vaccine candidates) • Broad-spectrum antivirals (e.g., remdesivir repurposing) • Nanoparticle delivery systems
Data Integration	• Fragmented global data • Reactive policies	• Federated learning for cross-border data sharing • Blockchain for outbreak traceability • WHO-backed collaborative platforms
Funding and Infrastructure	• Chronic underinvestment • Weak health systems	• Gavi-like vaccine alliances for neglected diseases • PPPs for diagnostics R&D • Decentralized manufacturing hubs

Abbreviations: AI, artificial intelligence; LMICs, low- and middle-income countries; NGS, next-generation sequencing; POC, point-of-care; PPPs, public–private partnerships.

effectiveness [46]. In addition, new drug delivery systems such as liposomes and micelles can improve the pharmacokinetic profiles of antiviral agents, enabling more efficient delivery to target tissues [47]. Incorporating these technologies into the design of treatments and diagnostics for neglected viruses could significantly improve clinical outcomes and reduce disease burden. Recent advances in AI, nanotechnology, and genomics are reshaping the neglected virus landscape. For example, CRISPR-based diagnostics and mRNA platforms—initially developed for other pathogens—are now being adapted for Lassa and Nipah viruses, demonstrating the transformative potential of cross-disciplinary technologies (Table 15.2). Similarly, decentralized surveillance networks leveraging portable sequencing and machine learning address longstanding gaps in outbreak detection, particularly in resource-limited settings.

15.4.2 Interdisciplinary Collaboration

Interdisciplinary collaboration is needed to engage policymakers, virologists, ecologists, and epidemiologists in developing comprehensive strategies for research and public health intervention. Interdisciplinary collaboration facilitates skill-sharing and resource integration, thereby improving the effectiveness of outbreak response and epidemic control [45]. For instance, ecologists can provide insight into environmental drivers of zoonotic spillover, while epidemiologists contribute to understanding transmission dynamics within and across populations. Interdisciplinary collaboration can generate holistic solutions to the complex challenges posed by neglected viruses.

The significance of applying One Health strategies, uniting the health of humans, animals, and the environment cannot be emphasized enough. The One Health model requires cross-disciplinary collaboration to enhance health outcomes across sectors [48]. For example, zoonotic outbreaks typically arise from complex interactions between wildlife, domestic animals, and human populations. Through adopting a One Health model, researchers and public health can more effectively understand transmission dynamics and implement targeted interventions to control and prevent outbreaks. Integrating this model into public health systems is essential for addressing and preventing future outbreaks of neglected viruses.

15.4.3 Educating and Inspiring Future Generations

Promoting and engaging young researchers and scientists in research on so-called neglected viruses is critical to the future of the profession in the longer term. Training programs that teach the global health relevance and significance of the 'neglected viruses' can inspire the next generation to pursue careers in infectious disease research and virology [49]. Through the generation of interest among early-career scientists, a skilled workforce can be developed to confront the challenges posed by neglected viral threats. Strengthening academic programs in virology and infectious disease research is essential to advance the field of neglected virus studies. An increase in funding is required support novel research areas, the development of emerging technologies, and opportunities for next-generation scientists [50]. Academic support is critical to building multidisciplinary research capacity by integrating diverse scientific domains into academic curricula. By embedding neglected virus topics into formal education, students can be better equipped to pursue impactful, long-term research in the field.

Globally, the future of virus research on neglected and underappreciated viruses' rests upon advancing technologies, interdisciplinary collaboration, and the proper education and empowerment of the next generation. It is through the application of AI, big data, and genomics that researchers can more effectively predict, monitor, and manage viral threats. Interdisciplinary strategies incorporating a One Health outlook will provide integrated solutions to the problems presented by the neglected viruses. Last, education and research investments will provide the future generation of scientists with the tools needed to address global health challenges effectively.

15.5 A CALL TO ACTION

As the global health profile continues to shift, governments, research institutes, and funders must prioritize research and response strategies for the so-called neglected viruses. These pathogens are often overshadowed by more prominent diseases, yet they pose a significant and growing threat to public health, particularly among vulnerable populations. It is necessary to urge the stakeholders to support neglected virus research aimed at developing effective prevention, preparedness, and response mechanisms. A relevant example is the establishment of contingency planning frameworks during dengue outbreaks, which could serve as a model for addressing other neglected viruses [51]. The strategies could involve the use of funds, training, and monitoring to prepare in case of possible outbreaks. The tiered prioritization in Table 15.3 highlights how Nipah and Lassa—despite differing transmission patterns—both demand Tier 1 attention due to their lethal potential in fragile health systems, whereas CHIKV's pandemic capacity warrants distinct mitigation strategies.

Encouraging international cooperation and PPPs is essential to sustaining research and innovation on neglected viruses. Through cooperation, the exchange of resources, knowledge sharing, and the development of novel solutions are significantly enhanced. The previously published research by Cuypers et al. [52] has been removed, as it does not substantiate claims regarding PPPs in neglected virus research, instead focusing on COVID-19 and nursing home outbreaks. Through the synergy between the public and the private sector, stakeholders can advance the development of diagnostics, therapeutics, and vaccines for neglected viral diseases.

TABLE 15.3
Evidence-based Prioritization Matrix for Neglected Viruses Using Modified CDC/WHO Pandemic Risk Criteria

Pathogen	Epidemic Risk (R_0 and Spillover Potential)	Clinical Severity (CFR/ Disability)	Globalization Potential	Countermeasure Gap (Tx/Vx/Dx)	Health Equity Impact	Priority Level	Rationale
Nipah	Moderate (0.4–0.8)*	Critical (40–75% CFR)	Emerging	Critical (no approved Vx)	High	Tier 1	High CFR with nosocomial amplification risk
Lassa	High (endemic)	Severe (15% hospitalized CFR)	Regional	Major (only ribavirin)	Extreme	Tier 1	Endemic burden + Congenital complications
CHIKV	High (urban cycles)	Debilitating (25% chronic pain)	Pandemic	Moderate (supportive care)	Significant	Tier 2	Explosive outbreaks with economic paralysis
Hantavirus	Low-moderate	Variable (1–40% CFR)	Niche	Major (no specific Tx)	Moderate	Tier 2	Occupational/ environmental exposure patterns

Note: Assessments incorporate: Epidemic Risk: Basic reproduction number (R_0) where available, or spillover frequency. Clinical: Case fatality rate (CFR) and DALY estimates. Globalization: Air travel connectivity, vector distribution shifts (for CHIKV). Countermeasures: Treatment (Tx), vaccine (Vx), diagnostic (Dx) availability. Equity: Burden on vulnerable populations.

Advocating public education and community engagement in outbreak preparedness and response is crucial in building resistance to neglected and underrecognized viral threats. Public health education campaigns that raise awareness about viral risks are essential for empowering communities to take preventive action. Community engagement in the development of health strategies fosters public trust and ensures that interventions are both effective and culturally sensitive. Carducci et al. [53] have been excluded, as the study does not support the claims related to community engagement or outbreak readiness. Raising public awareness about the potential for neglected viruses to evolve into future global threats can help generate support and funding for ongoing research. The COVID-19 global health emergency has exposed the interconnectedness of global health systems and the dangers of undetected viral emergence.

In summary, addressing the urgent challenges in neglected virus research requires a multi-pronged approach that includes research prioritization, international collaboration, and community engagement. Advocating for greater recognition of the importance of neglected virus research by governments, research institutions, and funders can strengthen global health security and improve outcomes in affected regions. Global partnerships and collaboration, and resource sharing can be advanced through coordinated efforts, while public engagement encourages community participation in outbreak preparedness and response. Last, sustained attention and long-term investment in this field depend on consistent reminders that neglected viruses have the potential to cause significant global health impacts. Moving from reactive to proactive management of neglected viruses demands synchronized efforts across five pillars (Table 15.4): (1) AI-powered surveillance to preempt outbreaks, (2) accelerated R&D pipelines leveraging COVID-19-era innovations, (3) decentralized health infrastructure to ensure equity, (4) policy reforms to secure funding, and (5) grassroots engagement to build trust. This framework's success hinges on adopting measurable targets, such as the 100-day vaccine goal championed by CEPI.

TABLE 15.4
Multidisciplinary Framework for Combating Neglected Viruses, Aligning Technological, Infrastructural, and Policy Interventions with Measurable Outcomes

Strategic Pillar	Core Actions	Expected Impact
Surveillance and Intelligence	• Deploy AI-driven outbreak forecasting tools • Establish One Health-integrated genomic surveillance	• Early spillover detection (e.g., pre-outbreak zoonotic signals) • Real-time cross-border alerts
Accelerated R&D	• Develop modular vaccine platforms (e.g., mRNA for Nipah) • Create antiviral libraries for repurposing	• 100-day vaccine target achievement • Reduced dependency on supportive care
Health Infrastructure	• Train frontline workers in LMICs on POC diagnostics • Strengthen regional BSL-3/4 lab networks	• 80% reduction in diagnostic turnaround time • Sustainable outbreak response capacity
Policy & Funding	• Launch neglected virus PPPs (e.g., CEPI model) • Advocate for WHO priority pathogen list inclusion	• $2B/year sustained funding by 2030 • Fast-track regulatory pathways
Global Coordination	• Implement blockchain-enabled data sharing • Community-led participatory surveillance (e.g., Ebola NGOs)	• Transnational trust in data governance • 50% higher outbreak reporting compliance

Abbreviations: AI, artificial intelligence; BSL, biosafety level; CEPI, Coalition for Epidemic Preparedness Innovations; LMICs, low- and middle-income countries; mRNA, messenger RNA; NGO, non-governmental organization; POC, point-of-care; PPP, public–private partnership; R&D, research and development; WHO, World Health Organization.

15.6 FINAL THOUGHTS

Upon reviewing the various threats of the so-called forgotten viruses, it becomes evident that sustained investment in virology research is essential. The world of infectious diseases is constantly evolving, with new viral threats emerging, reinforcing the need for continuous engagement in research, surveillance, and knowledge dissemination.

Investment in virus research is not merely a matter of scientific curiosity—it is a cornerstone of global health security. Improved surveillance networks, new diagnostics, and the development of effective treatments are all critical components of a global effort to counter neglected virus threats. Through interdisciplinary collaboration and the application of new technologies, we can establish a comprehensive framework to address these challenges effectively. Community involvement and public awareness of the importance of virus research in neglected disease are essential for empowering individuals to actively participate in disease prevention and control efforts.

In brief, we need to safeguard a vision of how scientific discoveries can transform the future of global health security. Rapid advancements in vaccine and diagnostic development during the COVID-19 pandemic demonstrate the power of science to address future major health threats. Through dynamism and strength in collaboration and innovation, we are well equipped to create a more resilient and healthier world, capable of combating the threats posed by neglected viruses. The path ahead may be long and complex, but with sustained commitment and investment in virus research, we are poised to lay the foundation for a safer global health future.

15.7 CONCLUSIONS

This chapter examined the threats posed by neglected viruses, with emphasis on zoonotic transmission and their potential to trigger global epidemics. It addressed the deficiencies in surveillance, the complexities of viral evolution, and the critical role of early warning systems in mitigating emerging

TABLE 15.5
Comparative Analysis of Intervention Effectiveness across Transmission Archetypes, Identifying Both Persistent Challenges and Emerging Technological Opportunities

Strategy	Spillover (Nipah/Hanta)	Vector-Borne (CHIKV)	Sustained H2H (Lassa)	Key Barriers	Innovation Opportunities
Predictive Surveillance	Environmental RNA monitoring	Mosquito genomics	Fever alert networks	Cross-border data policies	Wastewater surveillance adaptation
Prevention	Reservoir vaccination*	Wolbachia deployment	Rodent-proof housing	Ecological complexity	CRISPR-based vector modification
Diagnostics	Multiplex zoonotic panels	Dengue/Zika/CHIKV triage	Rapid antigen tests	Thermal stability requirements	AI-assisted image-based diagnostics
Therapeutics	mAb stockpiles	Antiviral repurposing	Lassa MATRx	Cold chain limitations	Nanoparticle delivery systems
Health Systems	BSL-4 readiness	Arbovirus task forces	Isolation protocols	Workforce attrition	Modular field hospitals

Abbreviations: BSL, biosafety level; mAb, monoclonal antibody; MATRx, multi-agent therapeutic regimen.

threats. Even in the face of improved diagnostic tools and immunotherapy strategies, neglected viruses continue to represent a significantly underrecognized public health risk.

A key takeaway from the above findings is the necessity of an integrated strategy incorporating epidemiological surveillance, cross-sectoral coordination, and investment in rapid response systems. Increased global readiness calls for sustained research on viral reservoirs, the development of advanced diagnostics, and equitable access to healthcare interventions.

Looking ahead, the synergy between climate change, human encroachment into wildlife habitats, and global travel will continue to influence the emergence of new pathogens. Mitigating the risks associated with them calls for a preventive strategy, in which scientific research and policy reform work in tandem to prevent future epidemics. Through cooperation among virologists, epidemiologists, and policymakers, we can forecast, prepare for, and respond to emerging threats, making the global health profile increasingly resilient. As Table 15.5 demonstrates, effective preparedness requires paradigm-specific adaptations—from environmental RNA monitoring for spillover viruses to Wolbachia deployments for arboviruses—while shared barriers like cold chain requirements highlight opportunities for unified technological solutions.

REFERENCES

1. Leinhos, M.; Qari, S.H.; Williams-Johnson, M. Preparedness and emergency response research centers: Using a public health systems approach to improve all-hazards preparedness and response. *Public Health Reports* 2014, *129*, 8–18. doi:10.1177/00333549141296s403.
2. Mantas, J. The importance of health informatics in public health during the COVID-19 pandemic. *Studies in Health Technology and Informatics* 2020, *272*, 487–488. doi:10.3233/shti200602.
3. Schmidt-Chanasit, J.; Hagen, R.M.; Frickmann, H. Chikungunya virus and armed forces: Relevance and vaccine prospects. *European Journal of Clinical Microbiology & Infectious Diseases* 2025. doi:10.1007/s10096-024-05028-x.
4. Rodríguez-Barraquer, I.; Solomon, S.S.; Kuganantham, P.; Srikrishnan, A.K.; Vasudevan, C.K.; Iqbal, H.S.; Balakrishnan, P.; Solomon, S.; Mehta, S.H.; Cummings, D.A.T. The hidden burden of Dengue and Chikungunya in Chennai, India. *PLOS Neglected Tropical Diseases* 2015, *9*, e0003906. doi:10.1371/journal.pntd.0003906.
5. Cain, M.; Ly, H. Oropouche virus: Understanding "sloth fever" disease dynamics and novel intervention strategies against this emerging neglected tropical disease. *Virulence* 2024, *15*. doi:10.1080/21505594.2024.2439521.

6. Doble, L.; Fèvre, E.M. Focusing on neglected zoonoses. *Veterinary Record* 2010, *166*, 546–546. doi:10.1136/vr.c2373.
7. Davies, S.E. National security and pandemics. *United Nations Chronicle/UN Chronicle* 2013, *50*, 20–24. doi:10.18356/0dfec716-en.
8. Pigott, D.M.; Howes, R.E.; Wiebe, A.; Battle, K.E.; Golding, N.; Gething, P.W.; Dowell, S.F.; Farag, T.H.; García, A.J.; Kimball, A.M. et al. Prioritising infectious disease mapping. *PLOS Neglected Tropical Diseases* 2015, *9*, e0003756. doi:10.1371/journal.pntd.0003756.
9. Kouadio, K.I.; Clement, P.; Bolongei, J.; Tamba, A.; Gasasira, A.; Warsame, A.; Okeibunor, J.; Ota, M.O.; Boima, T.; Gumede, N. et al. Epidemiological and surveillance response to Ebola virus disease outbreak in Lofa County, Liberia (March-September, 2014): Lessons learned. *PLOS Currents* 2015. doi:10.1371/currents.outbreaks.9681514e450dc8d19d47e1724d2553a5.
10. Wu, N.C.; Olson, C.A.; Du, Y.; Le, S.; Tran, K.; Remenyi, R.; Gong, D.; Al-Mawsawi, L.Q.; Qi, H.; Wu, T.T. et al. Functional constraint profiling of a viral protein reveals discordance of evolutionary conservation and functionality. *PLOS Genetics* 2015, *11*, e1005310. doi:10.1371/journal.pgen.1005310.
11. Laidlaw, P.P. Virus diseases and viruses. *Journal of the American Medical Association* 1939, *112*, 1531. doi:10.1001/jama.1939.02800150103033.
12. Luna, F. globalization and global health. *International Journal of Technoethics* 2015, *6*, 47–59. doi:10.4018/ijt.2015070104.
13. Rizzo, C.; Salcuni, P.; Nicoletti, L.; Ciufolini, M.G.; Russo, F.; Masala, R.; Frongia, O.; Finarelli, A.C.; Gramegna, M.; Gallo, L.G. et al. Epidemiological surveillance of West Nile neuroinvasive diseases in Italy, 2008 to 2011. *Eurosurveillance* 2012, *17*. doi:10.2807/ese.17.20.20172-en.
14. Michalski, A.; Niemcewicz, M.; Bielawska-Drózd, A.; Nowakowska, A.; Gaweł, J.; Pitucha, G.; Joniec, J.; Zielonka, K.; Marciniak-Niemcewicz, A.; Kocik, J. Surveillance of Hantaviruses in Poland: A study of animal reservoirs and human Hantavirus disease in Subcarpathia. *Vector-Borne and Zoonotic Diseases* 2014, *14*, 514–522. doi:10.1089/vbz.2013.1468.
15. Weiner, M.; Kubajka, M. Tularemia—Serious zoonotic disease. *Health Problems of Civilization* 2015, *1*, 39–46, doi:10.5114/hpc.2015.57104.
16. John, S.E.S.; Tomar, S.; Stauffer, S.R.; Mesecar, A.D. Targeting zoonotic viruses: Structure-based inhibition of the 3C-like protease from bat coronavirus HKU4—The likely reservoir host to the human coronavirus that causes Middle East respiratory syndrome (MERS). *Bioorganic & Medicinal Chemistry* 2015, *23*, 6036–6048. doi:10.1016/j.bmc.2015.06.039.
17. Morton, M.J.; Kirsch, T.D.; Rothman, R.E.; Byerly, M.M.; Hsieh, Y.H.; McManus, J.; Kelen, G.D. Pandemic influenza and major disease outbreak preparedness in US emergency departments: A survey of medical directors and department chairs. *American Journal of Disaster Medicine* 2019, *14*, 279–286. doi:10.5055/ajdm.2019.0341.
18. Kouroubali, A.; Kondylakis, H.; Kavlentakis, G.; Logothetides, F.; Stathiakis, N.; Petrakis, Y.; Tzikoulis, V.; Kostomanolakis, S.; Katehakis, D.G. An eHealth platform for the holistic management of COVID-19. *Studies in Health Technology and Informatics* 2020, *273*, 182–188. doi:10.3233/shti200636.
19. Sandhu, R.; Sood, S.K.; Kaur, G. An intelligent system for predicting and preventing MERS-CoV infection outbreak. *The Journal of Supercomputing* 2015, *72*, 3033–3056. doi:10.1007/s11227-015-1474-0.
20. Bucur, O. Emerging technologies for diagnostic pathology. *Discoveries* 2015, *3*, e46. doi:10.15190/d.2015.38.
21. Benner, S.A.; Kim, H.J.; Merritt, K.B.; Yang, Z.; McLendon, D.C.; Hoshika, S.; Hutter, D. Next-generation DNA in pathogen detection, surveillance, and CLIA-waivable diagnostics. *Proceedings of SPIE, the International Society for Optical Engineering/Proceedings of SPIE* 2015, *9490*, 94900K. doi:10.1117/12.2183481.
22. Wiedbrauk, D.L. Molecular methods for virus detection. *Laboratory Medicine* 1992, *23*, 737–742. doi:10.1093/labmed/23.11.737.
23. Ali, M.T.; Morshed, M.M.; Hassan, F. A computational approach for designing a universal epitope-based peptide vaccine against Nipah virus. *Interdisciplinary Sciences Computational Life Sciences* 2015, *7*, 177–185. doi:10.1007/s12539-015-0023-0.
24. Li, T.; Zhou, B.; Dong, H.; Lavillette, D.; Li, D. A human antibody derived from original SARS-CoV-2 infection effectively neutralizes omicron. *Advanced Biotechnology* 2024, *2*. doi:10.1007/s44307-024-00011-1.
25. Savaliya, R.; Shah, D.; Singh, R.; Kumar, A.; Shanker, R.; Dhawan, A.; Singh, S. Nanotechnology in disease diagnostic techniques. *Current Drug Metabolism* 2015, *16*, 645–661. doi:10.2174/1389200216666150625121546.

26. Backhus, L.; Adkar, S.; Klarin, D. Practical guide to multiomics big data sources. *JAMA Surgery* 2025. doi:10.1001/jamasurg.2024.6009.
27. Kong, L.X.; Perebikovsky, A.; Moebius, J.; Kulinsky, L.; Madou, M. Lab-on-a-CD: A fully integrated molecular diagnostic system. *Journal of Laboratory Automation* 2015, *21*, 323–355. doi:10.1177/2211068215588456.
28. Cooper, A.F.; Kirton, J.; Schrecker, T. Governing global health: Challenge, response, innovation. *Choice Reviews Online* 2007, *45*, 45–0909. doi:10.5860/choice.45-0909.
29. Farmer, P.; Kim, J.Y.; Kleinman, A.; Basilico, M. Reimagining global health: An introduction. *Choice Reviews Online* 2014, *51*, 51–6788. doi:10.5860/choice.51-6788.
30. Medact and the Global Equity Gauge Alliance. Global Health Watch 2: An alternative world health report. *Choice Reviews Online* 2009, *46*, 46–5645. doi:10.5860/choice.46-5645.
31. Taylor, A.L. Making the World Health Organization Work: A legal framework for universal access to the conditions for health. *American Journal of Law & Medicine* 1992, *18*, 301–346. doi:10.1017/s0098858800007322.
32. Utunen, H.; Attias, M.; George, R.; Ndiaye, N.; Piroux, C.; Farzi, M.R.; Sy, A.; Gamhewage, G. Global access to OpenWHO's online learning resources for COVID-19. *Studies in Health Technology and Informatics* 2020, *272*, 304–305. doi:10.3233/shti200555.
33. Carroll, D.; Daszak, P.; Wolfe, N.D.; Gao, G.F.; Morel, C.M.; Morzaria, S.; Pablos-Méndez, A.; Tomori, O.; Mazet, J.A.K. The Global Virome Project. *Science* 2018, *359*, 872–874. doi:10.1126/science.aap7463.
34. Carroll, D.; Morzaria, S.; Briand, S.; Johnson, C.K.; Morens, D.; Sumption, K.; Tomori, O.; Wacharphaueasadee, S. Preventing the next pandemic: The power of a global viral surveillance network. *BMJ* 2021, *372*, n485. doi:10.1136/bmj.n485.
35. Gray, B.; Eaton, J.; Christy, J.; Duncan, J.; Hanna, F.; Kasi, S. A proactive approach: Examples for integrating disaster risk reduction and mental health and psychosocial support programming. *International Journal of Disaster Risk Reduction* 2021, *54*, 102051, doi:10.1016/j.ijdrr.2021.102051.
36. Yang, Y.; Miskovich, T.A.; Larson, C.L. State anxiety impairs proactive but enhances reactive control. *Frontiers in Psychology* 2018, *9*. doi:10.3389/fpsyg.2018.02570.
37. Nachega, J.B.; Mbala-Kingebeni, P.; Rosenthal, P.J.; Rimoin, A.W.; Hoff, N.A.; Liesenborghs, L.; Vanlerberghe, V.; Andrei, G.; Rawat, A.; Wilson, L.A. The human Mpox global outbreak: available control tools and the opportunity to break a cycle of neglect in endemic countries. *The American Journal of Tropical Medicine and Hygiene* 2023, *109*, 719.
38. Nachega, J.B.; Mbala-Kingebeni, P.; Rosenthal, P.J.; Rimoin, A.W.; Hoff, N.A.; Liesenborghs, L.; Vanlerberghe, V.; Andrei, G.; Rawat, A.; Wilson, L.A. et al. The human Mpox global outbreak: Available control tools and the opportunity to break a cycle of neglect in endemic countries. *American Journal of Tropical Medicine and Hygiene* 2023, *109*, 719–724. doi:10.4269/ajtmh.23-0161.
39. Van Dijck, C.; Hoff, N.A.; Mbala-Kingebeni, P.; Low, N.; Cevik, M.; Rimoin, A.W.; Kindrachuk, J.; Liesenborghs, L. Emergence of mpox in the post-smallpox era-a narrative review on mpox epidemiology. *Clinical Microbiology and Infection* 2023, *29*, 1487–1492. doi:10.1016/j.cmi.2023.08.008.
40. Coburn, J.; Yaqub, O.; Chataway, J. Targeting research to address societal needs: What can we learn from 30 years of targeting neglected diseases? In *Handbook of Public Funding of Research*; Edward Elgar Publishing: 2023; pp. 156–171.
41. Melo, G.B.T.; Angulo-Tuesta, A.; Silva, E.N.D.; Santos, T.D.S.; Uchimura, L.Y.T.; Obara, M.T. Evolution of research funding for neglected tropical diseases in Brazil, 2004–2020. *PLOS Neglected Tropical Diseases* 2023, *17*, e0011134.
42. Soucat, A.; Khosla, R. Investing in public health systems is a global common good. *BMJ* 2022, *379*, o2475. doi:10.1136/bmj.o2475
43. Head, M.G.; Brown, R.J.; Newell, M.-L.; Scott, J.A.G.; Batchelor, J.; Atun, R. The allocation of US $105 billion in global funding from G20 countries for infectious disease research between 2000 and 2017: A content analysis of investments. *The Lancet Global Health* 2020, *8*, e1295–e1304.
44. Roux, S.; Hallam, S.J.; Woyke, T.; Sullivan, M.B.. Viral dark matter and virus–host interactions resolved from publicly available microbial genomes. *eLife* 2015, *4*, e08490.
45. Guajardo-Leiva, S.; Chnaiderman, J.; Gaggero, A.; Díez, B. Metagenomic insights into the sewage RNA virosphere of a large city. *Viruses* 2020, *12*, 1050.
46. Martinez-Hernandez, F.; Fornas, O.; Lluesma Gomez, M.; Bolduc, B.; De la Cruz Peña, M.J.; Martínez, J.M.; Anton, J.; Gasol, J.M.; Rosselli, R.; Rodriguez-Valera, F. Single-virus genomics reveals hidden cosmopolitan and abundant viruses. *Nature communications* 2017, *8*, 15892.

47. Geldenhuys, M.; Mortlock, M.; Weyer, J.; Bezuidt, O.; Seamark, E.C.J.; Kearney, T.; Gleasner, C.; Erkkila, T.H.; Cui, H.; Markotter, W. A metagenomic viral discovery approach identifies potential zoonotic and novel mammalian viruses in Neoromicia bats within South Africa. *PLOS One* 2018, *13*, e0194527.
48. Souza, J.V.C.; Santos, H.O.; Leite, A.B.; Giovanetti, M.; Bezerra, R.D.S.; Carvalho, E.; Bernardino, J.S.T.; Viala, V.L.; Haddad, R.; Ciccozzi, M. Viral metagenomics for the identification of emerging infections in clinical samples with inconclusive Dengue, Zika, and Chikungunya viral amplification. *Viruses* 2022, *14*, 1933.
49. Brinkmann, A.; Nitsche, A.; Kohl, C. Viral metagenomics on blood-feeding arthropods as a tool for human disease surveillance. *International Journal of Molecular Sciences* 2016, *17*, 1743.
50. Iani, F.C.M.; de Campos, G.M.; Adelino, T.E.R.; da Silva, A.S.; Kashima, S.; Alcantara, L.C.J.; Sampaio, S.C.; Giovanetti, M.; Elias, M.C.; Slavov, S.N. Metagenomic analysis for diagnosis of hemorrhagic fever in Minas Gerais, Brazil. *Microorganisms* 2024, *12*(4), 769. doi:10.3390/microorganisms12040769
51. Runge-Ranzinger, S.; Kroeger, A.; Olliaro, P.; McCall, P.J.; Sánchez Tejeda, G.; Lloyd, L.S.; Hakim, L.; Bowman, L.R.; Horstick, O.; Coelho, G. Dengue contingency planning: From research to policy and practice. *PLOS Neglected Tropical Diseases* 2016, *10*, e0004916.
52. Cuypers, L.; Keyaerts, E.; Hong, S.; Gorissen, S.; Menezes, S.M.; Starick, M.; Elslande, J.V.; Weemaes, M.; Wawina-Bokalanga, T.; Marti-Carreras J., Vanmechelen, B., et al. Immunovirological and environmental screening reveals actionable risk factors for fatal COVID-19 during post-vaccination nursing home outbreaks. *Nature* 2022, *3*(6), 722–733. doi:10.1038/s43587-023-00421-1
53. Carducci, A.; Arzilli, G.; Atomsa, N.T.; Lauretani, G.; Verani, M.; Pistelli, F.; Tavoschi, L.; Federigi, I.; Fornili, M.; Petri, D. Integrated environmental and clinical surveillance for the prevention of acute respiratory infections (ARIs) in indoor environments and vulnerable communities (Stell-ARI): Protocol. *PLOS One* 2024, *19*, e0309111.

Index